ENVIRONMENTAL POLICY AND PUBLIC HEALTH

ENVIRONMENTAL POLICY AND PUBLIC HEALTH

BARRY L. JOHNSON

CRC Press
Taylor & Francis Group
Boca Raton   London   New York

CRC Press is an imprint of the
Taylor & Francis Group, an informa business

CRC Press
Taylor & Francis Group
6000 Broken Sound Parkway NW, Suite 300
Boca Raton, FL 33487-2742

© 2007 by Taylor & Francis Group, LLC
CRC Press is an imprint of Taylor & Francis Group, an Informa business

No claim to original U.S. Government works
Printed in the United States of America on acid-free paper
10 9 8 7 6 5 4 3 2 1

International Standard Book Number-10: 0-8493-8434-6 (Hardcover)
International Standard Book Number-13: 978-0-8493-8434-9 (Hardcover)

Library of Congress Cataloging-in-Publication Data

Johnson, Barry L. (Barry Lee), 1938-
  Environmental policy and public health / Barry L. Johnson.
    p. ; cm.
  Includes bibliographical references and index.
  ISBN-13: 978-0-8493-8434-9 (Hardcover : alk. paper)
  ISBN-10: 0-8493-8434-6 (Hardcover : alk. paper)
  1. Environmental health--Government policy--United States. 2. Environmental policy--United States. 3. Medical policy--United States. I. Title.
    [DNLM: 1. Environment--United States. 2. Health Policy--legislation & jurisprudence--United States. 3. Environmental Health--United States. 4. Hazardous Substances--standards--United States. WA 33 AA1 J66e 2007]

  RA566.3.J64 2007
  362.10973--dc22   2006035545

Visit the Taylor & Francis Web site at
http://www.taylorandfrancis.com

and the CRC Press Web site at
http://www.crcpress.com

Dedication

This book is dedicated to the memory of Clay K. Johnson (1962–2005), a son, a friend, an environmentalist.

Foreword

We can chart our future clearly and wisely only when we know the path which has led to the present.

Adlai Stevenson, 1952[1]

We all benefit from potable water, clean air, food safe to consume, and sanitary disposal of wastes, all environmental conditions that were goals of our ancestors. Many of the horror stories of rivers afire, air full of choking chemicals, food unfit for consumption, and mismanaged municipal and industrial wastes are concerns of the past in the United States. The current issues pertain more to the preservation of improvements in environmental quality, human health protections, and the degree of risk posed by environmental hazards. Moreover, in the twentieth century we learned the importance of protecting our natural resources and improving environmental quality. These positive advances in environmental protection and public health did not occur by accident. They resulted from democratic processes that debated environmental health concerns and proposed legislative and other policy solutions.

This book describes how environmental health policies are developed, the statutes that have evolved to address public health concerns about specific environmental hazards, and the policy issues that impact how environmental health programs function within governmental structures. My focus is on those environmental hazards that have been associated with degraded human health. The hazards include air contamination, water pollution, unsafe food, and hazardous substances and waste. There are certainly more environmental hazards that could have been covered in this book, including the environmental consequences of the "built" environment (e.g., housing, roads) and the area of injury prevention, but they were omitted because they currently lack a foundation of environmental policies and practices.

I intend this work to be useful to students in academic programs of public health and environmental policy. Moreover, reflecting upon Adlai Stevenson's advice,[2] an appreciation is provided of the history of environmental health's evolution and legislative development. Policies and actions that help protect the public from the adverse consequences of environmental hazards did not appear without considerable struggle; knowing their history is vital if the protections they bring are to be maintained.

Many published sources have contributed to the content of this book, but some content represents the experiences and views of its author, whose public health career in environmental and occupational health began in 1960 and extends into the twenty-first century.

NOTES

1. Tripp, R.T., *The International Thesaurus of Quotations*, Thomas Y. Crowell, Publishers, New York, 1970, 281.
2. Adlai Stevenson (D-IL), former governor of Illinois, twice ran unsuccessfully for U.S. President against Dwight D. Eisenhower. Stevenson later served as the U.S. Ambassador to the United Nations.

About the Author

Barry L. Johnson, PhD, is adjunct professor, Rollins School of Public Health, Department of Environmental and Occupational Health, Emory University, Atlanta, Georgia. He is also editor-in-chief, *Journal of Human and Ecological Risk Assessment.* He received his doctorate from Iowa State University. In 1999 he retired from the U.S. Public Health Service's (PHS's) Commissioned Corps with the rank of assistant surgeon general. Dr. Johnson began his public health career at the PHS's air pollution program in Cincinnati, Ohio, conducting research on the neurotoxicity of criteria air pollutants. He continued his research when the newly formed U.S. Environmental Protection Agency was created, later joining the National Institute for Occupational Safety and Health as a research scientist and administrator. Dr. Johnson completed his public health career as director of the Agency for Toxic Substances and Disease Registry.

Dr. Johnson is author of *Impact of Hazardous Waste on Human Health* (1999, Lewis Publishers), and senior editor of *Hazardous Waste: Impacts on Human and Ecological Health* (1997, Princeton Scientific Publishing); *Hazardous Waste and Public Health* (1996, Princeton Scientific Publishing); *National Minority Health Conference Proceedings* (1992, Princeton Scientific Publishers); *Advances in Neurobehavioral Toxicology* (1990, Lewis Publishers); and *Prevention of Neurotoxic Illness in Working Populations* (1987, Wiley & Sons).

Acknowledgments

I would like to express my sincere appreciation to several persons who reviewed drafts of this book and made significant improvements in its contents. In particular, I express my appreciation to several graduate students at the Rollins School of Public Health, Emory University, who provided editorial comments: Libby Bower, Tiffany Dothard, Amy Funk, Caroline Hoffman, Jamie Howell, Shannon Jones, Kristen Mundy, Jeff Mutchler, and Amee Patel. Similarly, environmental health students at the Morehouse School of Medicine and the Georgia State University provided helpful editorial suggestions. A special note of appreciation goes to Dr. Charles Xintaras, a colleague in environmental health for more than four decades, who reviewed several drafts of this manuscript. I also thank Dr. Howard Frumkin, chair, Department of Environmental and Occupational Health, Rollins School of Public Health, Emory University and Associate Professor Melvin L. Myers, Rollins School of Public Health, Emory University, for helpful discussions; and Dr. Mark Bashor and Senior Engineer Morris Maslia, Agency for Toxic Substances and Disease Registry, and Dr. Maureen Y. Lichtveld, Centers for Disease Control and Prevention, for their helpful comments and materials shared during the preparation of this book.

Contents

1 Fundamentals of Environmental Policy

1.1 INTRODUCTION

Humankind's journey through the ages has been difficult. Our primordial ancestors faced threats to their survival in a hostile environment. Wild carnivorous animals abounded and natural disasters such as forest fires and floods surely presented grievous challenges to our ancestors. Over time the nature of the environmental hazards changed as humans passed from a nomadic, tribal existence to a more communal lifestyle in small villages and, later, large cites. As humans huddled together in increasingly large numbers, health problems magnified in both numbers and severity of disease. Perhaps no greater health calamity has befallen humankind than the bubonic plague (also called the Black Death). There were three major pandemics of the plague, occurring in the sixth, fourteenth, and seventeenth centuries. The death toll approximated 137 million victims. The pandemic of the fourteenth century was particularly devastating, causing the death of 25 million people. Ultimately, the plague killed about one-third of Europe's population over a five-year period, beginning in year 1347. The plague was eventually found to be caused by the bacterium *Yersinia pestis*, which is spread by fleas that infest animals such as the black rat [1]. The plague, an example of an extreme environmental health problem, illustrates the importance of environmental hazards as a public health concern.

A healthy environment promotes healthful conditions necessary to sustain living creatures. While this observation seems obvious, in practice, societies that have developed vigorous agricultural and industrial bases have found that pollution became a consequence of those activities. Air quality deteriorated, water purity diminished, and lands became fouled by chemical and other hazardous substances. As biomedical research on the effects of environmental hazards progressed, it became evident that environmental degradation was associated with adverse effects on the health of human populations and ecosystems.[1]

Policy: A definite course or method of action selected from among alternatives and in light of given conditions to guide and determine present and future directions [2].

In response to concerns about environmental hazards, the federal, state, territorial, and local governments in the United States have enacted various statutes meant to conserve the natural environment, assure environmental quality, and protect the public's health. Underpinning this effort are policies that shape the intent and implementation of the statutes. This chapter presents an overview of key fundamentals that shape the development of environmental policy. To be described are a summary of how environmental health has evolved, the fundamentals of public health, the role of government in environmental health, and public policies of relevance to environmental health. Moreover, how environmental policies have emerged in the United States and their relevance to the practice of public health is the focus of this book.

1.2 ENVIRONMENTAL HEALTH POLICY FRAMEWORK

Environmental health policy comprises actions that are intended to eliminate the effects of exposure to environmental hazards. One way to consider this kind of policy is to consider its uses, users, and nonusers, yielding the following five considerations.[2]

Directness. Some policies directly address environmental health. Examples include U.S. Environmental Protection Agency (EPA)[3] standards that regulate the levels of a contaminant in an environmental medium (e.g., levels of air pollutants in outdoor, ambient air). Other policies are primarily environmental policies, without a health focus, but they indirectly affect human health or environmental quality. An example would include the National Environmental Policy Act, discussed in chapter 4, wherein a national policy of environmental protection is articulated. And still other policies are not even environmental, but they incidentally have a major impact on environmental health. For example, national energy policy has an impact on which vehicle and heating fuels are used, which, in turn, can affect air quality and therefore human health. This book primarily addresses those environmental health polices that most directly affect human health, since they present a direct course of action in controlling the adverse consequences of environmental hazards on human health.

Level of Government. Environmental health policies span the spectrum of government. This book gives emphasis to federal government policies (e.g., the Clean Water Act and its attendant policies on controlling emissions of contaminants into bodies of water in the United States). However, state and local governments also develop environmental health regulations and enact legislation that addresses issues specific to a state's environmental conditions. States enact statutes that are necessary to comply with federal statutes and regulations. For instance, states will enact statutes and provide resources to meet the provisions of the Clean Air Act, which stipulates specific requirements of states. And local governments establish environmental health polices through ordinances, such as prohibitions on smoking of tobacco products in public facilities. In general, environmental health policies become more specific and targeted as they transition from federal to state to local government.

Primary Strategy. Policy makers such as legislators and government officials have made several primary strategies into environmental health policy. Some strategies aim directly at reducing the effects of hazards, some in a prospective manner (e.g., air pollution regulations), while others through retrospective action (e.g., cleanups of uncontrolled hazardous waste sites). Other policies do not directly regulate a hazard, but provide information to the public about the hazard, in effect relying on individuals to make informed health decisions. This is a kind of laissez-faire approach to controlling the effects of some environmental hazards. Examples include: health warnings on tobacco products; the Toxics Release Inventory, a public database compiled by EPA on the composition and amounts of pollution released from industrial facilities; and workers' right-to-know communications under the Occupational Safety and Health Act, wherein employers must provide employees with information on workplace hazards.

The Prime Actor in the Policy. There can be several prime actors in the development of environmental health policies. While this chapter emphasizes the role of government as the prime actor, private parties can also play a significant role. For example, the American Conference of Governmental Industrial Hygienists (ACGIH), a professional society, develops recommended exposure limits for substances found in workplaces. Private industry uses the ACGIH exposure limits as voluntary guidelines for workplace controls when government standards are not in effect. Similarly, the International Standards Organization develops recommended guidelines that industry and some government agencies adopt. As discussed later in the chapter, individuals can be prime actors in helping establish an environmental health policy through litigation of a government agency or a business. Consider the example of a person who litigated a restaurant chain when a cup of very hot coffee spilled on her legs while driving. The coffee's temperature was sufficiently high to cause severe burns. Litigation compensated the woman for her injuries and also contributed to the restaurant chain's voluntary decease in the temperature of the coffee served throughout the restaurant chain. As a consequence, one person's litigation contributed to control of an environmental hazard that was potentially faced by millions of people.

What Doesn't Get Regulated. This chapter focuses on policies that relate to regulations and standards as the primary means to control environmental hazards. Not described are important environmental issues for which regulatory policies do not exist. Examples of nonregulated environmental hazards include: indoor air of domiciles, which is not covered under the Clean Air Act; emissions of carbon dioxide, a major greenhouse gas implicated in global warming; and tobacco products for which product labeling is required, but product safety is not regulated. These three examples alone indicate that unregulated environmental hazards can exceed the public health impact of regulated hazards.

1.3 KEY DEFINITIONS

In order to understand and appreciate the complexities of establishing and maintaining environmental health policy, we need to have a common understanding of words and phrases. A common vocabulary is essential if communication and debate over environmental health policies are to occur in any productive manner. Some might say that meanings of words and phrases such as *policy, health, public health, environment, environmental health,* and *politics* are obvious and well known. This is not the case, however, because meanings of words reside in individuals themselves, not in any inherent properties of words themselves. Differences in how people understand words occur because of variations in individuals' cultural backgrounds, educational levels, home and business environments, and situational-specific settings. As aids to understanding meanings of words, dictionaries help us achieve partial common agreement on words' meanings, but even they must use more words in order to define meanings of specific words.

We can approach a common understanding of a word or phrase by accepting a definition chosen from a credible source (e.g., a dictionary or other creditable source) and then discussing the definition within the group needing a common definition (e.g., a group of students). With this approach in mind, the following definitions are proposed for key words and phrases pertinent to discussions of environmental health policy.

Meanings of words reside in people, not in any inherent property of words themselves.

1.3.1 POLICY

Policy: A definite course or method of action selected from among alternatives and in light of given conditions to guide and determine present and future directions [2]. More to the point of this book, policy is also defined as a plan that embraces the general goals and acceptable procedures in governmental action [ibid.]. Effective policy making will normally require choices among alternatives and will be based on conditions at hand. In a sense, making environmental health policy is no different from making family or business policies. Many families choose as a matter of policy to budget their expenditures. For businesses, some adopt a policy to service all customer complaints within a specified period of time. In both examples, alternatives were surely considered and an action selected to guide future actions.

Developing policy, according to our chosen definition, must involve the identification of alternatives that might be applied to specific situations. From the alternatives, policy makers (e.g., a legislative body, tribal council, or parent) determine the best (applying stated criteria) alternative, communicate their decision to interested parties, and apply the policy when future circumstances arise whose response must be based on policy.

1.3.2 HEALTH

Health: A state of complete physical, mental, and social well-being and not merely the absence of disease or infirmity [4]. This definition comes from the widely respected World Health Organization (WHO), headquartered in Geneva, Switzerland. The WHO is a component of the United Nations and its research, reports, and services are widely accepted by health agencies worldwide. It provides technical assistance and resources globally on programs of human health, including preventing the spread of AIDS, polio, and infectious diseases. Among its many contributions to global human health, the WHO led the campaign against smallpox as a global scourge to human health, announcing in 1981 that the disease had been eradicated globally.

The fundamental principle of public health is the prevention of disease and disability.

By the WHO's definition, a healthy individual, group, or population is free of physical and mental disease and infirmity, as well as being in a state of social well-being. As individuals, the absence of conditions such as bodily injury, cancer, heart disease, depression, or paranoia either is obvious or can be diagnosed by a medical doctor. Less obvious in the WHO's definition of health is what is meant by *social well-being*, certainly an altruistic component of the definition. What might be intended by the WHO? Several examples reflective of social well-being could include adequate housing, education, income, and living conditions; freedom from war, malnutrition, political abuse, and poverty; and ability to participate in political systems and public policy making.

1.3.3 PUBLIC HEALTH

Public Health: The process of mobilizing local, state, national, and international resources to solve the major health problems affecting communities [7]. This definition, one of several in existence [8], is appealing for use in a text on environmental health policy. This is because environmental hazards and problems are often community-based and, due to their complexity, require multiple resources for risk management and problem solution. Moreover, this definition of public health implies that major health problems must take priority over those of lesser consequence. Increasingly, risk assessment,[4] as described in chapter 11, is the tool used by environmental health specialists to separate major hazards to human and ecological health from those of lesser importance.

Public health can be understood as meaning "the public's health." Unfortunately, the U.S. public has an unclear concept of what public health agencies do and what their programs accomplish, and associate the term *public health* primarily with health services for indigent persons. In reality, the spectrum of U.S. public health programs and services encompasses such national efforts as childhood immunization, cancer research, mental health programs, lead exposure prevention, disease surveillance, physicians' education, and funding for local health centers. Local health departments conduct such programs as restaurant inspections, vector control, sanitation programs, immunizations, and activities to prevent the spread of infectious and chronic diseases. These examples show the broad impact of public health programs on the U.S. public.

1.3.4 ENVIRONMENT

Environment: The circumstances, objects, and conditions by which one is surrounded [2]. As an example, consider a student's classroom environment. Circumstances of a student's environment could include an assigned seat in the classroom, thus placing the student in the same location for all classes. Another circumstance could be whether the class was required or optional, which could determine which classroom the student occupies. Objects in the student's classroom environment could include other students, desks, tables, video equipment, and such. Conditions of the environment could include ambient air temperature, barometric pressure, relative humidity, lighting intensity, and noise levels.

1.3.5 Environmental Health

Environmental Health: Comprises of those aspects of human health, including quality of life, that are determined by physical, chemical, biological, social and psychosocial factors in the environment. It also refers to the theory and practice of assessing, correcting, controlling, and preventing those factors in the environment that can potentially affect adversely the health of present and future generations [5]. This verbose definition is no doubt the product of a committee. However, it bears the imprimatur of the WHO, which adds credibility and importance to the definition. Note that this definition is specific to human health and like the WHO's definition of health includes mention of physical, chemical, and social factors. The second half of the WHO's definition expresses the elements of both hazard assessment and risk management. Noteworthy in the definition is mention of *quality of life*, a subjective term. But given the overall environmental context of the definition, quality of life could include examples such as the adverse psychological consequences of living near a foul-smelling industrial facility or by living in a metropolitan area where a major highway has been constructed through a formerly well-established neighborhood, thereby exposing residents to more noise and air pollutants and fracturing social relationships due to neighbors' relocation.

Remarkably, the WHO definition is but one of many existing definitions of environmental health. One source collected twenty-six different definitions [6], which suggests insufficient effort has been given to achieving a consensus definition for use by environmental health specialists. Given increasing global commitment to reducing the impact of environmental hazards (e.g., the Kyoto Protocol to reduce greenhouse gases, which are chemicals that have the potential to increase global warming), a plea for a common definition seems over due.[5]

There is an alternative definition for the phrase *environmental health* that gives emphasis to the word *environmental*. In this definition, environmental health refers to the health of the environment, that is, considerations of environmental quality, ecosystems' well-being, and conservation of natural resources. For example, in this context, one could speak about the environmental health of equatorial ecosystems as affected by deforestation and human population growth.

1.3.6 Politics

Politics: a)The art of science of government, b) political affairs or business, c) the total complex of relations between people living in society [2]. Although most people associate politics with politicians and government, in fact, politics occur within families, businesses, civic organizations, schools, and other societal structures. In all these examples, politics must incorporate dialog, debate, negotiation, and, ultimately, compromise among the interested parties.

All politics must include discussion, negotiation, and, ultimately, compromise.

Politics permeates the development and execution of environmental health policy. Some persons may have a negative opinion of politics and politicians because the practice of politics necessarily involves negotiation and compromise, and some politicians have been poor examples of ethical behavior. Thus, to associate a somewhat unwholesome opinion of politics with an altruistic image of public health might seem contradictory to some persons. Moreover, if public health is about preventing disease and disabilities in human populations, should not something so important "be above" politics? The answer, of course, is no. Politics involve relationships among people, and the core of public health rests with people. How public health departments reach out to the public is a matter of politics, involving communication, negotiation, and compromise. Further, government public health organizations must compete with other government programs for budgets, personnel allocations, and authorities—all of which necessitates political acumen and wisdom.

This collection of definitions of *policy, health, public health, environment, environmental health*, and *politics* will help us better understand the development of environmental health policy in the United States and attendant actions resulting from specific policies.

1.4   EVOLUTION OF ENVIRONMENTAL HEALTH

An understanding of the evolution of environmental health is necessary for an appreciation of modern environmental policy. After all, as the Spanish-American philosopher George Santayana [10] commented, "Those who cannot remember history are condemned to repeat it." As discussed in this section, humankind long ago learned the importance of potable water and proper disposal of human wastes, perhaps dating to the time of the Neolithic Revolution, which occurred during the period 8,000–7,000 BCE [11]. During this period, humankind began changing from a hunter-gather society to a society that relied on agriculture and domesticated animals, forming small tribal settlements in the process. In fact, it can be asserted that modern public health has its historic roots in what we now call *environmental health*. Later, as human populations increased, clean air and safer food supplies were added to the environmental health agenda. Much of modern environmental health policy and practice in the United States. has roots in nineteenth century Europe, as will be subsequently discussed.

1.4.1 HISTORICAL ENVIRONMENTAL HAZARDS

The struggle by humans to overcome environmental problems is certainly not new. Archaeological research has revealed that some ancient civilizations developed ways to dispose of human wastes and to provide water to their expanding cities. As described by the public health historian George Rosen [12], archaeologists have found ancient ruins where bathrooms, flushing toilets, and water gutters were present (Table 1.1), some dating to 2100 BCE. The geographical diversity of these ruins is impressive—extending from northern India to the Incas in South America. Notable is the presence of water supply systems developed by the two major early European cultures: Greek and Roman. Both civilizations built elaborate systems of aqueducts and canals to bring water to the expanding cities of Athens and Rome, respectively.

The environmental health resources listed in Table 1.1 illustrate humankind's search for more healthful living conditions. Such conditions, then and now, include living with an ample, potable supply of water to meet daily needs and for sanitary disposal of human wastes. Maintaining these systems of water supply and waste disposal are constant challenges to modern policy makers because of increases in human populations and global climate change. The former puts added pressure on water resources and sewage systems; the latter will change geographic patterns of rainfall and land use.

1.4.2 HAZARDS TO SURVIVAL

Humankind's prosperity over the ages can be attributed to many factors, but surely meeting basic human survival needs must be the foremost factor. For human life to exist, there must be air, water, and food. Absence of any of these three is a death sentence. Another survival need is the sanitary disposal of human wastes, since improper man-

---

TABLE 1.1
Environmental Health Resources before the Common Era (BCE)

Location	Period	Environmental Health Resource
India: Indus Valley & the Punjab	2100 BCE	Bathrooms & drains found in excavated buildings
Egypt: Middle Kingdom	2100–1700 BCE	Water gutters found in excavated city
Troy	2000 BCE	Water supply system
Crete	2000 BCE	Flushing toilets in excavated palace
Incas	--	Sewage systems
Greece	600 BCE	Water supply system
Rome	312 BCE	Aqueduct to Rome

Note: Data compiled from Rosen, G., *A History of Public Health. Expanded Edition*, The Johns Hopkins University Press, Baltimore, 1993.

agement connotes disease and illness. The following sections describe the evolution of humankind's means to address these four basic survival challenges, presented in order of their likely historic development.

1.4.2.1   Sanitary Waste Management

There is, of course, no precise date in antiquity that demarcates humankind's awakening to the health hazards of their environment. But there were certainly environmental challenges faced by cave dwellers and other prehistoric peoples. Carnivorous animals, natural disasters, and emerging human diseases all surely took their toll on our earliest ancestors. However, one could postulate that diseases produced by unsanitary environmental practices and humankind's management of them could be called our first environmental health experience. More specifically, improved sanitation management of human wastes was a most important environmental health advancement as encampments grew into villages and then into cities. Too often human wastes was deposited into the residential environment, contaminating drinking water supplies. Cholera and dysentery were grievous outcomes of consumption of impure water.

Attempts to improve basic sanitation practices began during the middle ages in Europe. In the early Middle Ages, sanitary household practices were primitive to say the least. According to one source [13], "In much of medieval Europe, sanitation legislation consisted of an ordinance requiring homeowners to shout, 'Look out below!' three times before dumping a full chamber pot into the street." Because many houses were multistoried, dumping chamber pots literally caused a rain of human wastes on persons on the streets below. There the waste lay until rain washed it away to be deposited in lower lying areas or waterways. Later, larger cities began building sewers and reducing human wastes left on streets. Practices in China probably preceded anything done elsewhere. For instance, in rural China, "night wastes" have for centuries been routinely collected and used as fertilizer for crops and land, resulting in top soil thickness measured in feet, not inches as in the United States. As to the earliest environmental health intervention, some public health historians might attribute that to John Snow's removal of a pump handle in London, thereby preventing public access to a well contaminated with fecal coliform bacteria, which Snow associated with an ongoing cholera epidemic [14].

In more modern times, the United States has enacted federal statutes that control the levels of contaminants that can be released into water supplies and for management of human wastes. These are described in chapter 5.

1.4.2.2   Potable Water

Water quality was, and remains, an environmental health problem of great concern to many human populations. Over time, exposure to human wastes found in water gradually decreased by moving latrines, public toilets, and isolated privies away from such water supplies as wells, springs, lakes, and flowing streams. Some of these changes occurred when armies formed themselves into encampments. Military leaders knew the health importance of constructing latrines and requiring their use by troops. As a consequence, one can imagine troops returning to small villages with some experience on how to better manage the disposal of human wastes.

In the United States, as migration of immigrants increased in numbers, villages and cities sought better ways to protect their water supplies. In contrast, persons who lived on farms and in rural areas had to depend on wells, springs, and surface waters as sources of drinking water. For both urban and rural dwellers, avoidance of biologically contaminated water certainly was of concern, but without the population knowing how to protect themselves. Indeed, as discussed in chapter 5, sanitary practices and water contamination first came within the province of public health authorities in the early part of the twentieth century. Local sanitation authorities became involved with construction of sanitary sewers and location of waste facilities. The emergence of city and county health equipment occurred in the twentieth century. Sanitarians soon became integral members of local health departments.

1.4.2.3   Air Quality

As cities grew in size and complexity, air pollution resulting from burning coal for industrial purposes and for home heating became another environmental problem. In Europe and the United States, coal burning created huge amounts of carbon particulates that darkened the environment, fouled the air, and lowered the quality of life. The consequences of air pollution on environmental quality and public health are described in chapter 5. Suffice it to say here that death to residents of Donora, Pennsylvania, in 1948 and London, England, in 1952 from exposure to episodes of very polluted air had a major influence on enactment of federal clean air legislation. In more recent times, emissions from industrial plants and from vehicles powered by internal combustion engines have become of public health concern, as discussed in chapter 5, where other adverse public health effects of air contaminants are discussed.

1.4.2.4   Safe Food

Food, of course, is vital for human survival. In colonial America and well into the twentieth century, food was produced by farmers and ranchers. In villages and cities, food was purchased at local markets and prepared at home for consumption. Preventing foodborne illness was primarily the responsibility of those who prepared food. As the country passed from an agrarian society into an industrial economy, the U.S. food supply was increasingly produced by large agricultural enterprises, and imported supplies of food increased in volume and variety. As food sources became less familiar to consumers, food safety concerns increased.

Perishable food was a special problem for consumers. Methods were developed for canning vegetables, fruits, and some meat products. Canning involved placing cooked food into sterile, sealed containers, a process that killed microorganisms, thereby lessening the possibility of food poisoning. Other preservation methods included sun drying of some foods and the use of preservatives such as salt and the smoke from wood fires. These methods reduced the amount of moisture in the treated foods and thereby inhibited the growth of microorganisms. But it remained for a technological breakthrough to occur before perishable foods could be stored in large quantities for appreciable lengths of time. The invention of refrigeration equipment and its widespread distribution were responsible for increasing food safety. Beginning in the 1930s, perishable food could now be shipped in refrigerated trucks, stored in refrigerated warehouses, and sold to stores and restaurants for placement in freezers and other refrigerated equipment. The public's health was improved by this technology. However, as will be discussed in chapter 6, food safety concerns remain a major public health problem, given the large number of foodborne illnesses that occur annually in the United States.

1.4.3   EUROPEAN ROOTS

Modern environmental health systems and practices in the United States generally are dated from mid-nineteenth-century Europe, although this attribution may be wrongly based because of our lack of knowledge about conditions in other parts of the world. The evaluation of public health awareness and the sanitary movement in Europe in the early to mid-1800s had common roots: industrialization, unsafe and unhealthful working conditions, inadequate sanitation in crowded cities, and persons of vision who were committed to improving the public's health. These conditions were most evident in England, France, and Germany.

One source asserts that the devastating bubonic plague (also called the Black Death) that ravaged the globe during the mid-fourteenth century gave rise to the initial development of public health [13]. He notes, "The Black Death also played a major role in the birth of public health. One early innovation in the field was the municipal health board, such as those in Florence and Venice established in 1348 to oversee sanitation and the burial of the dead. Later the boards would grow more sophisticated. In 1377 Venice established the first public quarantine in its Adriatic colony of Ragusa (modern day Dubrovnik)." It is interesting to note that to some extent what we now call public health has some of its roots in humankind's struggle with a notorious pandemic plague.

In early nineteenth century England, the enclosure of common lands had the deleterious social consequence of creating huge numbers of rural poor. Their numbers exceeded the capacity of the existing relief system for the poor. These newly impoverished families migrated to the newly emerging industrial cities, where work, often hazardous and exploitive of children, was available [12]. Whole families were often crammed into dank basements and cellars, with inadequate or nonexistent sanitary facilities.

As workplace and community living conditions continued to worsen, social and health reformers emerged. Principal among them was Edwin Chadwick. The New Poor Law Act of 1834 created a new labor market, facilitating the immigration of the rural laboring poor into the harsh reality of urban factory work [ibid.]. Chadwick had been a primary author of the 1834 act. Later, in 1842, he and colleagues published the influential *Report on the Sanitary Conditions of the Labouring Population of Great Britain*. The report became the seminal work that reformed public health in England. Chadwick and others were convinced that prevention of epidemic disease (e.g., cholera) was less costly to the English economy than treating the consequences of unabated disease. The English model of disease prevention through improved living conditions and sanitary reforms found favor in France and Germany and also influenced public health policy in the United States.

In France, during the reign of Louis Philippe (1830–1848), the country's economy began to change from agriculture to industrialization. This change continued until the 1870s, according to Rosen [12].The French public health movement evolved during this roughly forty-year span. Terrible working conditions were mimicked by equally horrible living conditions, especially in rapidly expanding industrial cities. Overcrowded living conditions were but one symptom of urban distress. Lack of sanitary facilities, poor quality drinking water, and epidemic disease were the companions of impoverished communities.

Starting in 1841, with the passage of labor legislation regulating child labor in factories, a body of law and sentiment gradually emerged in support of a public health system in France. The outstanding figure in the French public health movement was Louis René Villermé, known for his study of textile workers' health, who aroused public opinion about hazardous workplace conditions [ibid.]. Earlier, in 1828, Villermé had published a report showing that mortality and morbidity rates were closely related to living conditions across social classes. Later, in 1848, a law created a network of local public health councils. These councils were not particularly effective, but did serve to commit the government to a national program of public health.

The public health movement in Germany emerged later than those in England and France. This was due in part to the fact that the modern German nation did not exist until late in the nineteenth century, when the Prussian leader Otto von Bismarck unified the German states into a nation. Similar to England and France, industrialization within the German states was evident by 1848, producing patterns of workplace hazards and unhealthful urban communities. Rosen [12] observes that two health reformers, Rudolph Virchow and Solomon Neumann, were leaders in shaping the German public health evolution. In 1848, Virchow advocated that government should provide public medical care for indigent persons. Although this and other social reform proposals foundered, the decades afterward led to a program of limited sanitary reform. During the 1860s and 1870s, public health reform again emerged. Focused efforts to improve sanitary conditions in Berlin and Munich contributed to the creation in 1873 of the Reich Health Office, the start of a unified, national public health system in Germany [ibid.].

1.4.4 MODERN TRENDS

Environmental factors play a central role in human development, health, and disease. Broadly defined, the environment, including infectious agents, is one of three primary factors that affect human health. The other two are genetic factors and personal behavior. Human exposure to hazardous agents in the air, water, soil, and food and to physical hazards in the environment are major contributors worldwide to disease, disability, and death. Furthermore, deterioration of environmental conditions in many parts of the world slows economic and social development. Poor environmental quality is estimated to be directly responsible for approximately 25 percent of all preventable ill health in the world, with diarrheal diseases and respiratory infections heading the list [15]. As discussed in chapter 9, ill health resulting from poor environmental quality varies considerably among countries. Poor environmental quality has its greatest impact on people whose health status already may be at risk.

Because the effects of the environment on human health are so great, protecting the public from exposure to environmental hazards has been a mainstay of U.S. public health practice, dating, perhaps to 1798 [16]. In that year, Congress enacted An Act for the Relief of Sick and Disabled Seamen, which established a loose network of marine hospitals, mainly in Atlantic seaboard port cities to care for sick and disabled mariners [17]. This care included issues of vessel sanitation and shipboard vermin. Later federal legislation, the Public Health Service Act of 1912, made more specific the link between federal public health responsibilities and protection from environmental hazards. In particular, the act specified, "The Public Health Service may study and investigate the diseases of man and conditions influencing the propagation and spread thereof, including sanitation and sewage and the pollution either directly or indirectly of the navigable streams and lakes of the United States, and it may from time to time issue information in the form of publications for the use of the public" [18]. Subsequent aments to the Public Health Service Act of 1912 increased and broadened the environmental health responsibilities of the federal Public Health Service.

National, tribal, state, and local efforts to ensure clean air and safe supplies of food and water, to manage sewage and municipal wastes, and to control or eliminate vector-borne illnesses have contributed significantly to improvements in public health in the United States. However, the public's awareness of the threat posed by chemical substances in the environment as a matter of public health is more recent. Events such as the publication in 1962 of Rachel Carson's book *Silent Spring*, in which the threat of pesticides to birds, with implications for human health, awakened the U.S. public to a new health concern. Another example of an event that raised public awareness was the well-publicized discovery that residents of Love Canal, in western New York, were being exposed to hazardous substances in their homes, which had been built over an abandoned chemical waste dump. Seepage of hazardous waste into local residents' homes brought national focus to a hitherto unrecognized threat to public health. A result of these and other similar events was an expanded environmental movement, which has led to the introduction into everyday life of such terms as Superfund sites, water quality, clean air, ozone, urban sprawl, and agricultural runoff [15].

As we begin humankind's journey through the 21st century, there may be the temptation to look smugly and confidently at our prospects. Global wars have not occurred for more than 60 years (which is not even a blink of an eye in time), scientific knowledge has proliferated (90% of all the scientists that the world has produced are alive now), life spans have lengthened for some groups in developed countries; and national economies and political stability have generally improved globally. There is, indeed, some basis for optimism that humankind will experience a better twenty-first century than the ones preceding it.

Yet, there are signs already that the twenty-first century could be a century of environmentally-caused turmoil and hardship. Increases in human population will continue to strain at, or deplete, natural resources—potable water and food supplies are examples. Technological solutions (e.g., cost effective desalination of sea water) as well as tough societal decisions (e.g., policy choices about urban development and water restrictions) will force tough policy choices. Perhaps the most daunting policy decision will be how to respond to global climate change and how to mitigate the effects. How these policy decisions are made on ways to share lesser and lesser amounts of natural resources will challenge democratic institutions and promote disharmony unless carefully managed.

The potential for environmental grief in the twenty-first century is real and potentially catastrophic. Already, bacteria and pests have evolved that are resistant to some chemical pesticides and therapeutic drugs. The result has been the re-emergence of cholera, tuberculosis, malaria, and other diseases as public health concerns. Moreover, chemical pollutants in air, water, and food remain as public health challenges. Given these conjectures about the twenty-first century, it is important to have a sense of public health fundamentals and government structure. Because public health programs are primarily government in nature, the structure of U.S. government is important to understand and will be discussed in chapter 3.

1.5 PUBLIC HEALTH FUNDAMENTALS

The fundamental principle of public health is the prevention of disease and disability. Prevention is as fundamental to public health practice as physiology and anatomy are to the practice of medicine. Preventing disease and disability brings with it elements of both idealism and practicality. Prevention is idealistic in the sense of altruistic conduct by persons concerned about the health and well-being of others. Disease and disability connote human suffering; prevention of suffering is an altruistic act, and an element of most religious beliefs and practices. What individual doesn't feel a sense of satisfaction when helping alleviate the suffering of another person? Moreover, some would consider prevention of suffering a necessary characteristic of what constitutes humanity.

Public health practitioners have developed several structures (i.e., models, paradigms) for preventing disease and disability. The elements of two models are shown in Table 1.2. The Disease Prevention Model consists of five elements [19], shown below. The application of the model to prevention of childhood lead intoxication is shown in parentheses as an example.

1. *Surveillance* for patterns of morbidity and mortality in at-risk populations (blood lead reporting systems administered by state or municipal health departments);
2. *Evaluation* of the factors underlying the observed patterns of morbidity and mortality (assessment by health officials to determine when reported blood lead levels exceed health-based guidelines);

TABLE 1.2
Two Public Health Models for Prevention of Disease and Disability

Disease Prevention Model	Industrial Hygiene Model
Surveillance	Anticipation
Evaluation	Recognition
Intervention	Evaluation
Infrastructure	Control
Impact	Assessment

3. *Interventions* or control strategies, including health education and risk communication (follow-up visits to homes of children with elevated blood lead levels
   to identify sources of exposure to lead);
4. *Infrastructure* at the federal, state, and local levels to implement interventions
   (grants from federal agencies to states and local health departments);
5. *Impact assessment* to assure that the interventions undertaken have been effective
   (evaluation by health officials to determine if blood lead levels have decreased
   in children at risk).

The Industrial Hygiene Model [20] (Table 1.2) differs in some important respects
from the Disease Prevention Model. Its four elements consist of the following:

1. *Anticipation*—Have in place mechanisms such as prospective risk assessment,
   basic research, surveillance systems that can identify potential morbidity or
   mortality.
2. *Recognition*—Identify specific patterns or instances of excess morbidity or
   mortality.
3. *Evaluation*—Assess the casual factors that account for the observed excess
   morbidity or mortality
4. *Control*—Implement strategies and actions that will reduce or prevent the identified patters of excess morbidity or mortality.

The Industrial Hygiene Model includes *Anticipation* as one of its elements, whereas
the Disease Prevention Model does not explicitly mention anticipation, although some
anticipation must be inherent in the latter model's *Surveillance* element in order to
provide a focus for surveillance. In contrast, the Disease Prevention Model contains
the elements *Infrastructure* and *Impact Assessment*, which are absent in the Industrial
Hygiene Model. However, both models serve useful purposes for protecting populations
at risk of adverse health effects.

Any system of disease and disability prevention will founder if the system is not
well planned and maintained. Dr. William Foege, former director, Centers for Disease
Control, identified what he called three enemies of prevention [20], to which a fourth

TABLE 1.3
Enemies of Prevention of Disease and Disability

Time

Distance

Greed

Ignorance

element, *Ignorance*, was added by the author. Shown in Table 1.3 are four enemies of disease/disability prevention: Time (e.g., diseases that have a long latency, loss of public health resolve, atrophy of funding), Distance (e.g., locally originating diseases can be transported to parts of the world that lack prevention programs, such as AIDS), Greed (e.g., tobacco companies continue to market a product that kills its users), and Ignorance (e.g., lack of knowledge of a disease or disability's causal factors make prevention a challenge).

In addition to the altruistic aspect of prevention of disease and disability, there is also a practical aspect in the sense of economic considerations and personal survival. Regarding economic considerations, persons who are sick or disabled are lost to contributing to family and other incomes. Tax bases are lessened when persons are unable to work, and monies spent on curative medicine are lost to other potential expenditures that impact national economies. Consider the horrific situation of the AIDS pandemic that has ravaged populations and national economies in certain African countries and elsewhere. Deaths from AIDS in such countries have removed potential new workers from contributing to national economies and industrial and agricultural development. Prevention of the spread of AIDS would have had obvious economic benefits. Left unchecked, disease pandemics have the potential to put at risk the survival of humans as a species. As to personal survival, national and local security can be compromised when defenders are sick or lost to death.

On a different perspective, long before modern public health programs were in place in the United States to prevent childhood diseases, families in colonial times through approximately the first third of the twentieth century experienced the death of many children lost to disease. Diphtheria, pertussis, malnutrition, measles, and polio all took their toll on children. Visit any old cemetery and observe the ages of those interred there. Many families chose to have large numbers of children, knowing that some would be victims of childhood diseases. This cycle of death was finally broken in many parts of the world through public health practices that included large scale vaccination programs and improvements in sanitation practices and water quality. Unfortunately, these improvements in public health remain quite unevenly distributed across nations. For example, the WHO [21] reports that polluted air and water and other environment-related hazards annually kill more than three million children under the age of five years, further observing that although just 10 percent of the world's population is under the age of five years, yet 40 percent of the environment-related disease burden falls on children in this age category. Moreover, the WHO elaborated on the global impact of environmental hazards by observing that:

1. "Unclean water causes diarrhea, which kills an estimated 1.8 million people worldwide each year, 1.6 million of whom are children under five. It's also responsible for many diseases including cholera, dysentery, guinea worm, typhoid and intestinal worms.
2. Eighty-six percent of all urban wastewater in Latin America and the Caribbean, and 65 percent of all wastewater in Asia, is discharged untreated into rivers, lakes and oceans.
3. Nearly one million children die each year from diseases caused by air pollution inside their own homes. Over 75 percent of households in most Asian and African countries cook with solid fuels, such as wood, dung, coal or crop waste, which produce a black smoke that, when inhaled, may give rise to, or worsen pneumonia and other respiratory infections" [ibid.].

These grim data on children's health illustrate the fact that saying that prevention of disease and disability is the keystone of public health is an easy statement to make. How to make prevention a reality is the tough act. How to make prevention a reality leads one to a consideration of public health practice. Fortunately, the concept and practice of public health have had the good fortune of thoughtful, dedicated, energetic practitioners. With the passage of time and from lessons learned, we now have guidelines that serve to drive modern public health practice. The following section discusses key elements of modern public health practice.

 "Children are the main sufferers of environmental hazards. It is unacceptable from every point of view that the most vulnerable members of a society should be the ones who pay the price for failures to protect health from environment dangers."   Dr. Lee Jong-wook, WHO Director General [21]

1.5.1 PUBLIC HEALTH PRACTICE

Modern public health practice has evolved to include the components shown in Figure 1.1, which was developed by the Centers for Disease Control and Prevention (CDC). The pyramid shown in this figure illustrates infrastructure components at the bottom, supporting scientific and technical capabilities in the middle tier, and public health programs in the apex [23]. The infrastructure shown in Figure 1.1 will depend on other existing social infrastructures. These include form of governance, communication systems, transportation, technology available, health care system, and people and economic resources. Put in different words, the public health infrastructure in the United States would differ somewhat from that in Great Britain. Although this diagram was

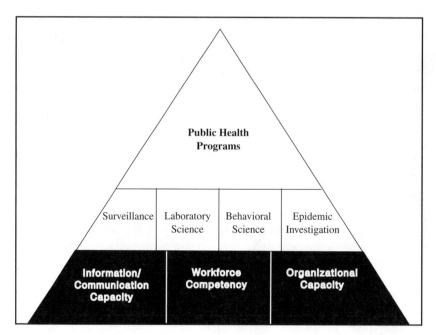

FIGURE 1.1   Elements of public health practice (Adapted from [22]).

developed by the CDC, the contents of this diagram also apply to state and local public health agencies, although the levels of investment of resources in a particular area (e.g., surveillance) will vary between governmental public health agencies. These elements are described in the sections that follow.

1.5.1.1   Organizational Capacity

This element represents the authority of an organization (e.g., EPA, CDC, Missouri Department of Health, Carroll County, Kentucky, Health Department) to receive and expend public monies. Legislative bodies such as the U.S. Congress and state legislatures are authorized by constitutional mechanisms to create organizational entities (e.g., a new public health agency), raise public funds through taxes and other means, and appropriate funds to governmental agencies. The created government organizations are organized into functional components (e.g., offices, divisions) and programs, based on authorizing legislation. Authorizing legislation at the federal level includes generic legislative acts, such as the Public Health Service Act of 1912, as amended, which authorizes specific federal public health activities, as well as specific legislative acts, and the Comprehensive Environmental Response, Compensation, and Liability Act of 1980, discussed in chapter 8, which authorized the creation of the Agency for Toxic Substances and Disease Registry, a federal public health agency.

1.5.1.2 Workforce Competency

Organizational capacity goes for naught if the workforce populating the organization is inadequate to do the tasks of the organization. The workforce of a public health organization must be adequate both in terms of numbers of persons as well as being professionally competent. The number of people within any organization depends on the organization's budget, which is appropriated by a legislative body. Also, some federal and state statutes will stipulate ceiling levels that specify the maximum number of agency employees. Another determinant of numbers of persons is how the organization executes its authorizing legislation. For example, some organizations are legislated to award and oversee grants awarded to academic institutions and individual researchers. The numbers of persons required to manage an agency's grants program are generally fewer than what is required to operate a heavily laboratory-oriented program.

Workforce competency also pertains to assuring that an organization's staff are well trained and professionally up to date. This obvious statement assumes that workforce competency spans the whole of an organization's workforce. For instance, professionalism in office administration is as important as professionalism in epidemiology in terms of the overall well-being of a public health organization. Maintaining workforce competency begins with hiring competent workers and proceeds through programs of internal and external training and performance reviews. Unfortunately, some organizations during periods of tight, constrained budgets will decrease or eliminate the funds directed to workforce training and development. This is a recipe for the downward spiral of an agency's effectiveness and performance. If workforce training is abandoned by an organization, long-term relevance of the organization will eventually be questioned by legislative bodies, the agency's parent organization, or the public, leading to questions of an organization's need to exist.

Workforce training and development programs are offered by various federal public health agencies, such as the CDC and FDA, state health departments, and private sector health providers. State health departments often provide training to county and municipal health departments and health care providers. In turn, local health departments offer training and other education opportunities for their own staff and members of the public that they serve. Another significant source of workforce training is provided by professional societies such as the American Medical Association, the American Public Health Association, the Society of Toxicology, and many others. These kinds of professional societies also serve an important purpose of credentialing those members who meet conditions of continuous education and testing.

1.5.1.3 Information/Communication Capacity

Silence or ineffective communication on important matters of public health is poisonous to the practice of public health. Given that prevention of disease and disability is the essence of public health, inability to communicate risks to health and well-being means that preventions won't occur. Even if organizational capacity and workforce competency are in place within a public health organization, lack of information and communication capacity will disserve these other two infrastructure elements.

A public health agency derives its communications to the public from science-based research findings, translated by communication specialists into effective public health messages. Public health communicators can range from an individual health care provider (e.g., attending physician) to large offices of media relations found in some federal public health agencies. For example, several federal, state, and local health departments actively outreach to the general public and populations at risk of contracting HIV. These programs of outreach and education have been credited with reducing the spread of HIV infection in the U.S. population. In an environmental context, EPA has effectively communicated the cancer risk of exposure to radon gas formed by the natural decay of uranium that is found in nearly all soils. Homeowners are advised to check for radon gas in their indoor ambient air, basements, and crawl spaces under houses.

1.5.1.4   Surveillance

Surveillance can be defined as a data-collection system that monitors the occurrence of disease (disease surveillance) or the distribution of hazard (hazard surveillance). Such systems are the eyes and ears of public health practice. Surveillance systems typically collect data from individual health care providers, hospitals, and entities such as health maintenance organizations. For example, state-based surveillance of birth defects and reproductive disorders has emerged, principally by way of federal grants. Other examples include surveillance of blood lead levels in children and, in some states, workers. These kinds of birth defects and blood lead data are typically collected by county and municipal health departments, reported to them by individual physicians, hospitals, and other health care providers.

Evaluation of surveillance data can reveal early, unusual patterns of disease or disability. In the case of disease outbreaks (and other health events), physicians' detection and reporting of sentinel cases is of great importance. Early detection provides public health agencies with an edge in developing and implementing any prevention programs. For example, identification of an unusual type of cancer, Kaposi's Sarcoma, led to the identification of what subsequently became known as Acquired Immunity Deficiency Syndrome (AIDS). The appearance of a rare cancer in numbers exceeding normal expectations was an alert that a possible health problem was developing. In time, programs of disease prevention evolved. Without an active disease surveillance system, the AIDS epidemic could have spread more quickly and with even more devastating effects on the public's health.

1.5.1.5   Epidemic Investigation

Referring again to Figure 1.1, epidemic investigation of suspicious patterns of disease or disability surfaced by surveillance systems is a key element of public health practice, whether at a federal, state, or local health department level. Epidemiologists are sleuths who examine patterns of morbidity or mortality and attempt to relate them to likely or plausible risk factors. Examples of epidemic investigations include those specific to risk factors in heart disease (e.g., the role of high density lipids in blood), Legionnaire's Disease, occupational injuries, mortality attributable to use of handguns, patterns of suicide in adolescents, and the spread of West Nile Fever through mosquito bites.

From epidemic investigations that identify specific risk factors (e.g., cholesterol levels or mosquito bites) can flow such prevention activities as public education efforts, vaccination programs, workplace redesign, and personal life-style changes (e.g., cessation of tobacco usage or choosing to reside in an area distant from sources of pollution). Findings from epidemic investigations, if of sufficient gravity for public health, can lead to legislative actions, which in turn, provide public health organizations with authorities and resources. A case in point is the legislative response to the AIDS epidemic, a response that has led to appropriating budgets and authorizing AIDS research, surveillance, and public education programs on AIDS prevention.

1.5.1.6 Laboratory Science

Laboratory science is defined here as those activities comprising laboratory research and laboratory practice. Bearing in mind that prevention is the central thesis of public health practice, findings from laboratory research can serve as powerful anticipatory data for prevention responses. While surveillance data are vital and indispensable to public health practice, such data nevertheless represent public health outcomes that have already occurred. In distinction, findings from laboratory research can sometimes serve as predictors of possible adverse health outcomes if interdictions are not taken. An example would be the laboratory testing for toxicity of new chemicals intended for use in consumer products, prior to introducing them into general commerce. Should toxicity be evident under laboratory conditions, appropriate interdictions could include abandoning the product or reformulating it and then retesting the modified substance or product.

> Bearing in mind that prevention of disease and disability is the central thesis of public health practice, findings from laboratory research can serve as powerful sentinel data for prevention responses.

Laboratory practice, as distinguished here from laboratory research,[6] is the establishment and application of laboratory services in support of public health programs to prevent disease and disability. Laboratory services can include measurement of toxicants in body tissues such as blood, bacterial levels in environmental media such as water supplies, viruses in human tissues or feral animals, and reference standards against which other laboratories can compare for accrediting purposes. In particular, the importance of laboratory practices as a means to help characterize exposure to chemical and biological agents cannot be overemphasized. Exposure characterization substantially strengthens epidemic investigations and surveillance programs. Moreover, exposure data can be used to hone public health prevention efforts. For example, measurements of blood lead levels in young children can aid public health interventions. Higher lead levels carry an urgency of medical intervention, while lower blood levels can trigger monitoring programs in order to identify and eliminate sources of children's contact with lead in the environment.

1.5.1.7   Behavioral Science

A relatively new element of public health practice is behavioral science. Traditional public health programs have relied on surveillance, laboratory science, and epidemic investigations, all of which have been used to build a data-based platform for specific public health prevention actions. Prevention efforts were largely focused on populations at risk of disease or disability. Individuals within the populations were given less attention, largely because of how surveillance systems and epidemic investigations yield their results. However, primarily from the public health experience in preventing the spread of HIV, public health agencies gradually recognized that too little was known about the personal behaviors of persons at risk of adverse health effects. For example, what are the personal determinants of why individuals choose to smoke cigarettes? Why do individuals still choose to smoke, given the overwhelming evidence of adverse health problems that are a consequence of smoking? Knowing answers to these kinds of question provides data for further refinement of anti-smoking public health endeavors. The addition of psychologists, sociologists, and behavioral scientists to the workforce of public health agencies holds the promise of better understanding individuals' risk-taking behaviors and subsequent refinement of prevention programs.

1.5.1.8   Public Health Programs

Categorical public health programs are at the apex of the public health practice pyramid developed by CDC [24]. These programs are the culmination of both legislated mandates as well as agency-determined public health needs. In both cases, funding must come from funds appropriated by a legislative body such as Congress or a state legislature. Categorical programs, as suggested by Figure 1.1, are built upon core public health infrastructure (base of pyramid) and supporting scientific and technical capacities (pyramid's middle level). Which categorical public health programs are expressed and exercised by a specific public health organization depends on the organization's authorities. This leads to differences in categorical programs between federal public health agencies, state health departments, and local health agencies.

As an example, categorical public health programs at the CDC include programs in immunization, sexually transmitted diseases, environmental health, chronic disease, birth defects, infectious diseases, bioterrorism, and injury prevention and control. These programs involve, at different levels of resources, surveillance, laboratory science, epidemic investigation, and behavioral science. State health departments' categorical health programs often mirror those of CDC, since the federal agency is a primary funding source for state and local health programs.

1.5.2   COMPARING PUBLIC HEALTH PRACTICE AND MEDICAL PRACTICE

While public health practice largely focuses on populations at risk of adverse health effects, and medical practice pertains largely to curing individual patients, this distinction should not be considered absolute. To elaborate, any public health practitioner must always remember that populations necessarily involve individuals. How an individual within a population will respond to a particular public health intervention (e.g., vaccination programs) must be of great importance to public health officials. Similarly, health care providers who provide medical treatment to individual patients should be alert to the possibility of applying their treatment methods to groups in need of health care. As an example, surgical procedures that are developed and administered to individuals can be generalized to provide relief to groups needing the same surgery.

Although both public health practice and medical practice have a common goal, healthy persons, there are important differences between the two disciplines. Some of the key conceptual differences are illustrated in Figure 1.2. Obviously, there are many more differences in addition to those illustrated in the figure. The following sections discuss the differences shown in Figure 1.2.

1.5.2.1   Focus of Attention

As previously noted, public health practice is focused on preventing disease and disability in populations (i.e., groups of people with common characteristics) who are at presumed health risk. The presumption of risk may come from surveillance, epidemic investigation, or laboratory science. Examples of populations presumed at risk for adverse health outcomes include persons with high cholesterol levels, workers exposed to workplace hazards, and children who lack vaccinations. Even as behavioral scientists attempt to understand the lifestyle behaviors of individuals within populations at risk, the focus nonetheless remains on preventing disease and disability in populations.

Medical practice, in distinction to public health practice, focuses on curing individuals' disease, mitigating disabilities, and relieving suffering. It is part science and remains part art. The practice of medicine, whether human or animal, draws upon contemporary scientific knowledge from physiology, anatomy, psychology, chemistry, and physics, just to name a few areas of science critical to medical practice. How and when to apply medical practices (e.g., how much of a therapeutic drug to administer) requires both scientific knowledge as well as personal experience. The art of medical practice stems from a practitioner's experience, the so-called physician-to-patient relationship. For example, for reasons not well understood, patients can improve their recovery from disease or disability by adopting a positive attitude about their prognosis. health care providers' reinforcement of positive attitudes can therefore contribute to the healing process.

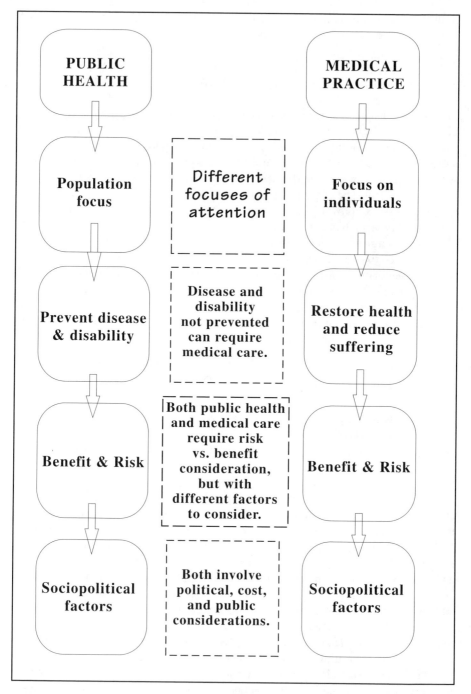

FIGURE 1.2   Comparison of public health and medical practice approaches.

1.5.2.2 Prevention and Restoration

When prevention of disease and disability does not occur, persons in ill health or diminished capacity seek medical care (Figure 1.2). Obviously, some persons will need medical care even when specific public health prevention programs have succeeded in reducing health risks. For example, persons who wisely chose not to smoke cigarettes may, however, experience a tobacco-related disease due to exposure to secondhand tobacco smoke from other persons.

For the practice of medicine, first, do no harm.

Medical practice is directed to improving individuals' health and reducing pain and suffering. The creed of medical practice is, "First, do no harm," an admonition that has philosophical, ethical, and practical implications—philosophical in the sense of ideals to which to strive, ethical in the context of human worth, and practical in the expression of health providers' practices. This creed, which is often erroneously attributed to the Hippocratic Oath, comes from other works of Hippocrates [25].

Public health practice does not have a manta similar to that practiced by providers of medical care. One could argue that "First, do no harm." should equally apply to the practice of public health. And indeed it does; no public health official would knowingly undertake prevention programs that might harm populations at health risk. Yet, public health programs sometimes *do* convey health risks (i.e., potential harm) to *some* individuals within a population. For example, vaccination against diseases can cause adverse reactions in a very small number of individuals. The kind of adverse reaction will depend on individual factors (e.g., age of person), kind of vaccine, how the vaccine was administered, and so forth. As another example, exercise programs to prevent obesity and to strengthen cardiovascular performance can produce additional wear and tear on body joints in some individuals, again depending on personal factors, kind and duration of exercise, and so forth. Public health officials must balance these kinds of health risks in individuals against the greater health benefits to a population. Therefore, the admonition, "First, do no harm" does not ring quite as true in public health as in medical practice.

For the practice of public health, first be and do good.

Given the preceding narrative, "First, be and do good." is proposed here as a creed for public health practice. Being good requires public health officials to be competent, ethical, altruistic, and effective. Doing good connotes actions that will knowingly improve the public's health and well-being. Moreover, mustn't one be good to do good? Between being good and doing good lies a challenging territory that spans: accumulating facts, considering relevant science, assessing benefits vs. risks, and coordinating outreach to public health partners (e.g., local health departments) and populations at known or hypothesized health risk.

1.5.2.3   Benefits and Risks

Both medical care providers and public health practitioners must assess benefits of a health action evaluated against potential health risks. Consider the surgeon who must decide if a patient is too ill or too elderly to merit undertaking a complicated surgical procedure (e.g., an organ transplant that could place the patient's life at risk). Simply put, will the patient survive the surgery is a question faced by the surgeon. Similarly, physicians and hospital administrators must assess the benefits of patients' extended post-operative stay in hospital versus the financial costs to the hospital and the patient.

These kinds of benefit vs. risk decisions are also made by public health organizations. Will the health benefits to a population at putative health risk outweigh any risks to the target population? Consider the public health official who must decide if spraying a mosquito-infested area to reduce the numbers of mosquito-borne infections outweighs the small risk of some persons being sensitive to the toxicity of the pesticides being sprayed. Consideration of benefits would include estimating the numbers of persons benefiting, nature of benefits (e.g., fewer mosquito bites), social improvements (e.g., decreased incidence of mosquito-borne disease), and impacts of benefits on other public health interests (e.g., decreased health care costs).

The astute reader will have observed that "benefit vs. risk" was the term of choice in the foregoing narrative. Considerations of *costs* associated with putative benefits of a specific public health program or an individual medical procedure have not been discussed. This "lapse" stems in part from historical beliefs that human health was too important to be reduced to economic bases. In a sense, a naive belief prevailed that "the money will always be there" to fund public health programs. Until the early 1980s, budgets for federal public health and environmental protection programs were generally moderately increased each year, with increases occurring mostly in support of new program initiatives (e.g., the development in the 1960s of new federal public health programs to research the effects on human health of environmental hazards such as air pollution). However, in the 1990s the U.S. public became less supportive of what they perceived as ineffective government programs. Further, these concerns were interwoven with those of escalating government budget deficits and rising taxes. As a result, federal budgets, in particular, came under greater congressional scrutiny as to justification, especially beginning in 1994 when Republicans gained control of both houses of Congress. From these changes that began in the 1990s have come the concepts of cost containment and cost-benefit analysis.

Cost-benefit analysis is a systematic assessment of whether the cost of an intervention is worth the benefit by measuring both in the same units; monetary units are usually used [26]. Its utility as a tool for policy formulation will be discussed in a subsequent section of this chapter. Cost-benefit analysis forces health care providers and public health officials to estimate the economic consequences of proposed programs and actions. What, for instance, would be the economic impact on a local health department if mandated by county commissioners to screen all young children for exposure to lead in the environment? Would the benefits of a small increase in IQ in children due to a community-wide program of removal of lead in the environment be worth the cost? Assume that the analysis affirms the need for a children's lead surveillance program. What kind of program would be the most effective? And, moreover, how would "effective" be defined by the local health department?

These questions are addressed through what is called cost-effectiveness analysis. Cost-effectiveness analysis measures the net cost of providing a service as well as the outcomes obtained [26]. Outcomes are reported in a single unit of measurement, e.g., per life saved, per life year gained, and per pain or symptom-free day [27]. Returning to our example of a local health department's decision on how to implement a children's lead prevention program, if resources were limited, would a more focused program on those children with greatest risk of exposure be more efficacious? This is the kind of question that now permeates public health and health care organizations.

1.5.2.4 Sociopolitical Factors

Some persons might think that preventing disease and disability in populations (public health practice) and caring for sick individuals (medical practice) are sufficiently altruistic and noble to be spared of sociopolitical influences. Moreover, shouldn't these practices be on a sufficiently high plane to be spared the rough and tumble of political negotiations, deals, and compromises? The answer is no. Why? This is because both public health and medical practice are to a considerable extent supported by public funds. The appropriation of any public funds is always a political exercise. Elected officials must decide funding priorities, and the fundamental decision on who gets what. Special interest groups attempt to influence public health appropriations, as also occurs with appropriations for medical care programs such as Medicare and Medicaid, hospital construction, and such. Sociopolitical considerations are therefore essential for facilitating decision-making processes in both practices of public health and medical care.

1.6 THE ROLE OF GOVERNMENT

A historic role of government is to promote the public good and to protect against threats to the public's well-being. Government by its very nature, represents a society's quest for protection of individuals in ways not always possible by individuals acting alone. Examples of government's protective role would include provisions for national defense and for assurance of public health. That is not to say that nongovernment organizations do not have a role in meeting certain societal needs. However, when authorized, government can provide resources and authorities not available to nongovernment organizations. This section will describe the roles of federal, state, and local governments in establishing environmental health policies and practices.

1.6.1 FEDERAL GOVERNMENT

Of the three branches of government in the United States, the federal government was the last to assume a significant role in protecting the natural environment and related consequences to public health. This will be discussed in chapter 3. However, suffice it to say here that local and state governments predated federal involvement in establishing various environmental policies. For example, small villages took action to protect water supplies before either state or federal policies emerged. The village well was both a source of drinking water and a social gathering place. Similarly, some states (e.g., California) developed air pollution control programs prior to federal action.

Over time, the U.S. federal government achieved primacy in developing national environmental health policies. This occurred because of a slow awakening in the U.S. public and members of Congress that environmental hazards (e.g., air pollution) were no respecter of local and state boundaries. Moreover, until the early twentieth century, the role of federal government was largely confined to areas specifically stated in the U.S. Constitution (e.g., national defense and foreign affairs). In the early twentieth century, federal legislation and laws emerged that were intended to protect the public's health. As presented in chapter 6, the Federal Meat Inspection Act and the Food, Drug and Cosmetic Act, both legislated in 1906, brought the federal government into the environmental health arena. In the mid-twentieth century, Congress enacted legislation that provided federal primacy, working in cooperation with states, to control such environmental hazards as air pollution, water contaminants, pesticides contamination of land, and improper disposal of solid and hazardous waste. These federal programs are increasingly being coordinated with international environmental organizations through the mechanism of national treaties, as described in chapter 9.

1.6.2 STATE GOVERNMENT

State governments have a significant responsibility for controlling environmental hazards that can impair their residents' health. Some state environmental health programs fulfill federal statutory responsibilities (e.g., implementing Clean Air Act regulations developed by EPA, but enforced by states with EPA approval). Other state environmental programs derive their authorities and resources from state legislatures. Such legislation addresses environmental hazards that are specific to state jurisdictions. State-based statutes that define the legal limit for blood alcohol levels in vehicle drivers are such an example. How states develop and implement their environmental health programs varies between the states. In some states, environmental health programs are placed in departments of health; in other states, similar programs are located in departments of environmental quality or similar entity. Environmental health programs will differ between states, but many programs address common environmental hazards, such as hazardous substances, sanitation, and emerging problems like the spread of the West Nile virus.

As an example of one state's environmental health programs, Georgia's Division of Public Health, which is part of the state's Department of Human Resources, administers a large public health effort that includes: epidemiological and outbreak investigations, maternal and child health programs, emergency medical services, vital records and health statistics, chronic disease prevention and health promotion, laboratory services, and infectious disease prevention, including sexually transmitted diseases. Within the Division of Public Health is located the Environmental Health and Injury Control Branch. This branch is further divided into four primary areas of coverage: environmental services, chemical hazards program, injury prevention, and emergency medical services. A brief description of these four programs follows [28]. The first two programs are those most germane to the purposes of this book.

Georgia Environmental Services. The mission of Environmental Services is stated to be "[t]o provide surveillance of environmental factors which may adversely affect the health of people and for compliance with state laws regulating the following." These programs include:

1. Food service establishments—More than 17,500 food establishments are monitored for compliance with food service rules and regulations to reduce potential problems of food borne illnesses. Evaluations are made by district and local environmentalists.[7]
2. Tourist courts—Approximately 1,300 tourist courts[8] are evaluated by district and local environmentalists to minimize risk of illness or injury.
3. Individual on-site sewage management systems—On-site evaluation of septic tank installations are conducted by district and local environmentalists.
4. Water Impoundments—Evaluated for permitting purposes while construction is in progress to control or prevent the incidence of malaria and other insect-borne diseases and to maintain the sanitary quality of water for public uses.
5. Occupational health—Upon request, investigation and monitoring is conducted of both workplaces and private residences.
6. Childhood lead poisoning prevention—Directs efforts to reduce and prevent children's exposure to environmental and occupational lead sources. This includes blood lead screening, laboratory support, and care coordination of services.
7. Related services—Other services under the branch's purview include swimming pools, evaluation of individual water supply systems, oversight of a rabies control program, and access of handicapped persons' access to food service establishments and tourist courts.

Georgia Chemical Hazards Program. The Chemical Hazards Program provides public health assessments and health consultations, technical assistance, community education, staff training, and referrals for district and local health departments, residents, educators, health care providers, and state and federal agencies.

1.6.3 LOCAL GOVERNMENT

As previously noted, environmental protection and public health programs, following the establishment of EPA, have been placed in separate organizations within federal and most state governments. However, this separation does not usually occur at local levels of government. County and city health departments ordinarily handle a suite of environmental problems, coordinating their programs with state health and environmental agencies. Local health agencies conduct myriad actions for their communities, including a range of environmental health responsibilities such as food and sanitation inspections, pest control, and audits of environmental hazards. There is no national catalogue of local health departments' environmental health programs, but one organization has provided helpful information that provides a useful perspective on such programs.

1.6.3.1 Environmental Health Responsibilities

The National Association of County and City Health Officials (NACCHO) provides technical assistance to county and city health departments. They develop and advocate policy and political positions for their membership. As noted above, city and county health departments conduct a host of public health programs, including those in environmental health [29].[9] The National Association of County Health Officials (NACHO) was interested in ascertaining a national picture of environmental health priorities, as expressed by their membership.

In 1990, NACHO conducted a national survey of local health departments to assess needs and resources for environmental health programs [29]. Their survey did not define environmental health. The purpose of their study was to identify: (1) various environmental health issues that challenge local health departments; (2) how these challenges are being met; and (3) the kinds of education, training, and other support local health departments need to adequately assess, communicate, and reduce environmental health risks. The NACHO survey of 1990 still represents the only national data on environmental health priorities specific to local health departments.

A questionnaire was used by NACHO to survey a stratified random sample of 670 of the 3,169 local health departments operating in the United States at the time of the survey. The sample was stratified according to the size of the population served, which NACHO used as an indicator of resources required by the department. Representative percentages of the total sample within each population range were selected to reflect a national picture.

From the survey, the most frequent environmental health services reported by local health departments are shown in Table 1.4. These six services are intended primarily

TABLE 1.4
Environmental Services Most Often Provided by Local Health Departments and Percentage of Departments Providing Them [29]

Food protection	91%
Nuisance control	88%
Sewage treatment	85%
Private well testing	83%
Swimming pool inspection	83%
Emergency response	80%

to protect the public from chemical and biological hazards in food, drinking water, swimming pools, sewage, and nuisances such as animal vectors of disease (e.g., rabid animals). Local health departments are also involved in providing directly, or coordinating with others, emergency response services. Given the terrorist threat to the United States, the role of local health departments in responding to chemical and biological threats will continue to increase in importance.

1.6.3.2   Case Example: DeKalb County, Georgia

Because each local health department tailors its programs, including environmental, to the needs of its geographic area of coverage, some variation in activities occurs between departments. An example of one local health department will illustrate the range of environmental health programs.

DeKalb County, Georgia, is an area northeast of the city of Atlanta, with a population of approximately 600,000. It is one of the counties that constitute metropolitan Atlanta. The mission of the DeKalb County Board of Health is stated to be, "Our mission is to promote and provide quality preventive and primary care. The prevention of disease, injury, disability and premature death is the primary purpose of the DeKalb County Board of Health. We unite with individuals, families and communities to serve the people who live, work and play in DeKalb" [30]. In support of this mission, the Division of Environmental Health services the environmental health needs of the county. Its programs are grouped into four main areas [ibid.].

Food Protection Services: The Food Protection unit reviews and approves plans for new food service establishments, issues permits, and conducts ongoing inspections. Approximately 1,800 food establishments and services are inspected by the county each year [30]. The results from restaurant inspections are made available to the public by: (1) posting a copy of the inspection report in a prominent place in each restaurant inspected, and (2) placing the inspection reports on the county's web site. As environmental health policy, providing the public with information with which to make personal health decisions is a right-to-know policy.

In addition, the unit evaluates and issues temporary event food service permits for festivals, carnivals, and fairs. Hotels and motels are evaluated and inspected for food safety. The unit also investigates all foodborne illness complaints and refers for follow-up any significant findings to disease surveillance programs operated by the state of Georgia's Division of Public Health.

Residential Services: The Residential Services team provides rodent control assistance such as investigating infestations and controlling rodent populations through baiting, as well as rabies control services including alerting citizens to areas of infection, enforcing home quarantine for dogs and cats and locating persons exposed to rabid animals. Unsanitary conditions such as the presence of garbage, trash, and dead animals are investigated and eliminated.

Land Use Services: The Land Use Services group is involved with several aspects of sewage removal. They issue permits for new septic tanks after reviewing plans, conduct soil analysis, and inspect sites, and inspect repaired septic systems. In addition, the group inspects and issues permits for sewage pump-out trucks, issues permits for commercial sewage disposal systems, and tests well water for possible use as drinking water.

Technical Services: The Technical Services team conducts a variety of regulatory and educational activities. It issues public pool and spa permits and evaluates water chemistry and pool safety. Efforts to protect against environmental hazards include coordinating in-home lead assessment and evaluating and testing homes for radon gas. Occupational health and safety services involve reviewing all occupational fatalities and making recommendations to prevent injuries and fatalities. The team also reviews and approves construction plans for commercial trash compactors. Educating the public is an important part of the scope of services.

1.7 PUBLIC'S POLICY EXPECTATIONS

Having now defined various key words and terms, such as health, public health, environmental health, and policy and explained the key differences between the practices of public health and medicine; it is time to discuss some public policies that have current relevance to environmental health policies. Public policies are referred to here as actions taken in accord with current expectations of the U.S. public. Some public policies, as will be illustrated, are the consequence of legislation; others have evolved as matters of public education or products of advocacy groups.

1.7.1 ACCOUNTABILITY

The notion of accountability is rooted in both ethics and law. For the former, human experience has evolved through religious teachings and secular wisdom to hold a person accountable for his or her actions. Whether based in religion or secularism, it could be argued that avoidance of chaos is a societal goal. Chaotic social structures are inherently unstable and don't have a good chance of long-term survival. As a matter of law, an individual's accountability comes into question when societal expectations, expressed as a body of law, are not met. This is true whether the law is based on English Common Law, Native American Tribal Law, or national law based on religious theology. For example, if murder of another human being is forbidden, the commitment of the act will bring some kind of specified societal response (e.g., imprisonment).

Holding government and corporations accountable for their actions is a relatively recent public policy in the United States. At the federal level, Congress enacted the Government Performance and Results Act (GPRA) in 1993, in part to "improve Federal program effectiveness and public accountability by promoting a new focus on results, service quality, and customer satisfaction" [31]. This act was meant to hold executive branches of government more accountable to the public. To the extent that the act is meeting its goal is currently unknown. On a corporate level, the financial irregularities associated with the Enron Corporation in year 2002 contributed to public skepticism and demands for more controls on how corporations manage their financial accountability to employees, stockholders, and the public. In regard to environmental health policy, the public expects environmental and public health agencies to respond to their concerns and to be accountable for protecting environmental quality and human health.

1.7.2 COMMUNICATION OF RISK

The communication of risk to the public goes hand-in-hand with the use of risk assessment as public policy (chapter 11). This is a corollary of the public's right-to-know. The emphasis here is on how to best communicate risk in order to enhance public understanding and achieve health actions. Indeed, risk communication has become a specialty discipline in some academic institutions, leading to research on how to more effectively communicate risk and evaluate the impact. As an example, how should information about potential threats by terrorists to public safety be communicated? Should every threat, bogus or creditable, be relayed to the public? If not, what are the criteria for selecting those threats that should not be communicated? These are difficult questions that have no easy answers. Experience and research will be needed if these kinds of particularly challenging risk communications are to be effective in preventing acts of terrorism, and at the same time, not unduly raise the public's anxiety.

1.7.3 COST-BENEFIT ANALYSIS

Although the U.S. public looks skeptically at personal health decisions that are based on their financial cost—consider the negative reactions of many persons to health maintenance organizations, where costs allegedly drive decisions on health care —the emergence of cost-benefit analyses and hazard management decisions have become policy within government agencies and business operations. Cost-benefit policy has occurred in government because of legislative directives and federal court decisions and within business operations that must have a sense of costs associated with their products, working conditions, and consumer affairs. On the surface, considerations of cost and benefits to the public are important and reasonable. But current cost-benefit analysis necessarily forces decisions about the worth of human life, forces estimates of technology costs by using projection models, and forces decisions when adequate information may be lacking on the costs and benefits to different races, cultures, and age groups. Cost-benefit will remain a public policy cloth, but with frayed edges until better data on benefits become available.

1.7.4 ENVIRONMENTAL JUSTICE

How environmental hazards are experienced by people of color became a matter of social justice in the 1970s, as described in chapter 10. Minority communities and persons of low income expressed their belief that toxic chemicals, in particular, were being deliberately released into their communities from hazardous waste dumps, incinerators, and pollution from industry. The resulting expressions of concern led to the establishment of environmental justice, defined by EPA as "[t]he fair treatment and meaningful involvement of all people regardless of race, color, national origin, or income with respect to the development, implementation, and enforcement of environmental laws, regulations, and policies" [32]. Federal government policies emerged in the early 1990s that were intended to address a number of environmental justice concerns. For example, offices of environmental justice were established at EPA, the U.S. Department of Energy (DOE), and elsewhere. State governments have emulated the federal example by creating offices that investigate environmental justice allegations and recommend corrective actions. While the effectiveness of federal, state, and some private sector actions to prevent environmental injustices continues to be debated, the public policy to prevent environmental injustices has become a cornerstone in the foundation of U.S. social justice.

1.7.5 FEDERALISM

Federalism is a kind of government that is structured around a strong central (i.e., federal) government, with specified authorities retained by lower levels of government, such as states and local governments. How power is shared among the central and subordinate governments must be specified in a constitution that binds the parties. The U.S. Constitution is an example, having been developed by the country's founders after a looser confederation of states was found to be ineffective. Our constitution specifies the powers of the federal government and delegates all other responsibilities to the states. State constitutions, in turn, specify the degree of power-sharing with local governments.

The sharing of power and authorities between the U.S. federal government and states and local governments is a public policy that crosscuts almost all social programs and discourse in this country, including environmental statutes. As an example pertaining to environmental health policy, as will be discussed in chapter 5, states have the primary responsibility for enforcing air pollution regulations developed by the federal government (i.e., EPA), with overall responsibility for the development of air quality standards vested with the federal government.

1.7.6 POLLUTER PAYS FOR CONSEQUENCES OF POLLUTION

In the 1970s, environmental groups adopted the strategy that those who cause pollution should pay the costs for its effects on the environment and remediation. The first federal expression of this theme is found in the Comprehensive Environmental Response, Compensation, and Liability Act of 1980 (chapter 8). In particular, companies and others that created hazardous waste dumps were liable for the costs of cleanup of the dumps and associated effects on human health and natural resources. A similar legal philosophy spread to the European Union[10] and elsewhere. Polluters paying for the consequences of their pollution is a kind of accountability policy, but is specific to environmental health. This policy has led to copious litigation between those who generate pollution and those, particularly government, who have the responsibility for enforcing remediation of polluted sites and attendant financial costs under cost recovery statutory authorities. This is discussed in chapter 8.

1.7.7 PREVENTION IS PREFERRED TO REMEDIATION

Common sense tells us that avoiding a problem is preferable to having to fix its conse-quences. Prevention of disease and disability is the cornerstone of public health. This cornerstone is policy at all levels of public health, from federal to state to local health departments and their programs. Disease and disability reduce a person's quality of life, can lead to costly health care, and lessen societal strength by eliminating or reducing ability to work and contribute to society. Although not always evident, environmen-tal statutes in general are expressions of the prevention policy. All such statutes are predicated on protection of human health and environmental quality. This protection is addressed through control of pollution levels in the environment, as pursued through regulations and standards and their enforcement.

1.7.8 PRODUCT SAFETY

The U.S. national economy has over the relatively brief existence of the country been based first on agriculture and trade, followed by industrialization, the manufacture of consumer goods, and most recently, information services and banking. Each change in economic engine has affected public policies of the time. The current U.S. economy is heavily based on what is called consumerism, which is the production, sale, and use of per-sonal products as diverse as personal computers, automobiles, clothes, and video games.

With consumerism has come public policy specific to product safety. Consumers do not want products that could harm them or their children. The result has been federal resources directed to the prevention of harmful consumer products' entry into commerce. The Consumer Product Safety Commission (CPSC), in particular, identifies commercial products (e.g., children's toys) that can be hazardous. For example, in 2003, a scooter manufacturer voluntarily recalled about 30,000 electric scooters and 55,000 electric mini-bikes due to eighty reports of injuries caused to young children. The injuries were associated with a malfunction in the products' electrical control circuits, causing the bikes and scooters to continue to run after the power had been cut off [33]. Other recent notable examples of products perceived unsafe by the U.S. public include automobile tires, sport utility vehicles, and canned baby food—all from specific manufacturers. These flawed products quickly attract the attention of newsmedia, public advocacy, attorneys, and elected officials; all of whom decry the product, its producer, and the failure to protect the public. Eventually, these public outbursts and private negotiations lead to specific products being recalled by the manufacturer and strengthened inspection systems by the producer, and where warranted, by the government.

1.7.9 PUBLIC'S RIGHT-TO-KNOW

It has become public policy in the United States that individuals have the right to know of conditions that may be hazardous to their health and well-being. The public's right-to-know is not absolute. For example, matters of national security and confidential business secrets are excluded from public view, unless ordered by a court. While the importance of right-to-know may seem self-evident, it has not always been the case, especially in the realm of information available to workers and consumers of commercial products. Not until the passage of the U.S. Occupational Safety and Health Act in 1970 were there general requirements to inform U.S. workers of on-the-job hazards (e.g., hazardous chemicals). Similarly, producers of consumer products had no requirement to inform U.S. consumers that flaws in product design or manufacture could be hazardous to them until enactment of the federal Consumer Product Safety Act of 1972. Other examples will be given later in chapters 4–8, where specific federal environmental laws are discussed.

1.7.10 RISK ASSESSMENT

How do we determine what level of risk is caused by a particular hazard (e.g., ozone in ambient air)? What scientific data should be examined and how should they be considered? And who should make these kinds of decisions about risk to human health and ecological systems? These questions are neither new nor confined to government. As individuals, we decide whether to use tobacco products, wear seatbelts while riding in vehicles, or reside in an area of dense automobile traffic, even in the face of information that advises different courses of action. Government agencies are directed to assess risk and take actions when risks are presumed in need of reduction. Businesses must pay attention to the risk posed to consumers by their products. So what's new about risk assessment and its adoption into policy by government? It is the public's gradual acceptance of a formal process, risk assessment, that has become a policy (chapter 11).

Risk assessment and risk management have become particularly dominant in the area of environmental health. As discussed in chapter 11, beginning circa 1980, federal regulatory agencies, primarily EPA and the OSHA, have implemented formal risk assessment procedures to estimate the degree of risk posed by individual environmental hazards. The impetus for this development came from federal court decisions which found that EPA and OSHA regulations had failed to establish degree of risk in certain proposed regulatory actions. In effect, the courts ruled that risk must be consequential, not trifling or unsupported, in any regulatory action.

The framework for the regulatory agencies' risk assessment procedures came from the U.S. National Academy of Sciences [9], which articulated the key components of risk assessment (toxicity assessment, dose response assessment, exposure evacuation, risk characterization) and recommended that risk assessment be kept separate from considerations of risk management. The academy's report and recommendations have had a profound impact on how environmental hazards are regulated.

As described in chapter 11, the formalization of risk assessment and risk management by federal agencies, international organizations, and business entities is a product of the 1980s. In particular, a Supreme Court decision in 1980, in effect, mandated that federal regulatory agencies use risk assessment when developing regulations and standards. The case brought before the Court was an industry challenge to an regulation developed by the Occupational Safety and Health Administration (OSHA) that would have lowered the exposure limit for benzene, a known carcinogen, from 10 ppm to 1 ppm in the ambient air of workplaces. The Court ruled that the Occupational Safety and Health Act required OSHA to establish that current allowed exposures to benzene posed a significant risk to workers' health, a decision that in effect told agencies to use risk assessment when developing workplace standards [34].

1.7.11 SOCIAL SUPPORT

Cultures have over eons of evolution learned that their survival depended on social support systems. These included hunting in groups for food, banding together to defend territory, and traveling in groups on trading expeditions. Living in close proximity, not as isolated individuals, gave protection to groups, eventually forming villages, cities, and nation states. Nations have developed social support structures that are intended to enhance the survival of both the state and the individual. Assistance to persons who need food, shelter, and education is commonplace, generally worldwide. In the United States, assistance programs include public education, subsidized housing, health services to the elderly and the indigent, and monetary subsidies to farmers, among many others. These are examples of public policy meant to provide social support to meet basic needs of members of a society. Unmet basic needs can lead to societal disruptions (e.g., political turmoil, chaos) that reduce a society's ability to protect itself, to foster economic gains, and enhance quality of life for individuals. How much of a society's resources should be devoted to social support systems is, and will remain, a legitimate debate.

1.8 CRITICAL THINKING

Establishing environmental health policy—indeed, any kind of policy—is a difficult undertaking. Much more is said in chapter 2 about the mechanics of policy making, but suffice it to say here that critical thinking is vital to the process of policy establishment. Critical thinking means different things to different people. However, for the purposes of this book, *critical thinking* means:

- Asking why and what if?,
- Looking beyond the obvious,
- Identifying interconnections,
- Understanding the players involved in policy making, their roles, and their motivations,
- Recognizing that complexity is the norm, not the exception.

As the five elements imply, critical thinking is an intellectual process that attempts to probe below the surface of contemplated actions in order to identify potential consequences.

Consider the following example. Assume that a community advocacy group has asked a member of a county's board of commissioners to support a proposed county ordinance that would require recycling of household waste. The proposed plan would require homeowners to separate their household waste into various categories of recyclable materials prior to pickup by trash crews. How should the commissioner proceed? The commissioner's critical thinking could proceed along these lines:

- What benefits of recycling household waste would accrue to homeowners? Moreover, how might the county's ordnance, if enacted, be perceived by individual homeowners? What would be the financial costs of the recycling program to the county? How would funds be found to operate the program? What if the program began operation and then became the target of citizens' discontent?
- If it is obvious that the community advocacy group has some citizens' support, just how extensive is their support? It is also obvious that the county's residents support environmental protection programs in general, but does this support extend to recycling of household waste?
- Interconnections between the state's environmental protection department and possibly the EPA would be required to effectuate any county-wide recycling program. What roles will these other layers of government assume in regard to recycling?
- Start-up and operation of a county-wide recycling program would be a difficult and complex operation. Does the community advocacy group understand the complexities?
- How will this get explained to the group and to the county residents should the board of commissioners approve the recycling proposal?

This kind of critical thinking should be applied during the course of any policy development, whether it be a personal policy or a federal environmental health policy.

1.9   SUMMARY

This chapter has summarized the evolution of environmental health, placing the evolution in a historical context. Humankind's search and aspirations for a less hazardous environment date from antiquity. Notable were ancient civilizations' attempts to acquire potable water and sanitary disposal of human wastes. The struggles of ancient peoples for healthful environmental conditions have continued into modern times. Only the means toward the end, control of environmental hazards, have changed.

Described in this chapter is one relatively modern development for controlling the effects of environmental hazards. The emergence of government as the primary force in hazard control is, over the long course of history, relatively new. In the United States, the triad of federal, state, and local governments all have roles to play in controlling environmental hazards. The traditional triad of government partners is essential for effective public health policies and programs. Without this triad, public health programs of disease and disability prevention, such as immunization and quarantine, would be far less effective. However, as will be discussed in the following chapter, this triad becomes strained when the subject becomes the control of environmental hazards. In effect, the triad must multiply itself in order to accommodate their federal, state, and local environmental protection partners.

The purpose of this chapter was to set the stage for what follows in the subsequent chapters. In order to navigate through the often turbulent waters of environmental health policy, a clear sense of direction and firm grip on resources are required. Sense of direction is predicated on an awareness of the linkage between environmental health policies (as often expressed in environmental statutes) and the control of specific environmental hazards (e.g., contaminants in drinking water). Resources required for environmental health policy navigation begin with a working knowledge of key definitions (e.g., *environmental health* and *policy*). Without a common and mutually understood vocabulary, effective communication is impossible. This lack of mutual understanding of terms often ferments disagreements between community groups and government officials when addressing concerns about toxic substances.

Another resource described in this chapter includes an appreciation of contemporary public policies that the U.S. public expects of elected officials and others who have the power to affect their health and well-being. This chapter described policies of accountability, communication of risk, cost-benefits analysis, environmental justice, federalism, polluters paying for the costs of their pollution, prevention being preferred to remediation, risk assessment, and social support. These public policies are important in their own right, but are all the more important when developing environmental health polices. Policy makers must be aware of these polices when developing new environmental policies or revising existing policies.

This chapter has set the course for an introduction to environmental health policy. The next chapter provides a discussion of the steps and processes involved with policy making.

1.10 POLICY QUESTIONS

1. Using the definition of *policy* provided in above, discuss: (a) Who are policy makers? Give examples of those who affect your daily life, (b) How do environmental polices (e.g., environmental protection) affect you? (c) What is your personal policy and what are its benefits to you.

2. Using the WHO definition of *environmental health* given earlier in this chapter, (a) give examples of *social and psychological factors* that could fit within the definition, and (b) discuss their importance to you in comparison with *chemical and biological factors*.

3. If *politics* is "the complex of relations between people living in society," discuss what you consider to be the positive aspects of politics, then discuss the negative aspects. Be specific and give examples based on your own life experiences.

4. Discuss how the judiciary can set environmental health policy. (a) Under what authority can courts change policies expressed in law by legislative bodies? (b) Give examples drawn from this text and other sources.

5. It is asserted in this chapter that for the practice of public health, "First, be and

do good." Do you agree, or disagree with this assertion? Discuss your reasoning.

6. Eleven elements of public policies in the United States of relevance to environmental health were presented in this chapter. Select any five and discuss each one's impact on your life. Give examples.

7. Do you think that the Public's Right-to-Know is absolute (i.e., government should *never* withhold any public health information from the public)? If not, discuss situations where, in your opinion, information should be withheld. Give specific examples and justify your reasons.

8. Assume that you are a newly-hired environmental health (EH) specialist working for a county health department in the United States. As the only EH specialist, the department's director asks you to provide a prioritized list of environmental health problems. (a) How would you proceed to develop the list? (b) What would be your key assumptions in developing the list? (c) How would you involve the public, or would you (involve the public)?

9. Given the chapter's discussion of historical environmental hazards, discuss the significance of historical data on modern-day environmental problems, as you have personally experienced them. Be specific.

10. Rachel Carson's book *Silent Spring*, published in 1962, is given much credit for enhancing concern in the U.S. public about environmental hazards. Discuss, in your opinion, (a) why the book had such a significant effect, and (b) opine whether such a book would today achieve the same sociopolitical prominence.

NOTES

1. Ecosystem—A biological community of interacting organisms and their physical environment [3].

2. The author expresses his gratitude to Dr. Howard Frumkin, Rollins School of Public Health, Emory University, for this contribution.

3. Lists of key definitions and abbreviations are found at the end of this book, as is a glossary.

4. *Risk assessment*—The characterization of the potential adverse health effects of human exposures to environmental hazards" [9].

5. The subject of greenhouse gases is discussed in Chapter 9.

6. This distinction, of course, is not absolute. In many instances research on laboratory methods must precede application of laboratory services.

7. The word environmentalist refers here to state, district, or local government employees.

8. *Tourist courts* is an older term for what are now called *motels*.

10. NACHO changed its name to NACCHO subsequent to the 1992 report. The organization is based in Washington, D.C.

11. The European Union is an organization of twenty-five European countries that cooperate through formal mechanisms on matters that include trade, common currency, environmental directives, and legislation. See chapter 9.

REFERENCES

1. BYU (Brigham Young University), The black death: Bubonic plague. Available at http://www.bye.edu/ipt/projects'middleages/LifeTimes/Plague.html, 1994.
2. *Webster's Ninth New Collegiate Dictionary*, Merriam-Webster Publishers, Springfield, MA, 1986.
3. Oxford Dictionary, Oxford Dictionary On-Line. Available at http://www.askoxford.com/?view=uk, 2005.
4. WHO (World Health Organization). Available at http//:www.who.int, 2000.
5. WHO (World Health Organization), Draft Definition Developed at a WHO Consultation in Sofia, Bulgaria, World Health Organization, Geneva, 1993.
6. Johnson, B.L., An Ensemble of Definitions of Environmental Health, U.S. Department of Health and Human Services, Environmental Health Policy Committee, Washington, D.C., 1998.
7. Detels, R. and Breslow, L., Current scope and concerns in public health, in *Oxford Textbook of Public Health*, 2nd ed, vol 1, Holland, W.W., Detels, R., and Knox, G., eds., Oxford Medical Publications, Oxford University Press, Oxford, 1991.
8. Johnson, B. L., *Impact of Hazardous Waste on Human Health*, CRC Press, Lewis Publishers, Boca Raton, FL, 1999.
9. NRC (National Research Council), *Risk Assessment in the Federal Government: Managing the Process*, National Academy Press, Washington, D.C., 1983.
10. Santayana, G., *The Life of Reason*, vol 1, Charles Scribner, New York, 1905.
11. Wikipedia, History of the Common Law. Available at http://en.wikipedia.org/wiki/Common_law, 2004.
12. Rosen, G., *A History of Public Health. Expanded Edition*, The Johns Hopkins University Press, Baltimore, MD, 1993.
13. Kelly, J., *The Great Mortality: An Intimate History of the Black Death, the Most Devastating Plague of All Time*, Harper Collins Publishers, New York, 2005.
14. CDC (Centers for Disease Control and Prevention), 150th anniversary of John Snow and the pump handle, *Mortality & Morbidity Weekly Report*, 53(34), 783, 2004.
15. DHHS (U.S. Department of Health and Human Services), Healthy People 2010, vol 1, Washington, D.C., 2000, 8–1.
16. CCPHS (Commissioned Corps of the U.S. Public Health Service). Available at http://www.os.dhhs.gov/phs/corps/direct1.html#history, 2001.
17. Mullan, F., *Plagues and Politics*, Basic Books, New York, 1989.
18. SGPHS (Surgeon General of the Public Health Service), Annual Report of the Public Health Service of the United States, Washington, D.C., 1913.
19. De Rosa, C.T. and Hansen, H., The impact of 20 years of risk assessment on public health, *Human and Ecological Risk Assessment*, 9, 1219, 2003.
20. Myers, M., 2002. Personal communication.
21. WHO (World Health Organization), Atlas of Children's Environmental Health and the Environment, Geneva, 2004.
22. CDC (Centers for Disease Control and Prevention), Surveillance for elevated blood lead levels among children—United States, 1997–2001, *Mortality & Morbidity Weekly Report*, 52, 1, 2003.
23. Lichtveld, M.Y., 2005. Personal communication.
24. CDC (Centers for Disease Control and Prevention), Homepage. Available at http://www.cdc.gov, 2001.

25. Gill, N.S., Is "first do no harm" from the Hippocratic Oath?. Guide to ancient/classical history. Available at http://ancienthistory.about.com/od/greekmedicine/f/HippicraticOath. htm, 2005.

26. Bandolier, E.M.B., Glossary. Available at http://www.jr2.ox.ac.uk/Bandolier, 2006.

27. Robinson, R., Cost-effectiveness analysis, *British Medical Journal*, 307(6907), 793, 1993.

28. State of Georgia, Rules and Regulations: Food Service. Available at http://www.ph.dhr. state.ga.us/publications/foodservice, 2002.

29. NACHO (National Association of County Health Officials), Current Roles and Future Challenges of Local Health Departments in Environmental Health, National Association of County Health Officials, Washington, D.C., 1992.

30. DeKalb County, Georgia, Information. Available at http://www.dekalbhealth.net/Information/inside/org-eh.html, 2002.

31. U.S. Congress, Government Performance and Results Act of 1993, Government Printing Office, Washington, D.C., 1993.

32. EPA (U.S. Environmental Protection Agency), Environmental justice. Available at http://www.epa.gov/ebtpages/environmentaljustice.html, 2005.

33. CPSC (Consumer Product Safety Commission), CPSC, Fisher-Price Announce Recall of Scooters and Mini-Bikes. Release # 04-031, Washington, D.C., November 13, 2003.

34. Merrill, R.A., The red book in historical context, *Human and Ecological Risk Assessment*, 9, 1119, 2003.

2 Steps in Environmental Health Policy Making

2.1 INTRODUCTION

Whether enacting federal legislation that covers national environmental issues such as clean air or a local business that decides to sell only lumber that comes from new growth forests, setting environmental health policies is typically a complicated undertaking. In particular, establishing environmental health policies is crucial to protection of human health, ecosystems, and environmental quality. These policies can result from legislation of new laws, changes in existing regulations and ordinances, and voluntary actions adopted by community groups, businesses, and individuals. Different points of view are always expected during consideration of a new or revised environmental health policy. As a democratic process, informed debate is vital, but necessarily lengthens the time to ensure a particular environmental health policy.

In order to understand the process of making environmental health policy requires an understanding of how government functions, since government is the primary source of environmental policies (discussed in chapter 3), and an appreciation of the influences that can influence policy making. This chapter provides a brief summary of key steps in environmental policy making, commencing with a description of factors that can have influence on establishing environmental health policies, followed by a simplified model of policy making. A particular kind of policy for controlling environmental hazards, called command and control, is discussed along with nonregulatory alternatives to command and control. The chapter concludes with a brief discussion of environmental ethics, since a framework of ethical behavior should accompany any policy of environmental protection and public health practice.

2.2 INFLUENCES ON ENVIRONMENTAL HEALTH POLICY

Establishing environmental health policy is a complicated political undertaking. This is true whether the policy is to be developed by government or an entity in the private sector (e.g., a corporation). Government policies must involve the public because of the Administrative Procedures Act, which will be described in chapter 3. Corporate or business policies are also complicated in their establishment, but usually receive less input from the public. The nineteenth-century German leader Otto von Bismarck is credited with saying, "There are two sights unfit for the human eye: making sausage and making legislation." Bismarck's observation surely applies to the messy and sometimes unpleasant making of environmental health policy. Why is this? Why should establishing environmental health policies that are intended to protect environmental quality and human health from hazards in the environment be a complicated, often passionate affair?

Setting government environmental health policy, be it to protect ecological or human health, normally forges lines of different public opinion, organizes bases of support and opposition, and energizes legislative machinery. When contemplating federal legislation (e.g., to protect the quality of the nation's supplies of drinking water), major issues quickly arise—not the least being the potential economic costs. Legislators are confronted with sorting out the impact issues that may accompany any enacted legislation. They must consider: What is the extent of the hazard (e.g., contaminated drinking water)? What would be the benefits to the public's health? What will be the burden on various business and government entities? How will the public react in general to the legislation? And will the enacted legislation meet specific environmental goals? Addressing these questions will energize special interest groups and often stimulate considerable passion. For instance, business groups will often allege that the economic costs of proposed regulations will be too great for them to bear. Environmental groups may argue that proposed legislation or regulations don't go far enough in protecting the environment.

There is no policy recipe or political cookbook that gives specific directions on how to successfully develop environmental health policy. Shown in Figure 2.1 are seven factors that can influence the establishment of environmental health policy. Not all are of equal consequence, and specific elements may be inconsequential at local and state government levels. These seven factors are discussed in the following sections.

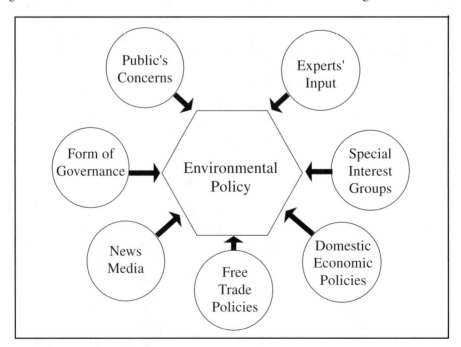

FIGURE 2.1   Factors that can influence environmental health policy.

2.2.1 THE PUBLIC'S CONCERNS

To establish environmental health policy through new legislation or refinement of existing regulations or ordinances requires concern expressed by the public—not all of the public, of course, but groups or individuals with special interests or specific concerns. These interests can take the form of an environmental organization, citizens group, business association, public health agency, or similar groups with a specific concern about an environmental hazard. Setting environmental health policy is seldom initiated by individual elected officials; few elected officials are willing to lead legislative efforts that they may perceive as politically risky.

On the other hand, elected officials are generally responsive to concerns from the public—if the pressure is great enough and the political risks are relatively low. Consider the example of uncontrolled hazardous waste releases into community environments. More specifically, take the example of Love Canal, New York, in the late 1970s. A community of homes had been constructed over an abandoned chemical dump. Over time, the chemical waste migrated into the overlying homes, leading to community concerns that birth defects, cancer, and other grievous health conditions were occurring. These health concerns became the material for national news reports, amplified by the discovery of other toxic waste dumps in Kentucky, Missouri, and other parts of the country. Pressure to act was eventually exerted on the Congress by national environmental organizations, Love Canal community groups, and public health advocates. In 1980, the Comprehensive Environmental Response, Compensation, and Liability Act was enacted and signed into law by President Carter. In retrospect, community concerns, well articulated by Love Canal spokespersons, had provided the impetus for new federal policy on managing uncontrolled hazardous waste.

2.2.2 SPECIAL INTEREST GROUPS

A necessary component of setting environmental health policy is the presence of special interest groups. These are organizations that have specific points of view relative to a particular environmental hazard or proposed policy. To be more specific, examples of special interest groups include national environmental organizations (e.g., World Wildlife Fund), business advocacy organizations (e.g., American Chamber of Commerce), and public health associations (e.g., American Public Health Association). As a particular environmental policy becomes of interest to a legislative body (e.g., Congress, state legislature, county commissioners), special interest groups will emerge to present their point of view and attempt to persuade legislators to support their view. This interaction occurs through presentations made by special interest groups in public meetings (e.g., congressional hearings) and private meetings with members of a legislative body and their staffs. Special interest groups are important for assuring a full range of debate and discussion on a proposed environmental health policy, but can be an impediment to democratic processes if they wield undue influence on policy makers.

2.2.3   NEWS MEDIA AND INTERNET COMMUNICATIONS

News media include newspapers, television, radio, cable, and other commercial and public outlets of news and information. Internet communications include web sites, e-mails, and blogs. Both the news media and the Internet serve a vital role in informing the public and shaping opinion. Much of what the U.S. public knows about environmental hazards is based on information from the news media, particularly from television sources. Readers and viewers are presented with words and images that depict the presence and consequences of specific environmental hazards. Over time, the news media have developed greater sophistication about which environmental reports to present to the public and how to convey their stories. What some would call "The Carcinogen of the Week" approach to environmental reporting has been replaced with a more cautious approach. Moreover, the amount of newsprint and television time devoted to environmental reports decreased during the 1990s, perhaps indicating a more pragmatic approach to reporting.

Few communication resources have proliferated as quickly and widely as the Internet. It began in the early 1960s at the Massachusetts Institute of Technology. University staff prepared concept papers that laid out the ideas of networking digital computers to exchanging packets of information. The World Wide Web technology was built upon the Internet infrastructure, allowing users to communicate globally [1]. One now has the capability of sending general information and personal communiqués around the globe, almost instantaneously, with but a stroke on a keyboard. It is now expected that any group that seeks public interest in its products, services, and agenda must have a web site. This, of course, includes environmental and public health groups and organizations. From such web sites can be obtained information on specific environmental hazards, copies of environmental laws and regulations, government policy positions, and special interest groups' stances. All this body of information can inform the public and thereby be used to protect an individual's health or become material for personal and group advocacy (e.g., letters to elected officials). However, there is a caution that must accompany use of information from the Internet. The credibility of a web site source can vary, particularly on matters of science, because there are relatively few constraints placed on web sites. On matters of science, the normal expectations of independent peer review[1] of research reports and their interpretations are not always followed. The consequence can be unsupportable scientific assertions that are intended to advocate a particular point of policy or other course of action.

In a policy context, news media and Internet resources can bring environmental problems to the attention of the public, leading to concerns in special interest groups and potentially affected members of the public. These concerns are soon brought to the attention of elected officials and policy makers, who in turn are expected by a concerned public to take action. Consider the spread of the West Nile virus within the United States. This virus, carried by mosquitoes, can infect birds and has caused loss of life in some persons bitten by infected mosquitoes. The presence of West Nile virus was first observed in the United States in 1999 in New York and other northeastern states. With the migration of birds, the virus was spread southward, causing additional loss of life. The news media reported the presence of West Nile virus in the United States, its public health consequences, and the spread of the virus into human populations. Reports of how to prevent human contact with infected mosquitoes were often presented by the news media and found on Internet sites operated by public health departments. This information and the availability of it raised public awareness and kept elected officials and other policy makers alert to their responsibilities with respect to mosquito control.

2.2.4 EXPERTS' INPUT

Environmental health policy can be influenced by the findings and recommendations from individual experts or groups of experts. Experts can be representatives from government agencies, universities, corporations, science councils, and such. Legislators and policy makers are often challenged to personally understand or appreciate the seriousness of a particular environmental hazard. In response to this challenge, legislative bodies will often turn to expert groups for their analysis and recommendations about a particular environmental hazard or issue. As an example, the U.S. Congress often asks (and funds) the National Academy of Science (NAS) to conduct an evaluation of an environmental issue and then provide their findings to the Congress and the public. These recommendations can help shape federal legislation. An example is the NAS report *Pesticides in the Diets of Infants and Children*, which reported on the public health hazard posed to young children from exposure to pesticides in the environment [3]. The report's recommendations for a greater level of protection for children had a major influence on the development and enactment of the Federal Food Quality Protection Act of 1998. This act requires EPA and other federal agencies to develop and implement regulatory actions that provide a higher level of protection for children exposed to pesticides.

2.2.5 DOMESTIC ECONOMIC POLICIES

Federal environmental health policy, and to a lesser extent, state and local policies, are influenced by domestic economic policies. For example, the U.S. economy is a free market, free enterprise economy, one that is very much undergirded by consumer spending. Consumer spending means the production of products and materials and the purchase of goods and services. This places money in general circulation, providing even more funds for development and investment. A part of this economic engine produces goods and services that are sold to international customers. This produces wealth for companies and can help economic development in other countries.

During times of weak national economies, it is difficult for elected officials and policy makers to impose laws and regulations that could result in further economic hardship. For instance, a particular proposed environmental policy that could harm international trade would be difficult to enact in times of economic austerity. Special interest business groups would argue that the contemplated environmental policy would result in job reductions, lessened corporate profits, stock devaluation, and such. Few elected officials would be willing to favor an environmental policy that might exacerbate a period of economic fragility.

2.2.6 FREE TRADE POLICIES

The trading of goods and services between nations and peoples is an ancient, even prehistoric, means of meeting personal and societal needs. Trading in this sense means selling goods from one person to another. Depending on the circumstances, each sale has the potential to produce an income for the seller, generating revenue. With sufficient revenue, regional and national economies can benefit. One way of enhancing the revenue from trade has been to place a tariff (i.e., a tax) on imported goods. This leads, predictably, to other nations placing tariffs on their own imported goods. This can lead to tariff, trade, and policy disputes between countries. Some disputes have had historic impact on a nation's growth and development. Consider the Boston Tea Party [4]. In 1773, the British Parliament allowed the East India Company to export a half million pounds of tea to the U.S. colonies, but without charging the company a tariff on the tea. This waiver of a tariff on tea placed U.S. tea merchants at an economic disadvantage, since they were still subject to paying a tariff on the tea they imported. On December 17, 1773, a band of U.S. merchants boarded the British ships that contained the tea and threw the cargo overboard into the Boston harbor, an event called the Boston Tea Party. This action, based on a tariff disagreement, furthered an already worsening relationship between the U.S. colonies and Britain.

Free trade policies remove trade barriers between nations and eliminate tariffs on goods. The European Union (EU), has no trade barriers between EU member states. In another example, Canada, Mexico, and the United States entered into the North American Free Trade Agreement (NAFTA) on January 1, 1994, a treaty between the three countries that removed barriers to trade across their borders. Some persons and groups argue that free trade translates into lowered environmental protection and public health protections. They assert that transnational corporations will relocate polluting and injurious (to workers) industries from countries with stringent environmental control to countries without such controls. A particular environmental health policy may founder on arguments that it would complicate free trade and economic well-being.

2.2.7 FORM OF GOVERNANCE

How a country, region, state/province, tribal nation, or locality chooses to govern itself can influence how environmental health policy is implemented. In the United States, the three branches of federal government (legislative, executive, judicial), discussed in chapter 3, are mirrored at the state and local government levels. Any federal or state environmental statute, regulation, or local government ordinance is subject to judicial processes if litigation is brought by a party that disagrees with some aspect of the statute, regulation, or ordinance. In the United States, the method of governance is a democratic republic, which means representatives are democratically elected and authorized to act for other persons who reside in a specific geographic area (e.g., a congressional district). As a democratic republic, serviced by elected officials, considerable input from the public on matters of environmental policy is both possible and encouraged.

Another form of governance is based on the election of a parliament as the national seat of legislation. Parliamentary governments are derived from the Parliament of Great

Britain, which in turn began in the Middle Ages as an advisory body to the monarch [5]. Beginning in the thirteenth century, the single advisory body evolved into the House of Lords (major landholders, chief nobles, clergy) and the House of Commons (knights, lower clergy, burgesses). In 1688, the Parliament succeeded in obtaining primacy over the monarchy for purpose of national government. In time, the House of Commons assumed the responsibility for enacting national legislation, with the House of Lords' power being limited to deliberations on legislation and initiating amendments to bills. Parliamentary government vests its legislative authorities with a parliament and its executive authorities with ministries (e.g., Ministry of Health). Courts are independent of parliament, except for receiving public funds through parliamentary appropriations.

Environmental health policy established within parliamentary systems is largely set by the relevant ministries of environment and health, operating under broad authorities of parliament. The amount of public input varies from one country to another, but is generally less than in the U.S. system of government.

2.2.8 OBSERVATION

Many factors will influence the development of environmental health policy, whether being made by public or private sources. The seven factors shown in Figure 2.1 are but one set of influences. There are others that come into play, depending on the specific policy and circumstances. More important than the specifics of individual influences (e.g., news media) is the recognition that establishing environmental policies is difficult and requires sustained support to achieve a particular policy. Knowing about this challenge gives even greater respect for those persons and organizations that have contributed to legislation and policies that have improved the quality of the air that we breath, the water that we drink, the food we consume, and the control of wastes that can threaten the public's health.

2.3 ESTABLISHING ENVIRONMENTAL HEALTH POLICY

The process of establishing environmental policy is usually a difficult proposition. This is because environmental issues in general will elicit divergence in opinion between special interest groups, especially if proposed policies might produce regulatory actions. Seldom do regulated organizations support new or expanded regulations that could have economic impacts. Rather, their opposition to a proposed environmental health policy will become part of the dynamic process that constitutes policy setting. Although policy setting is a difficult process, it can nonetheless be viewed as a structured process. For example, Rosenbaum [6] divides the policy cycle, which he defines as "the process of interrelated phases through which policy ordinarily evolves," into six phases:

All politics must include discussion, negotiation, and, ultimately, compromise.

---

1. Agenda Setting—It is quite difficult, but not impossible, for an individual to get a proposed policy issue made into an institutional policy. An exception can occur when an individual is a policy maker (e.g., county commissioner) and therefore has direct access to the policy-making machinery. However, a group effort is usually required, since policy makers are more easily persuaded when there are a large number of proponents for a proposed policy. The first step in achieving a group's policy-making aspirations is to get the desired policy onto the agenda of policy makers. This is because policy making is a political event, requiring political skill in the processes of negotiation and compromise. Examples of government agenda setters include local elected officials, state legislators, and members of Congress. In the private sector, agenda setters include company officials, boards of directors, and policy committees. Getting a desired policy into the hands of policy-setters occurs by lobbying, a time-honored process of political pressure brought upon elected officials and other policy makers.

2. Formulation and Legitimation—Getting a policy proposal onto the agenda of a policy maker is the essential first step in policy making. Rosenbaum's next step is formulation, which he states, "[i]nvolves setting goals for policy, creating specific plans and proposals for these goals, and selecting the means to implement such plans" (ibid., p 53). "Policies once created must also be legitimated—invested with the authority to evoke public acceptance." This usually is done through constitutional, statutory, or administrative procedures, such as voting, public hearings, presidential order, or judicial decisions upholding the constitutionality of laws—rituals whose purposes are to signify that policies have now acquired the weight of public authority" (ibid., p 54). Consider, for example, the formulation and legitimation of a municipal ordinance to prohibit tobacco smoking in public facilities. Policy makers would hold public hearings, meet with advocacy groups supportive or against the proposed anti-smoking policy, and consult with attorneys to ascertain the legality of various possible forms of the ordinance. Following the acquisition of such information, elected officials would vote in a public setting on whether to enact an anti-smoking ordinance.

3. Implementation—Policy implementation means acting upon a policy that has been formulated and legitimated by a policy-making body (e.g., a state legislature). Environmental health policy normally specifies the agencies that are responsible for affecting the enacted policy. For example, amendments to the federal Clean Air Act, discussed in chapter 5, specify EPA as the federal agency responsible for implementing the policy changes inherent in the amendments.

4. Assessment and Reformulation—Implemented policies always attract attention and oversight, particularly by those individuals and groups who advocated for the policies. For instance, groups that successfully lobbied for a municipal non-smoking policy in public facilities are likely to monitor the degree to which the policy has been effective. Implemented policies seldom go unchanged over time. Changes can occur as the result of court decisions, administrative experience, public dissatisfaction, and special interest intervention.

5. Policy Termination—Once in place, it is difficult to terminate a public policy, given the politicalness of the policy-making process. Policy makers are reluctant

to terminate a policy unless it has been convincingly shown to be detrimental to the public good. As described by Rosenbaum, "Terminating policies, environmental or otherwise, is such a formidable process that most public programs, in spite of intentions to the contrary, become virtually immortal. Policies usually change through repeated reformulation and reassessment" (ibid., pp 54–55).

6. Policy Making is a Combination of Phases—It is important for those who desire to set policy to know that policy is made and implemented through a combination of phases, commencing with agenda setting. Some phases can occur concurrently, or nearly so. For example, the phases of implementation and assessment and reformulation can occur at the same time, since implementation of a policy is one way to ascertain if it needs reformulation or termination.

The six phases of policy making, as described by Rosenbaum, can be simplified into a model that focuses on pressure, action, change, and modeling that occur during the policy-making process.

2.4   PACM MODEL

As prior narrative has stated or inferred, setting environmental health policies is a difficult, complex undertaking. It is a process that is thoroughly political—and should be, because one definition of politics is the "the total complex of relations between people living in society" [7]. Politics forces issues into the arena of discussion and debate, whether in the U.S. Congress or in one's family. Inevitably, all politics must include discussion, negotiation, and, ultimately, compromise. Just consider how politics shaped what became the U.S. Constitution. Representatives from the 13 states had many serious political disagreements over the content of the Constitution. For instance, smaller states had concerns that more populous states would take advantage because of greater numbers of representatives in a House of Representatives. This concern was settled by negotiating the creation of a U.S. Senate, a body that gave small states parity in representation.

Establishing any policy is political and the accompanying policy making and practice of politics are as intertwined as macaroni and cheese. Recognition of this reality leads to a simplified diagram of policy making, shown in Figure 2.2, which consists of four components: pressure, action, change, and monitoring (PACM). Figure 2.2 is arranged as a flow diagram, indicating how the four components interact. Establishing policy begins with pressure.

2.4.1   Pressure

Because setting any policy is a political event, and recognizing that all political systems have a certain amount of inertia, putting pressure on the system is required. This is because elected officials are often slow to support a proposed policy initiative until they have calculated the political implications of their support. The need to bring pressure

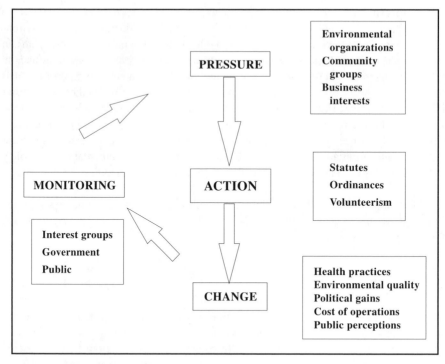

FIGURE 2.2   Simplified flowchart of environmental policymaking.

on elected officials and other senior policy makers is particularly important on matters of environmental health policy, where economic impacts of proposed policies (e.g., regulating the siting and operating of incinerators) often become controversial.

Bringing pressure on political systems to set federal environmental health policies occurs through lobbying by environmental organizations (e.g., Environmental Defense, Natural Resources Defense Council), business associations (e.g., chambers of commerce), trade associations (e.g., American Petroleum Institute), professional societies (e.g., American Public Health Association), state governments, and specialty experts (e.g., National Academy of Sciences). The pressure is exerted through meetings with policy makers and their staffs, testimonies given at public hearings, reports in the news media, and other outlets that can influence public opinion and motivate elected officials.

Pressure can be brought to bear on business enterprises to achieve environmental goals and changes in business practices. Like pressure directed to government policies and practices, meetings between environmental groups and senior policy makers in business can bring about debate, negotiation, and compromise. An example is the successful pressure applied to the McDonald's food chain to affect changes in how food was packaged. Specifically, environmental groups sought to replace Styrofoam™ containers with those fabricated of paper and cardboard. Environmentalists were concerned about the relatively lack of biodegradability of the polystyrene plastic materials, resulting in their presence in municipal landfills much longer than their paper counterparts.

2.4.2 ACTION

When sufficient pressure is brought to bear on a political system, action can occur. Action means simply that something occurs, gets done, or moves. Actions can take the form of enacted legislation by federal or state legislatures or ordinances promulgated by county and local governments. The enactment of the Food Quality Protection Act of 1996 represented action to improve how pesticides were regulated by the federal government. Action can also be new or revised federal or state environmental regulations or new policies. EPA's reassessment of the risk of exposure to dioxins changed that agency's procedures for conducting risk assessments by including for the first time mechanisms of cellular toxicity in their considerations. This action adds further science to how risk assessments of environmental hazards will be regulated (chapter 11).

2.4.3 CHANGE

Change, simply put, means that a different course of action will occur from what was previously done. For example, whereas the Air Pollution Control Act of 1955 was limited primarily to research and technical and economic assistance to state and local governments, the Clean Air Act Amendments of 1963 enlarged EPA's responsibilities for controlling air pollution. At the same time, the act continued congressional intent that "[t]he prevention and control of air pollution at the source is the primary responsibility of state and local governments" [8]. As will be discussed in chapter 5, amendments in later years further strengthened the federal government's primacy, working with state and local governments, in controlling air pollution.

Change can also occur when pressure is directed at executive branch agencies and departments. Such pressure is more narrowly focused and not always as visible to the public as pressure directed to legislative bodies. As an example, as noted in chapter 10, environmental justice advocates pressured the Clinton administration to issue an Executive Order (i.e., an action) that mandates several actions of federal government agencies in order to prevent the unjust imposition of environmental hazards on minority populations (i.e., change). This order changed risk management policies by requiring federal agencies to give attention to how environmental risks to minorities were characterized and managed. As a specific outcome, EPA created the National Environmental Justice Advisory Committee as a resource for advising the agency on communities' concerns regarding environmental injustice (see chapter 10). This committee has provided EPA with advice on agency environmental justice policies and guidance on implementing the policies.

2.4.4 MONITORING

Just because a change in a policy has occurred does not always mean that those who initiated the change will be satisfied. The changes may be perceived as being ineffective, too costly, or misdirected. Monitoring (or surveillance) of the changes is a common means to assess the impact of a change in policy. Monitoring data can, in turn, be used to refine policies or even get them withdrawn.

As an example of monitoring, Title III of the Superfund Amendments and Re-authorization Act of 1986 created the Toxics Release Inventory (TRI), which requires business entities to annually provide EPA with data on substances released into the environment and specifics on the quantities released. EPA is required to make TRI data available to the public. Environmental groups, nationally and locally, have evaluated TRI data to assess trends in releases of toxicants and have used the data to bring public attention to those facilities that release large amounts of pollution into the environment. The response of businesses has often been to voluntarily decrease their emission levels. Some persons have referred to this action as "regulation by shaming" [9].

In summary, an understanding of the PACM model is fundamental to policy development, be it a national environmental policy or a family budget. Application of this simple model must be tempered with the wisdom that politics will be a consideration at each step. No policy gets established without the art of debate, negotiation, and compromise.

2.5   POLICIES TO CONTROL ENVIRONMENTAL HAZARDS[2]

As described throughout this book, environmental hazards not sufficiently controlled can cause adverse effects on the public's health and environmental degradation. In the United States, over the course of many years, but particularly starting in the 1960s, policies have emerged that targeted the control of specific environmental hazards (e.g., chemical contaminants in drinking water supplies). The intent of the policies has been to prevent human exposure to those hazards that can cause human disease or disability. The most frequently used policy has been to legislate laws that required government agencies to develop and administer regulations to control the risk of exposure to select environmental hazards. However, alternative policies have emerged in support, or in lieu, of the regulatory approach.

This section will discuss some of the policy options that can be brought to bear on control of environmental hazards. At the core of these options is behavior—how individuals, public, and private entities act within a society toward attaining a common end. How behavior is expressed is the issue. For instance, society, acting through government, has controlled the behavior of polluters of the environment through the mechanisms of legislation, regulation, and enforcement. This is a kind of coerced behavior. But behavior can also be voluntary, as expressed by individuals, groups, and cultural entities such as corporations. Both coerced and voluntary behaviors are necessary for environmental policy purposes. The following sections will describe several policy choices for controlling environmental hazards.

2.5.1   COMMAND AND CONTROL

To regulate is to control, because a regulation is defined as a rule or directive made and maintained by an authority. Some control is vital to a society's well-being. This is true in application for both individuals and groups comprising a society. Control can be self-imposed as well as group-imposed. Rules that control behavior are purposeful for survival. Even our most distant human ancestors learned that individuals forming groups or tribes had better chances of survival than if each individual acted alone. Rules evolved that protected tribes and small groups (e.g., recognition that murder of tribal members generally weakened the group as there would be fewer hunters, warriors, child-bearers). As government evolved as the primary instrument for protecting large societies, control was passed from individuals and small groups to government.

All governments impose control over their society. One method of control is by establishment of laws and regulations. This is the way that U.S. environmental health policy has occurred. While some might consider environmental policies to be a product of twentieth-century U.S. government, in fact, environmental policies or practices have much deeper roots. Consider, for instance, the policy of indigenous people who took from the environment only what they could consume. An example is the reverence given to the bison by the indigenous people of what are now called the Great Plains of North America. The animal's body provided them with clothing, food, tools, and medicines. They took from the great bison herds only what they could consume. Contrast that environmental policy with the slaughter to near extinction of the bison herds by European settlers who immigrated westward from the eastern coast of North America. Herds were decimated for mere sport, pelts were made into "fashionable" clothing of the moment, and bison's grazing lands were divided into farms and ranches. As another early example of a people's environmental policies, leaders of the Massachusetts Bay Colony had to forbid the use of lead-lined pots in which to prepare rum for sale [10]. Environmental health practices and rules are not simply of recent origin in North America.

In the early twentieth century, the U.S. Congress began enacting legislation that was intended to control environmental hazards of public health consequence. As will be discussed in chapter 4, a considerable body of federal environmental legislation accrued during that century, peaking in the period between the mid-1950s through 1980. The enacted legislation, when signed into law by incumbent presidents, brought to bear the weight of law to control environmental hazards, protect the environment and natural resources, and protect the public's health.

In general, this body of environmental law gave federal agencies (e.g., EPA) the authority to *command and control* (also called *regulations and standards*) actions to control workplace and community sources of specific environmental hazards. Sources of pollution such as companies, businesses, and government agencies that are required to comply with specific regulations are called the *regulated community*. As an example, companies that generate electric power are a regulated industry under the provisions of the Clean Air Act.

Rosenbaum [6] observes that command and control comprises five phases through which pollution policy evolves: goals, criteria, quality standards, emission standards, and enforcement. These five phases apply to both federal and state environmental health policies.

1. Goals: Legislative bodies must state goals for statutes intended to control the release of pollutants or other environmental hazards. The goals are normally

stated in broad terms, (e.g., a goal of protecting public health and safety). However, specificity in achieving some goals is added by Congress when there is impatience with on-going regulatory processes. For example, slow progress by EPA in regulating hazardous air pollutants (HAPs) led Congress to specify in the 1990 Clean Air Act amendments more than 180 HAPs and timelines for which EPA was to develop regulations.

2. Criteria: "[t]he technical data, commonly provided by research scientists, indicating what pollutants are associated with environmental damage and how such pollutants, in varying combinations, affect the environment" [ibid.]. Criteria serve many purposes. Legislative bodies need criteria when contemplating environmental and other legislation. For instance, do criteria indicate or suggest that pollutants may cause adverse health effects to humans or ecosystems? Similarly, regulatory agencies require criteria during the development of risk assessments and risk management plans for specific environmental hazards (e.g., contaminants in drinking water). Agencies often prepare criteria documents in support of risk management actions, such as recommendations on permitted exposure levels to hazardous substances found in workplaces or in the ambient outdoor air of urban settings.

3. Quality standards: How pure do we want the air that we breathe, the water we drink, or the food we consume? This question goes to the regulatory concept of quality—air quality, water quality, food quality. *Quality standards* express the levels of pollution that can be present in an environmental medium (air, water, food) without causing harm to human or ecological health. For example, the primary air quality standard for carbon monoxide in ambient outdoor air is 10 mg/m^3 (8-hour average) (see chapter 5). Quality standards are legally enforceable. Quality standards are currently developed by U.S. regulatory agencies through risk assessment methodology, a structured process used to relate criteria to specific levels of pollutants in specific environmental media.

4. Emission standards: To achieve quality standards requires knowledge about, and control of, sources of pollution. *Emission standards* regulate the amount of pollution that can be legally released into specific environmental media over a specified amount of time. To achieve emission standards requires emission controls, technologies that will reduce or prevent emission of pollutants. As an example, automobile manufacturers must install emission controls (e.g., catalytic reactors) on vehicles sold in the United States in order to control vehicle emissions that contribute to air pollution.

5. Enforcement: Command and control regulatory policy, beginning with setting of goals, concludes with the enforcement phase. Simply establishing quality and emission standards is not sufficient to ensure that they will be implemented by the regulated community. Like other matters of law, regulatory standards must be supported by enforcement authority. Without enforcement authority, in effect, regulatory standards would become voluntary (i.e., sources of pollution could pick and choose those standards acceptable to them). Rosenbaum [6] observes, "[S]atisfactory enforcement schemes have several characteristics: they enable public officials to act with reasonable speed—very rapidly in the case of emergen-

cies—to curb pollution; they carry sufficient penalties to encourage compliance and they do not enable officials to evade a responsibility to act against violations when action is essential."

Command and control is a controversial policy. The regulated community often objects to the alleged economic impact of quality and emission standards, usually arguing that the costs to them far outweigh the benefits to the public. This debate in recent years has led to the requirement by Congress that regulatory agencies conduct a cost-benefit analysis of proposed regulations. Moreover, even after cost-benefit analyses are conducted and regulatory actions made final, litigation often occurs when the regulated community disagrees with the final regulatory action.

The ergonomics[3] standard developed in the 1990s by the Occupational Safety and Health Administration (OSHA) is an example of the politically thorny and difficult nature of current day attempts to develop and promulgate environmental regulations. As background, certain job activities, especially those that require repetitive body motion, can cause disorders in a person's musculoskeletal system. For example, persons who strenuously strike the keys of computer keyboards for data entry or typing can develop a painful wrist disorder called Carpal Tunnel Syndrome. Likewise, repetitive lifting of heavy objects (e.g., patients confined to hospital beds) can cause injuries to a hospital worker's back. These kinds of injuries are called musculoskeletal disorders (MSDs). They are almost always quite painful to afflicted individuals and can be debilitating, leading to job loss and workers' compensation demands. There is ample evidence to demonstrate that MSDs are prevalent, costly, and amenable to prevention strategies, such as better designed office equipment and mechanical aids for lifting and moving heavy objects.

OSHA spent the eight years of the Clinton administration developing an ergonomics standard (i.e., a set of regulations that would "command and control" how repetitive motion jobs and other physical labor would be performed in industry and other business enterprises). OSHA's Ergonomics Program Standard went into effect on January 16, 2001, which was four days before the Clinton administration ended. By the time the standard had gone into effect, opposition to its implementation had formed across much of the U.S. business sector. It was characterized by business interests as too broad in its coverage of work activities, too vague in what would be an ergonomic hazard, too costly in its requirements to redesign job conditions, and too burdensome in its paperwork and reporting requirements. As the result of these concerns, business interests brought the PACM model to bear on the U.S. Congress.

On March 6–7, 2001, the Senate and the House of Representatives voted to repeal OSHA's Ergonomics Program Standard. Congress's authority derived from the Congressional Review Act of 1996,[4] which requires federal agencies to submit regulations to Congress before they go into effect. Congress must act within sixty days on a proposed regulation. A proposed regulation can be set aside by a Joint Resolution of Disapproval by a simple majority vote of both houses of Congress [11]. Repeal of the OSHA ergonomics rule was the first, and to date only, such action by Congress under the Act. OSHA is currently revising the repealed standard, but has made no firm commitment on when a revised ergonomics standard might be issued [12].

Congress's authority to repeal a federal agency's regulation is relatively recent, and in a policy context, contradictory. It is contradictory in the sense that Congress mandates federal regulatory agencies to develop regulations, using the most current and relevant scientific data and judgment, to control specific environmental hazards, but can then overturn an agency's proposed regulations, based on political pressure from special interest organizations. This adds one more "hoop" for regulatory agencies to anticipate. On the more positive side, use of the Congressional Review Act can prevent the imposition of politically and, perhaps scientifically flawed, regulations before they take force.

2.5.2   ALTERNATIVES TO COMMAND AND CONTROL

The previous section described the Command Control approach for regulating environmental hazards. However, over time, alternatives have emerged to command and control as a regulatory policy. This has occurred in part because of dissatisfaction with the slowness of many federal regulatory actions and disagreement over whether regulatory policies should be risk-based. In fact, the wheels of government do generally turn slowly, given the political processes inherent in government policy making. Some voluntary actions, shown in Table 2.1, can proceed more quickly, as discussed in the following sections and the degree of personal control (e.g., litigation) can be greater than if government is proceeding along a regulatory approach for the issue at hand. However, some of the alternatives to Command and Control can themselves be rather time-consuming, with uncertain outcomes. Alternatives, therefore, must be carefully chosen and pursued.

2.5.2.1   Litigation

A substantial amount of environmental health policy has been established, or modified, through litigation. Courts have served as arbiters—and, by default, policy decision makers—on many wide-ranging environmental issues. For example, national environmental organizations have litigated EPA on whether the agency has complied with provisions

TABLE 2.1
Alternatives to Command and Control

Litigation

Market Power

Performance Incentives

Precautionary Approach

Public Education

Sustainable Development

Voluntary Action by Industry

of the Clean Air Act, such as allegations that EPA regulations on controlling fine particulate matter have been too lax. Similarly, EPA has litigated municipalities (e.g., Atlanta, Georgia) for failure to meet Clean Water Act standards. In these examples of "proactive" litigation, courts have become final authorities on how well government agencies have met their legal environmental responsibilities.

Private industry has also turned to courts for relief from alleged burdensome regulations. In perhaps the most significant litigation against a federal regulatory agency, in 1980, the U.S. Occupational Safety and Health Administration (OSHA) was litigated over their proposed regulation to control benzene levels in occupational settings (chapter 11). As related by Rodricks [13], "[O]SHA proposed simply to identify occupational carcinogens and to establish limits on worker exposure at the lowest technically feasible levels—in the absence of identifiable thresholds, technology should dictate the maximum allowable workplace exposures. When OSHA attempted to apply this regulatory philosophy to the leukemia-causing benzene, the affected industries mounted a legal challenge, based on the view that the law required OSHA to show explicitly the level of cancer risk the agency was attempting to control, and the level of risk reduction that would be achieved by the introduction of controls. The legal challenge made its way to the Supreme Court, and the nine judges, by a margin of 7 to 2, agreed with the industry position [Industrial Union Department, AFL-CIO v. American Petroleum Institute 1980]. They sent OSHA home with the assignment to engage in quantitative risk assessment when it attempted to regulate carcinogens." The court's decision completely changed the way that federal regulatory agencies assessed the risk from environmental hazards. Quantitative risk assessment became the status quo whenever it could be applied to a regulatory action.

Individuals can litigate private businesses when products are alleged to be unsafe or otherwise hazardous for personal use. Such litigation, called product liability suits, has often been used to address alleged harmful effects experienced by individuals. Examples of personal litigation include lawsuits against manufacturers of alleged defective automobile tires, toys, personal care products, tobacco products, and food products. Some lawsuits become "class action" litigation when many individuals join together as plaintiffs in a single lawsuit. The class action suits by asbestos-exposed workers against asbestos producers are an example.

Another example of a class action suit concerning an environmental issue pertains to the 1989 spill into the Prince William Sound in Alaska of crude oil from the ship *Exxon Valdez*. The ship had run aground, ruptured, and spilled 11 million gallons of crude oil, causing great damage to the local ecosystem and the region's economy. In the same year, a class action suit was filed against the ship's owner, the Exxon Corporation, by 32,000 fishermen and residents of the area [14]. In 2004, a federal judge awarded $4.5 billion to the plaintiffs, an action that is likely to be appealed by the corporation. Courts, by being drawn into individual and class action litigations, often set policies that linger long after the litigation ends. For example, manufacturers may redesign products to be less hazardous or change how their products are used by customers. This is achieved without resorting to enacting new legislation or regulations—in effect, serving as an alternative to environmental regulations and standards.

2.5.2.2   Market Power

In a society that is based on a consumer economy (i.e., one in which products are created, distributed, and sold at a profit), consumers of products and services can influence environmental policies through their marketing polices and practices. With adequate consumer education, individuals and groups who purchase environmentally sensitive products can help determine what products remain in commerce. As an example, the purchase by individuals of household detergents with low or zero phosphorus content helped reduce water pollution and algal growth in waterways. Similarly, companies, small businesses, and government agencies can voluntarily adopt polices on purchasing products that cause minimal harm to environmental quality.

Anti-smoking campaigns are an example of consumer power as environmental health policy in action. Anti-smoking activists have been quite successful in lobbying local governments and businesses to ban or restrict tobacco smoking in public buildings and in some private premises. A contemporary example is the restriction of tobacco smoking in restaurants. Many local governments have required restaurants and other food service establishments to either ban tobacco smoking altogether, or in some localities, provide areas where smoking is prohibited. Where anti-smoking ordinances do not exist, consumers can exert pressure by selecting restaurants that have voluntarily adopted a no-smoking policy, thereby promoting an increase in the number of such restaurants. Similarly, pressure from airline customers eventually led to no-smoking polices on domestic and international air travel, illustrating how market power can have a global impact.

A strategy used in market power to reduce the impact of environmental hazards is called Green Commerce. This strategy can be implemented by consumers who preferentially purchase products that protect the environment. The strategy can also be implemented by industry and businesses that operate in the arena of green commerce (e.g., companies that develop and market "environmentally-friendly" commercial products and services). It is a marketing strategy that intends to appeal to environmentally-supportive groups in the general population. In effect, the Green Commerce policy strives to link commercial entrepreneurialism with environmental advocacy. There is no regulatory apparatus that comes into force that requires a policy of Green Commerce, assuming, of course, that the green products or services do not violate regulations on safety or environmental quality.

Green Commerce strives to link commercial entrepreneurialism with environmental advocacy.

The introduction of Green Commerce products can be illustrated by changes occurring in the laundry, dry cleaning, and home cleaning businesses. In the dry cleaning business, less reliance on solvent-based cleansers means fewer hazardous pollutants released from dry cleaning establishments into the environment. Rather than dry cleaning, laundering garments by using phosphorous-free soap products is a Green Commerce advancement. Similarly, Green Commerce has resulted in a host of home cleaning products that are derived from citrus and other natural materials. Use of these products results in less water contamination than through use of products containing synthetic chemicals. Consumers who purchase products or services from environmentally-sensitive companies or from other business services can strongly influence positive environmental protections. Another Green Commerce example is the use of spent vehicle tires to make paving materials for roads. Old tires are ground into small pieces, combined with a petroleum mixture, and used in lieu of asphalt for repairing roadways. Recycling old tires is a major benefit to the environment and public health, since old tires left in waste dumps can become a most fertile breeding ground for mosquitoes, which in turn, can carry viruses such as West Nile that can cause human illness.

Sometimes changes in technology can bring about green marketing opportunities by giving customers more choices in purchasing green products. As an example, purchasing digital cameras rather than film cameras provides an environmental benefit, since the chemicals used in film processing are no longer necessary. In Sweden, water quality tests showed that silver levels have decreased by more than 50 percent in five years in the waters of the Stockholm archipelago. Swedish environmental officials attribute the dramatic decrease to the growing use of digital photography and the corresponding reduction in film processing laboratories [15].

An important market power initiative is called Green Communities, a five-year, $555 million initiative to build more than 8,500 environmentally healthful homes for low-income families. The initiative was created by the Enterprise Foundation, a national nonprofit organization based in Columbia, Maryland, that provides assistance to grassroots home ownership organizations in partnership with the Natural Resources Defense Council, a national environmental organization based in Washington, D.C. The Green Communities initiative intends to transform the way the United States thinks about, designs, and builds affordable communities. The initiative provides grants, financing, tax-credit equity, and technical assistance to developers who meet Green Communities criteria for affordable housing that promotes health, conserves energy and natural resources, and provides easy access to jobs, schools, and services. Projects are underway in the states of Massachusetts, Michigan, and Minnesota [16].

An advantage of Green Commerce policy is letting a free market loose to help eliminate specific environmental hazards. The creativity of free enterprise can be harnessed and applied to improving environmental quality and reducing public health impacts of hazards in the environment. Government involvement can be absent or minimal, a situation that appeals to many commercial interests.

Market power can also be used for environmental improvement through the use of government's use of *market-based instruments* (MBIs). The European Environmental Agency (chapter 9) [17] observes that much environmental pollution and depletion of natural resources occur from incorrect pricing of goods and services, because prices do not often reflect the true costs of production and consumption. In particular, the impacts on the environmental are not always correctly factored into prices of goods and services. Examples of hidden costs include the costs that come from responding to damage from air and water pollution, disposal of waste, loss of soils and species, and effects of climate change, floods, heat waves, and storms. The European Environmental

Agency (EEA) asserts that MBIs provide a stimulus to consumers and producers to change their behavior toward use of more ecologically-efficient use of natural resources by reducing consumption, by stimulating technological innovation, and by encouraging greater transparency on how much consumers and producers actually pay for products and services. The EEA specifies five main kinds of MBIs [17]:

1. *Tradable permits* that have been designed to achieve reductions in pollution (such as emissions of CO_2) or use of resources (such as fish quotas) in the most effective way through the provision of market incentives to trade.

2. *Environmental taxes* that have been designed to change prices and thus the behavior of producers and consumers, as well as raise revenues.

3. *Environmental charges* that have been designed to cover (in part or in full) the costs of environmental services and abatement measures such as waste water treatment and waste disposal.

4. *Environmental subsidies and incentives* that have been designed to stimulate development of new technologies, to help create new markets for environmental goods and services including technologies, to encourage changes in consumer behavior through green purchasing schemes, and to temporarily support achieving higher levels of environmental protection by companies.

5. *Liability and compensation schemes* that aim at ensuring adequate compensation for damage resulting from activities dangerous to the environment and provide for means of prevention and reinstatement. [ibid.].

According to the EEA, the use of MBIs in environmental policy has found increasing favor in Europe since the 1990s, particularly in the Scandinavian countries and The Netherlands [ibid.]. MBIs that address taxes, charges, and tradable permits have been those most often used in setting environmental policy. The long-term effectiveness of MBIs awaits analysis, but offers the promise of supporting policies of sustainable development, in particular.

2.5.2.3   Performance Incentives

Incentives are a powerful motivator of human behavior. In the sports world, some players' contracts contain performance goals that when achieved will result in extra salary. For example, a baseball pitcher who wins twenty games, has an earned run average less than 3.0, and is voted to the All-Star game would be paid more for achieving these goals if they were elements of his contract with his team. Similar in concept, salespersons who exceed sales goals are often paid extra. Incentives are metaphorically the "carrot and stick" approach to behavioral performance.

Pollution trading credits (PTCs)[5] are an example of a Performance Incentives policy. The idea is relatively simple. A regulatory body (e.g., EPA) grants individual polluting facilities (e.g., an electric power generating plant) an annual allocation of PTCs that equals at most the maximum amount of pollution that they can release into the environment. However, if a given facility releases less pollution than their emissions allocation, the difference in trading credits can become a commodity and sold to facilities that are not meeting their annual emissions limit. In other words, it makes good economic sense for a facility to overachieve in order to market their PTCs. This is the free enterprise system being used to drive gains in environmental quality. As discussed in chapter 5, the 1990 amendments to the Clean Air Act contain a provision that EPA implement a marketplace program to sell PTCs to control acid rain.

2.5.2.4 Precautionary Approach

In order to protect the environment, the precautionary approach should be widely applied by States[6] according to their responsibilities. Where there are threats of serious or irreversible damage, lack of full scientific certainty shall not be used as a reason for postponing cost-effective measures to prevent environmental degradation. (Principle 15. Rio Declaration on Environment and Development) [18]

The preceding words constitute the core of what is called the *precautionary principle*. They are relatively simple words, but loaded with imprecision and chockablock with deliberate ambiguity. Let's take a step back and consider how these simple words of public health significance came to be.

The prevention of disease and disability is at the heart of all public health programs and practice. For environmental health hazards, elimination of a hazard (e.g., elimination of tetraethyl lead in gasoline) or reduced exposure (e.g., lower levels of toxicants released into the environment through remediation of hazardous waste sites) are effective primary prevention policies. These examples of environmental preventive policies also engage the subject of risk assessment, discussed in chapter 11, and as such raise several essential questions. For example, what is the risk to human health posed by a specific hazard? And how is the risk determined? And by whom? What is a reasonable time upon which to reduce an environmental hazard to acceptable levels? And is there a better policy to prevent the consequences of environmental hazards than what is currently used by U.S. regulatory agencies?

The preceding questions have been—and continue to be—the subject of serious debate among government agencies, legislative bodies, and special interest groups (e.g., environmental organizations, business associations). From such debates, over many years, has emerged the precautionary principle. According to Kriebel and colleagues [19], "The precautionary principle has arisen because of the perception that the pace of efforts to combat problems such as climate change, ecosystem degradation, and resource depletion is too slow and that environmental and health problems continue to grow more rapidly than society's ability to identify and correct them." In other words, the command and control regulatory approach would be unnecessary in some circumstances if a precautionary approach were operative, since some environmental hazards would be interdicted prior to the need for regulatory action. Also, because it can be readily related to the public health core tenet of disease and disability prevention, the precautionary principle has found favor with public health and environmental groups.

The following sections will review its history, elements, policy issues, and policy position of the U.S. government.

2.5.2.4.1 History

The origin of the precautionary principle, much like the implementation of the principle itself, is a matter of interpretation. Many North Americans seem to link it as only a product of the Rio Conference on the Environment, convened by the United Nations Environment Programme in 1992 in Rio de Janeiro, Brazil, as elaborated in chapter 9. However, the precautionary principle has earlier roots in Europe, the date of origin being a subject of some historical disagreement. Let's examine three somewhat different interpretations.

The origin of the precautionary principle, according to Sand [20], was in Scandinavia circa 1970, in particular, the Swedish Environment Protection Act of 1969. According to him, this law was the first to translate the principle into a general legal rule. The act introduced the concept of "environmentally hazardous activities" and reversed the burden of proof required of Swedish regulatory authorities. More specifically, the mere, but not remote, risk presented by a hazard could trigger action by authorities; they did not have to demonstrate the certainty that an impact would occur. Precautionary action, not retrospective correction, was the gist of the Swedish Environmental Protection Act, an act still in effect in Sweden. Sand [ibid.] also describes the spread of precautionary principle legislation into Denmark, Norway, and East European countries during the 1970s.

Another source states, "The precautionary principle emerged during the 1970s in the former West Germany at a time of social democratic planning " [21]. They observed that at the core of early concepts of precaution was the belief that the state should seek to avoid environmental damage through careful forward planning. The German government used the precautionary principle in the 1980s to justify policies that addressed acid rain, global warming, and pollution of the North Sea. However, in order not to put German industry at a disadvantage because of the government's adoption of the principle, Germany pressed the European Union (EU) throughout the 1980s to adopt a similar principle. In 1993, precaution was adopted by the EU as a principle of the Union's environmental policy [ibid.].

A third source, the European Environmental Agency (EEA), located in Copenhagen, chose to date the precautionary principle's roots to 1896, when the first information emerged about the health hazard of radiation [22]. The agency based its date to the time when what is now called the precautionary principle could have been applied—in their belief, to prevention of exposure to radiation. The same agency produced an informative report *Late Lessons from Early Warning*, which is about "[t]he gathering of information on the hazards of human economic activities and its use in taking action to better protect both the environment and the health of the species and ecosystems that are dependent on it, and then living with the consequences" [ibid.]. The report is based on fourteen case studies that were selected by the EEA for their known environmental and human health impacts (Table 2.2). Each case study was conducted by scientists considered as experts in their respective subjects. Authors of each case study were asked

---

TABLE 2.2
Case Studies Evaluated by the EEA [22]

Antimicrobials	Hormones
Asbestos	Mad cow disease
Benzene	MTBE
DES	PCBs
Fisheries	Radiation
Great Lakes pollution	Sulfur dioxide
Halocarbons	Tributyltin

"[t]o identify the dates of early warnings, to analyse how this information was used, and to describe the resulting costs, benefits and lessons for the future." The primary findings from the asbestos case study will suffice to illustrate how the case studies in the report were developed.

The case study of asbestos [23] can be summarized by the quotation made in 1931 by Thomas Legge and which begins their article, "[L]ooking back in the light of present knowledge, it is impossible not to feel that opportunities for discovery and prevention of asbestos disease were badly missed." Legge was the ex-chief of England's Medical Inspector of Factories. Asbestos is a naturally-occurring mineral that was widely used for various industrial purposes, but primarily for its heat insulation properties. Inhalation of asbestos fibers is the main cause of asbestosis,[7] a lung disease caused by inhalation of asbestos fibers, and mesothelioma, an aggressive, almost always fatal, type of lung cancer. Workplace exposure to asbestos, in Europe alone, is expected to cause four hundred thousand cases of mesothelioma [ibid.]. The authors of the case study found that the earliest reports of asbestos as a workplace hazard occurred in 1898, when reported by an inspector of factories in England, who was concerned about the crystalline properties of inhaled asbestos dust. In 1897 had come the first British report of lung disease (what would now be called asbestosis) attributed to inhaled asbestos dust in a worker. In 1906, these lung disease findings were brought to the attention of the British government, but no action was taken. Additional articles published in the medical literature during the 1920s and 1930s described asbestosis in asbestos workers. In 1931, these findings led to the first asbestos dust control regulations in Great Britain, although they were widely ignored [ibid.]. In 1998–1999, both France and the European Union banned the use of all forms of asbestos. In the United States, using authorities in the Toxic Substances Control Act, §5, in 1989, EPA banned all new uses of asbestos; uses before 1989 are still allowed. Subsequent EPA regulations required school systems to inspect for asbestos and to eliminate or reduce the exposure by removal or containment in place. Asbestos released into air and water is regulated under the Clean Air Act and Clean Water Act, respectively [24]

The EEA asbestos case study identified several reasons why prevention actions were not implemented, even in the presence of considerable medical data. Economic interests were, of course, a significant weight used against taking a precaution approach toward asbestos. Asbestos was a valuable commercial product through the middle of the twentieth century in the industrialized countries. Asbestos companies paid local and other taxes and employed large number of workers, helping then to boost local economies. However, costs did not include the eventual health costs associated with health care for asbestos victims. Another reason for not taking a precautionary approach in the early decades of the twentieth century included the unprofessional involvement of some company physicians, who presented data and reports to government agencies, denying any asbestos health problems in their companies' workers.

The fourteen case studies in the EEA's report led to twelve general conclusions about implementing a precautionary approach to controlling environmental hazards (Table 2.3). Several of the recommendations address the need to have an adequate scientific basis upon which to predicate precautionary actions. Two lessons, numbers 10 and 12, merit further comment because of their policy implications.

Concerning Lesson 10, maintaining regulatory independence from economic and political special interests is consistent with similar advice from the U.S. National Research Council in 1982 [25]. They strongly recommended that federal government agencies keep risk assessment separate from risk management; whereas the former should be based on scientific data and judgment, the latter necessarily involves economic and societal considerations. Keeping this separation intact within U.S. regulatory agencies has become a difficult policy, owing in part to involvement of special interest groups during the preparation of specific risk assessments. This involvement occurs during public comment periods for a proposed risk assessment and from political lobbying of agency and legislative staffs.

Regarding EEA Lesson 12, preventing "paralysis by analysis" has become a sad reality for U.S. regulatory agencies, for which risk assessments may now require several years, particularly for high profile hazards. An example will suffice. EPA, as of May

---

TABLE 2.3
Lessons Learned from EEA Case Studies [22]

1   Respond to ignorance as well as uncertainty

2   Research and monitor for "early warnings"

3   Search out and address "blind spots" and gaps in scientific knowledge

4   Identify and reduce interdisciplinary obstacles to learning

5   Ensure that real world conditions are fully accounted for

6   Systematically scrutinize and justify the claimed "pros" and "cons"

7   Evaluate alternatives and promote robust, diverse, and adaptable solutions

8   Use "lay" and local knowledge as well as all relevant specialist experience

9   Take account of wider social interests and values

10 Maintain regulatory independence from economic and political special interests

11 Identify and reduce institutional obstacles to learning and action

12 Avoid paralysis by analysis

2006, has spent fifteen years in its reassessment of the health risk of dioxin. One reason for the delay in completing EPA's update assessment was what weight to give to a body of basic science about cellular mechanisms of dioxin toxicity. Using this basic science as a means to predict dioxin's carcinogenicity, rather than relying on experimental evidence, was a precedent, which became the focus of controversy between special interest groups—industry supporting, environmental groups opposing. This debate about the role of basic science in risk assessment helped contribute to a paralysis in the reassessment analysis of dioxin's risk.

2.5.2.4.2 International Charters

Regardless of how one interprets the history of the precautionary principle, it has been adopted as policy in several regional charters, although its definition varies between charters, and additional operational guidelines await. Applegate [26] notes the precautionary principle's presence in several international treaties and charters, including the 1993 charter of the European Union, the 1991 Bamako Convention on Hazardous Waste in Africa, the 1992 Convention on Biological Diversity, the 1992 United Nations Framework Convention on Climate Change (UNFCC), the Treaty on European Union in 1992, the Cartagena Protocol on Biosafety in 2000, and the 2001 Stockholm Convention on Persistent Organic Pollutants (chapter 9). Illustrative of the language found in these charters is wording in the Cartagena Protocol on Biosafety, "[I]n accordance with the precautionary approach contained in Principal 15 of the Rio Declaration on Environment and Development, the objective of this Protocol is to contribute to ensuring an adequate level of protection in the field of the safe transfer, handling and use of living modified organisms resulting from modern biotechnology that may have adverse effects on the conservation and sustainable use of biological diversity, taking also into account risks to human health, and specifically focusing on transboundary movements" [27]. This body of global charters is important because eventually the precautionary approach will be implemented through rules, regulations, and practices that flow from the treaties and charters.

Referring to the UNEP definition of the precautionary principle, placed at the beginning of this section, of note is the appearance of the words *cost-effective*. They are important words; they can be used to frustrate, or make more difficult, precautionary actions. For example, how is cost to be calculated? And who determines what is effective? The words *cost-effective* derive from actions attributable to the U.S. government. They first appeared in Article 7 of the Second World Climate Conference in 1990 at the insistence of the United States and were later added to the Rio Declaration text (chapter 9), but over the objections of the European Union and Japan [20]. The U.S. government's position of cost-effectiveness is consistent with congressional actions that have mandated cost-benefit analyses of federal regulatory proposals and other environmental health policies. Nevertheless, insistence on cost-effectiveness as a component of the precautionary principle seems to make it more difficult to operationalize the principle.

Notwithstanding the introduction of the precautionary principle into regional and international charters and conventions, it was the 1992 Rio Declaration of the United Nations Conference on Environment and Development that elevated attention and created a greater awareness of the principle in North and South American countries. The Rio Conference is significant because it was the first international environmental conference attended by heads of state, rather than representatives of ministerial or equivalent levels of national governments. President George H.W. Bush led the U.S. delegation.

2.5.2.4.3 Elements of the Precautionary Principle

As Applegate [26] observes, "Despite its wide acceptance as a foundation of international environmental law, the precise meaning of the Precautionary Principle remains surprisingly elusive. It has been defined variously over the last two or three decades in international legal instruments and by commentators, and the overall concept admits of varying degrees of environmental protection." Consistent with the diverse interpretations of the precautionary principle is the absence of consensus agreement on its elements. One source [ibid.] identified the four elements listed in Table 2.4, with actions associated with each element, but questions arise when trying to apply the elements to a specific environmental hazard. For example, when considering the trigger element, how does *serious damage* differ from *irreversible damage*, which is what the precautionary principle asserts? Moreover, what are the criteria for determining serious harm? Regarding timing, just how much of a scientific basis needs to exist before eliciting a prevention response? Again, are there criteria to determine the adequacy of key scientific findings? These are the kinds of questions that government agencies are debating when trying to make the precautionary principle operational (i.e., when trying to convert general policy into specific practices).

2.5.2.4.4 Environmentalists' Version of the Precautionary Principle

The precautionary approach has gained widespread, indeed, global favor with environmentalists, grassroots environmental organizations, and European governments. As previously noted, these and other organizations perceive that risk assessment-based,

---

TABLE 2.4
Elements of the Precautionary Principle [26]

Element	Action
Trigger	Potential serious or irreversible harm
Timing	Anticipatory action before causation can be scientifically identified
Response	Total avoidance Measures to minimize or mitigate harm Cost-effective regulatory measures Study alternatives, with an eye to prevention
Regulatory strategies	Bans and phase-outs Environmental effects assessment Pollution prevention Reversed burden of proof Polluter pays

regulatory approaches are now bound up with delay and inaction. Acting upon these concerns, on January 23–25, 1998, 35 academic scientists, grassroots environmentalists, government researchers, and labor representatives from the United States, Canada, and Europe met to discuss ways to formalize their version of the precautionary principle [28]. The meeting was held at the Wingspread Conference Center, Racine, Wisconsin, and has become known as the Wingspread precautionary principle, stating, "When an activity raises threats of harm to human health or the environment, precautionary measures should be taken even if some cause and effect relationships are not fully established scientifically" [19]. Although similar in theme to the Rio version of the precautionary principle, note that the words *cost-effective* are missing in the Wingspread version. One assumes that deletion of these words was intentional and for the purpose of removing a perceived impediment in applications of the Wingspread version of the precautionary principle.

As an aid to operationalizing the precautionary principle, the Wingspread's participants defined four components: preventative action should be taken in advance of scientific proof of causality; the proponent of an activity, rather than the public, should bear the burden of proof of safety; a reasonable range of alternatives, including a no-action alternative, should be considered when there may be evidence of harm caused by an activity; and for decision making to be precautionary it must be open, informed, and democratic and must include potentially affected parties [28].

The Wingspread interpretation of the Precautionary Approach, as defined by its four components, has gained the interest of sustainable agriculture advocates, who have used it in the State of Washington to protest the use of hazardous waste in the manufacture of fertilizers and in the response to the federal government's organic agriculture rule [ibid.]. Greater use of the precautionary principle by grassroots organizations is likely.

2.5.2.4.5   *Position of the U.S. Government*

There is strong support from environmentalists and public health groups for the U.S. government to adopt the precautionary principle as a matter of environmental health policy. In the supporters' opinion, required use of the principle would lead to prevention of more environmental health hazards because regulatory agencies could act more expeditiously than what is currently the case. Counterbalancing that support is opposition from the U.S. business community, who fear possible adverse economic consequences from adoption of the principle. These consequences are asserted to include more stringent environmental standards, greater numbers of banned or otherwise regulated commercial products, and possible greater product litigation.

In response to those who support, or oppose, the U.S. government's adoption of the precautionary principle as policy, no official position has been taken. EPA, which would be the U.S. agency most impacted by the principle as environmental policy, has put off a proposal for developing a position paper on it, stating that the time "[i]s not ripe for such a paper given the lack of a U.S. government position"[29].

It is unlikely that the United States will anytime soon take a position in support of adopting the precautionary principle as government policy. This assertion is based on practical considerations. First, a large body of U.S. law exists for the purpose of controlling environmental hazards. These laws have generally adopted the proposition that a substance is "safe" until proven otherwise. Contaminants are permitted, through emissions regulations, to be released into the environment, using risk assessment and risk management methods, to achieve "safe" environmental health conditions. As a matter of practicality, this body of environmental law, accompanying regulations, and court decisions is not likely to be significantly revised anytime soon. Second, the regulated community in the United States is generally opposed to adoption of the principle as government policy. There are too many uncertainties in how the principle would be implemented in ways that could impact them.

2.5.2.4.6   Position of the European Union

The EU (see chapter 9) has adopted the precautionary principle as policy. This is not surprising, given its origins in Sweden and Germany. Moreover, the slow adoption in Europe of quantitative risk assessment, in contrast to the United States, created something of a void in how to prevent adverse public health and environmental consequences of environmental hazards. In 2000, the European Commission, which is the administrative arm of the EU, published precautionary principle guidelines for the EU's member states [30]. The guidelines include the following:

1. The precautionary principle should be considered within a structured approach to the analysis of risk, which comprises three elements: risk assessment, risk management, risk communication. The precautionary principle is particularly relevant to the management of risk.
2. "Where action is deemed necessary, measures based on the precautionary principle should be, inter alia:
    *   proportional to the chosen level of protection.
    *   nondiscriminatory in their application,
    *   consistent with similar measures already taken,
    *   based on an examination of the potential benefits and costs of action or lack of action (including, where appropriate and feasible, an economic cost-benefit analysis),
    *   subject to review; in the light of new scientific data, and
    *   capable of assigning responsibility for producing the scientific evidence necessary for a more comprehensive risk assessment." [ibid.]

These guidelines make clear that the EU has adopted the precautionary principle as policy and is working on ways to implement it. Similar guidelines and policy adoption do not exist in the United States, since the federal government has taken no official position on the principle.

Adoption by the EU of the precautionary principle as policy has already had public health and economic consequences. An example of the former is found in an EU proposed program of chemical testing. In October 2003 the European Commission adopted legislation for a new EU regulatory framework for chemicals [31]. Called the Registration, Evaluation and Authorisation of Chemicals (REACH) framework [32], the proposal would require the chemical industry to test tens of thousands of chemicals that are used—or proposed for use—in the EU and for which toxicity data are lacking. The principle was the driving force behind REACH, which has been vigorously opposed by the U.S. government [33], asserting that the U.S. Toxic Substances Control Act (chapter 7) is adequate for testing chemicals that reach the United States The global chemical industry also opposes REACH, primarily because of the high cost of conducting toxicity tests. As environmental health policy, better toxicological databases of substances already in commerce will help make better regulatory decisions and provide improved programs of public health.

In November 2005, the European Parliament approved a modified version of REACH. The amended REACH program reduced the overall number of chemicals that would be required for testing by chemical producers. EU's member states (chapter 9) must approve the Parliament's legislation before a final REACH program is adopted throughout the EU [34].

An example of the economic impact of the EU's precautionary principle comes from its use to protect European trade interests. As related by Goldstein and Carruth [35], the European Community used the principle to justify a more stringent EU aflatoxin standard, one lower than that recommended by the World Health Organization's Food and Agriculture Organization and many other countries. The lower EU standard has been used to block the importation of aflatoxin-containing goods (e.g., peanuts) from African countries, thereby protecting farmers in EU member states in southern Europe.

In summary, the precautionary principle, depending on how it is made operational, could serve as an alternative to the Command and Control policy by its use to forego a risk assessment-based policy that has become litigious and time consuming. A challenge in the adoption of a precautionary approach would be to avoid its becoming another form of command and control.

2.5.3 PUBLIC EDUCATION

As another alternative to the Command and Control policy, an informed public can make a significant difference in reducing the effects of exposure to environmental hazards. For example, informed individuals can choose to purchase green products and services, patronize companies and business enterprises that have evidenced environmental contributions, and advocate for the importance of nurturing environmental health policies. Acting on the basis of environmental information is activism in practice. But how does an individual acquire such information? And from whom?

National environmental organizations such as the Sierra Club, Natural Resources Defense Council, Environmental Defense, and Physicians for Social Responsibility have effectuated Public Education policies. These and similar organizations provide scientific documents, news alerts, and policy recommendations of relevance to environmental health policies and practices. Because environmental organizations have achieved considerable credibility with the U.S. public, their Public Education actions and products have great political and personal impacts. Much of these organizations' educational materials are now made accessible on the Internet. As an example,

Physicians for Social Responsibility's environmental health program offers a brochure on reproductive disorders and birth defects that is available on their Web site (www. psr.org). Using the information in the brochure, individuals can make personal lifestyle choices that will reduce exposure to environmental hazards (e.g., solvent-based paint) to fetal development.

Many government agencies and academic institutions also have adopted Public Education policies. Environmental health information can be found on the Internet through the use of any Worldwide Web browser. For example, the Agency for Toxic Substances and Disease Registry, the EPA, and the National Institute of Environmental Health Sciences are federal government agencies that make environmental health information available to the public. Similarly, many schools of public health (e.g., the Harvard School of Public Health), publish environmental health information on their web sites.

Environmental health information from creditable sources can form the foundation for community activist groups upon which to build specific agendas and action plans. As an example, cleanup of uncontrolled hazardous waste is a matter of concern to many community and neighborhood groups. A host of Internet resources can readily be accessed that contain information about the human and ecological effects of exposure to hazardous waste. Armed with such information, community groups can use the PACM model to bring pressure on government agencies, business enterprises, and individuals to adopt beneficial environmental health practices.

A well-known example of a successful Public Education campaign is the designated driver program promoted by Mothers Against Drunk Driving (MADD). The goal of the campaign is to reduce the number of alcohol intoxicated motor vehicle drivers, thereby reducing car crashes caused by drunk drivers. Designated drivers are persons chosen in advance by persons planning to attend social events where alcohol will be served. By foregoing alcohol consumption, a designated driver can function as a sober provider of transportation for the other members of the group. The designated driver and similar Public Education efforts are most successful when they include societal responsibilities in their messages. An example is the sound bite, "Friends don't let friends drive drunk," which is a clear, effective, societal admonition that has become social doctrine. Other examples where public education has led to public health benefits include awareness of the adverse health effects of tobacco products, defects in a manufacturer's vehicle tires, and the safety problems caused by some sport utility vehicles that can tip over during vehicle operation.

2.5.4 Sustainable Development

The world's natural resources are not inexhaustible. Consider the loss of soil to erosion, oil lost to over reliance of human use of carbon fuels, forests lost to deforestation, fauna and wildlife disappearance because of human clearing of forests, and fishing of some fish populations to near extinction, amongst other examples. While some natural resources are renewable or can be supplanted by other resources (e.g., use of renewable sources of energy), current and future generations of humankind must consider how natural resources can be preserved and economic and societal development sustained.

One source observes that the main driving forces of human consumption of resources are population and economic development [36]. They note that the projected 50 percent growth in the global population over the next fifty years will put a significant strain on the environment. Further, the EEA observes that if the population of the developing countries achieves levels of material wealth like that in current-day industrialized countries, consumption of resources would increase by a factor ranging from two to five. Is there a policy solution to what appears to be a pending disaster of exhaustion of natural resources?

A policy that responds to the foregoing concerns and is gaining great international support is *sustainable development*. The concept is not new. Humankind has historically used principles of sustainable development upon which to exist. As examples, farmers long ago learned how to rotate crops annually in order to restore soil nutrients in farm fields, and nomadic peoples sustained themselves by domesticating animals, which then provided transportation, meat, milk, clothing, and commerce. A more current example is sustainable forestry, wherein lumber companies replace harvested trees with seedlings, achieving an overall balance of a constant or increased number of harvestable trees.

Humankind's ignorance (or unwillingness to practice) of sustainable development can be illustrated by a few examples. Overfishing of certain fish stocks in the North Atlantic Ocean (e.g., cod) has reduced them to near extinction numbers. Urban sprawl in large cities has led to destruction of local forests, contributing to loss of natural habitat for birds and other beneficial creatures. In industrialized countries, industrial waste was dumped directly into lakes, rivers, and other waterways, sometimes polluting them to the point of extinction of fish, shellfish, mammals, and other creatures. Currently, climate change is the most glaring example of failure to practice sustainable development. This is discussed further in chapter 9.

Sustainable development as global policy is relatively recent, dating to the 1972 United Nations Conference on the Human Environment, held in Stockholm, Sweden (chapter 9). The conference was convened for purpose of developing a global environmental protection policy and for enunciating common principles to preserve and enhance the human environment [37]. While the Stockholm conference was significant for focusing global attention on the interconnections between human development and the environment, it was fifteen years before a more precise focus was brought to bear on sustainable development.

Sustainable development is "Development which meets the needs of the present without compromising the ability of future generations to meet their own needs." [38]"

In 1987, the World Commission on Environment and Development, chaired by the Prime Minister of Norway, Gro Harlem Bruntland, published the report *Our Common Future* [38], which imprinted sustainable development on the international environmental health agenda. The report is often referred to as the Bruntland Report. It contains the most often quoted definition of sustainable development, "Development which meets the needs of the present without compromising the ability of future generations to meet their own needs." This definition can be used equally by governments and private sector entities. Indeed, the latter group may hold the greater promise for making sustainable development a practical reality. Citing three principles will provide an overall sense of the Declaration's thrust:

- Principle 2—The natural resources of the earth, including the air, water, land, flora and fauna and especially representative samples of natural ecosystems, must be safeguarded for the benefit of present and future generations through careful planning or management, as appropriate.
- Principle 5—The nonrenewable resources of the earth must be employed in such a way as to guard against the danger of their future exhaustion and to ensure that benefits from such employment are shared by all mankind.
- Principle 13—In order to achieve a more rational management of resources and thus to improve the environment, States[8] should adopt an integrated and coordinated approach to their development planning so as to ensure that development is compatible with the need to protect and improve environment for the benefit of their population.

As a global environmental health policy, sustainable development came into full flower at the 1992 UN Conference on Environment and Development, also called the Earth Summit, held in Rio de Janeiro, Brazil, to discuss how to achieve sustainable development. Under the auspices of the United Nations, the Earth Summit brought together more than 180 countries, represented by heads of state or national leaders. The Rio Declaration on Environment and Development builds upon the sustainable development recommendations in the Bruntland Report. The Rio Summit created agreements and conventions on critical issues such as climate change, deforestation, and desertification. In addition, the parties to the Rio Declaration drafted a broad action plan, *Agenda 21*, as the strategy for dealing with future global environment and development issues. Moreover, *Agenda 21*, which is discussed in chapter 9, includes commitments to: reduce global poverty, promotes women's rights, ban racism, and foster the welfare of children. These societal commitments are very much in the spirit of sustainable development.

The practice of the principles of sustainable development could obviate the need for regulatory control of some environmental hazards.

As a consequence of the Rio Declaration, regional and sectoral (e.g., business sectors) sustainability plans have been developed. Moreover, a host of groups have adopted the concept of sustainable development. These groups include businesses, municipal governments, and international organizations such as the World Bank (chapter 9). Indeed, the pervasiveness of sustainable development concepts portends a significant impact for better global environmental quality, resource management, and protection of human and ecological health. As a follow-up to the 1992 Earth Summit, the World Summit on Sustainable Development (WSSD)[9] was held from August 26 to September 4, 2002, in Johannesburg, South Africa, to elaborate on *Agenda 21*. A primary objective of the summit was to develop concrete steps and identify quantifiable targets for better implementation of *Agenda 21*. Two areas of focus at the summit were alleviation of global poverty and protection of the natural environment and human health. As a matter of environmental health policy, having an agenda, stated goals, and targets for global improvement is an important resource for long-range national and international planning. According to the UN [39], areas of commitments from the WSSD Plan of Implementation pertinent to environmental health policy include water and sanitation, energy, health, agriculture, biodiversity, and ecosystem management.

Although the United States is a signatory of the Rio Declaration and participated in the Johannesburg summit, little effort to support the principles of *Agenda 21* is evident. For example, two contemporary environmental issues that need to be vetted in the context of sustainable development include global climate change and deforestation of tropical rainforests. For the former, global policies (e.g., Kyoto Protocol) and actions must be adopted and implemented if projected massive socioeconomic and public health consequences of global climate are to be avoided. For the latter, unwise destruction of huge areas of tropic forests portends loss of oxygen-producing trees, unique plants that have medicinal potential, and species of animals both unique and irreplaceable.

In summary, practice of the principles of sustainable development could obviate the need for regulatory control of some environmental hazards (e.g., over use of pesticides and herbicides in agricultural and other uses). Problems that do not occur do not require command and control response. Moreover, sustainable development focuses on improving the quality of life for all the Earth's peoples, without using natural resources beyond the capacity of the environment to supply them indefinitely. As policy, sustainable development can be practiced without resorting to legislation and regulations, although some governments, particularly in Europe, have incorporated sustainable development into legal frameworks.

2.5.5 VOLUNTARY ACTION BY INDUSTRY AND BUSINESSES

Corporations and other businesses can adopt voluntary actions to eliminate environmental hazards in workplaces, communities, and homes. Voluntary actions are those not mandated by government agencies. A policy of Voluntary Action can reap benefits to business enterprises such as increased income, better community relations, and less litigation, depending, of course, on the nature of the voluntary action. As an example, in the fall of 2005 the Dannon Company announced that they would forego placing plastic overcaps on each container of yogurt, as had been their practice for many years, saving about 3.6 million pounds of plastic annually [40]. The result will lessen the amount of plastic that enters the waste stream, thereby lessening the volume of waste in sanitary landfills. As another example, producers of chemical stain repellants are redesigning their products to make them less hazardous to consumers and the environment. Stain repellants, used to ease the cleaning of carpets and clothing, are long-chain flurosurfactants, which can metabolize to a toxic compound. Replacement with shorter chain surfactants leads to a lesser hazard [41].

Other examples of Voluntary Action policies include a manufacturing plant's voluntarily decreasing the amounts of pollution released into the community environment beyond the levels permitted by environmental emission standards (e.g., clean water discharge standards). Sometimes, voluntary actions to reduce emissions are in response to public awareness of Toxics Release Inventory (TRI)[10] data required of a plant or facility. While reporting certain levels of emissions to EPA under the TRI regulations is mandatory, reducing the amounts of emissions beyond air and water quality standards is voluntary. Some companies have exerted extra effort to decrease their TRI emissions in order to improve their community image, an outcome called by some as "regulation by shaming" [9]. In 2005, EPA reported that TRI data showed that the amount of toxic substances released into the U.S. environment had declined 42 percent between the years 1999 and 2003 [42]. If emissions released into the environment are thought of as waste, and as such, an indicator of inefficient production, decreased emissions can have a positive economic benefit to a company.

Voluntary action can result from litigation by an individual or group against a company or other business enterprise. In effect, a single episode of a litigated environmental hazard's impact can result in a much broader prevention effort. Take the example of scalding hot coffee formerly sold by a fast food company. In one instance, a customer of McDonald's who purchased a cup of coffee placed it between her bare legs while driving a car. The coffee spilled onto her skin, causing severe burns. Later, a jury awarded the customer $2.7 million (later reduced by an appellate court to $470,000) for her medical costs and damages. While the customer's wisdom in how the coffee cup was held could be questioned, the fact remains that the beverage's temperature was quite high, and as such, was an environmental hazard. Subsequent to the litigation's outcome, the fast food company voluntarily lowered the temperature of coffee served to millions of customers. This act of primary prevention (i.e., elimination of a hazard) extended worldwide, owing to the thousands of food service establishments operated by the company. In effect, one response to person's injury was multiplied into a public health benefit for millions of people.

Another example of voluntary actions by business entities is an effort by the Green Building Rating System, which produced the Leadership in Energy and Design (LEED) framework for sustainable buildings. According to its developer, "LEED is a voluntary, consensus-based national standard for developing high-performance, sustainable buildings" [43]. According to the same source, LEED standards that are available or under development include: new commercial construction, homes, and neighborhood development. The extent to which LEED standards have been adopted by business entities is unknown, but the standards give promise of industry adoption, given that the standards are a consensus industry product. As a matter of environmental health policy, industry adoption of environmentally beneficial standards would be supportive of sustainable development goals.

2.5.6 POLICY CORNUCOPIA

The previous sections have described seven kinds of policies that have evolved in the United States and Europe for purpose of controlling environmental hazards. Of the seven policies, only one, Command and Control, bears the force of law. The other six policies require voluntary action by interested individuals and groups. The voluntary polices have emerged over time as means to replace or supplement existing policies perceived to have failed or otherwise been ineffective in controlling specific environmental hazards. To be more specific, many environmentalists and public health officials have become frustrated with the often painfully slow process that now comes with establishing government regulations and standards. To wit, new standards (e.g., water quality) are routinely challenged by special interest groups through protracted litigation and political pressure on legislative bodies. The rejected OSHA ergonomics standard is such an example.

It is doubtful that Command and Control will, or should, be replaced as the anchor policy in controlling environmental hazards. Without the weight of law to control environmental hazards, it is highly improbable that sources of pollution would be abated voluntarily in amounts sufficient to make a real difference in environmental quality. Further, regulatory frameworks "level the playing field" by treating all sources of pollution the same. That is to say, water and air quality standards apply equally to contaminants released by big corporations or by small businesses. Without regulations and standards, based on statutes, industrialized countries would become no different from developing countries, where pollution controls are largely nonexistent.

However, as history has shown, promulgating workplace and community environmental regulations has become increasingly challenging to regulatory agencies. The regulated community generally has the economic and political wherewithal to block or alter proposed or adopted environmental regulations. Moreover, the regulatory apparatus has become thoroughly political, as illustrated by Congress's repeal of the OSHA ergonomics standard. Given the difficulty of establishing regulations and standards, reliance on voluntary policies (e.g., sustainable development, Green Commerce, and market power) will increase in importance.

2.6 POLICY AS A MEANS TO DISEASE PREVENTION

The prevention of disease and disability is the essence and focus of public health. Prevention modes of action are sometimes grouped into three categories: primary, secondary, and tertiary [44]. The boundaries between these modes are sometimes not clear, nor is it important that they always be distinct. Elimination or reduction of conditions that cause adverse health effects in humans is called *primary prevention*. For example, some local health departments use larvicides[11] to kill mosquitoes that carry West Nile virus, thereby preventing human contact with the virus. *Secondary prevention* refers to the use of education, protective equipment, relocation away from a hazard, or other means to avoid contact with a hazard. For example, educational materials presented and discussed with residents of older houses where lead-based paint could exist, constituting a health hazard to young children, would be an act of secondary prevention. Removal of the paint would be primary prevention. *Tertiary prevention* relates to health care and consists of the measures available to reduce impairments and disabilities, and minimize suffering caused by existing departures from good health (adapted from ibid.). For example, health care that reduces the suffering in children exposed to quite large amounts of lead could be considered as tertiary prevention.

Given the foregoing discussion, do environmental health policies relate to public health programs of disease and disability prevention? One can assert that such policies, in fact, do constitute elements of public health disease prevention. Some examples of environmental health policies linked to primary and secondary prevention measures are shown in Table 2.5. Consider the primary prevention examples. Many environmental statutes (e.g., the federal Clean Air Act) contain provisions to control the levels of pollutants that can cause adverse health effects in humans. This statutory policy of pollution control comports with a strategy of primary prevention of disease (i.e., a reduction or elimination of a hazard). A similar line of reasoning would apply to the other primary prevention examples in Table 2.5.

As regards the secondary prevention examples shown in Table 2.5, each example pertains to some facet of education. For instance, consider the right-to-know policy. Individuals in the United States can now usually obtain environmental information of relevance to their personal health. Persons can now access databases on the Internet such as EPA's Toxics Release Inventory and thereby obtain information about pollution sources within their community. This information can then be used to make personal decisions such as whether to relocate from the community or create advocacy groups to lower community pollution emissions.

2.7   ENVIRONMENTAL ETHICS

Is it ethical to pollute the environment? If it is, under what conditions? And if it isn't, under what framework of ethical principles? What are the environmental ethics of national economies that are based on tenets of consumerism? What are the ethical dimensions of national and international policies regarding the environment? On an individual basis, is it immoral to purchase consumer products that are not "environmentally benign"

TABLE 2.5
Examples of Policies Linked to Disease Prevention

Kind of Prevention	Action
Primary	Environmental statutes
	Regulations and standards (R&S)
	Alternatives to R&S
Secondary	Right-to-know
	Public education
	Medical education

or, better, "environmentally friendly?" These are examples of serious and societally consequential questions that are being given increasing attention by ethicists who are turning their attention to environmental concerns under the rubric of environmental ethics, or ecoethics.

2.7.1 ECOETHICS

Before turning to what constitutes environmental ethics, it is useful to clarify terms. Even though the words *ethics*, *values*, and *morals* are often used interchangeably (and sometimes appropriately so), there are important practical differences in the words' meanings. *"Ethics* traditionally refers to the systematic framework of thought and analysis that deals with questions of right and wrong and the nature of the good and proper life" [46]. An ethical act is an action that is consistent within an ethical framework. *Values* refers more to a quality considered inherently worthwhile or desirable. *Morals* generally refers to that natural working-out of a personally affirmed ethical or value system. Morals become the conscious and visible basis for personal conduct and action. Morality is therefore concerned more with how a person acts than with the system that provides the framework for action [ibid.]. How are these concepts of ethics, values, and morality integrated into what is being called *environmental ethics*, or *ecoethics*?

According to Timmenman [47], "Many people date the rise of 'environmental ethics' as an academic discipline from a famous 1967 article in the journal *Science* by the U.S. historian of medieval science, Lynn White, Jr. In his piece White accused some aspects of Christianity of fostering an attitude toward nature as an object for exploitation and manipulation. This debate illuminated a number of hidden assumptions in medieval and modern history about the images of the human and the natural." *Environmental ethics* is concerned with the framework in which humankind relates to the natural environment. Although the intellectual underpinning of environmental ethics is being provided by philosophers—ecoethics is now considered a subdiscipline in ethics—these concepts will ultimately influence the laws and policies that pertain to the environment.

Ecoethics forces each of us to confront why we do things and the consequences of our actions to the environment. If one accepts the proposition that industrial, agricultural, consumer, and personal practices have resulted in pollution of the environment and caused adverse effects on human health and ecological systems; what accounts for the degraded state of environmental affairs in which we find ourselves? Lutzenberger [48] posits that an understanding to this question requires an examination of the basic philosophy of modern civilization. He asserts we do what we do because of the dogmas, premises, and postulates on which national and global economic activities are based. In particular, philosophical and religious thinking that has not integrated ecological concepts into their framework is seen by some as a major reason for current environmental problems that include deforestation, waste mismanagement, ozone depletion, and global warming.

Environmental ethics attempts to look at the world in a holistic view and through a different point of view, one that builds around a framework of the planet as an entity that must be shared by all human and natural occupants. In 1975, James Lovelock, a

British atmospheric chemist, and Lynn Margulis, a U.S. microbiologist, proposed that Earth be viewed as one enormous, complex ecosystem, which they called Gaia, after the Greek goddess of the Earth, and that humans constitute cells in a tissue of this supraorganism [46,48]. This has become known as The Gaia Hypothesis—Earth as a single, self-regulating organism—and has developed a very large following, especially within the activist and "deep ecology" segments of the environmental movement. (The Norwegian philosopher Arne Naess in the 1970s separated ecology into two divisions: "shallow" and "deep." In his scheme, shallow ecology advocates a continuation of current environmental and political paths with only certain revisions to modem lifestyles and ways of life. Deep ecology advocates deep and fundamental changes in human relationships with nature [46,47]). It is interesting to note that Margulis takes issue with how the Gaia hypothesis has been popularized by others. She affirms the notion that the Gaia hypothesis refers to Earth not as an individual organism, but an ecosystem [49]. Regardless of this controversy, it seems important to retain the essence of the Gaia hypothesis: Earth is a complex ecosystem and we as humans are but members of this system. We must function within the ecosystem and understand our relationships within it in ways that will sustain it.

2.7.2   ETHICS OF ORGANIZATIONS

Organizations such as government agencies, corporations, and environmental groups, like individuals, are expected by society to act ethically (i.e., to do "the right thing"). Laws and regulations are a formal, external means to express society's ethical expectations. For example, organizations that commit fraud are held accountable under applicable law. However, organizations' internal ethics are equally important to those imposed by external sources.

Internal ethics could include, for illustration of environmental issues, a commitment to protect the environment. From such an ethic could flow specific environmental policies, such as prohibition of environmental injustices, how waste is managed, support for employees' car-pooling, and purchasing of environmentally-benign products. Internal ethics must have the support of an organization's leaders, ideally originating with them, and be facilitated by rank-and-file employees.

2.7.3   ETHICS OF INDIVIDUALS

No person lives without impacting the ecosystem. In fact, the very act of life adds carbon dioxide and other bodily wastes to the environment. Ecoethics stresses the importance of a healthy and healthful environment that is capable of sustaining its quality. How we choose to protect the environment and minimize each person's ecological impact is a matter of individual choice. It is advisable to base one's personal ecoethics on a principle drawn from a framework of ethical conduct. Principles adhered to can serve as a long-standing basis for personal ethical conduct. In the case of personal ecoethics,

TABLE 2.6
Examples of Personal Ecoethics

Ethic	Illustrative Action
Water conservation	Reduce volume of water used in bathing & showering
Waste minimization	Recycle paper and paper products Reuse household products
Energy conservation	Drive hybrid automobiles Shut off electricity when not in use
Air quality support	Keep internal combustion engines in tune Use public transportation
Social support	Support conservation organizations Express environmental opinions to elected officials

principles such as sustainable development and the precautionary principle can serve as platforms for an individual's ecoethical behavior.

Shown in Table 2.6 are examples of a personal ecoethical framework. The framework used to construct the table is based on the practice of sustainable development, (i.e., using only those resources in amounts that can be renewed and thereby reducing an individual's ecological footprint). Consider the example of water conservation. It is an ethical act to help preserve water supplies, given contemporary global demands on water supplies because of such factors as increased human populations and climate change. Examples of ways to conserve water in everyday life could include reducing the amount of water used for showering and bathing, planting of plants that require less water for survival, and recycling of "graywater" (i.e., household water collected from showers, sinks, and laundries that can be minimally treated by municipal water systems and used for specified purposes, e.g., irrigation of crops and lawns).

Individuals evolve their own set of ethics, based on circumstances that include family experiences, religiosity, education, peer pressure, and societal expectations. Like organizations, individuals who violate ethics expressed in law run the risk of facing legal retribution. But beyond the ethics imposed on individuals by society, there are the ethics that persons can choose for themselves, including those that pertain to the environment and human health. For instance, an individual can choose to live a life of environmentally-nurturing ethics. Under such an ethical mode could follow personal policies such as purchasing green consumer products, conserving water usage, recycling household waste, lobbying local authorities for environmental protections, and becoming a member of groups that advocate environmental protection.

2.8 SUMMARY

An overview of the steps inherent in making environmental health policy was presented in this chapter. As described, many factors can influence the development and implementation of environmental policies, such as the public's concerns, special interest advocacy, news media reports, experts' inputs, and other influences. Policy makers need to know of, and respond to, these influences because they can have a powerful impact on the details of an environmental health policy. For example, special interest groups have traditionally had a powerful effect on environmental legislation and subsequent regulations and standards. Such groups have exerted their influence through lobbying of members of legislative bodies. A simplified model, called the Pressure-Action-Change-Monitoring model, was presented as a way to understand the rudiments of environmental policy making.

Controlling environmental hazards is the ultimate purpose of environmental legislation. The policy of command and control is characteristic of many environmental statutes. However, other policies have emerged that can supplement command and control. These alternatives include the precautionary principle, sustainable development, green commerce, and others discussed in the chapter. As presented, the precautionary principle has been viewed by some critics of the command and control policy as a preferable alternative. In particular, some public health groups and environmental advocates have contended that risk assessment, which undergirds many regulatory policies in the United States, takes too long to complete and has too many uncertainties in its development. However, there is a long experience with the use of risk assessment in the United States, which makes it uncertain if it will be replaced anytime soon, if at all.

Knowing the essence of making environmental health policy now opens the door to a description of the structure of the U.S. federal government, the major player in establishing environmental health policy, followed in later chapters by descriptions of the major federal environmental statutes and their public health implications.

2.9   POLICY QUESTIONS

1.   Using the PACM model shown in Figure 2.2, discuss how you, as a community environmental activist, would organize a campaign in your county to recycle household trash. Be specific.

2.   Using the strategy of critical thinking, discussed in chapter 1, apply it to a discussion of free trade policies as to the benefits and adverse effects on environmental health.

3.   Discuss how the precautionary principle comports with the public health strategy of disease and disability prevention. Be specific.

4.   Assume that federal, state, and local governments have no statutes or ordinances in place to control environmental noise sources. Further, assume that you are the leader of a grassroots environmental group. Choose any four of the seven elements in Figure 2.1 and discuss them in terms of advocating for a community ordinance that your group proposes for noise control.

5.   Discuss how you as an individual can promote the ideals of sustainable development. Include in your discussion those personal behaviors that would contribute to global sustainable development.

6.   Describe how you can use market power to achieve environmental protection goals. How would your actions contribute to public health?

7. What are the differences between ethics and morals? Develop your own list of personal ecoethics, using Table 2.6 as a guide.
8. Using elimination of tobacco smoking in public places as environmental policy, discuss how primary, secondary, and tertiary prevention programs could be used in support of the policy.
9. Contact your local health department. Describe their environmental health programs and how they affect you. Be specific.
10. Select a "green" product or service that you use and describe how its use benefits you personally and your community in general.

NOTES

1. *Peer review*—Evaluation of the accuracy or validity of technical data, observations, and interpretation by qualified experts in an organized group process [2].
2. The author expresses his gratitude to Prof. Melvin Myers, Rollins School of Public Health, Emory University, for his contributions to this section.
3. See Glossary.
4. More accurately referred to as Subtitle E, Title II, of the Small Business Regulatory Enforcement Fairness Act of 1996.
5. Also called Cap and Trade credits.
6. Note: *States* refers to member nations and other organizations comprising the United Nations.
7. "Breathing high levels of asbestos fibers for a long time may result in scar-like tissue in the lungs and its lining (pleural membrane) that surrounds the lung. This condition is called asbestosis" [24].
8. States refers to member states of the United Nations.
9. See chapter 9 for more details on the Rio and Johannesburg environment summits.
10. As discussed in chapter 8, under the provisions of the Emergency Planning and Community Right-to-Know Act of 1986, which is Title III of the Comprehensive Environmental Response, Compensation, and Liability Act (a.k.a. Superfund), releases of certain hazardous substances must be reported annually to EPA.
11. Larvicides are agents that prevent mosquito larvae from maturing [45].

REFERENCES

1. Internet Society, A brief history of the Internet. Available at http://www.isoc.org/internet/history/brief.shtml, 2002.
2. CRARM (Commission on Risk Assessment and Risk Management), Framework for Environmental Health Risk Management, U.S. Environmental Protection Agency, Washington, D.C., 1997.
3. NAS (National Academy of Sciences), Pesticides in the Diets of Infants and Children, National Academy Press, Washington, D.C., 1993.
4. BTP (Boston Tea Party), *History*. Available at http://www.bostonteapartyship.com/History.htm. 2004.
5. Grolier, *Multimedia Encyclopedia*, Grolier Interactive, Danbury, CT, 1998.
6. Rosenbaum, W.A., *Environmental Politics and Policy*, 4th ed., Congressional Quarterly, Washington, D.C., 1998, 52, 179, 184.

7. *Webster's Ninth New Collegiate Dictionary*, Merriam-Webster Publishers, Springfield, MA, 1985.
8. Fromson, J., A history of federal air pollution control, in *Environmental Law Review–1970*, Sage Hill Publishers, Albany, NY, 1970, 214.
9. Graham, M., Regulation by shaming, *Atlantic Monthly*, April, 2000, 36.
10. Johnson, B.L., and Mason, R.W., A review of public health regulations on lead, *Neuro-Toxicology*, 5, 1, 1984.
11. CRA (Congressional Review Act), Congressional Review Act of 1996. Available at http://www.usgovinfo.about.com/library/bills/blcra.htm, 2005.
12. OSHA (Occupational Safety and Health Administration), OSHA Announces Comprehensive Plan to Reduce Ergonomics Injuries, OSHA national news release, Washington, D.C., April 5, 2002.
13. Rodricks, J.V., What happened to the red book's second most important recommendation?, *Human and Ecological Risk Assessment*, 9, 1169, 2003.
14. Liptak, A., $4.5 billion award set for spill of Exxon Valdez, *New York Times*, January 29, 2004.
15. AP (The Associated Press), Swedish capital sees less silver pollution thanks to digital photos, November 25, 2005.
16. Enterprise Foundation, News article. Available at http://www.enterprisefoundation.org/resources/green/index.asp, 2005.
17. EEA (European Environmental Agency), Market-Based Instruments for Environmental Policy in Europe, Technical Report No. 8/2005, Publications of the European Communities, Luxembourg, 2005.
18. UNEP (United Nations Environment Programme), United Nations Conference on Environment and Development: Agenda 21, United Nations, New York, 1992.
19. Kriebel, D., et al., The precautionary principle in environmental science. *Environmental Health Perspectives*, 109, 871, 2001.
20. Sand, P.H., The precautionary principle: A European experience, *Human and Ecological Risk Assessment* 6, 445, 2000.
21. Jordan, A. and O'Riordan, T., The precautionary principle in contemporary environmental policy and politics, in *Protecting Public Health & the Environment*, Raffensperger, C. and Tickner, J., eds., Island Press, Washington, D.C., 1999, 15.
22. EEA (European Environmental Agency), Late Lessons from Early Warnings: The Precautionary Principle 1896–2000, Office for Official Publications of the European Communities, Luxembourg, 2001.
23. Gee, D. and Greenberg, M., Asbestos: From 'magic' to malevolent mineral, in Late Lessons from Early Warnings: The Precautionary Principle 1896–2000, European Environmental Agency, Office for Official Publications of the European Communities, Luxembourg, 2001, 52.
24. ATSDR (Agency for Toxic Substances and Disease Registry), *Division of Toxicology ToxFAQs. Asbestos*, Department of Health and Human Services, Public Health Service, Atlanta, GA, 2003.
25. NRC (National Research Council), *Risk Assessment in the Federal Government: Managing the Process*, National Academy Press, Washington, D.C., 1983.
26. Applegate, J.S., The precautionary preference: An American perspective on the precautionary principle, *Human and Ecological Risk Assessment*, 6, 413, 2000.
27. CBD (Convention on Biological Diversity), Cartagena Protocol on Biosafety–Text of the Protocol–Article 1. Objective. Available at http://www.biodiv.org/biosafety/articles.asp?Ig=0&a=bsp-01, 2002.

28. Raffensperger, C. and Tickner, J., *Protecting Public Health & the Environment*, Island Press, Washington, D.C., 1999, 3, 8.

29. SETAC (Society of Environmental Toxicology and Chemistry), EPA science advisers postpone workshop on precautionary principle, *Globe Newsletter*, March-April, 2002.

30. EC (European Commission), Communication from the Commission on the Precautionary Principle, COM(2000) 1 final, Commission of the European Communities, Brussels, 2000.

31. EU (European Union), The new EU chemicals legislation—REACH. Available at http://europa.eu.int/comm/enterprise/reach/overview.htm, 2005.

32. EC (European Commission), REACH in Brief. Brussels, Belgium. September 9, 2004.

33. *Wall Street Journal*, U.S. opposes EU effort to test chemicals for health effects, September 9, 2004.

34. Reuters, Parliament backs new EU law on toxic chemicals, November 17, 2005.

35. Goldstein, B.D. and Carruth, R.S., Implications of the precautionary principle: Is it a threat to science? *Human and Ecological Risk Assessment*, 11, 209, 2005.

36. EEA (European Environmental Agency), Sustainable Use and Management of Natural Resources, Report No. 9/2005, Office for Official Publications of the European Communities, Luxembourg, 2005.

37. UNEP (United Nations Environment Programme), Declaration of the United Nations' Conference on the Human Environment, New York, 1972.

38. Bruntland, G., ed., *Our Common Future*, Oxford University Press, Oxford, 1987.

39. UN (United Nations), Press summary of the Secretary-General's Report on Implementing Agenda 21, United Nations Department of Public Information, New York, September 4, 2002.

40. Dannon, Product label, Dannon Company, Inc., White Plains, NY, October, 2005.

41. Renner, R., Clean and green, *Scientific American.com*, January 16, 2006.

42. EPA Press Advisory, 2003 Toxics Release Inventory shows continued decline in chemical releases, Washington, D.C., May 11, 2005.

43. LEED (Leadership in Energy and Environmental Design). Available at http://www.usgbc.org/leed/leed_main.asp, 2005.

44. Karvonen, M. and Mikheev, M.I., *Epidemiology of Occupational Health*, WHO Regional Publications, European Series No. 20, Copenhagen, 1986, 388.

45. Raval-Nelson, P., Soin, K., and Tolerud, S., Analysis of *Bacillus sphaericus* in controlling mosquito populations in urban catch basins, *Journal of Environmental Health*, March, 28, 2005.

46. Miller, A.S., *Gaia Connections: An Introduction to Ecology, Ecoethics, Economics*, Rowman & Littlefield, Savage, MD, 1991, 32, 18.

47. Timmerman, P., Grounds for concern: Environmental ethics in the face of global change, in *Planet Under Stress*, Mungall, C. and McLaren, D., eds., Oxford University Press, London, 1990, 214.

48. Lutzenberger, J., Environmental ethics, *Ecodecision*, 1, 81, 1991.

49. Margulis, L., Gaia in science, *Science*, 259, 745, 1993.

3 Federal Government's Environmental Health Structure

3.1 INTRODUCTION

The previous two chapters described the fundamentals of public health and environmental policy, and presented the rudiments of environmental health policy making. Both chapters alluded to the U.S. federal government as a significant force in setting U.S. environmental health policy. In fact, since the 1960s, the U.S. federal government has been the big fish in the environmental health pool. The next chapter will describe the specific federal statutes federal statutes that constitute the core of U.S. environmental health policy. As preparation, this chapter lays out the federal government's environmental health structure, that is, a description of the federal agencies that have environmental health programs. In particular, attention is given to the establishment of the U.S. Environmental Protection Agency (EPA) and how its environmental health mandate evolved from the former responsibilities of the U.S. Public Health Service (PHS). Also described in this chapter is a summary of the structure of the U.S. government and the environmental health responsibilities of the PHS, the EPA, and other federal departments and agencies. Because many federal laws require the development of regulations and standards for control of environmental hazards, the chapter concludes with a summary of the federal rulemaking process. To have an appreciation of the federal structure requires some understanding of the three branches of the U.S. government, which commences this chapter.

3.2 CIVICS 101

One way to set environmental health policy is to enact laws, which in turn contain stated policies and purposes. For example, the federal National Environmental Policy Act of 1969, as discussed in chapter 4, contains the policy of individuals' responsibility to the environment, "[e]ach person should enjoy a healthful environment and that each person has a responsibility to contribute to the preservation and enhancement of the environment." Similarly, laws enacted by states and ordinances by counties and municipalities can contain, or be interpreted as containing, environmental health policies. States' environmental laws often emulate federal laws on corresponding environmental hazards (e.g., hazardous waste management). An example of a local environmental health policy is the application by county health departments of larvicides to control mosquito infestation. Because of the importance of legislated environmental policies—since they bear the weight of law—it is important to have some understanding of the basic elements of government and the U.S. federal government in particular, because federal environmental laws constitute much of the U.S. environmental framework.

One source defines *government* as "The set of legal and political institutions that regulate the relationships among members of a society and between the society and outsiders. These institutions have the authority to make decisions for the society on policies affecting the maintenance of order and the achievement of certain societal goals" [1]. There are several types of government worldwide, depending on such considerations as the nature of the ruling class, the kind of political institutions, the distribution of power, nature of national economy, and historical antecedents. Recognized types of government include monarchy, constitutional government, democracy, and dictatorship. The two forms of democratic government are the parliamentary and the presidential. In the parliamentary form, political power is vested in an elected parliament (an elected legislative body similar to the U.S. House of Representatives), where the prime minister is the leader of government and must be a member of parliament, along with members of his cabinet. This type of government is found (e.g., in Australia, Canada, Great Britain, India, and Israel). The presidential form of democratic government exists where voters elect a chief executive who is independent of the legislature, but whose responsibilities and actions are defined by constitutional authorities. Countries with presidential governments include France, Mexico, Russia, and the United States.

How a society determines how their government distributes political power and authorities with other governments is another important classification. There are three generally accepted forms of government power-sharing: federal systems, unitary states, and confederations [ibid.]. Federal systems, or *federalism*, is structured around a strong central (i.e., federal) government, with specified authorities retained by lower levels of government, such as states and local governments. The United States government is an example of federalism. In *unitary states* the national government performs all the governmental functions, but with subnational governmental units responsible for limited authorities within their jurisdictions. France is a country with unitary government, where strong central control is exercised over territorial administrative subdivisions of the country. *Confederations* are the weakest method of government power-sharing. In such a system of government, a central government has rather limited authority over its confederate states. Member states within a confederation retain their sovereignty, delegating only limited authorities to the central government. In a sense, it is government by handshake. As an example, following the collapse of the Union of Soviet Socialistic Republics (USSR), the Confederation of Independent States was formed, consisting of many of the member states of the former USSR.

The core structure of the U.S. federal government is shown in Figure 3.1. It derives from the U.S. Constitution. Of prime importance is the observation that the three primary branches are on the same organizational plane, that is, they are of equal rank and importance. This arrangement ensures a "check and balance" relationship among the three branches. In theory, there is supposed to be no "first among equals," but in reality, the legislative branch, which can enact legislation, appropriate monies to the executive and judicial branches, and override presidential vetoes of legislation, holds sway over

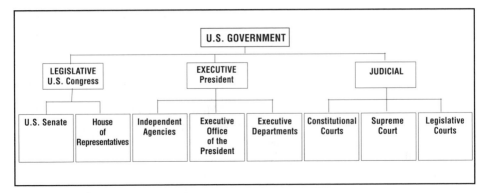

FIGURE 3.1 Core structure of U.S. government.

more policy-making power than the other two branches. Moreover, the legislative branch is the only branch authorized by the U.S. Constitution to impose taxes on the public, raise funds, and provide monies to both the executive and judicial branches; declare war; and approve (Senate) presidential appointments (e.g., secretaries of executive-rank departments, federal judges, ambassadors).

The following comments about the structure of the U.S. government are based on the authorities specified in the U.S. Constitution and illustrated with regard to setting environmental health policy. Moreover, much of what follows also applies to the structure of state government—and to a lesser extent—county and city governments.

3.2.1 Constitutional Basis of the U.S. Government's Structure

The U.S. government's environmental health statutes and the public health practices discussed in chapters 4–8 rest upon a foundation of government. That is, legislative bodies must enact statutes for the purpose of controlling environmental hazards. The executive branch of government is tasked with implementing the statutes enacted by legislative bodies. And the judiciary (i.e., the courts) must decide specific points of law when statutory authorities are disputed by parties affected by those authorities. In the United States, environmental health policy should be viewed not as a broth, but as a thick chowder. There are lots of policy ingredients in the chowder, with many cooks and seasoners.

Environmental policy making in the United States is heavily reliant on governmental action at all levels: federal, state, and local. The federal government's regulation of environmental hazards and other matters is an activity that requires the efforts of all three branches of government. Moreover, because of the importance of regulatory approaches for protecting human health and environmental quality, some details of the U.S. regulatory process are in order. The federal government now serves the primacy role in environmental policy making. It is important therefore to understand how the federal government is structured, based upon the U.S. Constitution, which defines the powers of each of the three branches of federal government. The following section provides a summary of the purpose and organization of the three branches of the U.S. federal government.

3.2.1.1 Legislative Branch

Article I, §1, of the U.S. Constitution states, "All legislative Powers herein granted shall be vested in a Congress of the United States, which shall consist of a Senate and House of Representatives." The framers of the U.S. Constitution were influenced by the two-body structure of the English Parliament: House of Commons and House of Lords. Moreover, in the U.S. Congress, to enact legislation requires that a particular bill (Senate) or resolution (House) be agreed to by both bodies. To be more specific, consider the legislative process that would be required to enact legislation to control an environmental hazard (e.g., environmental noise). First, pressure would be brought by public interest groups (e.g., environmental organizations such as the Sierra Club) on specific members of Congress to take action, specifically, to support legislation that would reduce ambient noise levels.

If sufficient interest results from the pressure, public hearings would be held and various special interest groups, expert groups, and government officials would present written and oral testimonies. Such hearings must be held within the rules of the House and the Senate. More specifically, standing committees and their subcommittees would convene the hearings. The usual sequence is to hold subcommittee hearings, which can lead to the subcommittee's preparation of a draft bill or resolution, which in turn, would be passed to the subcommittee's parent committee. Bills passed by committees are then taken by the committee's chairperson to the floor of the House or Senate for vote. This assumes that House and Senate leaders are willing to schedule the committee's bill for floor action. Bills passed by both the House and Senate but differing in content are referred to a conference committee that comprises members from each responsible committee. Conference committees typically produce compromise bills that are then referred back to the appropriate committees. Both houses of Congress then vote on the compromise bill, either passing or defeating it. When enacted, the bill becomes an act and sent to the president for approval or rejection. The U.S. Constitution requires that the president must act within ten days. If signed by the president, the act becomes public law. If rejected (i.e., vetoed by the president), Congress can override the veto if two-thirds of each body of Congress votes in the affirmative. If the president fails to sign the bill within the prescribed time, it becomes law without his signature if Congress is in session. Public laws are "codified" (i.e., combined into the numbering system for federal statutes known as the U.S. Code, which can be accessed on the Internet).

Legislation enacted by Congress contains language that mandates the Executive Branch to undertake specified actions. For example, EPA and the Food and Drug Administration (FDA) are authorized in the Food Quality Protection Act of 1996 (FQPA) to implement changes in how pesticides are registered in the United States and authorizes changes in risk assessment procedures in order to give greater protection to children exposed to pesticides (chapter 9). Authorizing legislation such as the FQPA must also contain language that authorizes the appropriation by Congress of public funds that can be used by a federal department or agency in support of its legislative mandates. Appropriations legislation authorizes specific amounts of public monies in the U.S. Treasury to be used by the executive branch in the conduct of authorized programs. Congressional appropriations committees are responsible for developing appropriations legislation, commencing with their consideration of the president's annual budget request to Congress.

3.2.1.2 Executive Branch

Article II, §1, of the U.S. Constitution states, "The executive Power shall be vested in a President of the United States of America." The executive branch, headed by the president, implements legislation enacted by Congress and signed into law by the president, or laws enacted by Congress through override of presidential vetoes. This agenda is accomplished by the various components of the executive branch (Figure 3.1). The primary components of the executive branch are the fifteen executive departments (also called the cabinet-rank departments) listed in Table 3.1. Also accountable to the president are independent agencies that are not part of any executive department. Such agencies include, among others, EPA, NASA, Federal Communications Commission, and the National Science Foundation. The third component of the executive branch is the Executive Office of the President, which has the responsibility for various offices, councils, and commissions that have been established by Congressional act or presidential appointment. These include, among others, the Office of Management and Budget, Council of Economic Advisers, National Security Council, and Office of Science and Technology Policy.

TABLE 3.1
Current Cabinet Rank Departments of the Executive Branch, U.S. Government, and Their Year of Establishment

Agriculture	1862
Commerce	1903
Defense (originally War Department)	1949
Education	1980
Energy	1977
Health and Human Services (originally Health, Education and Welfare)	1953
Homeland Security	2002
Housing and Urban Development	1965
Interior	1849
Justice	1870
Labor	1913
State	1789
Transportation	1966
Treasury	1789
Veterans Affairs	1989

A means to establish federal policy on environmental health and other matters is afforded to the president through issuance of executive orders, directives to the executive branch for purpose of achieving a particular action and policy. Executive orders are important because they have direct impact on not only the federal executive branch, but indirectly, the public. Two examples of executive orders that have had influence on environmental health policy are: (1) an environmental justice directive, and (2) energy policies for government facilities and operations. In the former, President Bill Clinton directed that all federal agencies develop and implement plans to prevent the unjust imposition of environmental hazards (e.g., location of polluting industries) on minority populations. In the latter example, President George W. Bush issued an executive order in 2001 that directs all federal agencies to reduce energy consumption by specified amounts.

Regarding environmental health policy, the executive branch must implement Congressional legislation that has been signed into law by the president or passed into law through Congress's override of a presidential veto. Using the example of the Food Quality Protection Act of 1996, the statute, upon its signature into law by President Clinton, directed EPA, the Department of Health and Human Services (DHHS), and the Department of Agriculture to revise specific polices and procedures bearing on the review and approval of pesticides and their potential impact on children. Upon receipt of such directives, affected departments and agencies must interpret the language in the new statute. Some legislation is deliberately written in vague terms, a product of failure by Congressional committees to negotiate more specific language. In such conditions, the executive branch must attempt its own interpretation of Congressional intent. Ultimately, the judicial branch is often required to interpret legislative intent and executive branch implementation. This occurs when concerned special interests litigate a federal department or agency, forcing courts to bridge the language in statutes and agencies' interpretation and implementation of the same language.

3.2.1.3   Judicial Branch

Article III, §1, of the U.S. Constitution states, "The Judicial Power of the United States, shall be vested in one supreme Court, and in such inferior Courts as the Congress may from time to time ordain and establish. The Judges, both of the supreme and inferior Courts, shall hold their Offices during good Behaviour, and shall, at stated Times, receive for their Services, a Compensation, which shall not be diminished during their Continuance in Office." The courts established under the powers granted by the Constitution are known as *constitutional courts*. Judges of these courts, who are appointed by the president, with the advice and consent of the U.S. Senate, are appointed for life. The constitutional courts consist of the Supreme Court, district courts, and the courts of appeal. The federal courts' jurisdiction, which is defined in the Constitution, covers litigation in which the U.S. government is a party, to controversies between the states, to disputes between a state, or its citizens, and foreign governments or their subjects, and to controversies between the citizens of one state and citizens of another state.

The Supreme Court is the highest appellate constitutional court and, as such, is the final judicial arbiter of federal constitutional questions and of the scope of federal laws. The court consists of nine justices, one of whom serves as the chief justice, who has overall administrative responsibility for the court. Other federal courts, derived from powers held to be implied in articles of the U.S. Constitution, are called *legislative courts*. These are courts created in law by the U.S. Congress and serve within the Judicial Branch's authorities and power. Legislative courts comprise the Claims Courts, the Court of International Trade, the Tax Court, and the territorial courts established in federally administered territories of the United States (e.g., Guam).

Federal courts have had a profound effect on interpreting environmental health policies. Because particular environmental laws (e.g., the Clean Air Act) require the federal government to regulate specific environmental toxicants, controversies can arise that become the fodder for litigation. Disputes arise over how a government agency (e.g., EPA) interpreted its authorities in law, or over a specific regulatory decision (e.g., a national ambient air quality standard), or questions of fairness in a regulation (e.g., environmental justice issues). Federal courts have concurred with specific environmental polices, thrown out some policies or practices, or referred some litigation back to lower courts for further review and action. In many situations, the affected federal agency (e.g., FDA) will be required to take some alternate path to regulating an environmental hazard.

The U.S. judicial system has further evolved from just what is specified in the U.S. Constitution. Shown in Figure 3.2 are the four general layers of U.S. law. Each layer builds upon the layer below it. More specifically:

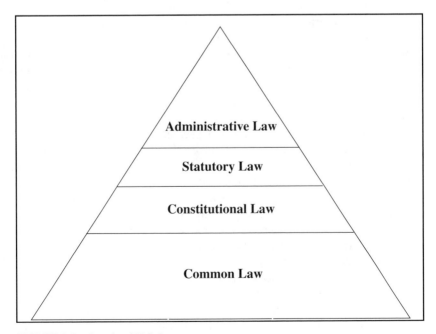

FIGURE 3.2 Levels of U.S. law.

Common law—The common law developed under the adversarial system that emerged in eleventh and twelfth century England in order to meet the legal needs of the times [2]. Common law was adopted by the English monarchies as the means to make law common, or uniform, across the country. Common law was devised as a means of compensating persons for wrongful acts perpetrated against them. As a colony of England, common law became a component of what became U.S. law. Common law applies in all U.S. states, except Louisiana, whose basis of law is of French origin, not English. An environmental example of common law would be litigation by an individual to recover damages to property when illegal dumping occurred.

Constitutional law—This type of law deals with the interpretation and implementation of the U.S. Constitution [3]. As such, it is concerned with issues between the states, issues between the federal government and the states, issues between the three branches of the U.S. government, and the rights of individuals in regard to rights specified in the Constitution. An environmental example was litigation in federal court that challenged the Comprehensive Environmental Response, Compensation, and Liability Act (CERCLA) as being unconstitutional on the basis of alleged violation of states' rights to control hazardous waste. (The court found the act to be constitutional.)

Statutory law—Statutes are defined as laws that are passed by the U.S. Congress, the various state legislatures and also includes enacted local ordinances [4]. Local ordinances are statutes passed by a county or city government to cover areas not covered by federal or state laws. Statutes form the basis for statutory law. Examples of statutes covered by statutory law include the federal Clean Air Act and the Food, Drug and Cosmetic Act. An environmental example would be litigation by northeastern states against the EPA, alleging that the agency had failed to adequately control air emissions from coal-burning electricity generating plants located in midwestern states.

Administrative law—This area of law encompasses laws and legal principles governing the administration and regulation of federal and state government agencies. The affected agencies are delegated power by a legislative body to act as agents for the executive branch of government [5]. Generally, administrative agencies are created to protect a public interest in distinction to protecting the rights of individuals. The Federal Administrative Procedures Act, which stipulates how federal government agencies must conduct their operations when effectuating the actions specified in federal statutes, is an example of administrative law. An environmental example would be litigation challenging whether a particular regulatory agency had conducted public meetings during the development of a proposed new regulation.

Environmental health policies can be subject to each of the four layers of U.S. law. For instance, common law may apply to the civil litigation of a person claiming harm from exposure to substances released from a landfill. Corporations may be litigated under statutory law, specifically the CERCLA, for failure to comply with that law's provisions for cleaning up uncontrolled hazardous waste sites. Constitutional law has been applied in some cases to question the constitutionality of some environmental statutes (e.g., provisions of the Clean Air Act). Administrative law has often been applied when parties allege that federal environmental regulations were developed without compliance with the Administrative Procedures Act. It is important to understand the distinctions among the four levels of U.S. law and how each might apply to a specific environmental issue.

3.3 DICHOTOMOUS ENVIRONMENTAL AND PUBLIC HEALTH STATUTES

Having discussed in chapter 1 the fundamentals of environmental health policy, we turn our attention to the roles of the U.S. federal government in establishing and implementing environmental health polices and programs. It is a story commencing in the early years of the twentieth century about ascendancy of federal public health authorities on matters of environmental hazards, gradual loss of primacy by PHS, and succeeded in the late 1960s by federal and state environmental agencies established within the federal and state governments. As noted by one source, "The driving force behind United States environmental law and regulation is the clear and explicit intent of the government to protect human health and the environment. [M]ost environmental laws [e]stablish environmental standards at levels that protect human health. [T]hese regulatory requirements provide the health professional with an extensive database [f]rom which preventive or responsive health practices can be developed"[6]. However, in the opinion of some public health specialists, this approach has resulted in an organizational dichotomy that has handicapped both environmental agencies and their public health counterparts in the United States, Canada, and some other countries.

The current environmental health dichotomy within the U.S. federal government will be discussed in this section. However, as prelude, the creation of EPA, postured by Congress to protect natural resources and human health through full regulatory authority, coexists with individual agencies of the DHHS that conduct research on environmental hazards and have some limited regulatory authority. Unfortunately, this arrangement has resulted in the creation of federal and state environmental protection agencies that possess public health responsibilities, but without sufficient depth in public health staff and practice. To complete the handicapped dichotomy, health agencies that conduct research on environmental hazards and offer services to states and the public often do so with too few staff with environmental experience, access to environmental databases, and insufficient practice in environmental science.

Beginning in the late nineteenth century, and continuing through the first half of the twentieth century, the U.S. federal government gradually became involved with protecting the public against environmental hazards, principally through congressional authorizations of authorities and resources. The existence of the PHS, created by Congress in 1796, provided a ready-made federal government structure to receive new congressional environmental health mandates. However, many of the federal government's efforts in controlling environmental hazards did not begin until the 1960s and afterwards.

In an analysis done in 1991, the Congressional Research Service (CRS), identified 13 federal statutes that comprised the major portion of the legal basis for programs administered by the EPA [7]. They are listed in Table 3.2, with the date of initial enactment of the statute or its antecedent and a brief statement of each statute's purpose. For

TABLE 3.2
Titles and Summaries of Major Federal Environmental Health Laws

Clean Air Act requires EPA to set mobile sources limits, ambient air quality standards, emission standards, standards for new sources, air quality deterioration requirements, and focus on areas that do not attain standards. (1955)

Comprehensive Environmental Response, Compensation, and Liability Act (or Superfund), establishes a fee-maintained fund to remediate abandoned hazardous waste sites. (1980)

Consumer Product Safety Act created the Consumer Product Safety Commission to protect against injuries from select consumer products. (1972)

Emergency Planning and Community Right-to-Know Act. Title III of Superfund Act. (1986)

Environmental Research, Development, and Demonstration Authorization Act provides authority for EPA research programs. (1976)**

Federal Insecticide, Fungicide, and Rodenticide Act governs the sale and use of pesticide products. (1947)

Federal Meat Inspection Act governs the production and use of meat and meat products. (1906)

Federal Water Pollution Control Act (known as the Clean Water Act) established sewage treatment construction grant programs, and a regulatory and enforcement program for discharges into U.S. waters. (1948)

Food, Drug, and Cosmetic Act governs the production and use of food additives, prescription drugs, and cosmetics. (1906)

Food Quality Protection Act amends the Food, Drug, and Cosmetic Act to provide a risk-based standard for pesticide residues in raw and processed foods and amends FIFRA. (1996)

Information Quality Act requires federal agencies to develop guidelines for determining the quality of scientific data used in decision-making. (2001)

National Environmental Policy Act established U.S. environmental policy and requires federal agencies to assess the environmental impact of major federal actions. (1969)

National Noise Control Act gave EPA the primary role in controlling noise in the general environment. (1972)**

Ocean Dumping Act regulates the intentional disposing of materials into ocean waters and establishes research on effects of, and alternatives to, ocean disposal. (1972)

Occupational Safety and Health Act requires the Department of Labor to develop and enforce workplace standards. (1970)

Oil Pollution Act expands oil spill prevention, preparedness, and response capabilities of the federal government and industry. (1990)

Pollution Prevention Act (PPA) seeks to prevent pollution through reduced generation of pollutants at their point of origin. (1990)

Public Health Security and Bioterrorism Preparedness and Response Act provides for planning, preparedness, and response to bioterrorism. (2002)

Resource Conservation and Recovery Act provides cradle-to-grave regulation of solid and hazardous waste. (1965)

Safe Drinking Water Act established primary drinking water standards, regulates underground injection practices, and establishes a groundwater control program. (1974)

Solid Waste Disposal Act. Title II of Clean Air Act of 1965, now generally referred to as the Resource Conservation and Recovery Act. (1965)

Toxic Substances Control Act regulates the testing of chemicals and their use in commerce. (1976)

NOTE: (Year of Enactment) ** Act is not currently funded by Congress.

the purposes of this book, to the CRS list have been added the Occupational Safety and Health Act, administered by the U.S. Department of Labor, which covers environmental conditions in workplaces; the Noise Control Act of 1972, and the Food Quality Protection Act, which are acts administered primarily by EPA; the Food, Drug and Cosmetic Act, administered by the Food and Drug Administration; the Federal Meat Inspection Act, which is administered by the U.S. Department of Agriculture; the Information Quality Act of 2001, which affects all federal regulatory agencies, and is administered by the White House's Office of Management and Budget; the Consumer Product Safety Act, administered by the Consumer Product Safety Commission; and the Public Health Security and Bioterrorism Preparedness and Response Act, which is administered by the DHHS and other federal departments. All the statutes listed in Table 3.2 will be discussed in chapters 4–8.

Several federal statutes have attendant regulations developed by EPA or other regulatory agencies, which are federal and state agencies that are authorized by specific statutes to develop, promulgate, and enforce regulations. Regulations are enforceable under the laws that mandate them. The reader is cautioned that the statutes cited in Table 3.2 are complex and their implementation by federal agencies is subject to change, because amendments to statutes occur occasionally and regulations change because of court decisions, agency updates, and legislative action. For instance, Congress enacted legislation in 1988 that gave itself the authority to review federal government regulations. Hence, caution should be exercised to ensure any contemplated action is consistent with current regulations and amendments to statutes. Consultation with the responsible federal, state, or local authority is advisable on any matter of regulatory compliance.

Of the various policy issues attending federal environmental statutes, none is more important to the public than the right-to-know requirements that are now found in several federal environmental laws. Right-to-know requires EPA and other federal agencies to make environmental hazards information available to the public. Environmental advocacy and community groups have used this information to pressure government and industry to take action to improve environmental quality. In other words, right-to-know policies have contributed to improved public awareness and better informed democratic processes. As examples, the Occupational Safety and Health Act of 1980 led the Occupational Safety and Health Administration to develop regulations that require private industry employers to make workplace safety and health information available to workers. The Emergency Planning and Community Right-to-Know Act, which was enacted as Title III of the CERCLA, includes the Toxics Release Inventory (TRI). The TRI requires businesses to report to EPA their emissions of hazardous substances, which the agency must then make available to the public.[1] Amendments in 1996 to the Safe Drinking Water Act require water suppliers to state the nature and levels of contaminants in drinking water. The Food Quality Protection Act of 1996 required federal agencies to provide families with information about pesticide levels in raw and processed food. Other policies of significance to public health practice are discussed throughout this chapter.

3.3.1 EMERGENCE OF THE U.S. PUBLIC HEALTH SERVICE

The role of the U.S. federal government in matters of public health dates from the late eighteenth century, the early days of the republic. As a young nation, the sea was a vital source of food, commerce, and security. Regarding commerce, mariners were essential to the nation's increasing prosperity. Ships transported goods and cargo between ports in North America and Europe. In these post-revolutionary years, mariners traveled widely, often became sick while at sea, and infrequently could find health care in port cities. Since they came from all the new states or former colonies, and could become burdens to port cities, the health of mariners became a problem for the nascent federal government. In response, in 1798, Congress established a loose network of marine hospitals, mainly in port cities to care for sick and disabled mariners. The hospitals comprised what was called the *Marine Hospital Service* (MHS) [8]. Funds to pay physicians and build hospitals were appropriated by Congress by taxing mariners 20 cents a month, collected by seaport customs officers, then paid into the federal treasury. The tax was abolished in 1884, succeeded until 1906 by a tonnage tax on vessels entering U.S. ports. From 1906 to 1981, when they were closed, the hospitals were funded out of general federal revenues.

In 1870, Congress reorganized the MHS into a centrally-controlled national agency with its own administrative staff, directed by a supervising surgeon, with headquarters in Washington, D.C., and assigned to the Department of Treasury [ibid.]. Dr. John Maynard Woodstock was the first supervising surgeon of the MHS and is therefore considered as the first surgeon general of what became the PHS. In 1876, the title was changed to supervising surgeon general of the MHS. The impetus for the reorganization of the MHS was the need for better accountability and improved medical services in what had been a loosely administered set of locally-operated marine hospitals [9]. Later, in 1889, a Commissioned Corps of medical officers was created by Congress, an action that provided the MHS with a cadre of professional, mobile officers who could be quickly assigned to disparate geographic areas in times of medical emergencies.

A key event in the history of U.S. public health occurred in 1887 when the MHS created a small bacteriology laboratory, called the Hygienic Laboratory, at the marine hospital on Staten Island, New York. It was later relocated to Washington, D.C. From this quite modest resource later sprang the National Institutes of Health, the world's premier biomedical research institution [ibid.].

In 1902, Congress enacted legislation that expanded the scientific work of the Hygienic Laboratory and appropriated a budget specific to the work of the laboratory. The act also changed the name of the Marine Health Service to the Public Health and Marine Hospital Service (PHMHS), directed by a surgeon general of the service. The change in name presaged further changes in the nation's emerging federal public health resources, changes that focused less on marine hospital services and more on biomedical research and containment of disease epidemics.

The Public Health Service Act of 1912 changed the name of the PHMHS to the Public Health Service (PHS), directed by a surgeon general, and again broadened its responsibilities by authorizing investigations into human diseases (e.g., tuberculosis, malaria, leprosy), sanitation, water supplies, and sewage disposal [8]. This act provided the first policy framework for federal research and services in public health, moving the nation's health needs past just those needed by mariners. This state of affairs lasted for approximately thirty years.

Cincinnati, Ohio, has held a special place in the development of PHS environmental health programs. The prominence of this city can be linked to the Ohio River steamboat trade in the mid- to late-nineteenth century. Cincinnati was a busy port during this period, a place where merchandise, farm products, livestock, and industrial goods were shipped and received on steamboats and barges [10]. All this occurred during a period of major growth in the city's economy and population. Regarding the latter, Cincinnati became the home of large numbers of German immigrants. The social and political impacts of the German migration were immediate and last even today. German social traditions of medical education, clinical services, public assistance, and support for cultural institutions were adopted by the city.

Cincinnati became the first port along the Ohio and Mississippi rivers to establish a medical service for steamboat crews. It seems that steamboats brought more than just goods and cargo to ports; their crews also brought sexually transmitted diseases, infectious agents, and vermin. The mariners' hospital quarantined disease-bearing crewmen, provided medical care, and generally tried to improve the health of both the city and that of individual crew members. The Cincinnati hospital was, in effect, an early occupational medicine facility of nineteenth century origin.

Beginning in the late 1950s, the PHS radiological health, water quality, air pollution, and occupational health research programs were conducted primarily in Cincinnati, Ohio. This location was attributable to several factors. One factor was the presence of the University of Cincinnati, home to an environmental health center and a radiological health program. Both university resources helped stimulate PHS programs in radiation protection, occupational health, and toxicology. Another factor was the construction of the Robert A. Taft Sanitary Engineering Center, which succeeded the Stream Investigation Station, which had been established under authorities in the Public Health Service Act of 1912 [9]. Although many of the former PHS programs and laboratories in Cincinnati were later moved to laboratories in North Carolina and Nevada, major NIOSH and EPA laboratories remain in Cincinnati.

3.3.2 DHEW AND DHHS

In 1942, some members of Congress perceived the need to further develop the PHS. There were two primary motivations for the congressional interest. First, the winds of war were beginning to stir in Europe, and the United States needed to have its public health capacity readied in case of war. Second, the Franklin D. Roosevelt administration desired a stronger, more active, PHS as part of the president's New Deal legacy [11]. The New Deal was the name for President Franklin D. Roosevelt's programs to combat the effects of the Great Depression. The Depression was devastating to the global economy. The stock market collapsed in 1929, reverberating in loss of jobs for millions of the U.S. population. At the depth of the Depression, between 10 to 15 million people were unemployed, which was about 25 percent of the available workforce

[12,13]. Roosevelt's New Deal programs of 1933 through 1935 created jobs through public works projects, home loans, and grants to individuals. Included in the New Deal were programs to stabilize and improve agriculture, business, and employment. New Deal policies and programs were unique in that the federal government assumed a primary social role in creating jobs and regulating private sector institutions in ways not previously experienced.

Real consolidation of the PHS had begun in June 1939, when it was transferred by President Roosevelt from the Department of Treasury to the newly formed Federal Security Agency (FSA), which combined a number of New Deal government agencies and services related to health, education, and welfare. All the laws affecting the functions of the services were also consolidated for the first time in the Public Health Services Act of 1944 [8].

The FSA was a noncabinet-level agency whose programs had grown to such size and scope that President Eisenhower, as head of the executive branch, submitted a reorganization plan in 1953 to Congress that called for the dissolution of the FSA and the transfer of its responsibilities to a newly created Department of Health, Education, and Welfare (DHEW). A major objective of this reorganization was to ensure that the important post-WWII programs of health, education, and social security be represented in the president's cabinet. In 1979, DHEW's education tasks were transferred to the new Department of Education and the remaining divisions of DHEW, including the PHS, were reorganized as the DHHS, which currently has administrative responsibility for several environmental health agencies and their programs.

The DHHS currently comprises several agencies and offices,[2] several of which have responsibilities for environmental health programs that represent a mix of research, services, and some narrow regulatory authorities. These programs are discussed in subsequent sections of this chapter.

3.3.3   ROLE OF PHS SURGEONS GENERAL

The federal government's current environmental protection and environmental health programs have evolved over many years. Originally, the programs were the sole responsibility of the PHS. The surgeon general of the PHS was the director of the early environmental health programs. These programs began in the early part of the twentieth century, undertaken by various surgeons general.

Two Surgeons General, Drs. Thomas Parran and Leroy Burney, were key leaders in shaping nascent federal environmental health programs.

The PHS is headed by the surgeon general of the PHS, more commonly referred to as the surgeon general. Actually, other components of the federal government (e.g., the U.S. Air Force) also have surgeons general. Over time, the duties and authorities of the surgeon general have changed. During the nineteenth century and up through the mid-twentieth century, the surgeon general had strong, independent authorities and resources to bring to bear on preventing cholera, tuberculosis, polio, and other epidemics.

However, the surgeon general's authorities over public health programs began to change in the mid-twentieth century and the position gradually transitioned to a "bully pulpit,[3]" that is, a position of advisor to the president on matters of public health and communicator to the public on health issues. Sometimes, as during the Clinton administration, the surgeon general concomitantly serves as the assistant secretary for health within DHHS. Currently, DHHS agencies and programs of environmental health, such as the National Institute of Environmental Health Sciences, report to parent Operational Divisions (e.g., the Director, National Institutes of Health), who in turn reports to the secretary, DHHS. The surgeon general now has no administrative authority over the DHHS operating divisions.

Through year 2006, there have been eighteen surgeons general. It is likely that all surgeons general have had some involvement in what is now called environmental health. For example, issues of human waste management, water pollution, vector control, and food contamination have long been challenges to the public's health, rising to the attention of surgeons general through the years. However, among those who served in that post, two have had the greatest effect on modern environmental health programs and policies.

Dr. Thomas Parran served from 1936–1948. He was a career PHS commissioned officer, a requirement for serving as surgeon general. In 1936, President Franklin Roosevelt appointed Parran as the nation's sixth surgeon general. Fortuitously for the public's health, the Social Security Act of 1935, Title VI, provided the PHS with funds and authority to build a system of state and local health departments [9]. Surgeon General Parran was in today's parlance an "activist" who used the authorities, particularly those given by Title VI of the Social Security Act, to develop a system of federal grants to state and local health departments. These grants required health departments to match federal funds and ultimately effected comprehensive public health programs within states and local health departments. The grants forged essential links between federal, state, and local health departments that began in 1936 and continue to the present.

The PHS grants to states that started in 1936 added programs in industrial hygiene and plague control to ongoing PHS and state programs in malaria control, privy construction, and mine-sealing activities [ibid.]. These efforts were therefore forerunners of later PHS and state health department efforts to reduce the human health toll of specific environmental hazards.

Later in his career, Surgeon General Parran became a key player in lobbying Congress to enact the Water Pollution Control Act of 1948. This legislation was a culmination of PHS's prior water quality investigations that dated from the Public Health Service Act of 1912 [ibid.]. The Water Pollution Control Act provided PHS with authorities and resources that extended investigations of fecal contamination in water supplies and streams to include new water quality problems from chemical contaminants.

As the PHS's environmental health programs and budgets expanded in the late 1950s, expectations grew, especially within environmental organizations, that the PHS would become the federal government's leader in protecting the public against specific environmental hazards. However, as early as 1954, decreases in the PHS water pollution budget led to criticism by environmental organizations. Unfortunately, hope exceeded reality. Seeds of discontent with PHS's leadership of environmental health programs were being sowed.

The other surgeon general who had a major responsibility for environmental health problems was Dr. Luther E. Burney, the eighth to serve in the position. Burney was appointed to his position in 1956 by President Dwight Eisenhower and served until 1961. Like Surgeon General Parran, Dr. Burney was a career PHS commissioned officer. He was at the PHS helm during a period of time, the mid-1950s, when evidence began to mount that chemicals in water supplies, poor ambient air quality, and increased radiation levels in the atmosphere from above ground nuclear weapons testing had become potential threats to the public's health. Surgeon General Burney, like Parran a decade earlier, took assertive steps to position the PHS to respond to these new environmental challenges.

Surgeon General Burney foresaw the need for research to elucidate the effects on the public's health of radiation sources, contaminated water supplies, and polluted air. Using his authorities in the Public Health Service Act of 1912, as amended, Dr. Burney mobilized talented, multidisciplinary researchers in order to pursue answers to basic questions on the human health consequences of environmental hazards. These teams were composed of physicians, toxicologists, epidemiologists, engineers, chemists, physicists, radiation biologists, and statisticians, among others. Laboratory studies and measurement of environmental contamination levels were prominent activities whether in radiation, water quality, or air pollution research programs. The emphasis on conducting basic (e.g., hypothesis testing) and applied (e.g., collecting data on environmental quality) research fit within the traditional disease prevention model used by public health practitioners. Knowing your opponent (i.e., disease) before you engage it with regulations and forced responses was thought preferable to an attitude of predicting harm and later assessing impacts. In this public health approach, education and hazard control followed the establishment of sufficient scientific data to say that a public health problem could occur.

As related by PHS historian Mullan [9], Dr. Burney appointed a national advisory committee on radiation, resulting in the creation of the Division of Radiological Health, located in Cincinnati, Ohio, thereby consolidating all PHS radiation programs. In 1957, he convinced Congressman John Fogarty, a powerful member of Congress, to add language to the PHS appropriations bill in order to implement new PHS environmental programs and to expand existing programs. At the same time, a committee of experts was proposed for the purpose of advising the PHS and Congress on environmental health problems. As it turned out, this new congressional interest and largesse to the PHS had set in motion events that would lead to an eclipse of PHS environmental health programs and leadership.

By 1960, Surgeon General Burney and PHS leadership were faced with a proposal in Congress to remove water pollution control from the PHS. This proposal was based on the argument that water pollution involved conservation and wildlife concerns in addition to sewage treatment, the area of PHS emphasis [9]. Of note, PHS engineers were apparently behind-the-scene proponents of the proposal being considered by Congress. Their advocacy was predicated on frustrations with the leadership of PHS, comprised largely of medical doctors.

Dr. Burney, in a report to Congress, presaged the eventual PHS acquiescence of environmental health primacy. He wrote, "When we are dealing with the possible harmful effects of the byproducts of industry and the wastes of nuclear technology, our goal is not *conquest* (emphasis added) but *containment* (emphasis added) [ibid.]." In retrospect, this is a quite revealing comment from the leader of the PHS. Burney, who was one of the great surgeons general, implies that primary prevention (i.e., hazard elimination) could not, or would not, be applied to environmental health problems. Would he have made such a statement about an infectious disease? Of course not. Surgeon General Burney's comment gives insight into the disease-focused orientation of the PHS.

By the early 1960s, PHS leadership of environmental health programs had begun to wane.

With President John F. Kennedy's election in 1960 came changes that quickly impacted the nascent PHS environmental health programs begun by surgeons general Parran and Burney. Burney was not reappointed and was succeeded by Dr. Luther Terry. In August 1961, Dr. Terry established the Committee on Environmental Health Problems and asked them to develop long-range objectives for the environmental health programs of the PHS. This committee was the realization of Surgeon General Burney's efforts to convene an expert committee to advise the PHS and Congress on matters of environmental health.

The Committee on Environmental Health Problems [16] delivered its report to Surgeon General Terry on November 1, 1961. The committee's many conclusions included:

1. "The Committee believes that immediate action should be taken to establish a center where the operational research, and training programs of the Service (i.e., Public Health Service) in environmental health can be brought together.
2. A major national effort, both government and nongovernment, must be started if the environmental health problems resulting from the rapid growth of our highly technological civilization are to be adequately understood and if measures for their control and ultimate prevention are to be developed.
3. The focus of this national effort should be centered in the U.S. Public Health Service."

These quotes portray the committee's strong support for PHS primacy in federal environmental health programs. Their aspirations were to be only partially realized. The PHS did not seize the opportunity and build momentum to establish its leadership role in environmental health policy and programs. Why did this lapse occur? The answers are complex and, in hindsight, predictable.

The PHS was established by Congress as a federal government resource to protect the public's health, primarily to prevent the spread of infectious disease (e.g., sexually transmitted diseases). Although engineers and sanitarians have always been key members of the PHS, physicians had dominated the leadership and programs of the PHS by reason of their numbers and influence. Regrettably, they were not, by virtue of their medical training and professional experience, well attuned to environmental health problems.

Other forces, in addition to the PHS leaders' discomfort with environmental health problems, contributed to Surgeon General Terry's failure to embrace recommendations from the Committee on Environmental Health Problems. The most significant force was an increasingly active environmental lobby, which saw environmental health in a larger dimension than just human health concerns. In particular, conservation groups and ecology organizations lobbied Congress for legislation that would add more emphasis to improving environmental quality and protecting natural resources. These lobbying efforts were not lost on the PHS leadership, which gradually relinquished any substantive efforts to provide the nation's leadership and serve as the primary resource in environmental health. Lack of environmental health vision within the PHS and leaders who lacked environmentally-relevant backgrounds contributed to an eventual loss of PHS primacy in environmental health. The die had been cast for creation of EPA as the nation's principal environmental authority.

The role and authority of surgeons general began to change in 1967 during the Johnson administration, when the position of Assistant Secretary of Health (ASH) was created, reporting directly to the Secretary of the DHEW. This position added another political appointee, along with the surgeon general, to the department's leadership. In 1968, DHEW Secretary Wilbur Cohen redefined the PHS organizations structure and the surgeon general's role by placing Assistant Secretary for Health Dr. Philip Lee, in charge of the PHS, with the Surgeon General as his deputy. With the stroke of a pen, PHS leadership changed so that all PHS agencies (e.g., NIH, FDA, CDC) reported to the ASH. As a second consequence, the surgeon general's authorities became largely advisory, and somewhat ceremonial, and duties focused on being a spokesperson on public health, preparing reports on major health problems (e.g., tobacco use), advocating for public health programs, and representing the U.S. government at international health meetings.

Prior to EPA's establishment, an event occurred that created a lasting schism between public health programs and environmental protection organizations. In 1961, all authority for water pollution control was transferred from the surgeon general to the PHS's parent leader, the secretary of DHEW, later to become DHHS. In 1965, the Federal Water Pollution Control Administration was established by Congress; shortly thereafter this new administration was transferred from the DHEW to the Department of Interior (DOI). Upon this transfer, the DOI decreed that all PHS commissioned officers assigned to the water pollution program would have to resign their commissions and convert to civilian (i.e., Civil Service) appointments. This policy had an immediate and lasting chilling effect on a key PHS resource, its commissioned officers. The result was a hemorrhage of key personnel, particularly medical officers and some engineers who preferred to remain PHS officers, where retirement benefits and pay generally exceeded what were available to federal civil service employees. A personnel policy crafted by federal bureaucrats had led to a setback in public health protection.

Federal water pollution control efforts continued, but with a significantly different focus: pollution monitoring and application of engineering controls to reduce water pollution loads. Unwittingly, the DOI's personnel policy decision to discourage the presence of PHS officers was the genesis of a dichotomy in environmental health: environmental protection, standards, and regulations vs. public health, consensus recommendations, and voluntary action.

3.3.4   ESTABLISHMENT OF EPA

The story of the environmental movement in the 1960s and 1970s has much to do with two men: Senator Edmund Muskie of Maine and President Richard M. Nixon. Both were central to efforts to improve environmental quality and protection of public health from specific environmental hazards. Both men exhibited tenacity and, ultimately, wisdom in how to deal with the U.S. public's growing concern in the 1960s about environmental hazards. Some context is required. In order to understand what led to the establishment of EPA as a component of the federal government requires some background. This context can best be understood in terms of the social, legislative, and governance climates of the 1960s and 1970s.

3.3.4.1   Societal Climate

The 1960s and 1970s were times of great, bitter societal turmoil. It was a time of war on the battlefields of Vietnam and protesters of the war on U.S. streets. Looking back, these years saw the greatest change in United States social progress since the Civil War. These events included the awakening and national commitment to correcting civil rights abuses against African Americans and other minority populations. Other notable, and lasting, changes in the U.S. social fabric included emergence of the environmental movement, feminism, peace activists, outer space exploration, and assassinations of national leaders, including a U.S. president. These events strained the nation's social fabric, which frayed at the edges, but did not tear.

Perhaps the pivotal event that precipitated a tidal wave of later social change was the assassination of President John F. Kennedy (D-MA)[4] on November 22, 1963, in Dallas, Texas. Kennedy's election in 1960 was seen by many persons to presage an era of progressive change, including improved civil rights for minorities, a national vision for the country's future, and a commitment to science and education. It was a brief period of national innocence and optimism. These feelings ended with the president's assassination..

Upon Kennedy's death, Vice President Lyndon B. Johnson (D-TX) became president. Prior to his becoming Kennedy's vice president, Johnson had been the powerful leader of the U.S. Senate. He was justly renowned as a politician who got what he sought. He wielded absolute control over all Senate legislative matters. His intimate knowledge of the congressional legislative process would be needed as the nation's attention became focused on correcting civil rights abuses.

Led by the Rev. Martin Luther King, Jr., demands mounted on the U.S. Congress to correct civil rights injustices imposed on African Americans in the southern states, such as voting restrictions, lack of unfettered access to public places, and denial to housing of choice. The major impediment in passage of civil rights legislation in Congress were senators and congressmen from southern states who chaired key congressional committees. Their opposition meant that civil rights legislation would founder in committee inaction. President Johnson was able to lead efforts that culminated in the landmark Civil Rights Act of 1964, which outlawed racial discrimination in public accommodations and by employers, unions, and voting registrars. This act was soon challenged in federal court, ultimately being decided by the U.S. Supreme Court, which held that the commerce clause of the U.S. Constitution applied, thereby overriding states' rights. In 1965, the Voting Rights Act suspended (banned in later legislation) the use of literacy or other tests of voter qualifications. In the aftermath of the assassination of Dr. King on April 4, 1968, Congress passed the Fair Housing Act of 1968, which banned racial discrimination in housing financed by federal funds. Overall, this body of civil rights legislation reinvigorated a proactive policy of making social change through federal law. One of these changes was a federal response to protecting the natural environment.

Lyndon Johnson's leadership in getting civil rights legislation enacted into law would have been enough to make for a positive place in United States history. However, this place was denied him because of his insistence on America's winning the war in Vietnam. This policy was widely unpopular, leading to widespread demonstrations against the war by thousands of street demonstrators and some acts of violence against people and property. Johnson chose not to seek reelection and Richard M. Nixon (R-CA), was elected president in 1968. Ultimately, the war ended during Nixon's administration. The U.S., South Vietnam, and North Vietnam governments signed the Paris Peace Accord, which went into effect on January 17, 1973, and set into motion the end of U.S. military action in Vietnam [17]. In 1975, the last U.S. military personnel withdrew from South Vietnam. Shortly thereafter, North Vietnam overthrew the South Vietnam government, an action that unified the two Vietnams into one country. A lesson from the U.S. experience in the Vietnam war was the power of public protests on social policies, a lesson not lost on the Congress, including pending federal environmental legislation.

3.3.4.2   Legislative Climate

The 1960s and 1970s saw the emergence of environmentalism as an engine of social change, although, as noted in chapter 4, some laws and programs were already in existence. Prior to 1960, these laws were primarily focused on conservation of natural resources and, secondarily, on public health research and services pertinent to water quality, food safety, solid waste disposal, and radiation hazards. An opportunity to strengthen existing environmental laws, and develop new ones, began with the election in 1958 of Edmund S. Muskie (D-ME) (Figure 3.3) to the U.S. Senate. Senator Muskie, prior to his election to the Senate, was governor of Maine, a state heavily dependent on an economy based on timber, fishing, and recreation. As governor, he became concerned about protecting the state's natural resources. He carried these con-

FIGURE 3.3   Senator Edmund S. Muskie (D-ME), circa 1985.

cerns to the Senate, along with his support for preserving the authorities of the states in environmental affairs.

As related by Landy and colleagues [18], Muskie's role in environmental legislation began inauspiciously. As a freshman senator he had offended Senate Majority Leader Lyndon Johnson. As a result, Muskie was assigned to the Senate's Public Works Committee, a committee that was not held in high regard by other senators because it dealt with projects of special interest to members of the Senate (e.g., bridge repairs, highway construction, and canal dredging). In 1963, Muskie was appointed chairman of the committee's newly created environment subcommittee, a position he kept for the next seventeen years before he left the Senate to campaign for the presidency.

Muskie sponsored a series of seminal water and air quality bills during the 1960s that had great impact on the nation's environmental policies. In 1963, he led the development of amendments to the Water Pollution Control Act, which transferred authority for water pollution from the PHS surgeon general to the secretary of DHEW. The secretary was provided authority to establish water quality standards for interstate waterways if states' standards were deemed unprotective of public health. The 1963 amendments became the Water Quality Act of 1965 [ibid.]. In 1966, he sponsored the Clean Water Restoration Act, which provided states with federal funds for sewer construction.

Air quality legislation was also an interest and product of Muskie's subcommittee. In 1967, he sponsored an air quality bill that brought sweeping change. States were required to establish and enforce air quality standards that were to be based on scientific data from the federal government. Further, the act required the federal government to develop air quality standards for automobile emissions [ibid.].

Although some critics of Muskie thought his body of air and water quality environmental legislation was too dependent on states' actions, his contributions to improving environmental quality were both undeniable and vital.

3.3.4.3 Governance Climate

Another key player in the nation's emerging commitment to environmental protection was Richard M. Nixon (Figure 3.4). The election in 1968 of Nixon soon led to quite significant changes in the U.S.'s environmental policies and resources [18]. Nixon perceived that the Republican Party needed to expand its voting base. He correctly understood the political implications of the migration, started in the 1950s, of persons from urban areas to the suburbs, areas where the outdoor ambient air was cleaner, green spaces were available, and housing was less expensive. Republican Party strategists thought suburbanites were more aware of good environmental quality and therefore ripe for Republican Party outreach to persons who supported improved environmental quality—and an expanded voting base of Republicans in the suburbs. Moreover, the inaugural Earth Day celebration of April 22, 1970, had drawn attention to environmental issues and raised public concern for the environment, situations ripe for political cultivation.

For political gain, Nixon became a supporter of a stronger federal role in protecting the environment. He signed into law the National Environmental Protection Act of 1969, the Clean Air Act Amendments of 1970, and the Occupational Safety and Health Act of 1970. Further, in May 1969 he appointed the Environmental Quality Council, a cabinet-rank committee, and tasked it with preparing a strategy for addressing environmental

FIGURE 3.4 Richard M. Nixon (R-CA), 37th U.S. president, circa 1970.

issues. The committee failed to deliver its report, which led to the appointment of a task force to produce the report. In February 1970 a preliminary report was produced. It recommended the establishment of a new Department of Environment and Natural Resources (DENR) [ibid.].

Nixon initially gave his approval to creating a DENR and asked an advisory committee on government reform, chaired by Roy Ash, formerly the head of Litton Industries, to consider the proposal. The committee became known as the Ash Council. Staff within the Council soon split over the merits of a DENR, in part because it would create problems within the existing congressional committee structure that had environmental responsibilities and in part because a DENR was seen as too large and unwieldy. The president withdrew his support for a DENR and settled, instead, on establishment of EPA, a more focused and visible federal agency than what a DENR would have been. Notably, the secretary of DHEW supported the plan for an EPA, aware that DHEW's public health environmental programs would be transferred into EPA.

On July 9, 1970, Nixon submitted to Congress his reorganization plan to establish EPA. According to an existing statute on executive branch reorganizations, Congress had sixty days to react to the proposal. Since neither the House nor Senate expressed opposition to the plan, it went into effect on September 17, 1970 [ibid.]. However, the White House disagreed with environmental groups and their congressional allies over the mission of EPA and how the agency would relate to Congress. According to Landy and colleagues [18], the White House expected EPA to pursue its mandate so as not to hinder industrial expansion and resource development. In contrast, the environmental community wanted EPA to champion environmental values via statutes that bound the executive branch to enforcing strict limits on environmental hazards.

The debate about EPA's mission and how it would function within the executive branch was settled to a considerable extent by the agency's first administrator, William Ruckelshaus. He chose to emphasize enforcement of air and water quality regulations as policy [ibid.]. This was a natural choice, given Ruckelshaus's background as Attorney General of Indiana, where he had aggressively litigated entities that had broken the state's environmental statutes. His decision cast EPA's future as a regulatory and enforcement agency. Moreover, his orientation of EPA as a regulatory agency met with support from environmental groups, which had become dissatisfied with the nonregulatory approach taken by the PHS. But with enforcement and the establishment of regulations and standards under specific environmental statutes came the inevitable opposition from companies and other entities that would be targets of environmental regulations and enforcement.

This new agency was constructed of several existing environmental programs within DHEW and other federal departments, subsuming PHS programs in air pollution, solid waste, pesticides, drinking water, aspects of radiological health [9], and water pollution control from the DOI. EPA did not repeat the personnel policy of the DOI by excluding PHS commissioned officers, although few medical officers chose to transfer to EPA. In fact, there remain today a substantial number of PHS officers on loan from the PHS to EPA. Some would assert that EPA's policy of retaining PHS officers has had a beneficial effect by infusing greater public health perspective into environmental protection, however, the relatively few number of medical doctors in

EPA remains a problem in regard to getting a medical and public health perspective infused into regulatory decisions.

Since its establishment, EPA has had federal primacy in environmental protection and protection of human health against specific environmental hazards. Its current organizational structure is shown in Figure 3.5. Indeed, the current stated purpose of EPA is "[t]o protect human health and the environment" [19]. Protection of human health is pursued by controlling—using risk assessment policies and procedures—individual environmental hazards (e.g., contaminants in drinking water supplies) and managing the risks through "command and control" regulations. These endeavors are intended to reduce or prevent human contact with specified environmental hazards. Less exposure means a reduced potential for adverse human health effects and improved environmental quality.

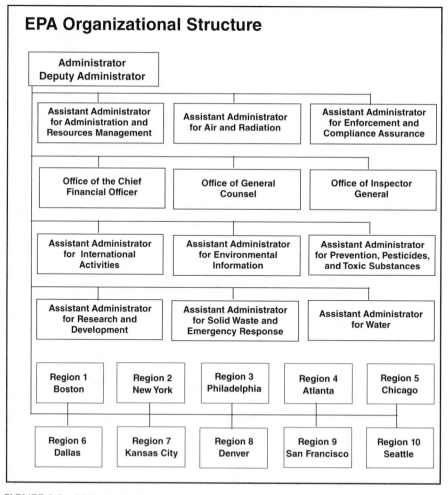

FIGURE 3.5 EPA organizational chart [19].

In addition to creating EPA, Nixon signed into law the Occupational Safety and Health Act of 1970. This act established the Occupational Safety and Health Administration (OSHA), a regulatory agency within the Department of Labor. As described in chapter 4, OSHA has the responsibility for controlling workplace environmental hazards through control of workplace conditions, conduct of workplace inspections, and enforcement of workplace regulations and standards. The act also created the National Institute for Occupational Safety and Health, an agency now within the DHHS, for the purpose of conducting health and safety research on workplace hazards and other public health duties that are discussed later in this chapter.

3.4   DHHS AGENCIES WITH ENVIRONMENTAL PROGRAMS

The DHHS is home to several environmental health agencies and programs that date from the mid-twentieth century. The current programs are conducted through the use of the traditional public health approach of science, consensus, and services. As intended here, science comprises basic biomedical research, health surveillance systems, epidemiology, and applied research. Stakeholders include state and local health departments, expert advisory groups, and public health organizations. Consensus refers to the process of stakeholder dialog on the public health significance of scientific findings. Services flow from the consensus achieved. The services can include, depending on the nature of the identified hazard, such activities as public health education, resources for health care providers, immunization programs, and assistance to local health departments.

Even with EPA's emergence as the federal government's principal regulatory authority on environmental health hazards, environmental health programs have remained, and grown, as components of DHHS, comprising agencies and offices. Agencies are organizational entities that administer programs authorized by law and specific to the purposes of each agency. Offices, in distinction to agencies, administer functions that support the administrative needs of the secretary. Not all DHHS agencies and offices administer programs and activities of relevance to environmental health. Shown in Figure 3.6 are those agencies and offices that administer the bulk of the DHHS programs of environmental health research and services. Following, in alphabetical order, are brief descriptions of each DHHS agency and office that has environmental health responsibilities.

3.4.1   AGENCY FOR TOXIC SUBSTANCES AND DISEASE REGISTRY

In the late 1970s the United States gradually became aware through intense and frequent news media reports of a new environmental health hazard: uncontrolled hazardous waste sites. Uncontrolled is used in the sense of abandoned or not under regulatory control. The most prominent case was a small suburban community, Love Canal, New York, located near Niagara Falls, New York, which had been built over a chemical waste dump [20]. As the buried drums of chemical waste gradually deteriorated, chemical fumes migrated through the soil and permeated the houses above. Residents became concerned that the

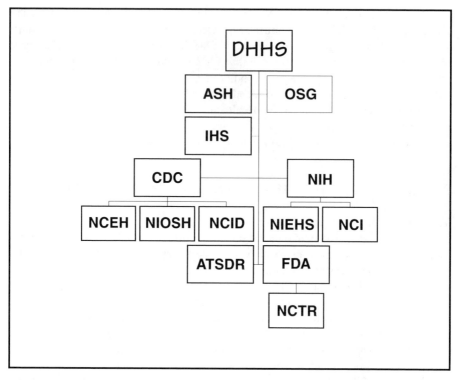

FIGURE 3.6   Organizational chart of the DHHS agencies with environmental health programs. ASH - Assistant Secretary for Health; ATSDR - Agency for Toxic Substances and Disease Registry; CDC - Centers for Disease Control and Prevention; DHHS - Department of Health and Human Services; FDA - Food and Drug Administration; IHS - Indian Health Service; NCEH - National Center for Environmental Health; NCI - National Cancer Institute; NCID - National Center for Infectious Diseases; NCTR - National Center for Toxicological Research; NIEHS - National Institute of Environmental Health Sciences; NIH - National Institute of Health; NIOSH - National Institution for Occupational Safety and Health; OSG - Office of Surgeon General.

vapors were associated with adverse health effects in their children. During the same time period, other news reports of hazardous waste sites were broadcast (chapter 8).

As news media coverage of hazardous waste sites grew, the environmental lobby made waste site cleanups and protection of human health into a legislative priority. Two organizations, the Environmental Defense Fund (now called Environmental Defense) and the Natural Resources Defense Council, lobbied energetically for enactment of federal legislation, postulating that the public's health was at risk. Chemical companies and waste generators lobbied equally energetically to discourage hazardous waste legislation, arguing that the environmental and human health consequences had been exaggerated. The scene was set for a vigorous congressional debate over hazardous waste site management.

Congressional debate divided generally along partisan political lines. Democrats generally favored federal action to identify and remediate (i.e., cleanup or fix) uncontrolled hazardous waste sites. Republicans generally favored state-based actions that might lead to state legislative action. In the House of Representatives, Congressmen

James Florio (D-NJ) and John Dingell (D-MI) led the drafting of a bill that would become the Comprehensive Environmental Response, Compensation, and Liability Act (CERCLA). Similar efforts in the Senate were led by Senator Robert Stafford (R-VT). The CERCLA is described in chapter 8.

The ATSDR's primary mission is to conduct the public health agenda specified in the CERCLA.

A key policy problem facing Congress was how to respond to pressure from community groups who represented people residing near waste sites and the environmental lobby, both of which demanded "victims' compensation" for persons residing near uncontrolled hazardous waste sites. In concept, this proposal was similar to workers' compensation, arguing that one's residential location, through no fault of their own, had caused financial (lower property values) and health harm (cancer and other health problems). In a social justice context, residents asserted they had become victims, thereby deserving compensation. The Senate's CERCLA bill contained language to provide victims' compensation; the House version did not. In such instances of disagreement, a conference committee of House and Senate members is appointed and tasked with trying to find a consensus bill that then goes back to the House and Senate for vote. Resolving the differences over victims' compensation was among the conferees' challenges.

House and Senate conferees eventually abandoned the issue of victims' compensation [21]. Conferees dropped the idea, fearing such a policy would be abused. Moreover, the policy was thought premature in light of considerable scientific uncertainty about actual health effects in persons residing on or near uncontrolled hazardous waste sites. Their solution was to create a new federal agency specifically tasked with assessing the health of persons potentially impacted by hazardous substances released from uncontrolled waste sites. This organization, ATSDR, was therefore the congressional prescription to avoid victims' compensation. The agency would, in theory, develop the health effects database that could be used by waste site area residents in private litigations against waste site owners. ATSDR's public health responsibilities under the CERCLA are discussed in chapter 8.

On December 10, 1980, President Jimmy Carter signed the CERCLA into law. The ATSDR was made an agency of the DHHS. However, the incoming Reagan administration was philosophically opposed to any further growth in federal government programs and therefore chose not to provide resources to create ATSDR as a new federal organization. Rather, ATSDR's responsibilities to conduct health assessments of the CERCLA waste sites were assigned to CDC's Center for Environmental Health. In 1983, a lawsuit in federal court, litigated by strange bedfellows (Environmental Defense Fund, Chemical Manufacturers' Association, American Petroleum Institute), sued EPA and the DHHS for failure to implement various sections of the CERCLA. One of the points of litigation was failure of the federal government to fully establish ATSDR as a new federal agency. A settlement between the litigants led to agreement to organize ATSDR and provide it with resources. However, DHHS elected to tether the new agency to CDC, designating the CDC director to also serve as the ATSDR administrator. Both ATSDR and CDC are headquartered in Atlanta, Georgia. Dr. James O. Mason served as ATSDR's first administrator. As a matter of environmental health policy, the arrangement between CDC and ATSDR was a marriage of necessity, given the antipathy of the Reagan administration to the creation of new government agencies, as well as an animus toward the CERCLA itself. This is an example of a conflict over environmental health policy that occurred between the executive and the legislative branches of the U.S. government, requiring the third branch, the judiciary, to settle the issue over how and when to structure the ATSDR.

Following its establishment, the Superfund Amendments and Reauthorization Act of 1986 gave additional responsibilities to ATSDR The agency's public health programs grew to include health assessments of communities impacted by hazardous waste sites, the preparation of toxicological profiles on substances released from waste sites, health surveillance of persons exposed to hazardous substances, epidemiological investigations of populations exposed to hazardous substances, toxicological studies, and education programs for health care providers. These programs are conducted in collaboration with EPA, CDC, and other federal agencies, and through grants awarded to state health departments.

3.4.2 CENTERS FOR DISEASE CONTROL AND PREVENTION

The Centers for Disease Control and Prevention (CDC), promotes itself as " [t]he lead federal agency for protecting the health and safety of people—at home and abroad, providing credible information to enhance health decisions, and promoting health through strong partnerships" [22]. The agency's mission is stated to be, "To promote health and quality of life by preventing and controlling disease, injury, and disability." CDC's roots were planted in World War II when control of malaria in U.S. troops was of great concern because of the large concentration of military bases in the southern United States. A military hospital was established in Atlanta to provide medical care for troops who had contracted malaria. After the war, public health authorities continued with efforts to prevent malaria and typhus in civilian populations residing in the southern states and Caribbean countries. The authority for these efforts was the Malaria Control in War Areas (MCWA) program, which was developed in anticipation of troops returning to the United States, having contracted various tropical diseases unfamiliar to U.S. clinicians [9].

A visionary PHS officer, Dr. Joseph Mountin, foresaw the need for a public health agency that would monitor infectious disease outbreaks, provide education and services to state and local health departments, and conduct epidemiological investigations. His vision became reality when in 1946 the Communicable Disease Center was established as part of the PHS, later to become the Communicable Disease Centers, then the Centers for Disease Control, and now named the Centers for Disease Control and Prevention. CDC's programs cover infectious disease, sexually transmitted disease, chronic disease, injuries, environmental health, occupational health and safety, bioterrorism prevention, and immunization. The core public health disciplines that serve as the spine of CDC are surveillance, epidemiology, laboratory research, education, and health services. Within the CDC structure are three organization components that have environmental health responsibilities, which are described in the following sections.

3.4.2.1 National Center for Environmental Health, CDC

The Center for Environmental Health was created in 1980 in response to a reorganization of the then Communicable Diseases Center, soon to become the Centers for Disease Control. Leaders of CDC developed a long-term strategic plan that was based on mortality and morbidity statistics. They asked the question, what are the leading causes of premature death and preventable disability in the American population? From this exercise, environmental hazards, traumatic injuries, and chronic diseases (e.g., heart disease) were added to the list of existing CDC programs, principally focused at the time on the prevention of infectious diseases.

The National Center for Environmental Health's programs include surveillance of environmental illnesses, prevention of lead exposure, epidemiological investigations of environmental hazards, and laboratory assessment of exposures to hazardous substances in the U.S. population.

The newly formed Center for Environmental Health (CEH) assumed the responsibility for an existing federal vector control program, commenced the development of a laboratory to measure environmental toxicants in human tissues, initiated a program to reduce children's exposure to lead in the environment, and took responsibility for a nascent public health program created by the CERCLA, even though the law had created the Agency for Toxic Substances and Disease Registry (ATSDR) for that purpose. In 1983, litigation brought against DHHS forced the Reagan administration to organize and fund ATSDR, which removed the hazardous waste program from CEH. In 1987, CEH was given responsibility for nonocupational injury control programs and was renamed the Center for Environmental Health and Injury Control, and the word *National* was added in 1991 [23]. At about the same time, the center assumed the responsibility for an existing vessel (i.e., cruise ships, primarily) sanitation program. Although no regulatory authority accompanies vessel sanitation, the cruise ship industry looks to NCEH to provide inspections of vessels and give advice on how to improve sanitary practices. As environmental policy, this is a good example of government and industry cooperation that improves public health.

NCEH's organizational structure and programs have changed over the years. In 1992, the injury control program was transferred from NCEH and became CDC's National Center for Injury Prevention and Control (i.e., its own center). A similar change in NCEH's structure, resulted from the Children's Health Act of 2000, which transferred NCEH's reproductive effects program into a new CDC center, the National Center for Birth Defects and Developmental Disorders. The current NCEH programs include surveillance of environmental illnesses, prevention of lead exposure, epidemiological investigations, cruise ship inspections, and laboratory assessment of exposures to hazardous substances in the U.S. population

3.4.2.2 National Center for Infectious Diseases, CDC

The National Center for Infectious Diseases (NCID), CDC, conducts surveillance, epidemic investigations, epidemiological and laboratory research, training, and public education programs to develop, evaluate, and promote prevention and control strategies specific to infectious diseases.

The National Center for Infectious Diseases, CDC, conducts surveillance, epidemic investigations, epidemiological and laboratory research, and other programs for prevention of infectious diseases.

Among the NCID programs of infectious disease prevention are two programs of relevance to environmental health: food safety and potable water. The Foodborne Diseases Active Surveillance Network (FoodNet) is the principal foodborne disease component of CDC's Emerging Infections Program (EIP). FoodNet is a collaborative project of the CDC, ten EIP states[5], the U.S. Department of Agriculture (USDA), and the Food and Drug Administration (FDA) [24]. The project consists of active surveillance for foodborne diseases and related epidemiological studies designed to help public health officials better understand the epidemiology of foodborne diseases in the United States. Three other surveillance systems track outbreak data, which are derived from three different reporting systems: (1) foodborne disease outbreaks, as reported to NCID by state epidemiologists, (2) *E. coli* outbreaks, as reported by the states or regulatory agencies, and (3) *Salmonella enteritidis* outbreaks reported to CDC by the states or regulatory agencies. In addition to operating these four surveillance systems, NCID investigates foodborne disease outbreaks and establishes both short-term control measures and long-term improvements to prevent similar outbreaks in the future.

NCID's Safe Water System (SWS) is a water quality intervention that uses simple, inexpensive technologies appropriate for the developing world [ibid.]. The objective is to make water safe through disinfection and safe storage at the point of use. The intervention consists of three steps: (a) point-of-use treatment of contaminated water using sodium hypochlorite solution purchased locally and produced by a local manufacturer or in the community from water and salt using an electrolytic cell; (b) safe water storage in plastic containers with a narrow mouth, lid, and a spigot to prevent recontamination; and (c) behavior change techniques, including social marketing, community mobilization, motivational interviewing, communication, and education. These activities increase awareness of the link between contaminated water and disease, the benefits of safe water, and hygiene behaviors, including the purchase and proper use of the water storage vessel and disinfectant. The Safe Water System is pertinent to the discussion of the Clean Water Act, discussed in chapter 5, and global environmental public health policies, as found in chapter 9.

3.4.2.3 National Institute for Occupational Safety and Health, CDC

In 1970, after a decade of effort, organized labor was successful in lobbying Congress to enact legislation to protect U.S. workers' health and safety. The Occupational Safety and Health Act of 1970 (OSHAct) created a framework for regulatory control of workplace hazards and vested the responsibility with a new organization, OSHA, which was placed within the Department of Labor. OSHA was given the responsibility to develop and enforce workplace standards, conduct workplace investigations, and conduct educational programs on workers' health and safety.

The OSHAct also created the National Institute for Occupational Safety and Health (NIOSH) and placed it within the DHEW. Placement of NIOSH and OSHA in different federal departments was necessary in order to gain the political support of Sen. Jacob Javits, a liberal Republican from New York. His support was needed to ensure enough Republican votes to gain Senate passage of the OSHAct. He believed that NIOSH's public health and science responsibilities should be separate and independent from the influences of regulatory actions assigned to OSHA. The policy of organizationally separating environmental regulatory agencies from public health agencies was repeated in law when the CERCLA created ATSDR as an agency separate from EPA, which was given the overall responsibility for administering the law.

NIOSH's programs include health hazard evaluations of workplaces, substance-specific criteria documents, and surveillance and epidemiological investigations of workplace hazards.

Placing regulatory agencies administratively apart from health agencies is a significant environmental health policy, with both positive and negative consequences. On the positive side of separate organizations, if public policy issues such as regulating workplace levels of specific toxicants or community levels of air pollutants are to be based on firm scientific findings, one can argue that scientific research and its interpretation should be left to the purview of scientists unbiased by regulatory imperatives. Such imperatives would include taut timelines to publish a regulatory action to control toxicants and political pressures from business organizations and environmental groups to establish regulations that comport with their individual interests.

Further, independence of health agencies has traditionally been based to a great extent on investigator-initiated research, such as the biomedical research grants program at the NIH and NIOSH. Researchers are less bound to agency priorities than is the case with regulatory agencies, where priorities are more driven by regulatory interests. As such an example, EPA has linked its research program to the needs of the agency for risk assessment data and prioritized its research support on the basis of comparative risk assessment of highest-ranked hazards to the environment and human health [25].

Arguments against a policy of separating regulatory agencies from public health agencies would include the same arguments in favor of separation, but with an opposite perspective. For example, some would argue that greater efficiency in resource expenditures would occur if public health research was more focused on priority environment and workplace hazards and the data needs of risk assessment. Moreover, risk assessors would have closer contact with agency scientists who might be more steeped in a specific issue of science that is relevant to a particular risk assessment. For example, in the first years of NIOSH, criteria documents were written to a considerable extent by institute scientists actively engaged in NIOSH research. As the number of criteria documents increases within an agency (e.g., NIOSH and ATSDR), the demands of preparing them outstrips the ability and willingness of intramural scientists to help prepare them. As a result, persons with scientific backgrounds and graduate degrees are hired expressly to write the documents, with some documents written by contract consultants.

NIOSH was given authority to conduct health hazard evaluations of workplaces, develop substance-specific criteria documents, do surveillance and epidemiological investigations of workplace hazards, and offer training courses in workplace safety and health. NIOSH was created from the PHS Bureau of Occupational Safety and Health, based in Cincinnati, Ohio, thus building upon an administrative structure that had existed for many years. The first NIOSH director, Dr. Marcus Key, had served as the bureau's director.

NIOSH was first placed in the DHEW's Health Services and Mental Health Administration, and was later made a center within the CDC structure. At its establishment, NIOSH's headquarters were located in Rockville, Maryland, in order to be close to Washington-area resources and political contacts with organized labor and congressional sources. The main research facilities were located in Cincinnati, Ohio, Morgantown, West Virginia, and Salt Lake City, Utah. Cincinnati was the location of NIOSH's surveillance, field studies, toxicology, analytical laboratories, and training programs. Morgantown was the site of NIOSH's respiratory research programs, and Salt Lake City was the site of industrial hygiene and safety programs. In the 1970s, all of NIOSH's Salt Lake City programs were transferred to Morgantown, West Virginia, a relocation that was promoted by the West Virginia congressional delegation and organized labor. In 1981, with the election of Ronald Reagan, the incoming administration moved NIOSH's headquarters to Atlanta, Georgia, with the purpose of lessening contacts with organized labor and those Members of Congress concerned about occupational health and safety issues. In 1992, the Clinton administration acceded to labor's request that NIOSH's headquarters be returned to Washington, D.C.

In an environmental health policy context, NIOSH has always been sensitive to the interests of organized labor, sometimes to the institute's detriment, because industrial and business groups perceived some NIOSH programs were not in their best interests. As an example, business interests became concerned about specific programs of research (e.g., ergonomics research and recommendations to OSHA for the development of workplace standards to control ergonomic hazards that cause musculoskeletal disorders). During the first Reagan term, NIOSH was targeted for elimination or major reduction in responsibilities, an agenda that was thwarted through the intercession of organized labor and professional societies such as the American Industrial Hygiene Association. On the other hand, organized labor's health and safety concerns helped develop a realistic agenda for NIOSH research and services, not to mention a stable political base of support for the institute. During the Clinton administration, NIOSH achieved more cooperative relationships with corporate and business interests (e.g., in developing a long-range strategic research plan in occupational health and safety).

3.4.3 FOOD AND DRUG ADMINISTRATION

The Food and Drug Administration, like the NIH, can trace its history to a small laboratory established in the nineteenth century. In 1862, the newly established Department of Agriculture established a laboratory to analyze samples of food, soils, fertilizers, and other agricultural substances [26]. This was in response to public concerns that would today be called product safety. Some goods had been found to have been deliberately adulterated by merchants in order to increase their profits. An early question investigated by the laboratory was whether adding sugar to fermenting wine to increase alcohol content was food adulteration. (The laboratory concluded that adding sugar was legitimate.) In time, the laboratory grew into the Bureau of Chemistry. The public health problem of food adulteration was the responsibility of states until 1906, when the federal Food and Drugs Act was enacted. The Bureau of Chemistry enforced the 1906 law until 1931, when the Food, Drug, and Insecticide Administration was formed, to be renamed in 1931 as the Food and Drug Administration (FDA) [26]. In 1940, to prevent conflicts between food producers and consumers, FDA was transferred from the U.S. Department of Agriculture to the Federal Security Agency, which in 1953, became the DHEW [ibid.], as previously noted. FDA's headquarters are located in Rockville, Maryland.

As discussed in chapter 6, FDA has a significant responsibility for enforcement of the provisions of the Federal Food, Drug, and Cosmetic Act, along with provisions of various other laws pertaining to public health [27]. FDA regulates products that the U.S. public encounters daily. These include regulation of food (e.g., dietary supplements, product labeling), drugs, therapeutic devices, biologics (e.g., vaccines, blood products), animal feed and drugs, cosmetics, and radiation-emitting products (e.g., microwaves). Research on the potentially toxic properties of substances of interest to FDA is conducted at the agency's National Center for Toxicological Research (NCTR), located near Jefferson, Arkansas.

3.4.3.1 National Center for Toxicological Research, FDA

As background to the establishment of NCTR,[6] during the Nixon administration the U. S. agreed to stop production of biological weapons. This policy was in reaction to international efforts to forego egregious weapons of human destruction. With great fanfare, President Nixon announced that all U.S. biological weapons facilities would be closed. As it turned out, how these newly surplused facilities were to be used was influenced by a report to the secretary of the DHEW, prepared by a consultant, Dr. Emil Mrak, who at the time was chancellor, University of California, Davis campus.

In 1969, the DHEW established the Commission on Pesticides and Their Relationship to Environmental Health, which was tasked to conduct the first assessment of pesticide risks [28]. In December 1969, Dr. Mrak delivered the *Report of the DHEW Secretary's Commission on Pesticides and Their Relationship to Environmental Health* to the Secretary, DHEW [29]. The development of this report had been stimulated by Rachel Carson's 1962 book *Silent Spring* [30]. Mrak was well known for developing the food science department at the University of California, Davis, and was also active in the work of the National Research Council.[7] In 1968, at the time that the commission was created, pesticides regulation was the responsibility of three different federal departments (Agriculture, Interior and DHEW). The Commission operated during a period when the U.S. public had become increasingly concerned about the public health effects of pesticides in the environment.

The Mrak Commission's report recognized the need for increased research to develop better ways to assess the inherent risks associated with the use of hazardous chemicals. The Mrak Commission also recognized the need for an increased role of the federal government in developing methods to test chemicals for their potential to produce toxic effects in humans [28]. The commission's report had a major effect on toxicology and the federal government's regulation of hazardous chemicals in the environment.

With the need to redeploy surplused buildings and personnel that had previously been part of the U.S. chemical weapons program, two important centers, Ft. Detrick, Maryland, and the Pine Bluff Arsenal, Arkansas, were repurposed. The former was transferred to the National Cancer Institute. The Pine Bluff Arsenal had been built with thick concrete walls and with state-of-art air handling equipment as a high security biological containment facility. It was used by the U.S. Army for its biological warfare materials production in the 1950s and 1960s. Rather than demolishing the facility, President Nixon continued his environmental health epiphany by announcing that the Pine Bluff weapons facility was to be converted into an environmentally-relevant research laboratory. Federal departments were encouraged to propose concepts for the facility's use. With the closing of the Pine Bluff facility, the Arkansas congressional delegation, eager to protect 350 jobs at the Pine Bluff Arsenal, asked the U.S. Army to find another use for the facility. However, the Army declined. The Departments of Interior and Agriculture also declined because the facility was mainly a large, chemical production facility in disrepair, along with a few animal housing rooms and microbiology laboratories.

The mission of NCTR is to conduct peer-reviewed scientific research that supports and anticipates the FDA's current and future regulatory needs.

The FDA successfully proposed that the facility become a toxicology laboratory. The new laboratory was to conduct basic and applied research on the toxicity of drugs of interest to FDA, and through an arrangement with EPA, would conduct studies on environmental substances of interest.

EPA was established on December 2, 1970, and given the responsibility to regulate pesticides and other chemicals in the environment. FDA retained the responsibility to regulate chemicals in food. In January of 1971 the Office of Science and Technology, Executive Office of the President, issued a news release announcing the creation of the NCTR. The news release stated, "new major project aimed at investigating the health effects of a variety of chemical substances found in man's environment .. [i]n the surplus facilities of the Pine Bluff Arsenal, Pine Bluff, Arkansas."

The NCTR began operations in early 1971 as a joint venture between the FDA and EPA. The NCTR program was provided oversight by a joint FDA-EPA Policy Board. NCTR was administratively organized as a bureau of the FDA. The director of NCTR reported to the Commissioner of the FDA. In 1971, Dr. Morris Cramer, an EPA scientist on detail to FDA, was appointed the first NCTR director. Over time, EPA's interest in the NCTR's toxicology programs decreased as their own intramural toxicology laboratories increased in number, resources, and toxicological specialty.

As stated by the FDA, the current mission of the NCTR is "[t]o conduct peer-reviewed scientific research that supports and anticipates the FDA's current and future regulatory needs. This involves fundamental and applied research specifically designed to define biological mechanisms of action underlying the toxicity of products regulated by the FDA. This research is aimed at understanding critical biological events in the expression of toxicity and at developing methods to improve assessment of human exposure, susceptibility and risk" [31]. It is interesting to reflect on the societal good achieved by converting a chemical weapons facility into a facility dedicated to preventing the harm done by substances found to be toxic to human health.

3.4.4 Indian Health Service

The Indian Health Service (IHS) was established in 1955. Its headquarters are located in Bethesda, Maryland, with regional offices in western states and Alaska. The IHS is responsible for providing federal health services to Native Americans and Alaska Natives. The provision of health services to members of federally-recognized tribes grew out of the special government-to-government relationship between the U.S. federal government and Native American tribes [32]. This relationship, established in 1787, is based on Article I, §8, of the U.S. Constitution. The IHS is the principal federal health care provider and health advocate for Native Americans and Alaska natives, including providing environmental health services.

The Indian Health Service is responsible for providing federal health services to Native Americans and Alaska Natives.

According to the IHS [ibid.], the principal legislation authorizing federal funding of health services provided to recognized Native American tribes is the Snyder Act of 1921. It authorized funds "for the relief of distress and conservation of health...[and]...for the employment of...physicians...for Indians tribes throughout the United States."

In 1993, Congress enacted the Indian Self-Determination and Education Assistance Act, which provided tribes the option of either assuming from the IHS the administration and operation of health services and programs in their communities, or to remain within the IHS administered direct health systems. This is an interesting and significant policy option. In other words, Native American tribes and Alaska natives that are recognized as such by the U.S. government have the option of operating their own health care system, with federal funding, or opting to have the IHS provide their health care. In 1994, Congress enacted the Indian Health Care Improvement Act, which is a health-specific law that supports the options in the 1993 Act. The goal of the 1994 Improvement Act was to provide the quantity and quality of health services necessary to elevate the health status of Native Americans and Alaska Natives [ibid.].

The IHS currently provides health services to approximately 1.5 million Native Americans and Alaska Natives who belong to more than 557 federally recognized tribes in thirty-five states [32]. IHS services are provided either directly and through tribally contracted and operated health programs. In 2004, the federal system consisted of thirty-six hospitals, sixty-one health centers, forty-nine health stations, and five residential treatment centers. In addition, thirty-four urban Native American health projects provide a variety of health and referral services.

Of relevance to environmental health policy, the IHS notes that since 1960, more than 230,000 Native American homes have benefited from IHS funding of water and sewerage facilities, solid waste disposal systems, and technical assistance for operation and maintenance organizations. The IHS also observes that the age-adjusted death rate from gastrointestinal disease for Native Americans and Alaska Natives has decreased by more than 91 percent since 1955, the year the IHS was established [ibid.]. Approximately 93 percent of Native American and Alaska Native homes have been provided sanitation facilities since the inception of an IHS sanitation construction program. The IHS also funds construction of new and replacement hospitals and ambulatory care facilities and staff quarters. Moreover, the IHS provides technical assistance to Native American tribes on such environmental problems as hazardous waste removal, water purification, and food safety.

3.4.5   NATIONAL INSTITUTES OF HEALTH

The National Institutes of Health (NIH), based in Bethesda, Maryland, is the nation's premier resource in biomedical research. The NIH can trace its history to 1887, when Surgeon General John B. Hamilton established the Hygienic Laboratory, a one room facility in the Marine Hospital on Staten Island, New York [9]. The early work of the laboratory was in the areas of bacteriology and pathology. In 1891, the Hygienic Laboratory was moved to Washington, D.C., near the U.S. Capitol [33]. In 1902, Congress enacted legislation that increased the Hygienic Laboratory's authorities and resources, authorizing the establishment of divisions of chemistry, zoology, and pharmacology.

These changes established a firm and necessary link between scientific research and public health practice. In 1930, Congress formally created the NIH, renaming the Hygienic Laboratory, and in 1938 the construction began of buildings on 45 acres of land in Bethesda, Maryland, acreage donated to the federal government.

The current mission "[i]s science in pursuit of fundamental knowledge about the nature and behavior of living systems and the application of that knowledge to extend healthy life and reduce the burdens of illness and disability" [34]. The NIH currently comprises twenty-seven institutes and offices [35]. The 20 institutes comprise the National Cancer Institute, National Eye Institute, National Heart, Lung, and Blood Institute, National Human Genome Research Institute, National Institute on Aging, National Institute on Alcohol Abuse and Alcoholism, National Institute of Allergy and Infectious Diseases, National Institute of Arthritis and Musculoskeletal and Skin Diseases, National Institute of Biomedical Imaging and Bioengineering, National Institute of Child Health and Human Development, National Institute on Deafness and Other Communication Disorders, National Institute of Dental and Craniofacial Research, National Institute of Diabetes and Digestive and Kidney Diseases, National Institute on Drug Abuse, National Institute of Environmental Health Sciences, National Institute of General Medical Services, National Institute of Mental Health, National Institute of Neurological Disorders and Stroke, National Institute of Nursing Research, and the National Institute of Medicine. As indicated by their titles, institutes have a focus on specific diseases or disorders.

3.4.5.1   National Cancer Institute, NIH

The National Cancer Institute (NCI) is the foremost cancer research and services organization in the United States. It was established by the National Cancer Institute Act of 1937, which was signed into law by President Franklin D. Roosevelt on August 5, 1937. The purpose was stated to be, "To provide for, foster, and aid in coordinating research relating to cancer; to establish the National Cancer Institute; and for other purposes" [36]. An appropriation of $700,000 for each fiscal year was authorized by the 1937 act, an amount which is now less than many NCI grants awarded to individual cancer researchers. The NCI is headquartered in Bethesda, Maryland.

Of historical note, the NCI was the centerpiece of a presidential political platform. In 1971, President Richard Nixon declared his initiative, *War on Cancer*, and made it a priority for his administration. Other presidents have made similar political declarations (e.g., President Franklin Roosevelt's *New Deal*, President Kennedy's *Great Society*, and President Johnson's *War on Poverty*); however, Nixon's *War on Cancer* was the first and only declaration that focused solely on a public health issue. As background, on December 23, 1971, Nixon signed into law the National Cancer Act of 1971 [37], which initiated the National Cancer Program; authorized the establishment of fifteen new research, training, and demonstration cancer centers; established cancer control programs as necessary for cooperation with state and other health agencies in the diagnosis, prevention, and treatment of cancer; and provided for the collection, analysis, and dissemination of all data useful in the diagnosis, prevention, and treatment of cancer, including the establishment of an international cancer data research bank. This act gave a major boost to the nation's fight against all forms of cancer.

The mission of NCI is "to eliminate the suffering and death due to cancer" [38]. In support of its mission, NCI supports a broad range of research to expand scientific discovery at the molecular and cellular level, within a cell's microenvironment, and in relation to human and environmental factors that influence cancer development and progression [39]. Regarding environmental health, NCI supports investigations ranging from basic behavioral research to research on the development and dissemination of interventions in areas such as tobacco use, dietary behavior, sun protection, decision making, and counseling about testing for cancer susceptibility and participation in cancer screening.

Of special note to environmental health policy, NCI's Human Genetics Program provides an expanded focus for interdisciplinary research into the genetic determinants of human cancer, including research to explore and identify heritable factors that predispose to cancer, including studies of gene-environment interactions. This kind of research will provide a better understanding of why some persons contract cancer from an environmental hazard (e.g., a carcinogen), while other persons exposed to the same hazard do not express disease.

To the extent that scientific knowledge accrues about the causes of human cancer, environmental health policy will benefit through more focused legislation and health services.

3.4.5.2   National Institute of Environmental Health Sciences, NIH

In 1966, the NIEHS began as the Division of Environmental Health Sciences, NIH, by action of the surgeon general. The division was housed on the main NIH campus in Bethesda, Maryland. That location was changed the next year when the state of North Carolina donated 509 acres within the Research Triangle Park, located in an area between Durham, Raleigh, and Chapel Hill, North Carolina, as the home for the fledgling division. The park was created during the administration of Gov. Luther Hodges, who later became secretary of commerce in the Kennedy administration. Through political lobbying by North Carolina State officials, including Hodges, the Division of Environmental Health Sciences was proposed as a key occupant of the new park. Environmental health lobbyists also were active in promoting the establishment of a federal government entity that could help develop a scientific database that would be needed to reduce the effects of environmental hazards.

The NIEHS awards research grants to investigate environmental hazards, and conducts intramural research focused mainly on mechanisms of toxicity and related matters of basic science.

In 1969, the Division of Environmental Health Sciences was raised to institute status within the NIH structure and became the NIEHS [40]. Dr. Paul Kotin, the division director, was appointed as the first director of NIEHS, serving through part of 1971. NIEHS was, and remains, the only institute of NIH that is located outside the Bethesda, Maryland, area. Its headquarters are located in Research Triangle Park, North Carolina.

The NIEHS administers a broad-based program of research grants to universities and other eligible organizations. The grants are principally investigator-initiated and selected through the traditional NIH approach of ranking of grant applications by expert committees composed of non-government scientists. The NIEHS also conducts intramural research focused mainly on mechanisms of toxicity and related matters of basic science (e.g., the environmental genome project that is investigating human genes possibly responsible for how individuals react to specific environmental toxicants).

In addition, the NIEHS provides the scientific and administrative leadership within the DHHS for the National Toxicology Program (NTP). The NTP had begun as a program conceived and administered by the NCI, which was reacting to environmental and congressional pressures to investigate the carcinogenicity of chemicals found in the general environment. NCI's response was a program largely devoted to testing specific toxicants for carcinogenicity, using laboratory animals under controlled exposure conditions. The testing was conducted by commercial toxicology testing laboratories, using a study protocol designed by NCI. Unfortunately for the NCI, one of the major contractors was found inadequate and their alleged poor quality work became the subject of critical news media reports and articles in prestigious scientific journals such as *Science*. Weary of the negative publicity, the secretary of DHHS transferred the NTP to the NIEHS for administration.

In 1981, under NIEHS's administration, the NTP became the federal government's principal program for assessing the toxicity of substances found in the environment. As a matter of policy, the NTP receives scrutiny and advice from standing extramural committees comprising experts in toxicology and related disciplines. A major activity of the NTP is to coordinate the preparation of a biennial report for DHHS on substances judged to be carcinogenic by government scientists.

3.4.6 OFFICE OF THE ASSISTANT SECRETARY FOR HEALTH, DHHS

From the establishment of the PHS in 1798 through the middle of the twentieth century, surgeons general were the service's directors. Throughout this span of time, presidents appointed career PHS commissioned officers to the post of surgeon general. Although career officers, strictly speaking, surgeons general were appointed by the president with the advice and consent of the U.S. Senate. In a policy sense, political appointees are persons whose authorities are enhanced by such appointments. They are expected to do their designated duties, and at the same time, have access to political advice, support, and resources within the political structure of the administration in which they serve. At the same time, political appointees must not use political allegiance to override the duties and responsibilities of their office. As an example of the latter, during the first term of President Reagan's administration, political appointees were selected for purpose of *not* implementing the newly enacted the CERCLA. Congressional hearings led to perjury charges brought against one political appointee, who subsequently was incarcerated.

The ASH has had responsibility for PHS environmental health programs in both the context of organizational line authority as well as program advice. Regarding the former, PHS agencies' budgets, program progress, and policy issues were subject to review and approval by the ASH. The ASH's line authority was in effect from 1967 through 1995, when a Clinton administration reorganization of DHHS resulted in all PHS agencies reporting directly to the Secretary, DHHS. The ASH became the primary health advisor to the secretary, along with the surgeon general. Sometimes the position of surgeon general was not filled, rather, the duties and responsibilities were assumed by the ASH.

Regarding environmental health policy, in 1973, ASH Charles Edwards created the *DHEW Committee to Coordinate Toxicology and Related Programs* (CCTRP).[8] It was established as a multi-agency group to provide a forum to assure exchange of information on toxicology and related programs, to coordinate these programs, to enhance sharing of resources, and to provide advice to the Department. Membership of the Committee was composed of the ASH; director, Office of Program Operations, Office of the ASH; director of NIH as well as directors of several NIH components (National Cancer Institute, National Institute of Environmental Health Sciences, National Institute of Mental Health, National Library of Medicine); director of CDC, Commissioner of FDA and the director of FDA'S National Center for Toxicological Research. In 1978, Assistant Secretary for Health Julius B. Richmond expanded the membership of CCTRP to include liaison representatives from the CPSC, EPA, National Science Foundation, OSHA, Council for Environmental Quality, Library of Congress, Department of Energy, National Oceanographic and Atmospheric Administration, and the following components of the Department of Defense: Department of Army, Edgewood Arsenal, Navy Department, U.S. Army Environmental Hygiene Agency, and Wright-Patterson Air Force Base.

In 1979, Assistant Secretary Richmond changed the committee's name to DHEW Committee to Coordinate Environmental and Related Programs. With the new name came an expansion in purpose that included "[i]nformation on environmental health, toxicology and related programs, to coordinate these programs [..]." The Coordinating Committee shall interface these activities with other components of the federal government in areas of mutual interest and concern. In 1985, DHHS Secretary Margaret Heckler changed the name to the DHHS Committee to Coordinate Environmental Health and Related Programs (CCEHRP). Once again the role of CCEHRP was expanded, "[t]o coordinate and promote the exchange of information; to provide advice, review and, where needed, carry out processes and efforts which encourage a balanced, objective consensus on all environmental health-related research efforts, exposure assessments, risk assessments, and risk management procedures, and to serve as the primary focal point within the DHHS for information coordination within and outside the Department for all environmentally related issues."

In 1989, Assistant Secretary for Health James O. Mason established CCEHRP as a standing committee of the PHS, to be chaired by the ASH, with a vice chair. In 1992, the committee was renamed the Environmental Health Policy Committee, but with the same membership, purposes, and advisory function.

Over the years, the Environmental Health Policy Committee and its predecessors provided a useful structure and forum for the debate of DHHS environmental policies and for consideration of emerging environmental issues. However, the committee did not undertake performance reviews of agencies' environmental health budgets, programs, and authorities. Moreover, each agency's coordination with other DHHS agencies' environmental health programs and with the EPA were, and remain, largely a matter of individual initiative. Although the committee has been inactive since year 2000, several of its reports of significance to environmental health policies remain available. These reports include *Environmental Diseases from A to Z* (released in 1999), *Review of Fluoride: Benefits and Risks* (1999), *A Primer on Health Risk Communication* (1998), *Drinking Water and Human Health* (1997), and *Multiple Chemical Sensitivity* (1998), all of which can be obtained from DHHS [41].

3.4.7 PERSPECTIVE

The foregoing material about the historical role of the PHS (and the role of surgeons general), the creation of EPA, and the current environmental health programs of the agencies of DHHS was meant to provide perspective on how environmental health programs evolved at the federal level of U.S. government. As was described, the emergence of concerns for protection of natural resources and improved environmental quality moved legislators and policy makers to create new policies that ultimately reduced the environmental health role of the PHS, but preserving a role of biomedical research and public health services for federal public health agencies.

Whether or not the bifurcation of environmental health programs into those of environmental protection and public health research and services is the best arrangement for serving the U.S. public is a topic that remains current and somewhat contentious. Some environmental organizations have argued for a greater public health perspective at EPA, particularly in regard to the adoption of risk assessment as the agency's policy to determine priorities and management strategies for environmental hazards. Other groups have questioned the role of PHS agencies in not being more involved in environmental risk management actions. This kind of debate is useful, if for nothing more than trying to best arrange federal resources meant to protect both the environment and human health.

What is currently in place with respect to coordination between regulatory environmental programs (e.g., EPA, OSHA) and PHS agencies (e.g., NIH, CDC, ATSDR) can best be described as collegial coordination, when it exists, on matters of mutual interest. For instance, considerable cooperation between EPA and CDC exists on matters of lead abatement in order to reduce children's exposure. Similar collaborative efforts on the CERCLA hazardous waste sites exists between EPA and ATSDR, particularly on matters of community health concerns over releases of toxicants from waste sites into community environmental media. In addition to "collegial coordination," matters of policy are taken to various White House offices and committees for development and resolution. This more formal coordination can serve to bridge differences between federal agencies when needed, but this desirable outcome is not always achieved. The White House Council on Environmental Quality (CEQ), which was established under the National Environmental Policy Act (chapter 4), reports to the president and can, in theory, influence how policy differences are resolved between agencies. However, the CEQ is only as effective as a president wants it to be, and its power to resolve policy differences is influenced accordingly.

3.5 LEGISLATED ENVIRONMENTAL HEALTH PAIRINGS

Given the need for environmental regulatory agencies and public health agencies to work on issues of mutual interest, Congress has devised a means to achieve greater coordination between the two kinds of agencies. Specifically, some federal legislation has required such coordination, usually with statutory language that stipulates agencies' responsibilities and the methods of interagency coordination.

In theory, placing regulatory responsibilities in an agency separate from public health agencies isolates the science and scientists from the political pressures of developing environmental regulations and standards. There are both benefits and detriments to this kind of bureaucratic arrangement. Concerning benefits, in theory, scientific research should be conducted objectively and spared the pressures of regulatory timelines and influence. In this theory, investigator-initiated research would be encouraged and scientific inquiry would prevail. In reality, both government conducted and sponsored (via grants) research is seldom without the constraint of agencies' research budgets and priorities. Concerning disadvantages, regulatory agencies need the best, contemporary scientific findings to bolster risk assessment and risk management decisions. Having a research program that is integrated with a regulatory agency's needs improves both the efficiency and effectiveness of risk-based decisions. In fact, EPA has over the years developed an effective research program, primarily in toxicology and ecology, that has met many of the agency's data needs in risk assessment. Following are two examples of this kind of "legislated cooperation" between environmental and public health agencies.

3.5.1 NIOSH AND OSHA ROLES AND CONNECTIONS

The OSH Act of 1970 was the first comprehensive federal legislation intended to protect U.S. workers from workplace hazards. Two federal agencies were created for the purposes of the act. OSHA was established as a regulatory agency within the Department of Labor. Its primary duties were to develop and enforce workplace standards. The act gave OSHA the authority to enter workplaces to inspect work conditions and to impose penalties when standards were violated.

The act also established the National Institute for Occupational Safety and Health (NIOSH) as a component of DHEW. NIOSH's responsibilities included conducting health hazard evaluations of workplace conditions, surveillance of workers' injuries and health problems, development of criteria documents, biomedical research, and certification of respiratory protection devices. NIOSH is not a regulatory agency and functions generally as a traditional public health organization (i.e., development and assessment of science, consensus formation, the delivery of services intended to prevent workplace injuries and disease, and safety education programs).

NIOSH's research findings and health hazard evaluation findings are relevant to OSHA's programs of standards development and workplace enforcement. The two agencies have workgroups and committees that meet and coordinate information and program plans (e.g., status of on-going research projects). These interactions are largely in the form of collegial relationships; their effectiveness depends in substantive measure on the interests of OSHA and the political climate of incumbent administrations. For example, contact between OSHA and NIOSH was infrequent and sometimes acrimonious during the first Reagan administration. More productive relationships were evident during subsequent administrations.

3.5.2 ATSDR AND EPA ROLES AND CONNECTIONS

A second example of a legislated relationship between a federal public health agency and a federal environmental protection agency is found in the CERCLA, also called the Superfund Act, which will be discussed in chapter 8. The act was the result of public concern over large amounts of uncontrolled hazardous waste that had been discovered in residential communities (e.g., Love Canal, New York). Two concerns were paramount: alleged effects of the waste on human health, and removal of the waste from the environment. Under the Act, EPA's primary responsibility is to identify those uncontrolled hazardous waste sites that require remediation (i.e., cleanup) and, where possible, recover the costs of remediation from those parties that sent waste to the site to be remediated. ATSDR is directed by the act to conduct various public health actions of significance to communities residing near waste sites and to develop toxicological and epidemiological data for use by EPA, states, and communities. Other CERCLA responsibilities for EPA and ATSDR are discussed in chapter 8. However, an example will illustrate the close yoking of these two federal agencies under the CERCLA.

First, the CERCLA, as amended, requires ATSDR to conduct health assessments of every site that EPA places on its National Priorities List (NPL), which are the sites that score highest on EPA's Hazard Ranking System, which is a numerical ranking system for ranking the hazard of uncontrolled hazardous waste sites [42]. Public health assessments are ATSDR's evaluations of a site's potential harm to persons potentially exposed to substances released from NPL sites. ATSDR uses EPA data as an important resource for characterizing the extent and impact of exposure. This cooperation between different federal agencies is facilitated by clear authorities stated in law (i.e., the CERCLA).

Various committees and workgroups have been established between the two agencies in order to implement applicable provisions of the CERCLA, as amended. Without the establishment of ATSDR in law, it is highly doubtful that its various waste site responsibilities would have been pursued by other public health agencies, using general authorities under the Public Health Service Act of 1912, as amended, because of what would have been perceived as higher priorities, e.g., infectious disease prevention.

3.6   OTHER FEDERAL ENVIRONMENTAL HEALTH PROGRAMS

In addition to DHHS, other federal departments also have environmental activities of relevance to public health. These programs are derived from federal laws specific to individual federal departments. As will be described, some environmental health programs, under a particular federal law, are shared between federal departments. All programs to be described are headquartered in Washington, D.C.

3.6.1   DEPARTMENT OF AGRICULTURE (USDA)

The agriculture department was established in 1862, thereby being one of the federal government's oldest departments [43]. The current USDA administers programs and services that include concerns for the economic well-being of farmers and the health of consumers of farm products. These activities include farm price supports, food stamps for low-income citizens, loans to farmers, soil conservation, biological research, the grading and inspection of meat and other products, crop forecasting, crop insurance, and negotiations with foreign governments for trade in agricultural products [44].

The USDA is responsible for some programs pertinent to environmental quality and, thereby, the public's health. For example, USDA's Pesticide Data Program provides data on pesticide dietary exposure, food consumption, and pesticide usage. This program is of great public health importance, because it provides essential background exposure data that epidemiologists and others can use in health research on U.S. populations of interest. Food products such as fruit and vegetables, fruit juices, whole milk, grain, and corn syrup have been tested for the presence of approximately forty pesticides [44].

Two other USDA programs, the Center for Animal Health Monitoring and the Food Safety Research program [45], are also pertinent to public health practice. The center collects national data, and conducts studies, on interactions among animal health, welfare, production, product wholesomeness, and the environment. Poor health of domestic and feral animals can be an important sign of potential human health problems. An example is the finding that fish and waterfowl in the Great Lakes region displayed abnormal sexual development, attributed to chemical contaminants in lake water, which led to investigations of human populations for signs of adverse health effects [46]. Regarding the Food Safety Research program [47], the USDA provides funds to universities to conduct research on laboratory and epidemiological methods that can be used in assuring food safety. The USDA also offers services to farmers on managing wastes from farms, use of pesticides and other farm chemicals, and conducts market basket surveys of food contaminants.

3.6.2   DEPARTMENT OF COMMERCE (DoC)

Created in 1903 as the Department of Commerce and Labor, the department was reorganized into the DoC in 1913. The department promotes economic growth, advancements in technology, negotiates trade agreements with other countries, and provides the public with information on weather conditions and business conditions. The DoC contains the Bureau of Census, which is responsible for conducting the census of the U.S. population every ten years, as required by the U.S. Constitution [48] and the Patent and Trademark Office, which performs the services implied by the office's title. The Department's International Trade Administration monitors, investigates, and evaluates foreign compliance with trade agreements, and provides advice and services to U.S. companies that have, or desire, international sales of goods and services. Of particular note for environmental purposes, is the work of the National Oceanographic and Atmospheric Administration (NOAA), established in 1970. NOAA has numerous statutory responsibilities that pertain to oceanic and atmospheric science and services. The responsibilities include coastal zone management, the management and conservation of resources within two hundred miles of the U.S. coast, issuance of weather forecasts and warnings, the preparation of nautical aeronautical charts and other navigational aids; and the management of NOAA laboratories [49]. The work of NOAA has significant import for public health. For example, warnings of severe weather help prepare public health and emergency responders for their delivery of public services and for preparing hospitals and other health care providers.

3.6.3 DEPARTMENT OF DEFENSE (DoD)

Following the end of World War II, the DoD was organized in 1949 as a result of the National Security Act of 1947. It replaced the Departments of War and Navy. The DoD has environmental responsibilities and programs that involve managing the natural resources under DoD stewardship, remediating DoD contaminated waste sites, developing pollution prevention programs, and implementing occupational safety and health programs for the department's civilian and uniformed personnel [50]. Individual military services have specialized environmental programs that meet specific needs in toxicology, industrial hygiene, radiation biology, and environmental health. For example, the U.S. Army's Ft. Detrick laboratory performs toxicological testing of substances of military interest. Other DoD programs deal with environmental hazards at military bases such as housing, repair shops, and weapons testing facilities. Also, the DoD gets involved with the environmental consequences of military actions. For example, the effects of chemical defoliants (e.g., Agent Orange) on Vietnam war veterans remains a subject of debate and final resolution as a matter of public health. Similarly, research on the adverse health effects experienced by some Gulf War veterans and health services for them is an active program in the department.

3.6.4 DEPARTMENT OF ENERGY (DOE)

The DOE was established in October 1977. It consolidated the activities of the Energy Research and Development Administration, the Federal Power Commission, the Federal Energy Administration, and elements of other agencies. The DOE has wide-ranging powers to set energy prices, enforce conservation measures, and allocate fuel. It is also empowered to engage in research on new sources of energy and direct nuclear-weapons research and development. The DOE conducts and sponsors research on alternative sources of energy (e.g., solar power) and is responsible for assessing and remediating DOE hazardous waste sites [51].

DOE's National Institute for Global Environmental Change supports work on human-induced influences on the environment. The institute was created by the Energy and Water Development Act of 1990. This effort is pursued by dividing the United States into six regions in order to study environmental change on different geographical and geological systems. Each region has a regional center, usually a university, that develops and administers research programs conducted by individual investigators through competition for grants. For example, carbon levels in Great Plains grasslands and coastal margin research in the West are two kinds of regional research studies [52].

The Center for Excellence for Sustainable Development provides educational materials to communities facing problems of congestion, urban sprawl, pollution prevention, and resource overconsumption. The materials describe how sustainable development can provide a framework under which community resources and infrastructure are used more efficiently, and economic development is enhanced.

The U.S. Human Genome Project, composed of the DOE and NIH Human Genome Programs, is the national effort to characterize all human genetic material by determining the complete sequence of the DNA in the human genome. As stated by the DOE, "[t]he ultimate goal is to discover all the more than 30,000 human genes and render them accessible for further study" [53]. The DOE Human Genome Program supports research projects at universities, DOE national laboratories, and other research organizations. Information from the Project will dramatically change almost all biological and medical research.

In another environmentally important program, DOE's Carbon Sequestration Program is addressing environmental problems caused by CO_2 emissions from widespread use of fossil fuels. To stabilize and then reduce this particular greenhouse gas's emissions and atmospheric levels will require the sequestration of carbon. This includes carbon capture, separation, storage, and reuse. This nascent DOE program will study processes to capture CO_2 and carbon separation, possible storage of sequestered CO_2 in geological formations, injection of CO_2 into oceans, and other sequestration concepts. As discussed in chapter 9, CO_2 is the most important greenhouse gas and therefore is strongly associated with concerns for global warming. Methods to reduce CO_2 emissions will have great significance for global environmental health and human health consequences. Lessened emissions will equate with improved environmental quality (e.g., lesser arid areas) and protection of human health (e.g., fewer heat-related illnesses).

3.6.5 Department of Homeland Security (DHS)

The various cabinet level federal departments, with one exception, have evolved as products of normal growth in the U.S. sociopolitical structure. Some departments, such as the State Department, date from the early beginnings of the U.S. government. Others, e.g., the Department of Energy, came into existence when the United States became more heavily involved in the development of energy policies and the need to control the production of nuclear weapons. In one instance, in a crucible of fear, a department was created relatively quickly, based on reaction to a catastrophic event. In 2002, the Department of Homeland Security was created in reaction to terrorists' flying high-jacked airplanes into the twin towers of the World Trade Center in New York City on September 11, 2001, causing approximately 3,000 fatalities and total destruction of the massive towers. This horrific event changed overnight the U.S. policy on protection against terrorists, changing from a policy of reaction to terrorist events to a policy of preemptive strikes against individual terrorists and groups and countries that support them.

The Homeland Security Act of 2002 created the Department of Homeland Security (DHS). It consists of agencies, resources, and programs formerly in other federal departments, including the U.S. Coast Guard, Secret Service, Customs Service, Border Patrol, Transportation Security Administration, and parts of the Immigration and Naturalization Service. The stated mission of the department is, "Prevent terrorist attacks within the United States; reduce America's vulnerability to terrorism; and minimize the damage and recover from attacks that do occur" [54]. The emphasis of the department is clearly the prevention of terrorist acts against the U.S.

Regarding environmental health, the U.S. Coast Guard[9] has responsibility under the CERCLA, as amended, for coordinating cleanups of emergency releases of hazardous substances into internal and external waterways of the United States. They also are a member of the National Response Team and provide personnel to operate the National Response Center. The Coast Guard also provides education to mariners and recreational boaters on environmental hazards and management.

Specific environmental health policies and programs in the DHS have not yet been formally announced. However, in a general sense, preventing the presence of biological, chemical, and radiological weapons of terror in the environment will be required of the department's programs.

3.6.6 DEPARTMENT OF HOUSING AND URBAN DEVELOPMENT (HUD)

HUD is the agency principally responsible for federal programs relating to housing and urban improvement. It was created by Congress in 1965. HUD's programs include mortgage insurance for home-buyers, low-income rental assistance, and programs for urban revitalization that are developed in conjunction with state and municipal authorities. HUD also administers the Lead Hazard Control program, which awards grants to states in order to eliminate lead hazards in low-income housing, promote educational programs, and conduct research [55]. This program includes funds for blood lead testing of children living in low-income housing, removal of lead-based paint from low-income housing, inspection of low-income housing for detection of lead hazards, community education and outreach, and job training for lead hazard control workers. HUD also administers a Healthy Homes Program, which funds local projects that address a multitude of health hazards in houses. According to HUD, the grants help develop cost effective methods for assessing and controlling hazards in low-income housing.

3.6.7   DEPARTMENT OF INTERIOR (DOI)

Created in 1849, the DOI administers conservation programs, manages fish and wildlife resources, operates national parks and historic sites, assesses mineral resources and directs their management on federal lands. The DOI also administers programs for the interests of Indian and Alaskan Native Americans and the inhabitants of Pacific island territories that are under U.S. administration. The DOI's U.S. Geological Survey (USGS) conducts several programs of direct relevance to the public's health [56]. For example, they have conducted surveys on the environmental occurrence and distribution of organic chemicals known to adversely affect human health. One USGS survey pertains to drinking water quality, collecting data on potential contamination sources and strategies to protect sources of drinking water. An example is the USGS survey of arsenic levels in groundwater supplies in southeastern Michigan.

Although U.S. drinking water supplies are generally safe and do not adversely affect the public's health, except when water management safeguards fail, having data on contaminants in drinking water supplies is vital to decisions on water quality standards and policies to ensure potable water supplies.

3.6.8   DEPARTMENT OF JUSTICE (DOJ)

The DOJ, established in 1870, is headed by the attorney general. The department is the federal government's legal office. It is responsible for the enforcement of federal laws, represents the federal government in litigations, and gives legal advice to other federal departments and agencies. The DOJ represents the federal government in litigations that involve environmental health and protection law suits, e.g., representing EPA when that agency seeks to recover costs for remediating uncontrolled hazardous waste sites.

3.6.9   DEPARTMENT OF LABOR (DOL)

The DOL was established in 1913 to "foster, promote, and develop the welfare of the wage earners of the U.S., to improve their working conditions, and to advance their opportunities for profitable employment." OSHA was established in 1970 by Congress to enforce safety and health standards in industry [57]. The Bureau of Labor Statistics compiles the Consumer Price Index and indexes of wholesale prices and publishes information on employment and earnings. The DOL, through its OSHA, has statutory responsibilities under provisions of the Occupational Safety and Health Act of 1970 for developing workplace standards and conducting workplace inspections.

OSHA has regulatory authority over workplaces covered under the Act, generally excluding small businesses, self-employed individuals, and various government agencies. OSHA develops, promulgates, and enforces (e.g., workplace inspections) regulations and standards to control hazards in U.S. workplaces. The hazards include chemicals, physical agents, biological agents, and workplace procedures such as construction work.

Another DOL component, the Mine Safety and Health Administration (MSHA), derives its authorities from the Federal Mine Safety and Health Act of 1977 (more commonly called the Mine Act) [58]. MSHA develops mandatory safety and health standards applicable to both surface mines and underground mines. They are required to inspect mines two (surface) to four (underground) times annually, unless regulations direct otherwise; investigate mine accidents; review for approval mine operators' mining plans; and provide technical assistance and training to mine operators.

3.6.10   DEPARTMENT OF STATE (DOS)

The DOS is the oldest federal department, created by Congress in 1789 [59]. The department's current responsibilities include negotiating treaties between the United States and other countries, representing the United States in foreign countries, and developing and implementing policies on international affairs. Of note to environmental issues, the U.S. Agency for International Development [60] assists other countries in developing their national economies and improving quality of life. Of importance to environmental policy, USAID encourages nations to take "[a]n integrated approach to natural resources management. Land and water must be managed skillfully so that they are able to maintain our basic ability to produce food for the nine billion people that the world is expected to have by 2050." Without calling it such, this is a statement of the policy of sustainable development.

3.6.11   DEPARTMENT OF TRANSPORTATION (DOT)

The DOT was established in 1966. It is responsible for policies aimed at environmental safety and an efficient national transportation system that can also facilitate national defense. The major divisions of the DOT are the Federal Aviation Administration, the Federal Highway Administration; the Federal Railroad Administration; Urban Mass Transportation Administration, Maritime Administration, St. Lawrence Seaway Development Corp., and the National Highway Traffic Safety Administration. The DOT has responsibilities for interstate shipment of hazardous cargo, emergency response programs, and remediation of DOT hazardous waste sites.

The DOT administers several environmental programs of note. These programs reside in DOT's Federal Aeronautics Administration (FAA), the Office of Hazard Materials Safety, and the National Response Center. The Office of Hazardous Materials (HMS) develops and recommends regulatory changes in the transportation of hazardous materials (hazmat) and implements guidance for approved polices on hazmat transportation. The HMS Office also provides technical support regarding hazmat classes and their containment and packaging. It also supports hazmat research and development programs and develops hazmat safety training polices and programs [61].

The National Response Center serves as the sole national point of contact for reporting of all oil, chemical, radiological, biological, and etiological (i.e., agents of disease) discharges into the environment anywhere in the United States and its territories. Among several duties, the NRC: (1) receives and relays reports of incidents reportable under the Hazardous Materials Transportation Act; (2) receives incident reports under the Federal Response System, which is supported under the Clean Air Act, Clean Water Act, Title III of the CERCLA, as amended; and the Oil Pollution Act of 1990, relaying incident reports to EPA for response; (3) serves as a 24-hour contact point to receive earthquake, flood, hurricane, and evacuation reports for the Federal Emergency Management Agency; and (4) releases of etiological and biological agents are received and referred to the Centers for Disease Control and Prevention for response.

The Federal Aeronautics Administration's (FAA) primary mandate is aircraft and flight safety. The FAA's Office of Environment and Energy develops models to assess airport air quality from aircraft engines. Another environmental problem, noise from aircraft landing or departing from airports, is not an FAA responsibility per se because aircraft schedules are the responsibility of the airlines. However, the FAA does have responsibility for certifying engine noise performance.

3.6.12 NATIONAL AERONAUTICS AND SPACE ADMINISTRATION (NASA)

NASA is the federal agency responsible for the development of advanced aviation, space technology, and space exploration. The agency's origins can be traced to 1915, with the establishment of the National Advisory Committee for Aeronautics (NACA). Through the 1940s, this committee established centers for aeronautical research and development. In order to better coordinate federal civilian aeronautics programs, the National Aeronautics and Space Act of 1958 transformed NACA into NASA and provided administrative structure for space research and exploration [62].

NASA's environmental programs of relevance to public health include agency activities to reduce environmental pollution from NASA activities; pursuing new technologies using environmentally benign substances and processes; and expanding the use of environmental monitoring systems in NASA programs. The agency conducts efforts to remediate hazardous waste sites that are their responsibility. NASA conducts intramural programs of research and supports extramural grant programs of relevance for improving environmental conditions. For example, the Ultra-Efficient Aircraft Engine Technology program has the goals to: increase performance of a wide range of revolutionary aircraft; address local air quality concerns by developing technology for reducing NO_x emissions from aircraft engines; and increasing engine performance to enable reductions in CO_2. Reductions in NO_x and CO_2 emissions from aircraft engines will contribute to lower ozone levels and reduced global warming, respectively. Both kinds of reduction will contribute to improved public health; the former to reduced respiratory disease and the latter to fewer heat-related illnesses.

3.6.13 NATIONAL SCIENCE FOUNDATION (NSF)

The NSF is an independent federal agency, established in 1950, which reports to the U.S. president [63]. The agency supports basic and applied research through grants and contracts to universities and other research organizations. NSF activities related to environmental research and education involve support of basic disciplinary research, except biomedical research, which is funded by NIH; focused interdisciplinary research; and other environmentally-relevant programs. The NSF supports a broad range of educational, international, and outreach functions that span a wide spectrum of environment and natural resources scientific interests.

One of the NSF's priority areas is biocomplexity in the environment (BE), a program that promotes new approaches to investigating the interactivity between biota and the environment. As stated by the NSF, "The key connector of BE activities is complexity—the idea that research on the individual components of environmental systems provides only limited information about the behavior of the systems themselves" [64]. Grants are awarded to research institutions in response to NSF applications for grant proposals. Another NSF program of relevance to human health is the U.S. Global Change Research Program. This program supports activities that range from international collaborative field programs for collection of data critical to the development, testing, and application of improved models encompassing various geographic and temporal scales, and research on human contributions and responses to global change.

Having discussed the environmental health programs within departments and agencies of the executive branch of government, it is time to consider how federal agencies develop rules and regulations that are mandated by individual environmental statutes.

3.7 ADMINISTRATIVE PROCEDURES ACT

Before discussing the federal government's process for developing environmental health regulations, it is important to understand how the public is involved in the process. When developing regulations and standards, regulatory agencies such as EPA and OSHA must follow the terms and conditions specified under the Administrative Procedures Act of 1946, as amended. This little known federal law, which is more than a half century old, establishes specific citizen rights to access government information of relevance, to participate in government decisions affecting them, and to legal accountability for the agencies' decisions [65]. This access provides citizens with the ability to obtain copies of federal government documents, using the Freedom of Information provisions of the Administrative Procedures Act, except for documents that are exempted from distribution, such as military or security documents. Even in those circumstances, a citizen can challenge the government through litigation in a federal court.

Prior to proceeding, two definitions are in order [66]. Rule means the whole or a part of an agency statement of general or particular applicability and future effect designed to implement, interpret, or prescribe law or policy, and rulemaking means agency process for formulating, amending, or repealing a rule. More specific to the subject of regulations and standards, the Administrative Procedures Act requires that all federal agencies must comply with rulemaking procedures specified in the act, which include:

1. Make documents available to the public,
2. Conduct meetings open to the public, having publish in advance of the meeting the date and location of such meetings,

3. Accept data from the public on matters relevant to rulemaking; and
4. Publish proposed, revised, and final documents in the *Federal Register.*[10]

The Administrative Procedures Act of 1946, as amended, provides citizens and other interested parties a powerful access to government rulemaking procedures and the documents and public meetings that accompany rulemaking. Moreover, rulemaking that is not in compliance with the act can be litigated in federal courts.

3.8 FEDERAL GOVERNMENT'S REGULATORY PROGRAMS

Federal agencies such as EPA, OSHA, and FDA and at least fifty others are called regulatory agencies because they are authorized by specific federal laws enacted by Congress to develop, promulgate, and enforce rules (i.e., regulations) that carry the full force of a law. Moreover, regulations can be the subject of judicial review and decision. In other words, Congress provides the regulatory tool chest from which executive branch agencies choose the tools to use; judicial review can decide if the tool chest and the ways the tools are used are lawful. Most federal government regulations fall into one or more of the five categories shown in Table 3.3 and discussed below [66].

Process requirements—These are regulations that control emissions from sources of pollution and usually require the use of performance standards for pollution control technologies. For example, the Clean Water Act specifies that those who generate water contaminants must use the "best available technology (BAT)" that is economically feasible before releasing the contaminants into U.S. bodies of water.

Product controls—These regulations control certain commercial products, including a product's design and potential uses. For example, the Food, Drug, and Cosmetic Act authorizes the FDA to control prescription drugs and medical devices.

Notification requirements—These regulations require notification of government regulatory agencies and/or the public about a company's actions. For example, under Title III of the CERCLA, companies that release pollutants into environmental media in excess of specified limits must report the nature and quality of contaminants to EPA and the public.

Response requirements—These regulations require that a federal regulatory agency be notified when emergency conditions occur. For example, the Clean Water Act requires that emergency releases of chemical contaminants into bodies of water be

---

TABLE 3.3
Basic Forms of Federal Regulations [68]

Process requirements

Product controls

Notification requirements

Response requirements

Compensation requirements

---

reported to the EPA or the U.S. Coast Guard, depending on whether the spill occurred on land or a body of water.

Compensation requirements—These are regulations that require an individual or business entity to reimburse the federal government for actions detrimental to the public's welfare. For example, as discussed in chapter 8, the CERCLA authorizes EPA to recover costs of remediating uncontrolled hazardous waste sites from those parties responsible for creating the waste site.

According to the Office of Management and Budget (OMB) [67], an office of the White House, the history of federal regulation in the United States of certain areas of commerce and other social endeavors can be viewed as having occurred in four periods. In the first period, which preceded the mid-nineteenth century, the federal government had little to no involvement in what would be called today *regulatory authority*, other than the establishment of tariffs on goods brought into the United States and the imposing of taxes on various commercial activities. The prevailing philosophy was that commerce, in particular, should not be impeded by government control, and that states should have the primacy in determining where control (i.e., regulation) should be exerted on business, transportation, and other commercial enterprises.

In the second period of regulatory development in the United States, which could be called the *commerce and banking* period and extending through the first three decades of the twentieth century, the federal government began regulating specific commercial activities, in particular, in the areas of banking and securities exchange. The third period occurred during the late 1960s and lasted for approximately a decade. This period might be referred as the *quality of life* period of government regulation. During this period, regulations emerged that provided consumer protections, improved environmental quality, and better workplace protections.

The fourth period could be called *regulatory relief*, or deregulation. It is difficult to pinpoint when this period began, but perhaps the deregulation of the airlines during the Carter administration could be cited as the nascent beginnings of deregulation, i.e., the repeal or substantial rollback of an existing regulation. In this example, airlines in the United States, which had had their ticket costs and air routes regulated by the Federal Aviation Administration, were permitted to set their own ticket costs and participate in commercial activities that had previously required federal government review and approval. The regulatory relief period perhaps began in earnest with the election in 1980 of President Ronald Reagan, whose campaign had stressed the need to lessen the "burden of government rules and regulations." A brief history follows of how federal government rules and regulations have evolved in the United States. [ibid.].

3.8.1 History of Federal Regulations

According to OMB [67], the oldest federal regulatory agency still in existence is the Office of the Comptroller of the Currency, established in 1863 to charter and regulate national banks. However, federal regulation is usually dated from the creation in the late nineteenth century of the Interstate Commerce Commission (ICC), which was charged with protecting the public against excessive and discriminatory railroad rates.

The regulation was economics in nature, setting rates and regulating the provision of railroad services. The Commerce and Banking period of U.S. regulatory development began in the early twentieth century, with the creation of the Federal Trade Commission (FTC) in 1914, the Water Power Commission in 1920 (later the Federal Power Commission), the Federal Radio Commission in 1927 (later the Federal Communications Commission), the Federal Reserve Board in 1913, the Tariff Commission in 1916, the Packers and Stockyards Administration in 1916, the Commodities Exchange Authority in 1922 and Food and Drug Administration in 1931.

The Quality-of-Life regulatory period began in the late 1960s with the enactment of comprehensive, detailed legislation intended to protect the consumer, improve environmental quality, enhance workplace safety, and assure adequate energy supplies.

The Franklin D. Roosevelt administration was actively engaged in creating a variety of new regulatory programs [ibid.]. Some of Roosevelt's New Deal economic regulatory programs were implemented by the Federal Home Loan Bank Board created in 1932, the Federal Deposit Insurance Corporation (FDIC) created in 1933, the Commodity Credit Corporation created in 1933, the Farm Credit Administration created in 1933, the Securities and Exchange Commission (SEC) created in 1934, and the National Labor Relations Board created in 1935. In addition, the jurisdictions of both the Federal Communications Commission (FCC) and the Interstate Commerce Commission were expanded to regulate other forms of communications (e.g., telephone and telegraph) and other forms of transport (e.g., trucking). In 1938, the role of the Food and Drug Administration (FDA) was expanded to include prevention of harm to consumers in addition to corrective action. The New Deal also called for the establishment of an agency to enforce the Fair Labor Standards Act of 1938 by the Department of Labor, which is now called the Employment Standards Administration.

The Quality-of-Life regulatory period began in the late 1960s with the enactment of comprehensive, detailed legislation intended to protect the consumer, improve environmental quality, enhance workplace safety, and assure adequate energy supplies. In contrast to the pattern of economic regulation adopted before and during the New Deal, new social regulatory programs tended to cross many sectors of the economy (rather than individual industries) and affect industrial processes, product designs, and by-products (rather than entry, investment, and pricing decisions). The consumer protection movement of that era led to creation of several agencies designed to improve transportation safety. They included the Federal Highway Administration (created in 1966), which sets highway and heavy truck safety standards; the Federal Railroad Administration (1966), which sets rail safety standards; and the National Highway Traffic Safety Administration (1970), which sets safety standards for automobiles and light trucks. Regulations were also authorized pursuant to the Truth in Lending Act, the Equal Credit Opportunity Act, the Consumer Leasing Act, and the Fair Debt Collection Practices Act. The National Credit Union Administration (1970) and the Consumer Product Safety Commission (1972) were also created to protect consumer interests [ibid.].

In 1970, EPA was created as part of an Executive Branch reorganization plan by President Richard M. Nixon in order to consolidate and expand environmental programs. This effort to improve environmental protection also led to the creation of the Materials Transportation Board (created in 1975) (now part of the Research and Special Programs Administration in the Department of Transportation) and the Office of Surface Mining Reclamation and Enforcement (1977) in the Department of the Interior.

OSHA (created in 1970) was established in the Department of Labor (DOL) to enhance workplace safety. Major mine safety and health legislation had been passed in 1969, following prior statutes reaching back to 1910. The Pension Benefit Guaranty Corporation and the Pension and Welfare Administration were established in 1974 to administer and regulate pension plan insurance systems [ibid.].

Also in the 1970s, the federal government attempted to address the problems of the dwindling supply and the rising costs of energy. In 1973, the Federal Energy Administration (FEA) was directed to manage short-term fuel shortage. Less than a year later, the Atomic Energy Commission was divided into the Energy Research and Development Administration (ERDA) and an independent Nuclear Regulatory Commission (NRC). In 1977, the FEA, ERDA, the Federal Power Commission, and a number of other energy program responsibilities were merged into the Department of Energy (DOE) and the independent Federal Energy Regulatory Commission (FERC).

Another significant regulatory agency, the Department of Agriculture (USDA), has grown over time so that it now regulates the price, production, import, and export of agricultural crops; the safety of meat, poultry, and certain other food products; now including the regulatory work of the U.S. Forest Service (created in 1905), the Natural Resources Conservation Service (1935), the Farm Service Agency (1961), the Food and Consumer Service (1969), the Agricultural Marketing Service (1972), the Federal Grain Inspection Service (1976), the Animal and Plant Health Inspection Service (1977), the Foreign Agricultural Service (1974), the Food Safety and Inspection Service (1981), and the Rural Development Administration (1990) [ibid.].

As stated by OMB, "The consequence of the long history of regulatory activities is that federal regulations now affect virtually all individuals, businesses, State, local, and tribal governments, and other organizations in virtually every aspect of their lives or operations. It bears emphasis that regulations themselves are authorized by and derived from law. No regulation is valid unless the Department or agency is authorized by Congress to take the action in question. In virtually all instances, regulations either interpret or implement statutes enacted by Congress. Some regulations are based on old statutes; others on relatively new ones. Some regulations are critically important (such as the safety criteria for airlines or nuclear power plants); some are relatively trivial (such as setting the times that a draw bridge may be raised or lowered). But each has the force and effect of law and each must be taken seriously" [ibid.].

3.8.2 FEDERAL GOVERNMENT'S RULEMAKING

The federal government's process of making regulations (i.e,, rulemaking) proceeds along the following course.

1. Congress must enact legislation, signed into law by the president (or by way of overturning a presidential veto), that requires a federal agency to develop and promulgate rules (i.e., regulations) specific to a particular congressional concern. Such legislation is called *authorizing legislation* or *enabling legislation.* For example, the Clean Air Act requires EPA to issue regulations for controlling levels of outdoor, ambient air pollutants. Congressional legislation often results from pressure that individuals or special interest groups place upon members of Congress. The formal proceedings of Congress, including bills (i.e., draft legislation being considered by the House and/or Senate) that have been proposed or enacted by Congress, are published in the *Congressional Record,* the official journal of all actions taken by Congress.

2. The designated executive branch agency develops proposed regulations, following the requirements of the Administrative Procedures Act. In general, the agency must publish in the *Federal Register,* the federal government's journal of daily announcements, its intent to issue regulations in compliance with a specific statute. Meetings held during the rulemaking process must generally be open to the public. The public has the opportunity to submit data to the agency during the rulemaking process. Draft and final rules must be published in the *Federal Register* and on the agency's Web site.

3. Final rules take effect upon publication in the *Federal Register,* but often provide a time schedule for compliance by the regulated community. For example, EPA's regulations on underground storage tanks gave owners ten years to inspect, repair, or replace their tanks. Final regulations are incorporated into the *Code of Federal Regulations,* which can be accessed through the Internet. Failure to be in compliance with a regulation can result in penalties that are specified in the enabling legislation.

3.9 SUMMARY

This chapter has described the relationship between public health organizations and environmental protection agencies. In particular, the roles of the PHS and the EPA are discussed. The PHS, dating from the Public Health Service Act of 1912, was the U.S. government agency first tasked by Congress to investigate health hazards in water, air, food, and sanitation. During most of the first half of the twentieth century, states had primacy in controlling pollution and other environmental hazards within their borders. The PHS provided states with advice, services, and guidelines for controlling levels of contaminants in outdoor air, water supplies, and municipal waste. By the late 1960s, critics of the PHS approach of laissez-faire response to environmental hazards had begun to advocate for a regulatory approach to hazard control.

In 1970, EPA was established by the Nixon administration, in effect, supplanting the PHS as the federal government's primary agency for protection of the environment and public health. This created a dichotomy of federal and state environmental protection agencies (with public health ethos) and public health agencies (with environmental hazards of interest). This awkward dichotomy has been replicated as policy by most

U.S. states and some countries. The awkwardness derives from environmental regulatory agencies that have public health responsibilities, but without the requisite public health resources and experience. On the other side, some public health agencies have environmental health responsibilities, but without the necessary resources in environmental science and experience. This kind of dichotomy can produce uncoordinated programs of environmental research and services to the public.

As described in this chapter, in addition to the PHS and the EPA, many other federal departments have environmental programs that impact segments of the U.S. public. These programs are agency specific and generally respond to statutory requirements such as hazardous waste management. Because federal statutes on control of environmental hazards are generally based on the development of environmental standards and regulation of environmental hazards, the chapter concluded with a discussion of the federal rulemaking process that must be followed when federal agencies develop regulations.

This and the preceding chapter have discussed how humankind has come to grips with the nature of environmental hazards, the evolution of environmental health, and the emergence of government as the key player in the control of environmental hazards and the process of policy-making. Given this background, the next chapter will begin a discussion of basic federal environmental statutes.

3.10 POLICY QUESTIONS

1. Assume that the traditional public health approach toward preventing disease and disability is through science (i.e., problem identification), consensus-formation (i.e., problem resolution), and services (to affected organizations and at-risk populations). Using this paradigm, discuss the current roles of any two agencies of the Department of Health and Human Services in preventing adverse health effects of environmental hazards.

2. Discuss bioterrorism as a modern-day environmental health problem in the United States. Be specific as to the environmental dimensions of the threat. Using Internet resources, describe the role of local and state health departments in protecting against bioterrorism events.

3. Given the discussion about the historical role of the surgeon general, discuss whether or not the surgeon general's former primacy in public health leadership, including environmental hazards, should be restored.

4. Select any of the cabinet-rank departments of the executive branch of the federal government (Figure 3.1) and describe its key environmental programs and environmental health policies.

5. The Occupational Safety and Health Act of 1970 created two new agencies within the federal government. Discuss them and their differences in purpose.

6. Why did the U.S. Public Health Service (PHS) lose it primacy in leading the nation's environmental health programs? What could the surgeon general have done to preserve PHS primacy?

7. EPA's first Administrator, William Ruckelshaus, made enforcement of EPA's federal regulations as his top priority. Assume that he had made science and

services his first priority. Speculate on how EPA might have evolved as a federal regulatory agency.

8. Referring to Figure 3.1, select any one of the three branches of the U.S. government and discuss some of its policy-making implications for environmental health policy.

9. In regard to environmental health, discuss the key differences between statutory law and administrative law.

10. Why do governments care about the public's health?

NOTES

1. As described in chapter 8, the Pollution Prevention Act of 1990 requires companies and businesses to also report data on recycling of wastes and provide other information on pollution prevention plans and actions.

2. As of 2006, the DHHS agencies comprise the Administration for Children and Families, Administration on Aging, Agency for Healthcare Research and Quality, Agency for Toxic Substances and Disease Registry, Centers for Disease Control and Prevention, Centers for Medicare and Medicaid Services, Food and Drug Administration, Health Resources and Services Administration, Indian Health Service, National Institutes of Health, Program Support Center, and Substance Abuse and Mental Health Services Administration [14].

3. *Bully pulpit*: a prominent public position (as a political office) that provides an opportunity for expounding one's views [15].

4. It is common practice to denote elected officials by political party and state. Kennedy (D-MA) signifies Democrat from Massachusetts and Nixon (R-CA) denotes Republican from California.

5. California, Colorado, Connecticut, Georgia, New York, Maryland, Minnesota, Oregon, Tennessee and New Mexico

6. The author is indebted to Dr. Morris Cranmer, first director of NCTR, for materials and insights on the establishment of NCTR.

7. The National Research Council is a component of the National Academies. Other components are the National Academy of Sciences, the National Academy of Engineering, and the Institute of Medicine. The National Academies is located in Washington, D.C.

8. The author is indebted to Ronald Coene, FDA, and former Executive Secretary of the CCEHRP, for supplying much of the information about the CCEHRP and its antecedent committees.

9. The Homeland Security Act of 2002 transferred the U.S. Coast Guard from the Department of Transportation to the Department of Homeland Security.

10. The *Federal Register* is the executive branch's newsletter. It contains announcements of federal agencies activities, such as public meetings. It is available on the Internet.

REFERENCES

1. Grolier, *Multimedia Encyclopedia*, Grolier Interactive, Danbury, CT, 1998.
2. Wikipedia, History of the Common Law. Available at http://en.wikipedia.org/wiki/Common_law, 2004.

3. LII (Legal Information Institute), Constitutional Law: An Overview. Available at http://wwwsecure.law.cornell.edu/topics/constitutional.html, 2004.

4. Essortment, What is the Statutory Law? Available at http://ar.essortment.com/whatisstatutor_rita.htm, 2004.

5. LII (Legal Information Institute), Administrative Agencies: An Overview. Available at http://wwwsecure.law.cornell.edu/topics/administrative.html, 2004.

6. Klees, A.A., The interface of environmental regulation and public health, *Occupational Medicine*, 11, 173, 1996.

7. CRS (Congressional Research Service), Summaries of Environmental Laws Administered by the Environmental Protection Agency, The Library of Congress, Washington, D.C., 1991.

8. NIH (National Institutes of Health), Images from the history of the Public Health Service, National Library of Medicine, Bethesda, MD, 2003.

9. Mullan, F., *Plagues and Politics*, Basic Books, New York, 1989, 20, 142.

10. Silberstein, I.H., Cincinnati Then and Now, The Voters Service Education Fund of The League of Women Voters of the Cincinnati Area, Cincinnati, OH, 1982, 29.

11. Snyder, L.P., A new mandate for public health—50th anniversary of the Public Health Service Act, *Public Health Reports*, July-August, 1994.

12. Patterson, J.T., The American Presidency: New Deal. Encyclopedia Americana. Available at http://gi.grolier.com/presidents/ea/side/newdeal.html, 2000.

13. DeLong, J.B., 1997. Slouching Towards Utopia?: The Economic History of the Twentieth Century. XIV. The Great Crash and the Great Slump. Available at http://econ161.berkeley.edu/TCEH/Slouch_Crash14.html, 1997.

14. DHHS (U.S. Department of Health and Human Services), Agencies and offices. Available at http://www.dhhs.gov>, 2004.

15. Merriam-Webster,. *Merriam-Webster Online*. Available at http://www.m-w.com/cgi-bin/mwwod.pl, 2001.

16. Committee on Environmental Health Problems, Committee Report to Surgeon General Luther Terry, U.S. Department of Health, Education, and Welfare, Public Health Service, Washington, D.C., 1961.

17. Freelao, Paris Peace Accord. Available at http://freelao.tripod.com/id87.htm, 2004.

18. Landy, M.K., Roberts, M.J., and Thomas, S.R., *The Environmental Protection Agency: Asking the Wrong Questions from Nixon to Clinton*, Oxford University Press, New York, 1994.

19. EPA (U.S. Environmental Protection Agency), Homepage. Available at http://www.epa.gov, 2002.

20. Johnson, B.L., *Impact of Hazardous Waste on Human Health*, CRC Press, Lewis Publishers, Boca Raton, FL, 1999.

21. Moore, C. 1986. Personal communication.

22. CDC (Centers for Disease Control and Prevention), Homepage. Available at http://www.cdc.gov, 2001.

23. NCEH (National Center for Environmental Health). Available at http://www.cdc.gov/nceh/history/history.htm, 2001.

24. CDC (Centers for Disease Control and Prevention), National Center for Infectious Diseases. Available at http://www.CDC.gov, 2004.

25. Patton, D. and Huggett, R., The risk assessment paradigm as a blueprint for environmental research, *Human and Ecological Risk Assessment*, 9, 1337, 2003.

26. FDA (Food and Drug Administration), History of the FDA. Available at http://www.fda.gov/oc/history/historyoffda/default.htm, 2002.

27. FDA (Food and Drug Administration), Laws enforced by the FDA & related statutes. Available at http://www.fda.gov/opacom/laws/lawtoc.htm, 2002.
28. Van Steenwyk, R. and Zalom, F., Editorial, *California Agriculture*, Jan-Mar, 2005.
29. DHEW (Department of Health, Education and Welfare), Report of the Secretary's Commission on Pesticides and Their Relationship to Environmental Health,. Superintendent of Documents, U.S. Government Printing Office, Washington, D.C., 1969.
30. Carson, R., *Silent Spring*, Houghton Mifflin Co., New York, 1962.
31. FDA (Food and Drug Administration), NCTR's mission, Center for Food Safety and Applied Nutrition, Washington, D.C. Available at http://www.fda.gov/nctr/overview/mission.htm, 2005.
32. IHS (Indian Health Service), Programs of the Indian Health Service. Available at http://www.ihs.gov, 2004.
33. NIH (National Institutes of Health), A short history of the National Institutes of Health. Available at www.nih.gov/od/museum/exhibits/history/main.html, 2002.
34. NIH (National Institutes of Health), Mission statement. Available at http://www.nih.gov/about/almanac/index.html, 2005.
35. NIH (National Institutes of Health), About NIH: Fast facts. Available at http://www.NIH.gov/about/, 2004.
36. NCI (National Cancer Institute), The National Cancer Institute Act of 1937. Available at http://www3.cancer.gov/legis/1937canc.html, 2005.
37. NCI (National Cancer Institute), The National Cancer Act of 1971. Available at http://www3.cancer.gov/legis/1971canc.html, 2005.
38. NCI (National Cancer Institute), Mission. Available at http://www.nih.gov/about/almanac/organization/NCI.htm, 2005.
39. NCI (National Cancer Institute), NCI research programs. Available at http://www.nih.gov/about/almanac/organization/nci.htm#programs, 2005.
40. NIEHS (National Institute of Environmental Health Sciences), Brief history of NIEHS. Available at http://www.niehs.nih.gov/external/history.html, 2000.
41. EHPC (Environmental Health Policy Committee), U.S. Department of Health and Human Services, Washington, D.C., 2005.
42. EPA (U.S. Environmental Protection Agency), Introduction to the HRS, Office of Solid Waste, Washington, D.C. Available at http://www.epa.gov/superfund/programs/npl_hrs/hrsint.htm, 2002.
43. USDA (U.S. Department of Agriculture), History. Available at http://www.fsis.usda.gov/About_FSIS/Agency_History/index.asp, 2005.
44. USDA (U.S. Department of Agriculture), Quick Facts about PDP. Pesticide Data Program. Available at http://www.ams.usda.gov/science/pdp/quick.htm, 2002.
45. USDA (U.S. Department of Agriculture), Quick Facts about PDP. Pesticide Data Program. Available at http://www.ams.usda.gov/science/pdp/quick.htm, 2002.
46. Colborn, T., Dumanoski, D., and Myers, J.P., *Our Stolen Future: Are We Threatening Our Fertility, Intelligence, and Survival? A Scientific Detective Story*, Penguin Books USA, New York, 1997.
47. USDA (U.S. Department of Agriculture), Food Safety Research. Available at http://www.reeusda.gov/pas/programs/foodsafetyresearch/index.htm, 2002.
48. DoC (U.S. Department of Commerce), Department of Commerce—Milestones. Available at http://www.commerce.gov/milestones.html, 2005.
49. NOAA (National Oceanographic and Atmospheric Administration), A history of NOAA. Available at http://www.history.noaa.gov/legacy/noaahistory_13.html, 2004.
50. Goodman, S.W., Environmental issues: A top priority for Defense leadership, *Defense*

Issues, vol 13, no. 20–Environmental Issues, U.S. Department of Defense, Washington, D.C., 1998.

51. DOE (U.S. Department of Energy), Energy and Environmental Quality. Available at http://www.energy.gov/environ/index.html, 2002.

52. NIGEC (National Institute for Global Environmental Change), About NIGEC. Available at http://nigec.ucdavis.edu/about/content.htm, 2002.

53. DOE (U.S. Department of Energy), Energy and Environmental Quality. Available at http://www.energy.gov/environ/index.html, 2002.

54. DHS (U.S. Department of Homeland Security). Available at http://www.whitehouse.gov/deptofhomeland/ 2003.

55. HUD (U.S. Department of Housing and Urban Development), Homes and Communities. Available at http://www.hud.gov, 2002.

56. USGS (U.S. Geological Survey), Environmental and Human Health. Available at http://www.usgs.gov/themes/environment_human_health.html, 2002.

57. OSHA (Occupational Safety and Health Administration), OSHA's 30th Anniversary. Available at http://www.osha.gov/as/opa/osha-at-30.html, 2000.

58. MSHA (Mine Safety and Health Administration), History of Mine Safety and Health Legislation. Available at http://www.msha.gov/MSHAINFO/MSHAINF2.htm, 2005.

59. DoS (U.S. Department of State), State Department History. Available at http://www.state.gov/r/pa/ho/faq/#department, 2005.

60. USAID (U.S. Agency for International Development), Our Work: A Better Future for All, Washington, D.C., 2005.

61. OHM (Office of Hazardous Materials), Functions of the Office of Hazardous Materials. Available at http://hazmat.dot.gov/ohhms_fn.htm, 2002.

62. NASA (National Aeronautics and Space Administration), History. Available at http://history.nasa.gov/brief.html, 2005.

63. NSF (National Science Foundation), About the National Science Foundation. Available at http://www.nsf.gov/about/, 2005.

64. NSF (National Science Foundation), Environmental Research and Education. Available at http://www.nsf.gov/home/crssprgm/be, 2002.

65. Andrews, R.N.L., *Managing the Environment, Managing Ourselves*, Yale University Press, New Haven, 1999, 66.

66. LII (Legal Information Institute), Subchapter II—Administrative Procedure. Available at http://www.law.cornell.edu/uscode/htm/uscode05/usc_sup_01_5_10_1_30_5_40_11.html, 2005.

67. OMB (White House Office of Management and Budget), Report to Congress on the Costs and Benefits of Federal Regulations, Washington, D.C. Available at http://www.whitehouse.gov/omb/inforeg/chap1.html, 2002.

68. Chemalliance, Environmental regulations. Available at http://www.chemalliance.org/Handbook/background/back-detail.asp, 2001.

4 General Environmental Statutes

4.1 INTRODUCTION

Commencing in the 1950s, a gradual awareness arose in the U.S. public's mind that environmental hazards might be harmful to human health, an awareness that helped in the enactment of various federal and state laws to control environmental hazards. The enactment of individual laws that now constitute the main body of federal environmental law has occurred over a half century of public concern, legislative enactment, executive branch implementation, and judicial interpretation. This body of law did not occur without heated debate and impassioned feelings. In particular, environmental groups and business organizations were often at odds as how to deal with specific environmental hazards, usually leading legislative bodies like the U.S. Congress to seek legislative language that negotiated the differences between business and environmental groups.

Over the decades, environmental laws were enacted for reasons that differed across time. To be more specific, the impetus for legislation in the earlier years differed from what occurred later in the twentieth century. Four conditions that contributed to the passage of specific environmental laws are listed in Table 4.1. The individual statutes identified primarily by abbreviations will be described in subsequent sections, but examples of the operative conditions shown in Table 4.1 are presented in the following sections.

Public health tradition The earliest federal environmental legislation was drawn from the traditions of public health. This tradition comprised: conducting scientific research on specific environmental hazards and assessing the public health significance of scientific data, fostering consensus on the public health significance of scientific information, and providing services to states tasked with controlling environmental hazards. Legislation for the Clean Air Act (CAA); Clean Water Act (CWA); Safe Drinking Water Act (SDWA); Federal Insecticide, Fungicide, and Rodenticide Act (FIFRA); and Resource Conservation and Recovery Act (RCRA) was developed through application of the public health tradition. For example, consider the example of the Clean Air Act. As will be described in the following chapter, in 1955, the U.S. Congress enacted the Air Pollution Control Act, which provided the Public Health Service (PHS) with funds and authorities to research the effects of air pollutants on human health and to provide technical assistance to states' air pollution programs.

Over the next decade, Congress gradually gave the surgeon general, as head of the PHS, additional authorities and resources to further investigate the effects of air pollution on human health and to commence the development of air quality standards. An advantage of this public health tradition as a foundation for environmental legislation is the development of a body of science that in turn guides specific statutory language. As policy, this is an early legislative example of the precautionary principle (discussed

149

TABLE 4.1
Impetus for Various Federal Environmental Statutes

Impetus	Environmental Statutes
Public health tradition	CAA, CWA, SDWA, FIFRA, RCRA, FDCA, FMIA, Noise Control
Fear of catastrophic events	CERCLA, Oil Pollution Act, Medical Waste Tracking Act, Bioterrorism Preparedness Act
Opportunistic conditions	FMIA, NEPA, Title III of CERCLA (as amended), Information Quality Act
Confluence of special interests	FQPA, TSCA, OSHA

in chapter 2). A disadvantage is the time required to develop both the body of science and infrastructure for providing services, most often to state health programs as part of a public health federalism compact.

> Few concerns motivate elected officials as quickly as fear. No legislator or policy maker wants to be accused of ignoring conditions that could produce a catastrophic event.

Fear of catastrophic events   Few concerns motivate action by elected officials as quickly as fear. No legislator or other policy maker wants to be accused of ignoring conditions that could produce a catastrophic event. This is particularly true of catastrophes of anthropogenic causes. Consider the body of legislation that almost immediately followed the September 11, 2001, terrorist events that occurred in New York City, Washington, DC, and Pennsylvania. No member of Congress wanted to be accused of delaying legislation that would strengthen airport security, identification of terrorists, and improve the nation's resources in managing biological and chemical agents that could be used by terrorists.

A product of public fear can therefore be quickly enacted legislation that attempts to control a hazard before feared events occur. Perhaps the premiere example of federal environmental legislation is the CERCLA, also called the Superfund law. This law was enacted in response to the public's concerns that toxic waste was leaking into homes and environmental media (e.g., groundwater), exposing human populations to substances that could cause cancer, birth defects, and other fearful outcomes. These concerns were broadcast and amplified by news media across the U.S. landscape. Because toxic waste was viewed as insidious, silent, and a threat to children, it was feared and characterized as such, even in the absence of causal scientific research. This means of legislative operation was clearly different from the public health tradition approach, where policy was based on a persuasive body of existing scientific data. However, using the limited scientific data available, Congress, in effect, acted upon the precautionary principle, concluding that the small amount of scientific data was adequate to enact precautionary legislation.

Opportunistic conditions   Sometimes legislation occurs simply because the conditions were opportune. An example of such legislation is the Emergency Response and Community Right to Know Act of 1986, which is Title III of the CERCLA. This act established community-based requirements for emergency response teams, requiring in particular, that local responders (often fire departments) be provided information by companies and others about hazardous materials stored on their premises. Such information provides emergency responders with advance information about chemicals that could be harmful to them and to communities if released into the environment. Further, Title III established the Toxics Release Inventory (TRI), an EPA database of substances released by industrial sources into the environment. Moreover, the act stipulates that the TRI be made available to the public, thereby providing the public with information that they could use in managing their own exposure to environmental hazards.

Passage of the Emergency Response and Community Right to Know Act had failed to gain Congressional support until 1986. While the act was strongly supported by national environmental groups and local government, business interests energetically opposed the legislation, objecting in particular to reporting of chemical stocks and releases to government agencies and the public. These objections were couched as trade secret information that if released could give competitors an edge in producing commercial products. However, supporters of the act were able to overcome these objections by attaching the act to the CERCLA amendments of 1986. The public at that time remained very supportive of preventing releases from uncontrolled hazardous waste sites and emergency chemical releases were considered a similar chemical threat, albeit acute in nature. Support for the CERCLA legislation was strong within Congress, and the conditions were opportune for attaching the Emergency Response and Community Right to Know Act to the Superfund Amendments and Reauthorization Act of 1986. What wasn't previously possible became possible when the act was piggy-backed onto popular legislation that had broad-based support in 1986.

Confluence of special interests   In 1980, the election of Ronald Reagan commenced two decades of political separation between the political party of the incumbents of the White House and the political party controlling the U.S. Congress. Republican Reagan served during a period when Democrats controlled both houses of Congress, as was the situation with President George H. W. Bush, a Republican. Similarly, Democrat Bill Clinton, commencing in 1994, found himself working with a Republican-controlled Congress, a condition that continued for President George W. Bush until year 2001, when Democrats regained control of the U.S. Senate. President George W. Bush is the exception, working with a Congress controlled by the Republican Party. A political consequence of this method of "separation of power" is the necessity for negotiation and compromise if legislation is to successfully navigate the political waters. Confluence of special interests can assist in the legislative process.

The Food Quality Protection Act of 1996 is an example of legislation that resulted from the confluence of special interests. Environmental groups and children's health advocates were concerned that the country's young children were experiencing exposure to pesticides in food in amounts deleterious to their health. At the same time, producers of pesticides and related substances had long sought relief from the Delaney Clause,[1] a provision in the Food and Drug Act that required the Food and Drug Administration

(FDA) to ban any carcinogen found added to food. The Delaney Clause became, in effect, an environmental health policy of zero cancer risk from carcinogens in food. The producers' prevailing attitude was that the Delaney Clause was inflexible and out of date, given the emergence of risk assessment as the principal method for determining the risk of environmental hazards. Although the interests of pesticide producers and environmental and children's health advocates were different, there was a confluence of desire to change how pesticides were regulated by the U.S. government. This desire led to enactment, without a dissenting vote in either the House of Representatives or Senate, of the Food Quality Protection Act of 1996. As will be subsequently described, this act materially altered federal pesticide statutes, which are described in chapter 7.

4.2 REGULATIONS AND STANDARDS

Before launching into a presentation of the major federal environmental—or environmentally-relevant—statutes and policies, it is useful to repeat some material from chapter 2. In particular, the primary environmental health policy, command and control (also called regulations and standards), merits reiteration, since this policy is operative in many of the statutes to be discussed. In particular, the distinction between quality and emission standards needs elaboration.

As reminder, *quality standards* are the maximum levels of contaminants to be permitted in specific environmental media, such as air, water, and food. Quality standards are established by regulatory agencies through a process of review of scientific literature, risk estimation, and health impact. In distinction, *emission standards* prescribe the amount of contaminant discharges that can be released from significant emission sources, such as industrial facilities, municipal discharges into water, and landfills. It is important to understand the meanings of both kinds of standards and their essential interrelation.

Emissions standards and quality standards are essential partners for controlling environmental quality. Consider the simple act of showering. Assume that the temperature of the water is the quality factor to be controlled. Some persons prefer a tepid temperature; others find a hot shower to be preferable. The quality of the showering event can be controlled through setting of the water's temperature. A comfortable water temperature is therefore the quality standard. The quality standard in this example is based upon a set of data (i.e., previous showering) that was evaluated by a knowledgeable expert (i.e., the person showering). Water coming from the showerhead can be considered as the emission source. Its volume and temperature can be controlled to achieve the quality standard. As this example suggests, a quality standard is only a goal if an emissions standard is lacking.

While the showering example may be a simple illustration of the relationship between quality and emission standards, application of these standards to large geographic areas is very complex and challenging. Consider the problem posed by a regional airshed (i.e., a geographic area such as a city and its outlying metropolitan areas). Assuming that quality standards have been established for individual air pollutants (e.g., particulate matter), how can emission standards be established to control sources of pollution within the airshed? Setting emission standards to apply to an airshed requires authorities, normally state agencies acting under federal law (e.g., the Clean Air Act), to identify sources of air pollution, derive emission standards for polluting sources, and work through a regulatory apparatus to implement the emission standards. The process is complicated by uncertainties in characterizing sources of pollution and computer modeling of airshed pollutants. Moreover, producers of pollution will resist those emission standards they consider as too costly. Protracted litigation is often the result.

4.3   ENFORCEMENT AND PENALTIES

Federal statutes that require federal agencies to develop regulations and standards and control of environmental hazards in general also contain enforcement authorities. The enforcement authorities are normally accompanied by penalties for failure to comply with specific statutory requirements. For example, under the Clean Air Act, §113, EPA is authorized, in ascending order of severity, to: (1) issue an administrative penalty order, (2) issue an order of compliance, (3) bring a civil action, or (4) request the U.S. attorney general to commence a criminal action. Examples of EPA enforcement actions and penalties are found in chapters 5 through 8. The examples are given for the purpose of showing the seriousness of federal enforcement authorities and illustrate the civil (monetary penalties) and criminal (felony actions) consequences of failure to meet statutory directives.

Having discussed how federal environmental legislation has evolved according to the conditions of the times and a brief discussion of regulations and standards, the following sections will describe some of the federal statutes that contain broad environmental health purposes and policies, together with statutes that are important for understanding larger policy issues such as information quality. Subsequent chapters will describe the key federal environmental statutes, organized into air and water, food safety, pesticides and toxics, and wastes. Each statute will be described in terms of its history, key provisions pertinent to public health, and public health impacts.

4.4   CONSUMER PRODUCT SAFETY ACT

4.4.1   History

Of the four major federal regulatory agencies, the smallest in resources and perhaps least visible to the public is the Consumer Product Safety Commission (CPSC), headquartered in Bethesda, Maryland, and with regional offices in New York City, Chicago, and Oakland, California. Other regulatory agencies (e.g., EPA, FDA) are arguably better known to the U.S. public, with broader environmental responsibilities. However, the CPSC fills an important public health role, protection against hazardous consumer products.

The CPSC was created by Congress under the Consumer Product Safety Act of 1972 (CPSA). The legislation occurred in the early 1970s, a period of energetic congressional and White House environmental activism, although reasons for the activism differed between the two major political parties. Republicans, particularly President Nixon, were interested in expanding their political base by attracting suburban voters

---

The Consumer Product Safety Act's purpose is to protect the public "against unreasonable risks of injuries associated with consumer products." [1]

---

interested in environmental protection. Democrats supported environmental legislation as being consistent with their view of the use of federal government to advance social programs. The act directs the CPSC to protect the public "against unreasonable risks of injuries associated with consumer products" [1]. The CPSC is an independent federal agency, not part of any federal department or agency. Three commissioners head the Commission. They serve seven-year terms, appointed by the President and confirmed by the U.S. Senate. The Chairman of the Commission serves as the principal executive officer of the Commission. In 2002, the CPSC had a staff of approximately 480 persons, with a budget request in 2003 of $56 million [2]. Compared to its public health responsibilities, CPSC's resources are modest, to say the least.

The purposes of the act are stated to be [3]:

1. To protect the public against unreasonable risks of injury associated with consumer products,
2. To assist consumers in evaluating the comparative safety of consumer products,
3. To develop uniform safety standards for consumer products and to minimize conflicting state and local regulations, and
4. To promote research and investigation into the causes and prevention of product-related deaths, illnesses, and injuries.

Of these four purposes, the first and third are particularly germane to prevention of injuries and deaths from consumer products, as discussed below.

4.4.2   KEY PROVISIONS OF THE CPSA RELEVANT TO PUBLIC HEALTH

While the CPSC has statutory authority to develop safety standards for products under their jurisdiction, voluntary standards are mandated by the act, which states, "The Commission shall rely upon voluntary consumer product standards rather than promulgate a consumer product safety standard prescribing requirements [w]henever compliance with such voluntary standards would eliminate or adequately reduce the risk of injury addressed and it is likely that there will be substantial compliance with such voluntary standards" [1]. According to the commission, since 1990, their cooperative work with industry has resulted in 214 voluntary standards. During the same period, 35 mandatory rules were issued [2]. By law, the CPSC can issue a mandatory standard only when a voluntary standard has been determined by the commission not to have eliminated or adequately reduced the risk of injury or death or if a voluntary standard is unlikely to have substantial compliance. In 2002, the commission issued 950 corrective actions that included 387 recalls involving about fifty million consumer product units that either violated mandatory standards or presented a substantial risk of injury to the public [ibid.]. This is an impressive body of public health accomplishment, particularly in view of the modest amount of federal resources made available to the commission.

4.4.3   PUBLIC HEALTH IMPLICATIONS OF THE CPSA

Consumer products can injure some of those who use them. According to CPSC data, product-related deaths and injuries in the United States average 23,900 deaths and 32.7 million injuries annually [2]. This is a heavy public health burden on the U.S. public, a burden not reflected in the commission's authorities and resources. The deaths, injuries, and property damage associated with consumer products cost the nation more than $700 billion annually. These kinds of injuries are investigated by the CPSC when the weight of evidence supports a follow-up. The commission has jurisdiction over more than 15,000 kinds of consumer products used in and around the home, in sports, recreation, and schools. However, the CPSC has no jurisdiction over many of the most hazardous consumer products. The excluded products include automobiles, other on-road vehicles, aircraft, tires, boats, alcohol, tobacco, firearms, food, drugs, cosmetics, pesticides, and medical devices [ibid.]. While some of these products are covered by other federal statutes (e.g., Food, Drug and Cosmetic Act, Toxic Substances Control Act), many escape regulatory coverage (e.g., firearms, tobacco) due to pressure brought on Congress by special interest groups. In a public health context, exclusion of hazardous products like tobacco from CPSC jurisdiction is not good environmental health policy.

The commission uses a variety of tools to reduce the risk of hazardous consumer products. "[T]he tools include: (1) developing and strengthening voluntary and mandatory safety standards, (2) initiating recalls and corrective actions of hazardous products and enforcing existing regulations, and (3) alerting the public to safety hazards and safe practices" [2]. The CPSC has authority to direct recall of specific products that have been found to present an "unreasonable hazard" to consumers. Under recall manufacturers are required to remove the designated products from commerce until revisions are made to remove the product's hazardous features. The act also requires the Commission to maintain an Injury Information Clearinghouse to "[c]ollect, investigate, analyze, and disseminate injury data, and information, relating to the causes and prevention of death, injury, and illness associated with consumer products" [1]. The CPSC's recalls are notices available to the public and are archived on their Web site (www.cpsc.gov).

The CPSC relies heavily on injury data submitted by consumers. This leads to post hoc investigations and regulatory actions such as recall of products. While such hazard elimination or reduction can serve as prevention actions, they are triggered only after injuries have already occurred.

As a matter of policy, the CPSA encourages the development of voluntary standards through industry and CPSC cooperation. This approach can expedite the development of voluntary standards, hastening the introduction of actions that can prevent injury or death caused by consumer products. However, implementing voluntary standards is largely a matter of industry responsibility. Although the commission has statutory authority to convert a voluntary standard into a mandatory standard if compliance with a voluntary standard is inadequate, mandatory standards are subject to the same challenges and litigation that often delay regulatory actions undertaken by other federal regulatory agencies. Moreover, as a relatively small agency, the CPSC lacks the resources to aggressively pursue voluntary or mandatory standards' implementation by industry.

4.5   ENVIRONMENTAL RESEARCH, DEVELOPMENT, AND DEMONSTRATION AUTHORIZATION ACT

4.5.1   HISTORY

Many federal statutes that pertain to environmental protection and concern for effects of environmental hazards on public health contain provisions for research and development. The intent of Congress has been to encourage research that advances scientific knowledge about specific environmental hazards and their consequences. According to an analysis by the Congressional Research Service, EPA's statutory mandate for research and development grew piecemeal from parts of many environmental protection laws, enacted and amended over the years, involving at least 12 separate federal environmental statutes [4].

The Environmental Research, Development, and Demonstration Authorization Act authorizes all EPA research and development programs.

The Environmental Research, Development, and Demonstration Authorization Act of 1976 (ERDDAA) coalesced EPA's research and development programs under one authorization statute. The ERDDAA was enacted annually through 1980, ending in 1981 when Congress did not enact an authorization for fiscal year 1982. The lack of current authorization means that, in the House, bills appropriating funds for those programs are potentially open to objections because they do not comply with the rule that money cannot be appropriated without prior authorization. This problem has not been unique to the ERDDAA. During the 1980s, authorization for appropriations for many of EPA's programs expired for a time. In the absence of the ERDDAA, EPA's current and continuing authority for research and development derives from the combination of authorization provisions in basic environmental protection statutes, requirements and precedents established by the laws that authorized appropriations for EPA's overall R&D program annually, and annual (unauthorized) appropriations for EPA [ibid.].

Although not covered by the ERDDAA, several other federal departments conduct research and development on environmental hazards and conditions. This includes the Department of Health and Human Services (DHHS), Department of Energy, Department of Defense, and Department of Interior. These departments' appropriations are authorized through various authorizing statutes that are specific to each department.

Of policy note, amendments in 1978 to the ERDDAA established the EPA Science Advisory Board (SAB) [5], an independent (i.e., nongovernment) committee of scientists and other specialists. The SAB provides advice to EPA on matters of science that include reviews of EPA draft criteria documents, risk assessment methods, emerging environmental problems, international environmental hazards, and risk communication issues. The SAB's work has influenced EPA's environmental policies that rely on science-based risk assessment and judgment.

As policy, many government agencies have formed advisory committees to advise them on matters of science and other issues. Federal government agencies must construct their advisory committees (e.g., EPA's SAB) in compliance with the Administrative Procedures Act (chapter 3), which requires that committee meetings must be announced in advance and held as public meetings. (Committees, however, can convene in executive session, which is not open to the public.) Agencies are encouraged to create advisory committees that contain racial, gender, and geographic diversity. Committee members from academic institutions often constitute the majority of advisory committees, owing to their perceived objectivity on matters of science.

Federal agencies that choose not to form an advisory committee for a specific issue of science or public health practice often turn to the National Academies for advice. The academies comprise the Institute of Medicine, National Academy of Sciences, and the National Academy of Engineering. The academies can create committees to advise federal agencies on matters of medicine, engineering, and science. Federal agencies can (e.g., EPA) commission the academies to create a committee to address a particular matter of interest. As an example, the National Academy of Sciences has often been asked to advise federal agencies on such subjects as risk assessment, toxicology, and epidemiology.

Agencies are not obligated to accept the advice proffered by advisory committees, since it is the agency, not the committee, that is legally accountable under a particular federal or state law. However, agencies that ignore the advice from advisory committees run the risk of criticism from the lay public, special interest groups, and science organizations, depending on the science issue at hand. It is good environmental health policy for government agencies to accept advice and recommendations from their advisory committees, and document to the public why any specific recommendations were not adopted.

4.5.2 PUBLIC HEALTH IMPLICATIONS OF THE ERDDAA

Understanding the effects of environmental hazards on human and ecological health has been, and continues to be, a matter of "catch-up." Unlike the considerable body of human health data that clearly, and unequivocally, causally links specific infectious agents (e.g., HIV) with corresponding diseases (e.g., AIDS), research on the effects of specific environmental stressors (e.g., climate change) with potential health consequences (e.g., famine) is often equivocal and fraught with uncertainties. Some of the complicating factors in ascertaining associations between the environment and human health impacts include: uncertain exposure regimens, nonhomogeneous study populations, missing basic mechanisms of toxicity or other biological action, poorly established records of environmental contaminants, and reliance on risk assessment methods that can be blunt instruments when characterizing human risk.

Notwithstanding the previous pessimistic comments, a considerable body of research now exists on the effects of environmental hazards and their remediation or control. This can be attributed, in part, to the ERDDAA, which was enacted as an answer on how to authorize EPA's need to conduct research on environmental hazards and protection of the environment. This was a statement from Congress that environmental research was important and would be pursued by federal agencies. EPA and many other federal agencies currently sponsor or conduct serious and sometimes ambitious programs of basic and applied research on the effects of the environment on humans and ecological systems. Although the ERDDAA has not been reauthorized since 1981, it remains as a reminder that environmental research programs must accompany programs of environmental protection and public health.

4.6   INFORMATION QUALITY ACT

4.6.1   History

The federal statutes discussed in this chapter are, with the exception of what is called the Information Quality Act of 2001, the products of congressional hearings, public debates, and congressional authorizing committees' action. This process makes public the business of making legislation. For environmental legislation, the outcome is a statute that authorizes federal agencies to take regulatory or other actions on a specific environmental problem (e.g., air pollution). Authorizing legislation such as the Clean Air Act contains language that permits appropriating funds from the U.S. Treasury in support of agencies' programs. Without congressional appropriations that are specific to an authorizing statute, federal agencies can take no action. Each year Congress must enact appropriations bills for each federal department and agency.

> The Information Quality Act's purpose is to enhance the quality of information used by federal agencies in science-based decision-making.

Appropriations bills have sometimes contained language that some argue is actually authorizing language. Such language is called a *rider*, since it rides along with a bill that has a primary purpose different from the rider. Appropriations bills are a favorite vehicle for members of Congress to place riders, given that such bills are seldom vetoed by the president, since that could lead to federal agencies going without funds, in effect, a shut down of government. Adding a rider to a bill typically requires only the assent of the chairperson of the responsible congressional committee. Hearings and public debate do not usually occur, thereby denying the visibility and democratic embrace that accompany authorizing legislation.

In 2000, Congress enacted legislation that addressed the alleged problem of inadequate information quality used in federal regulations and other policy decisions. Although the Information Quality Act is not an environmental law, its implications for regulatory agencies such as EPA are profound, which is why the act is included in this chapter. As described by Weiss [6], the Information Quality Act[2] was the brainchild of a Washington, D.C., corporate lobbyist working for the Center for Regulatory Effectiveness, who had worked in the White House's Office of Management and Budget (OMB) during the Reagan administration. The lobbyist drafted what became the Information Quality Act and gained the support of a congresswoman who agreed to insert the draft into a massive appropriations bill. The congresswoman was a former lobbyist and former director of communications for the National Republican Congressional Committee. No hearings by Congress occurred to discuss the act, nor was the public informed of the act and its implications. As policy, this was the legislative process at its worst. It is an example of stealth legislation.

Further background information on the evolution of the Information Quality Act was provided by Baba and colleagues. [7], who discovered the role played by the tobacco giant Philip Morris. The company, which was concerned about EPA's risk assessment of the carcinogenicity of secondhand tobacco smoke, retained the Center for Regulatory Effectiveness, a Washington, D.C., advocacy firm, to draft language that eventually became the backbone of the act. This link between a major cigarette company and its promotion of government policies to frustrate scientific rigor in public health standards was discovered in court documents in litigation brought against the tobacco industry.

The Information Quality Act of 2001 was quietly enacted as twenty-seven lines in the Treasury and General Appropriations Act for Fiscal Year 2001. Section 515 of the act states the following:

SEC. 515. (a) IN GENERAL—The Director of the Office of Management and Budget shall, by not later than September 30, 2001, and with public and Federal agency involvement, issue guidelines under sections 3504(d)(1) and 3516 of title 44, United States Code, that provide policy and procedural guidance to Federal agencies for ensuring and maximizing the quality, objectivity, utility, and integrity of information (including statistical information) disseminated by Federal agencies in fulfillment of the purposes and provisions of chapter 35 of title 44, United States Code, commonly referred to as the Paperwork Reduction Act.

(b) CONTENT OF GUIDELINES- The guidelines under subsection (a) shall—
  (1) apply to the sharing by Federal agencies of, and access to, information disseminated by Federal agencies; and
  (2) require that each Federal agency to which the guidelines apply—
    [A] issue guidelines ensuring and maximizing the quality, objectivity, utility, and integrity of information (including statistical information) disseminated by the agency, by not later than 1 year after the date of issuance of the guidelines under subsection (a);
    [B] establish administrative mechanisms allowing affected persons to seek and obtain correction of information maintained and disseminated by the

agency that does not comply with the guidelines issued under subsection (a); and

[C] report periodically to the Director—

    (I) the number and nature of complaints received by the agency regarding the accuracy of information disseminated by the agency; and

    (II) how such complaints were handled by the agency.

The Information Quality Act rode along with the rest of the appropriations bill and was signed into law by President Clinton. OMB's final guidelines were published in the *Federal Register* in January 2002, with a corrected version appearing in February 2002 [8].

The act met with the U.S. Chamber of Commerce's approval and that from other business interests. In support of the act, the Chamber of Commerce stated, "Federal law has not historically required that information used by agencies to support regulations meet particular quality standards. This omission has frequently resulted in politically motivated regulations that would not, if tested, withstand scientific or statistical scrutiny. However, as a result of the Information Quality Act, which became effective on October 1, 2002, a new and exciting opportunity exists to bring to a close the days of such poorly supported regulations" [9]. In essence, the chamber asserted that federal regulations have frequently been based on flawed science. As policy, alleged "imperfect or unnecessary" regulations have long been a target of regulated enterprises, primarily because of alleged economic impacts.

In compliance with the Information Quality Act, on October 15, 2002, EPA published its *Guidelines for Ensuring and Maximizing the Quality, Objectivity, Utility, and Integrity Disseminated by the Environmental Protection Agency* [10]. The guidelines state that EPA will use a graded approach to establish the appropriate level of quality, utility, and integrity of information based on the intended use of the information. The more important the information (e.g., an epidemiology investigation used in a proposed water quality standard), the higher the quality standards to which it will be held by the agency. EPA guidelines specify how persons who believe EPA information does not meet the requirements of the Information Quality Act can submit a Request for Correction (RFC) to the agency. Presumably, EPA's denial of a RFC can be litigated.

◊ ◊ ◊

In December 2004, OMB released its final guidelines on how federal agencies must conduct peer reviews of scientific documents [11]. In the Information Quality Act (§515a), Congress directed OMB to issue guidelines to "[p]rovide policy and procedural guidance to Federal agencies for ensuring and maximizing the quality, objectivity, utility and integrity of information" disseminated by federal agencies. The OMB guidelines note that *peer review* "[i]s a form of deliberation involving an exchange of judgments about the appropriateness of methods and the strength of the author's inferences." Peer review is conducted on draft documents for purpose of identifying errors and flaws prior to the document becoming final. Two forms of peer review are acknowledged: *internal*, where members of a federal agency who were not involved with the preparation of a draft document conduct the peer review, and *external*, where the reviewers are not affiliated with the agency that has prepared the draft document.

The OMB guidelines exempt various types of information from their requirements for peer review, leaving the decision to the agency that is developing a scientific document. For example, time-sensitive health and safety determinations could be excluded from peer review, depending on an agency's determination that the public's welfare would be disserved by the time delay inherent in any peer review. In contrast, "highly influential scientific assessments" are required to be peer reviewed and under specific conditions. EPA's development of a risk assessment for an air pollutant would be an example of a highly influential scientific assessment, which would be required to follow OMB's peer review guidelines.

Under the OMB guidelines, "[i]n general, a federal agency conducting a peer review of a highly influential scientific assessment must ensure that the peer review process is transparent by making available to the public the written charge to the peer reviewers, the peer reviewers' report(s), and the agency's response to the peer reviewers' report(s). The agency selecting peer reviewers must ensure that the reviewers possess the necessary expertise. In addition, the agency must address reviewers' potential conflicts of interest (including those stemming from ties to regulated businesses and other stakeholders) and independence from the agency" [ibid.].

The OMB guidelines are required policy for all federal agencies that produce scientific information. The peer review requirements, since they cut across all federal agencies, will bring about a more uniform approach to the preparation, review, and dissemination of scientific information. Better science translates into better public policy.

◊ ◊ ◊

When environmental legislation gets enacted into law, opponents often shift their attention to combating agency regulations that flow from specific environmental statutes (e.g., the Clean Air Act). This opposition can include lawsuits that take issue with how an agency has developed or implemented a particular regulation. Another kind of opposition includes providing agencies with written comments in response to an agency's published, proposed regulations. One can argue that changing how an agency develops its regulations is also a path to opposing environmental statutes. Some will assert that the Information Quality Act is such an impediment.

Whether delays in effectuating government decision-making are a problem depends on whose policy ox is being gored. Those interests that have eschewed government regulations will likely be pleased with delays in promulgating—and possibly achieving less onerous—regulations. Some organizations will trumpet that the Information Quality Act has lessened the government's reliance in rulemaking on the now tiresome term "junk science." However, other groups will use the Information Quality Act to delay or overturn government decisions they think are not stringent enough. Recall that the mature ox has two horns and is adept at goring with either or both horns. And that's the predictable impact of the Information Quality Act: more litigation, more lawyers, delayed public policies, more contentious debates over science, and heightened cynicism about government's failure to protect the environment, environmental quality, ecosystems, and public health.

4.6.2 Public Health Implications of the Information Quality Act

Given the newness of the Information Quality Act, it is not clear if the act will deleteriously impact the public's health. As environmental health policy, it would be unfortunate if the act were used simply by vested interests to delay or otherwise impede the issuance of federal regulations and standards that would decrease environmental risks and improve public health. Some would argue that delay of flawed regulations is a contribution to public health and environmental policy. Critics would argue that regulations based on alleged poor science impose an unjustified economic burden on the public, decreasing resources that could be used for higher social priorities. The allegation of poor science inherent in federal government regulations has historically not been verified when outside scientific reviews have been conducted (e.g., EPA science reviews). For example, the National Academy of Sciences (NAS) has often been asked by Congress to review the science that underpins specific proposed federal regulations. Examples of NAS reviews include the toxicity of arsenic and mercury. Both reviews supported the EPA analysis of the published literature pertaining to the two toxicants.

Just how the Information Quality Act will impact the traditional practice of public health is not yet known. Presumably, such actions (e.g., federal quarantine of ocean vessels with passengers presenting symptoms of illness) might be subject to provisions of the Information Quality Act. The possible impact would be delay of public health interventions and second-guessing of public health policies. Moreover, the very core of public health practice, which is prevention of disease and disability, might be severely hampered if information quality considerations must precede taking public health actions. Also, it is uncertain to what extent the Information Quality Act might euthanize the adoption of the Precautionary Principle in U.S. environmental policy, which requires acting on incomplete information if a public health threat is sufficiently compelling.

Any benefits of the Information Quality Act will lie in the eye of the beholder—special interest groups, in particular. What is undeniable is that the act will affect how federal agencies do their business. One effect already apparent is litigation, real or threatened. The sources of litigation are likely to represent a wide spectrum of organizations and individuals. Three examples will suffice. In 2002, the Center for Regulatory Effectiveness, in a letter to the White House Office of Science and Technology Policy, expressed concern about computer models used in modeling regional climate change [12]. In 2003, the Public Employees for Environmental Responsibility filed a lawsuit against the Army Corps of Engineers [13]. The suit alleges the Corps used flawed data in a report. In 2003, the Environmental Working Group challenged the Food and Drug Administration's (FDA's) document on mercury levels in fish [14], questioning the data used by FDA in preparing a document made available to the public. Regardless of the merits, or demerits, of these three examples, it seems clear that the Information Quality Act will spawn more litigation and agency reviews, leading to even more delay in government decision making.

4.7 NATIONAL CONTINGENCY PLAN

Of relevance to several federal statutes to be subsequently discussed, the National Oil and Hazardous Substances Pollution Contingency Plan, more commonly called the

National Contingency Plan (NCP), is the federal government's blueprint for responding to both oil spills and hazardous substance releases. The NCP resulted from efforts to develop a national response capability and promote overall coordination among the hierarchy of government and private sector responders and contingency plans [15].

The National Contingency Plan is the federal government's blueprint for responding to both oil spills and emergency releases of hazardous substances.

According to the EPA [ibid.], the first National Contingency Plan was developed and published in 1968 in response to a massive oil spill from the oil tanker *Torrey Canyon*, which had ruptured off the coast of England the year before. More than 37 million gallons of crude oil spilled into the water, causing massive environmental damage. To avoid the problems faced by European response officials involved in the incident, U.S. officials developed a coordinated approach to cope with potential spills in U.S. waters. The 1968 plan provided the first comprehensive system of accident reporting, spill containment, and cleanup, and established a response headquarters, a national reaction team, and regional reaction teams, which are now called the National Response Team and Regional Response Teams.

Congress has broadened the scope of the National Contingency Plan over the years. As required by the Clean Water Act of 1972, the NCP was revised the following year to include a framework for responding to hazardous substance spills as well as oil discharges. Following the passage of the CERCLA in 1980, the NCP was broadened to cover releases from those hazardous waste sites that require emergency removal actions. Over the years, additional revisions have been made to the NCP to keep pace with the enactment of legislation. The latest revisions to the NCP were finalized in 1994 in order to reflect the oil spill provisions of the Oil Pollution Act of 1990. The details of the NCP can be found elsewhere [ibid.], but include provisions such as: establishment of the National Response Team, establishment of the Regional Response Teams, establishment of general responsibilities of Federal On-Scene Coordinators, the provision of funding for responses under the Oil Spill Liability Fund, and authorization of the lead agency to initiate appropriate removal action in the event of a hazardous substance release.

As environmental health policy, the NCP provides the means and resources to quickly respond to emergency conditions that involve the release of oil or hazardous substances. The NCP, because it is based on law, brings together the coordination between federal government agencies and others in order to protect the public's health and the well-being of ecosystems. Yoking government agencies in legally binding ropes of coordination serves the public well, since such cooperation is often difficult to achieve in the absence of law.

4.8   NATIONAL ENVIRONMENTAL POLICY ACT

4.8.1   HISTORY

The National Environmental Policy Act (NEPA) is a relatively brief, concise statute that articulates U.S. national policy on the environment and establishes goals for federal programs in terms of their impact on the environment. The act is an administrative procedures act, that is, it directs federal executive agencies to do specified actions. One source considers the NEPA to have had a profound impact in the sense that the act focused the federal government's attention on environmental consequences and forced government agencies to assess the impact of their actions on the environment [16]. Of note, the NEPA's concepts have been emulated by many state statutes and other national governments.

The National Environmental Policy Act (NEPA) created the Council on Environmental Quality (CEQ) and requires federal agencies to prepare environmental impact statements.

According to one source, "Few statutes of the United States are intrinsically more important and less understood than is the National Environmental Policy Act of 1969" [17]. The basis for this assertion is the fact that the NEPA firmly established the nation's endorsement of the importance of environmental quality and preservation. Further, NEPA established the policy of individuals' responsibility, "[e]ach person should enjoy a healthful environment and that each person has a responsibility to contribute to the preservation and enhancement of the environment" [ibid.]

According to Caldwell [17], the enactment of the NEPA was the culmination of 10 years of effort in Congress, which began in 1959, when Sen. James E. Murray (D-MT) introduced the Resources and Conservation Act in the 86th Congress. Murray's bill contained several provisions that were ultimately included in the NEPA of 1969, including a declaration of national environmental policy, the creation within the Executive Office of the President of an advisory council on environmental quality, and the preparation of an annual report on the nation's environmental quality. Congressional hearings in 1960 on Murray's bill were important because they brought forward dialog on protecting the environment.

As noted by Caldwell [17], "environment," as understood today, had a very limited meaning prior to the 1960s. The prevailing perspective was conservation of natural resources, with recognition of an endangered environment only slowly emerging in the early 1960s. Murray's bill was opposed by the Eisenhower administration, by many federal agencies, and by business trade associations. This opposition stemmed from the Eisenhower administration's general reluctance to expand the role of federal government, federal agencies' fear of loss of authorities and resources, and recognition by business groups that the Murray bill might have an economic impact on them [ibid.].

Following the failure of Murray's bill to get sufficient Congressional support to enable its passage, alternative bills on the environment were introduced during the decade of the 1960s. During this period, the environmental movement began coming to the forefront, supporting congressional efforts to protect the environment. Moreover, a scientific literature, which had begun in the 1930s, increased in amount and gravity of findings through the 1960s. This literature, which included *Silent Spring* by Rachel Carson [18] and *The Quiet Crisis* by Stewart Udall (D-AZ) [19], helped raise an environmental awareness in the U.S. public and the public's political representatives. In particular, *Silent Spring* made environmentalism a middle-class issue. Her book described in detail how DDT enters the food chain and accumulates in the fatty tissues of animals and humans. She concluded that DDT and other pesticides had poisoned the world's food chain. The book was widely circulated and awakened environmental concerns in people who had moved from cities to suburban areas in search of better environmental conditions. Environmentalism was no longer simply a concern only to conservation groups; the new environmental movement was focused on environmental quality and protection. This wider interest soon became apparent to elected officials and contributed to legislative support for new or more protective environmental policies.

In 1966, Senator Henry Jackson (D-WA) and Congressman John Dingell (D-MI) prepared bills on environmental policy that included several features of Murray's bill. These legislative efforts came to fruition in December 1969 when both houses of Congress passed a House-Senate conference bill, resulting in enactment of the National Environmental Policy Act of 1969. President Richard Nixon signed the act into law on January 1, 1970. With Nixon's signature came fulfillment of a decade's labor in Congress to make environmental protection a matter of national policy.

The NEPA states four important purposes: "(1) to declare a national policy that will encourage productive and enjoyable harmony between man and his environment, (2) to promote efforts that will prevent or eliminate damage to the environment and biosphere and stimulate the health and welfare of man, (3) to enrich the understanding of the ecological systems and natural resources important to the nation, and (4) to establish a Council on Environmental Quality" [20].

Regarding national policy the act states, "The Congress, recognizing the profound impact of man's activity on the interrelations of all components of the natural environment, particularly the profound influences of population growth, high-density urbanization, industrial expansion, resource exploitation, and new and expanding technological advances and recognizing further the critical importance of restoring and maintaining environmental quality to the overall welfare and development of man, declares that it is the continuing policy of the Federal Government, in cooperation with state and local governments, and other concerned public and private organizations, to use all practicable means and measures, including financial and technical assistance, in a manner calculated to foster and promote the general welfare, to create and maintain conditions under which man and nature can exist in productive harmony, and fulfill the social, economic, and other requirements of present and future generations of Americans" [20].

4.8.2 Key Provisions of the NEPA Relevant to Public Health

The NEPA comprises two titles.

Title I—*Congressional Declaration of National Environmental Policy*—states, "[i]t is the continuing responsibility of the Federal Government to use all practicable means, consistent with other essential considerations of national policy, to improve and coordinate Federal plans, functions, programs, and resources to the end that the nation may—

(1) fulfill the responsibilities of each generation as trustee of the environment for succeeding generations;[3]

(2) assure safe, healthful, productive, and aesthetically and culturally pleasing surroundings for all Americans;

(3) attain the widest range of beneficial uses of the environment without degradation, risk to health and safety, or other undesirable and unintended consequences;

(4) preserve important historic, cultural, and natural aspects of our national heritage, and maintain, wherever possible, an environment which supports diversity and variety of individual choice;

(5) achieve a balance between population and resource use which will permit high standards of living and a wide sharing of life's amenities;[4] and

(6) enhance the quality of renewable resources and approach the maximum attainable recycling of depletable resources.

Title I also requires that an environmental impact statement (EIS) must be prepared by all agencies of the federal government to "[i]nclude in every recommendation or report on proposals for legislation and other major federal actions significantly affecting the quality of the human environment." The EIS is to include: "[a] the environmental impact of the proposed action, [b] any adverse environmental effects which cannot be avoided should the proposal be implemented, [c] alternatives to the proposed action, [d] the relationship between local short-term uses of man's environment and the maintenance and enhancement of long-term productivity, and [e] any irreversible and irretrievable commitments of resources which would be involved in the proposed actions should it be implemented" [20]. The EIS process involves participation and review at all levels of federal government and allows interested public groups to be involved.

Title II—*Council on Environmental Quality*—creates the Council on Environmental Quality, which is placed in the Executive Office of the President, and defines its responsibilities. The CEQ is to assist and advise the President in the preparation of an annual Environmental Quality Report to Congress. The report must include: (1) the status and condition of the major natural, manmade, or altered environmental classes of the nation; (2) current and foreseeable trends in the quality, management, and utilization of such environments; (3) the adequacy of available natural resources; (4) a review of government and nongovernment programs as to their effect on the environment and on the conservation, development and utilization of natural resources; and (5) a program for remedying the deficiencies of programs, together with recommendations for legislation [20]. In addition, CEQ advises the President on policies and legislation, gathers timely ᴬation concerning trends in environmental quality, and conducts studies related ᴵmental quality and ecological systems.

Technical amendments to the NEPA occurred in 1970, 1975, and 1982. Of note, the Environmental Quality Improvement Act of 1970 established the Office of Environmental Quality within the Executive Office of the President. The director of the Office is stipulated to be the Chairman of CEQ. The Office is authorized to conduct administrative actions and information collection attending the responsibilities of the CEQ.

In a more practical context, the White House's Council on Environmental Quality provides political leadership on environmental issues, including environmental health. The council interacts with federal agencies and private sector organizations to help shape environmental policies and represents the administration in interactions with the Congress. In theory, the CEQ is an advocate for an administration's policies and practices on environmental affairs.

4.8.3 Public Health Implications of the NEPA

In a policy sense, the public health implications of the NEPA are found in statements of purpose in the act's Title I:

"(2) Assure safe, *healthful* (emphasis added), productive, and aesthetically and culturally pleasing surroundings for all Americans;

(3) Attain the widest range of beneficial uses of the environment without degradation, *risk to health* (emphasis added) and safety, or other undesirable and unintended consequences."

These two statements codify in law the nation's commitment to preventing adverse health effects from environmental hazards. The public's health, it can be argued, must be preserved and the NEPA serves as an anchor for this protection.

4.9 NOISE CONTROL ACT

4.9.1 HISTORY

Unique among the federal environmental statutes is the Noise Control Act of 1972. It is unique because it is an existing law, but not funded by Congress since 1981. Lack of funds means an agency is prohibited from conducting programs in support of authorities in law. This renders a statute impotent. How did this happen in regard to federal noise control?

The Noise Control Act authorizes EPA to conduct and coordinate noise research, review noise control regulations, and set standards for major sources of noise.

As background, the Noise Control Act of 1972 was signed into law by President Nixon on October 28, 1972. The act gave EPA the primary role for controlling environmental noise. It had been submitted in 1971 to Congress as part of the Nixon administration's environmental package [21]. Under the provisions of the 1972 statute, EPA has the responsibility for coordinating all federal programs in noise research and control. EPA must be consulted by other federal agencies prior to publishing new regulations on noise. If the agency feels that any proposed new or existing federal regulations do not adequately protect the public health and welfare, the agency can call for public review of them. Further, EPA has the authority to set standards for any product or class of products that have been identified as a major source of noise. Categories of equipment covered by the Noise Control Act include construction, transportation, motors or engines, and electrical and electronic devices.

Under the act, EPA created the Office of Noise Abatement and Control (ONAC). It had the overall responsibility for administering EPA's responsibilities under the Noise Control Act of 1972. In 1978, Congress amended the 1972 act to require coordination between federal agencies on matters of noise control and abatement, primarily with the intent of facilitating better coordination between EPA and the Federal Aviation Administration on issues of aircraft noise regulations [21]. The 1978 amendments, called the Quiet Communities Act of 1978, also authorized EPA to provide grants to state and local governments for noise abatement.

In 1981, funding for the ONAC was eliminated by Congress in response to a proposal from the Reagan administration. The proposal was part of the Reagan administration's anti-regulatory agenda, an agenda that included the proposed elimination of EPA, OSHA, the Department of Education, and the downsizing of the Department of Interior. This agenda soon foundered because of adverse public reaction and eventual opposition by the Democrat-controlled Congress. However, the administration's proposal to eliminate noise control regulations did not result in Congressional repeal of the Noise Control Act and its amendments. Rather, Congress left the act stand, but eliminated funding of the ONAC. In 1997, a bill to restore funding for the ONAC was introduced in Congress. Called the Quiet Communities Act of 1997, it failed to generate sufficient support for passage by either body of Congress.

As an environmental policy matter, this action by Congress to not repeal the Noise Control Act makes it relatively easy to restart the EPA program of noise control and abatement. Restored funding provided through the appropriations process would be all this is required, since the existing law already provides the necessary authorizing language that appropriations committees require.

4.9.2   PUBLIC HEALTH IMPLICATIONS OF THE NOISE CONTROL ACT

There are few environmental hazards as pervasive as noise pollution. Noise is defined as unwanted sound, which in some instances may be considered desired sound by others (e.g., loud music). One source observes that noise affects millions of people globally on a daily basis [22]. The same source cites highway noise alone affecting more than 18 million people in the United States and 100 million worldwide. In the United States it is estimated that community noise levels have increased more than 11 percent during the decade of the 1990s [23]. Most urban noise is caused by automobile traffic, an environmental problem on the rise globally. Internationally, noise control and abatement has gained more attention as countries set their own national noise standards [22]. No similar federal government actions are underway in the United States. However, some local governments in the United States have established noise control ordinances.

The adverse health effects of noise have been summarized by the World Health Organization [24]. WHO considers the health significance of noise pollution to include: noise-induced hearing impairment, interference with speech communication, disturbance of rest and sleep, effects on residential behavior and annoyance, and interference with intended activities. For public health purposes, two adverse health effects are further described.

Noise-induced hearing loss is no doubt the first recognized adverse effect of noise pollution. Noise in workplaces was early recognized as a major occupational hazard and remains so in both industrialized and developing countries. The principal social consequence of hearing loss is an inability to understand speech. Interpersonal communication is made more difficult, contributing to frustration and stress in hearing-impaired individuals. More recently, community noise pollution has become a concern in regard to hearing loss [24]. One source estimates that 120 million people worldwide suffer from disabling hearing difficulties [ibid.].

Effects on the cardiovascular system are the second public health effect of note due to exposure to noise pollution. The WHO notes, "Many studies in occupational settings have indicated that workers exposed to high levels of industrial noise for 5–30 years have increased blood pressure and statistically significant increases in risk for hypertension, compared to workers in control areas. In contrast, only a few studies on environmental noise have shown the populations living in noisy areas such as airports and noisy streets have an increased risk for hypertension. The overall evidence suggests a weak association between long-term environmental noise exposure and hypertension, and no dose-response relationships could be established" [ibid.]

Data are limited on the health costs associated with exposure to excessive noise. Research in Germany has estimated that the annual cost of noise on public health is approximately $500 to $1900 million ECU (approximately $725 million to 2.7 billion U.S. dollars) per year for road noise and $100 million ECU per year for rail noise (cited in [24]). In summary, noise in the workplace and community environment is an environmental hazard of global importance. High levels can impair hearing, cause hypertension, and other health consequences. Moreover, noise levels in the United States. and other countries have continued to increase. Many countries, both industrialized and developing, have developed noise control and abatement programs, while in the United States the federal government noise control program is frozen in time, owing to failure of Congress in 1981 to fund the provisions of the Noise Control Act of 1972.

4.10   OCCUPATIONAL SAFETY AND HEALTH ACT

4.10.1   History

Workers' health and safety have been concerns of long standing. Since antiquity, the nature of human activity called work has always had the potential to cause harm to those who do it. Our earliest human ancestors, whether stalking feral animals for food or growing grain for consumption, were subject to injury from predatory animals, wildfires, and traumatic events. As agriculture became the principal commercial human endeavor, farm workers fell victim to injuries caused by domestic animals, farm equipment, and weather hazards.

The Occupational Safety and Health Act requires the Occupational Safety and Health Administration to develop and enforce workplace standards and the National Institute for Occupational Safety and Health to research and investigate workplace hazards.

The Industrial Revolution replaced agriculture as the main economic engine in most countries. This development brought new kinds of hazards to workers, who found themselves subjected to chemicals (e.g., tars, solvents, metals), physical agents (e.g., heat, noise), particulates (e.g., soot, silica), and musculoskeletal trauma (e.g., loss of limbs, lower back disorders). As the Information Age gradually succeeded the Industrial Age, computer-based technology and information management brought their own kind of workplace problems, including musculoskeletal disorders (e.g., carpal tunnel syndrome) associated with repetitive body motion from work at computer keyboards, eyestrain from staring at video displays, and stress from challenging work schedules and economic pressure.

In the United States, workplace safety became an issue with trade unions in the 1930s. Improved workplace conditions, better pay, and other benefits were cornerstones for organized labor. Organized labor, as it grew in national political influence, lobbied Congress to enact legislation to protect workers' safety and health. These efforts, usually opposed by trade associations and business interests, culminated in the passage into law of the Occupational Safety and Health Act of 1970 (OSHA).

The OSHA was the first federal legislation to deal comprehensively with health and safety problems in the workplace. The declared purpose and policy of the statute is "[t]o assure so far as possible every working man and woman in the nation safe and healthful working conditions and to preserve our human resources" [25]. To effectuate this goal, the OSHA created three new government agencies. The Occupational Safety and Health Administration, which is located in the Department of Labor, is directed to develop workplace safety and health standards, conduct workplace inspections, enforce regulatory actions developed by the agency; help set up state occupational safety and health programs and monitor their effectiveness, and conduct education and training programs for safety and health professionals. As discussed in chapter 3, the National Institute for Occupational Safety and Health (NIOSH), which was placed in DHHS, is tasked with conducting scientific research, performing workplace health evaluations, and developing criteria documents that contain recommendations for safety and health exposure conditions. The Occupational Safety and Health Review Commission was established by the OSHA to adjudicate disputes arising from enforcement of the act.

The most consequential OSHA responsibility under the act is the development and promulgation of safety and health standards. This is done through regulatory action, taking into account all relevant scientific information and public concerns. OSHA standards affect most workplaces, large and small. Employers convicted of willful violation of OSHA standards may face civil or criminal penalties, depending on the seriousness of the infraction. OSHA's inspectors conduct inspections of workplaces to assess compliance with OSHA safety and health standards. Since 1977, OSHA has attempted to direct most of its inspections toward high-hazard industries—construction, petrochemicals, and manufacturing [26]. Where no specific standards exist for a workplace condition, the act directs each employer to provide "[a] place of employment which is free from recognized hazards that are causing harm to employees." This "general duty" clause is used by OSHA to control workplace hazards that are obvious and for which no specific standard exists [25,26].

An OSHA standard of particular importance is the Hazard Communication Standard (HCS). The standard covers more than thirty-two million U.S. workers exposed to an estimated 650,000 hazardous chemicals in all industrial sectors [27]. It requires that the hazards of all chemicals imported into, produced, or used in U.S. workplaces are evaluated and that the hazard information is disseminated to affected employers and exposed employees. The hazard information is conveyed by means of labels on containers and material safety data sheets (MSDSs). All employers covered under the HCS must have a communications program that includes labels on containers, MSDSs, and employee training [27]. One organized labor source considers it the most significant job safety and health regulatory action ever adopted [25]. The HCS is known as the "Right-to Know" standard and its concept has been widely adopted by numerous states and some local governments.

Language related to waste management is found in the 1986 amendments to the CERCLA, which directs OSHA to promulgate standards for the health and safety protection of employees engaged in hazardous waste operations. The CERCLA amendments stipulate that OSHA regulations pertaining to hazardous waste workers must include the following: (1) site analysis (each hazardous waste site is to have a specific plan for worker protection), (2) training requirements for contractors to provide workers with training in hazardous waste operations, (3) medical surveillance of workers, (4) protective equipment requirements, (5) engineering controls concerning the use of equipment, (6) maximum exposure limits for workers, (7) information programs informing workers of hazards, (8) handling of hazardous waste, (9) new technology that would maintain worker protections, (10) decontamination procedures, and (11) emergency response requirements [20]. The subject OSHA regulations were promulgated in final form in 1990.

The OSHA of 1970 was amended in 1990 by Public Law 101-552 and in 1998 by Public Laws 105-198 and 105-241. The 1990 amendments raised OSHA fines seven-fold over their corresponding amounts specified in 1970 [28]. The amendments in 1998 required: (1) the secretary of labor to establish a program under which employers may consult with state officials in respect to compliance with occupational safety and health requirements, and (2) directed the secretary of labor not to use the results of enforcement activities to evaluate employees directly involved in enforcement activities under the act or to impose quotas or goals with regard to the results of such activities.

4.10.2 PUBLIC HEALTH IMPLICATIONS OF THE OSHA

Job-related deaths are normally separated into two categories: (1) deaths due to work-place injuries, and (2) deaths attributable to occupational diseases. The former includes deaths due to causes such as electrocutions, motor vehicle accidents, falls, homicides, and machinery-related events. The latter category includes deaths due to diseases such as asbestosis, silicosis,[5] and cancer. One source estimates that in 1992 there were about 6,500 deaths from workplace injuries and 60,300 deaths due to occupational diseases [29].

Herbert and Landrigan [30] summarized occupational mortality data from several sources. They cite data showing a total of 88,622 deaths from work-related injuries during the period 1980 through 1994. The number of deaths per year decreased from 7,405 in 1980 to 5,406 in 1994, with the annual death rate declining from 7.5 per 100,000 workers in 1980 to 4.5 per 100,000 workers in 1994. More recent fatality data indicate a lesser decline in death rates: 5.3 per 100,000 in 1994 and 4.5 per 100,000 in 1998 [ibid.] These figures are based on injury surveillance systems; no similar systems exist in the United States on deaths from occupational diseases.

The Bureau of Labor Statistics, U.S. Department of Labor, conducts surveys annually to estimate the incidence rates of occupational injuries and illnesses. The data in Figure 4.1 indicate that between the years 1973 through 2000, injury incidence rates have declined over the years, while illness rates decreased from 1973 through 1985, then increased until 1994, declining thereafter (Figure 4.1). It is unclear why the illness

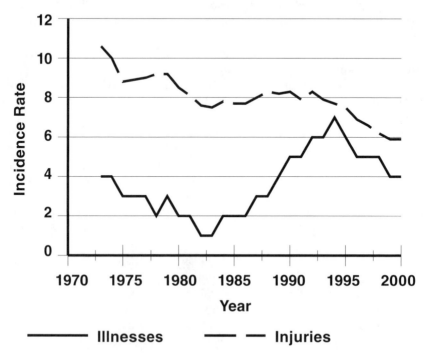

FIGURE 4.1   Injury and illness (total cases) incidence rates per 100 full-time workers. Note: Illness rates are shown as 10 times their value [31].

incidence increased from 1985 onward, but may be attributable to improved reporting systems for occupational illnesses. For example, the year 2000 incidence rate for injuries and illness combined is 6.1 per 100 full-time workers, a decrease of approximately 45 percent from year 1973. As a matter of environmental policy, do decreases of this magnitude portray changes in U.S. workplaces due merely to the nature of work (e.g., fewer manufacturing jobs or how work is performed)? Or are they due to the effectiveness of the act? Or, most likely, due to both? This dilemma illustrates the difficulty in policy analysis of changes that occur over long periods of time, in this case, 30 years.

4.11 PUBLIC HEALTH SECURITY AND BIOTERRORISM PREPAREDNESS AND RESPONSE ACT

4.11.1 HISTORY

The United States has been fortunate to have sustained only a few homeland attacks on its national security. In substantial measure this has been due to the nation's geography—sheltered by two major oceans—and good relations with its hemispheric neighbors. That is not to say that attacks on the U.S. homeland have not occurred. In 1814, British forces burned the public building in Washington City (now Washington, D.C.) during the War of 1812. In 1941, Japanese forces bombed the U.S. Naval base at Pearl Harbor, Hawaii, precipitating U.S. entry into World War II. And on September 11, 2001, terrorists high-jacked four commercial airplanes and flew two planes, with their passengers, into the twin World Trade Towers in New York City; one plane was flown into the Pentagon in Washington, D.C., and the fourth plane crashed in Pennsylvania. In total, approximately 3,000 persons were killed by this act of terrorism [32]. In each of these three examples, the nation came together, united in response to threats to its national security.

> The Bioterrorism Preparedness and Response Act of 2002 is a comprehensive public health plan for use should acts of bioterrorism occur within the U.S.

America's response to the terrorists' attacks of September 11, 2001, was swift and multifaceted. Of relevance to environmental health policy was the enactment of federal legislation to strengthen the nation's public health structure in order to be prepared for any future acts of terrorism on the United States. In concert with Congress, the George W. Bush administration participated in the enactment of the Public Health Security and Bioterrorism Preparedness and Response Act of 2002 (hereafter shortened to Bioterrorism Preparedness Act). The act's titles are listed in Table 4.2. They portray an act that provides protections, through federal and state public health agencies, of food, drugs, and drinking water supplies through enhanced programs of advance planning and preparedness. A discussion of each of the act's titles follows.

TABLE 4.2
Public Health Security and Bioterrorism Preparedness and Response Act Titles

Title	Name of Title
I	National Preparedness for Bioterrorism and Other Public Health Emergencies
II	Enhancing Controls on Dangerous Biological Agents and Toxins
III	Protecting Safety and Security of Food and Drug Supply
IV	Drinking Water Security and Safety
V	Additional Provisions

4.11.2 KEY PROVISIONS OF THE BIOTERRORISM PREPAREDNESS AND RESPONSE ACT RELEVANT TO PUBLIC HEALTH

Title I—National Preparedness for Bioterrorism and Other Public Health Emergencies

The aptly named Title I is built upon the traditional public health infrastructure, which requires federal, state, and local public health authorities to coordinate in programs to prevent disease and disability. Title I is focused on preparedness in response to acts of bioterrorism and related public health emergencies.

Title I directs the Secretary of DHHS to develop and implement a coordinated health-related plan for use in responding to acts of bioterrorism. The plan must be coordinated with states and local governments. A new position, Assistant Secretary for Public Health Response Emergency Preparedness, appointed by the president, is established under Title I, with responsibility for the act's newly established National Disaster Medical System, which is to provide health care services, health-related social services, other human services, and auxiliary services needed to respond to the needs of victims of a public health emergency. Title I represents a policy decision to utilize the traditional public health approach of addressing public health hazards through coordinated efforts of federal, state, and local health authorities. Five subtitles specify the actions to be taken under Title I.

Subtitle A—National Preparedness and Response Planning, Coordinating, and Reporting–§101 [§2801(a) of Public Health Service Act] "The Secretary shall further develop and implement a coordinated strategy, building upon the core public health capabilities established pursuant to Section 319A, for carrying out health-related activities to prepare for and respond effectively to bioterrorism and other public health emergencies, including the preparation of a plan under this section. The Secretary shall periodically thereafter review and, as appropriate, revise the plan." §2801(a)(2) "In carrying out paragraph (1), the Secretary shall collaborate with the States toward the goal of ensuring that the activities of the Secretary regarding bioterrorism and other public health emergencies are coordinated with activities of the States, including local governments." §2801(a)(3) "Developing and maintaining medical countermeasures (such as drugs, vaccines and other biological products, medical devices, and other supplies) against biological agents and toxins that may be involved in such emergencies."

§2801(a)(5) "Enhancing the readiness of hospitals and other health care facilities to respond effectively to such emergencies."

Subtitle B—Emergency Preparedness and Response—§2811(a)(1) "There is established within the Department of Health and Human Services the position of Assistant Secretary for Public Health Response Emergency Preparedness. The President shall appoint an individual to serve in such position. [..]." §2811(b)(1) "The Secretary shall provide for the operation in accordance with this section of a system to be known as the National Disaster Medical System. The Secretary shall designate the Assistant Secretary for Public Health Emergency Preparedness as the head of the National Disaster Medical System [..]." §2811(b)(2)(A) "National Disaster Medical System shall be a coordinated effort by the Federal agencies specified in subparagraph (B), working in collaboration with the States and other appropriate public or private entities, to carry out the purposes described in paragraph (3)." (B) "The Federal agencies referred to in subparagraph (A) are the Department of Health and Human Services, the Federal Emergency Management Agency, the Department of Defense, and the Department of Veteran Affairs." (§2811(b)(2)(A) "The Secretary may activate the National Disaster Medical System to (I) provide health services, health-related social services, other appropriate human services, and appropriate auxiliary services to respond to the needs of victims of a public health emergency [..]." §121(1) "The Secretary of Health and Human Services [i]n coordination with the Secretary of Veteran Affairs, shall maintain a stockpile or stockpiles of drugs, vaccines and other biological products, medical devices, and other supplies in such numbers, types, and amounts as are determined by the Secretary to be appropriate and practicable [t]o provide for the emergency health security of the United States [..]."

Subtitle C—Improving State, Local, and Hospital Preparedness for and Response to Bioterrorism and Other Public Health Emergencies—§131[§319C-1(a) of Public Health Service Act] "To enhance the security of the United States with respect to bioterrorism and other public health emergencies, the Secretary shall make awards of grants or cooperative agreement to eligible entities to enable such entities to conduct the activities described in subsection (d)." §319C-1(d)–"Use of Funds (1) To develop Statewide plans [f]or responding to bioterrorism and other public health emergencies. [...] (3) To purchase or upgrade equipment. [...] (4) To conduct exercises to test the capability and timeliness of public health emergency response activities. [...] (6) To improve training or workforce development to enhance public health laboratories. [...] (9) To enhance communication to the public of information on bioterrorism and other public health emergencies, including through the use of 9-1-1 call centers. [...] (17) To improve surveillance, detection, and response activities to prepare for emergency response activities." [...]

Subtitle D—Emergency Authorities; Additional Provisions—§142 Streamlining and Clarifying Communicable Disease Quarantine Provisions.

Subtitle E—Additional Provisions—§153 "The Secretary of Health and Human Services [a]cting through the Director of the National Institute of (sic) Occupational Safety and Health, shall enhance and expand research as deemed appropriate on the health and safety of workers who are at risk for bioterrorist threats or attacks in the workplace." [...] §154(1) "The Secretary of Veterans Affairs shall take appropriate actions to enhance the readiness of Department of Veterans Affairs medical centers to protect the patients and staff of such centers from chemical or biological attack." [...] §159 [§312 of Public Health Service Act] "The Secretary shall award grants to States, political subdivisions of States, Indian tribes, and tribal organizations to develop and implement public access defibrillation programs." [...]

Title II—Enhancing Controls on Dangerous Biological Agents and Toxins

This title provides authorities to the Departments of Health and Human Services and Agriculture for purpose of controlling dangerous biological agents and toxins through registration and enforcement requirements of those persons in possession of biological agents and toxins. The following four subtitles contain Title II's principal authorities.

Subtitle A—Department of Health and Human Services (DHHS)—§351A(a)(1)(A) "The Secretary shall by regulation establish and maintain a list of each biological agent and each toxin that has the potential to pose a severe threat to public health and safety." (b) "The Secretary shall by regulation provide for—(1) the establishment and enforcement of safety procedures for the transfer of listed agents and toxins." [...] c) "The Secretary shall by regulation provide for the establishment and enforcement of standards and procedures governing the possession and use of listed agents and toxins." [...]

Subtitle B—Department of Agriculture (USDA)—This subtitle parallels Subtitle A, but with the Secretary of Agriculture responsible for developing a list of biological agents and toxins, with the requirement to develop regulations for transportation and use of the listed agents and toxins.

Subtitle C—Interagency Coordination Regarding Overlap Agents and Toxins—221(a)(1) "The Secretary of Agriculture and the Secretary of Health and Human Services shall in accordance with this section coordinate activities regarding overlap[6] agents and toxins."

Subtitle D—Criminal Penalties Regarding Certain Biological Agents and Toxins—§231(a)(1) "Whoever transfers a select agent to a person who the transferor knows or has reasonable cause to believe is not registered as required by regulations under subsection [b] or [c] of section 351A of the Public Health Service Act shall be fined under this title, or imprisoned for not more than 5 years, or both." C)(1) "Whoever knowingly possesses a biological agent or toxin for which such person has not obtained a registration required by regulations under section 31A of the Public Health Service Act shall be fined under this title, or imprisoned for not more than 5 years, or both."

Title III—Protecting Safety and Security of Food and Drug Supply

This title established extra protections for the U.S. food and drug supplies. Title III requires the development of a communications and education strategy when there are bioterrorism threats to the U.S. food supply. Further, the Secretary of Agriculture is directed to register food production facilities and to increase the number of food inspections of such facilities. Similar to authority to register food production facilities, foreign producers of drugs must be registered. The following three subtitles contain the primary public health authorities under Title III.

Subtitle A—Protection of Food Supply–§301 "The President's Council on Food Safety (as established by Executive Order No. 13100) shall, in consultation with the Secretary of Transportation, the Secretary of the Treasury, other relevant Federal agencies, the food industry, consumer and producer groups, scientific organizations, and the States, develop a crisis communications and education strategy with respect to bioterrorist threats to the food supply." [...] §302(h)(1) "The Secretary shall give high priority to increasing the number of inspections under this section for the purpose of enabling the Secretary to inspect food offer for import at ports of entry into the United States, with the greatest priority given to inspections to detect the intentional adulteration of food." §305 Registration of food facilities— "The Secretary shall by regulation require that any facility engaged in manufacturing, processing, packing, or holding food for consumption in the United States be registered with the Secretary." [...].§317R "Food Safety Grants— "The Secretary may award grants to States and Indian tribes [t]o expand participation in networks to enhance Federal, State, and local food safety efforts, including meeting the costs of establishing and maintaining the food safety surveillance, technical, and laboratory capacity needed for such participation." §313 "The Secretary of Health and Human Services, through the Commissioner of Food and Drugs and the Director of the Centers for Disease Control and Prevention, and the Secretary of Agriculture shall coordinate the surveillance of zoonotic diseases."

Subtitle B—Protection of Drug Supply–§321 Annual registration of foreign manufacturers; shipping information; drug and device listing— "[o]n or before December 31 of each year, any establishment shall, through electronic means in accordance with the criteria of the Secretary, register with the Secretary the name and place of business of the establishment, the name of the United States agent for the establishment, and the name of each person who imports or offers for import such drug or device to the United States for purposes of importation."

Subtitle C—General Provisions Relating to Upgrade of Agricultural Security— §331 "The Secretary of Agriculture [m]ay utilize existing authorities to give high priority to enhancing and expanding the capacity of the Animal and Plant Health Inspection Service to conduct activities to (1) increase the inspection capacity of the Service at international points of origin; [E]nhance methods of protecting against the introduction of plant and animal disease organisms by terrorists."[...] §332 "The Secretary of Agriculture may utilize existing authorities to give high priority to enhancing and expanding the capacity of the Food Safety Inspection Service to conduct activities to [...] (1) enhance the ability of the Service to inspect and ensure the safety and wholesomeness of meat and poultry products; (2) improve the capacity of the Service to inspect international meat and meat products, poultry and poultry products, and egg products at points of origin and at points of entry." [...]. §335 "The Secretary of Agriculture [m]ay utilize existing research authorities and research programs to protect the food supply of the United States by conducting and supporting research activities." [...].

Title IV—Drinking Water Security and Safety

This title provides added protections for public water supplies and for their customers. At the heart of these protections are vulnerability assessments, which are required of all public drinking water systems that serve 3,300 or more persons. The assessments, which are required to be secured by the EPA Administrator, must be used by water suppliers to design and implement an emergency response system to protect against acts of terrorists, as stipulated in §1433.

§1433(a) "Vulnerability Assessments—Each community water system servicing a population of greater than 3,300 persons shall conduct an assessment of the vulnerability of its system to a terrorist attach or other intentional acts intended to substantially disrupt the ability of the system to provide a safe and reliable supply of drinking water. The vulnerability assessment shall include, but not be restricted to, a review of pipes and constructed conveyances, physical barriers, water collection, pretreatment, treatment, storage and distribution facilities, electronic, computer or other automated systems which are utilized by the public water system, the use, storage, or handling of various chemicals, and the operation and maintenance of such system." [...] §1433(a)(2) "Each community water system referred to in paragraph (1) shall certify to the Administrator that the system has conducted an assessment complying with paragraph (1) and shall submit to the Administrator a written copy of the assessment." [...] §1433(a)(5)[C] "no copy of an assessment or part of an assessment, or information contained in or derived from an assessment shall be available to anyone other than an individual designated by the Administrator." §1433(b) "Each community water system serving a population greater than 3,300 shall prepare or revise, where necessary, an emergency response plan that incorporates the results of vulnerability assessments that have been completed." [...] §1433(d) "The Administrator shall provide guidance to community water systems serving a population of less than 3,300 persons on how to conduct vulnerability assessments, prepare emergency response plans, and address threats from terrorist attacks or other intentional actions designed to disrupt the provision of safe drinking water or significantly affect the public health or significantly affect the safety or supply of drinking water provided to communities and individuals." §1435(a) "The Administrator, in coordination with the appropriate departments and agencies of the Federal Government, shall review (or enter into contracts or cooperative agreements to provide for a review of) methods and means by which terrorists or other individuals or groups could disrupt the supply of safe drinking water or take other actions against water collection, pretreatment, treat, storage and distribution facilities which could render such water significantly less safe for human consumption," [....]

Title V—Additional Provisions

This title has, in truth, little to do with the purpose of the Bioterrorism Preparedness Act. One subtitle provides the FDA with extra funds that can be used to further reduce the time required by FDA to approve new drug applications from pharmaceutical companies. This is obviously a benefit to such companies in that they can get their drugs into commerce more quickly. Another subtitle relates to the allotment of digital television channels, a benefit to the television industry. Title V is a prime example of how special interest groups can get extraneous legislation grafted onto a bill whose primary purpose is focused elsewhere. The two subtitles of greatest public health importance follow.

Subtitle A—Prescription Drug User Fees. §501(2) "the public health will be served by making additional funds available for the purpose of augmenting the resources of the Food and Drug Administration that are devoted to the process for the review of human drug applications and the assurance of drug safety" [...].

Subtitle C—§531(a) "In order to further promote the orderly transition to digital television, and to promote the equitable allocation and use of digital channels by television broadcast permittees and licensees, the Federal Communications Commission, at the request of an eligible licensee or permittee, shall, within 90 days after the date of enactment of this Act, allot, if necessary, and assign a paired digital television channel to that licensee or permittee" […].

4.11.3   Public Health Implications of the Bioterrorism Preparedness and Response Act

The Public Health Security and Bioterrorism Preparedness and Response Act was enacted by Congress in response to terrorist acts against the United States. The act provides authorities and resources to federal government agencies, in coordination with state and local agencies, to prepare for, and respond to, acts of bioterrorism. The act requires actions to protect food, drug, and drinking water supplies and registration of food production facilities and foreign drug manufacturers of drugs imported into the United States. Further, the act directs the development of the National Disaster Medical System, which would be called into deployment in instances of bioterrorism. The beneficial implications to the public's health are obvious.

The challenge to public health authorities will be to maintain the act's preparedness programs and resources over extended periods of time, since the potential for bioterrorism is not likely to diminish for many years. Experience shows that today's public health crisis can become tomorrow's humdrum activities unless care is taken to refresh the responses to the crisis. This is why public health campaigns to reduce tobacco smoking, prevent the spread of HIV infection, and abolish children's exposure to lead all need periodic reinforcement in terms of public education and awareness raising.

4.12   PUBLIC HEALTH SERVICE ACT

4.12.1   History

The Public Health Service Act of 1912 and subsequent amendments constitute the basis for several environmental health programs. As background, prior to 1912, the Marine Hospital Service, formed in 1870, provided health care for merchant mariners, and gradually expanded its services over the next thirty years. These expanded services included the control of infectious disease and quarantine responsibilities. In 1912, Congress enacted the Public Health Service Act, which brought together the various federal health authorities and programs under one statute. As described by a PHS historian, "The Public Health Service Act of 1912 made explicit what had been an increasingly important element of the work of the Service from 1887 when the Hygienic Laboratory opened—the all-out exploration of disease in the laboratory and in the field" [33]. The Public Health Service Act of 1912 and its subsequent numerous amendments are the bedrock federal public health legislation, whose various authorities impact the full breadth of the U.S. society. The act of 1912 consisted of only two brief sections. Section 1 of the act is relevant to environmental health concerns. The language of that section follows [34].

"AN ACT To change the name of the Public Health and Marine-Hospital Service to the Public Health Service, to increase the pay of officers of said service, and for other purposes.

The Public Health Service Act of 1912 and its subsequent numerous amendments are the bedrock federal public health legislation, whose various authorities impact the full breadth of the U.S. society.

Be it enacted by the Senate and House of Representatives of the United States of America in Congress Assembled, That the Public Health and Marine-Hospital Service of the United States shall hereafter be known and designated as the Public Health Service, and all laws pertaining to the Public Health and Marine-Hospitals Service of the United States shall hereafter apply to the Public Health Service, and all regulations now in force, made in accordance with law for the Public Health and Marine-Hospital Service of the United States shall apply to and remain in force as regulations of and for the Public Health Service until changed or rescinded. The Public Health Service may study and investigate the diseases of man and conditions influencing the propagation and spread thereof, *including sanitation and sewage and the pollution either directly or indirectly of the navigable streams and lakes of the United States* (emphasis added), and it may from time to time issue information as publications for use by the public."

The PHS Act of 1912 was significant in regard to environmental health policy and practice. For example, the act authorized surveys and studies of the impact of water pollution on human health. The act further directed the PHS to develop the first national water standards [35], which materialized in 1914. The standards introduced the concept of maximum contaminant limits in drinking water supplies. However, the standards applied only to water supplies that served interstate transportation because they were intended to protect the traveling public [ibid.]. These fledgling steps to establish national water quality standards did not lead to true national standards for another sixty years when Congress enacted the Safe Drinking Water Act of 1974.

The same act also initiated the first federal policy concerning the disposal of human wastes. The act provided technical advice and assistance to communities and commenced federal support for research and technical studies on the sanitary disposal of human wastes. Sanitary disposal of human wastes, like protecting the quality of drinking water, was a public health success, eliminating one major source of disease-producing pollution in the environment. The policy of federal involvement in reducing or preventing environmental hazards in states and local communities was wise and timely, although not without some opposition—some maintained that such responsibilities lay exclusively with states. That federal-state argument over jurisdiction remains today, a product of the U.S. Constitution, which limits the role of federal government and assigns to states those responsibilities not specifically assigned in the Constitution. Of course, federal court decisions over the years have elaborated and further defined the role of federal government, generally holding that the federal government has primacy on matters of environmental pollution.

The 1912 legislation remained intact until the years preceding World War II. Some members of Congress and the PHS leadership—in particular, Surgeon General Thomas Parran—foresaw the need to reorganize the country's health resources in advance of a likely global war. Two acts of Congress reshaped the Public Health Service and the nation's health needs [36]. Congressman Alfred Bulwinkle (D-NC) and Senator Elbert Thomas (D-UT) were the leaders in Congress who spearheaded the legislation. The first Act, signed into law in November 1943, organized the PHS Commissioned Corps along military lines and assembled PHS functions into four subdivisions: the National Institute of Health, the Office of the Surgeon General, the Bureau of Medical Services, and the Bureau of State Services.

The second Act, signed on July 3, 1944, was of great significance for the PHS's future. It codified all of the PHS's responsibilities and further strengthened the role of the surgeon general in public health policy making [ibid.]. According to Snyder [36], the act of 1944 established the PHS Commissioned Corps as the leadership cadre of the PHS and included "[f]inancial, technical, and advisory support to State and local health departments, the funding of extramural research through grants-in-aid, the provision of construction funds for hospitals and other facilities, and continued clinical services for a wide range of Federal beneficiaries." The act also expanded the PHS tuberculosis program to include support to state and local health departments.

The Public Health Service Act of 1944 became the framework upon which the modern public health programs in the United States are built upon.

Various environmental health programs are authorized and funded through the Public Health Service Act, as was discussed in chapter 3. These include programs of biomedical research at the National Institute of Environmental Health Sciences, National Cancer Institute, and Food and Drug Administration; the Centers for Disease Control and Prevention's surveillance of environmental exposures to toxicants and health effects of environmental hazards; and toxicity testing programs under the auspices of the National Toxicology Program. Environmental education programs, physician education credits, and grants to states for their environmental health programs (e.g., exposure to lead prevention efforts) are other examples of environmental health programs funded through the Public Health Service Act, as amended.

4.13 SUMMARY

Twenty-three federal environmental statutes will be described in this and subsequent chapters. They constitute the skeleton for the body of U.S. environmental health programs, policies, and practices. Concern for public health is a characteristic of most of these laws. The 1970s were a watershed period for Congressional legislation to control environmental contaminants in outdoor air, bodies of water, and drinking water supplies. Much of this legislation was predicated on human health concerns and enacted without an existing body of causal science and public health data. In a very real sense, this was an act of Precautionary Action (i.e., legislating without complete scientific data and information). As noted in this chapter, a scientific body of published reports has reinforced the importance of having in place environmental statutes that protect human health and quality of environment.

The statutes discussed in this and the chapters that follow have often been emulated in other countries. Moreover, some of the U.S. environmental health policies, (e.g., polluters pay for the costs of their pollution) have been adopted by regional governments and individual nations. In particular, the United Nations is playing an increasingly important role in developing policies for controlling environmental hazards, especially in developing countries, as described in chapter 9. As the United Nations and the European Union continue to implement environmental health policies, such as the management of hazardous waste, they will impact how the United States interacts with global partners in trade, commerce, and risk management of environmental hazards.

4.14 POLICY QUESTIONS

1. Let us assume that you are working for a member of your state's legislature. As an elected official, she has been asked by several community groups to get a law enacted that would regulate traffic noise, which the groups believe has gotten out of control and is deleteriously affecting their quality of life and possibly causing adverse health effects. (A) Your assignment is to provide the following: (1) Name of the proposed act, (2) Policy statement to be included in the act, (3) Purpose of the act, (4) Titles (or subtitles if you prefer) in the act, (5) Key provisions in each title (or subtitle). (B) Discuss what you anticipate to be the key issues in getting the proposed legislation adopted by the state legislature. *Note*: Assume a "command and control" regulatory structure, and assume the traffic noise comes from only three sources: vehicle tires, vehicle mufflers, and vehicle radios. *Suggestion*: A useful document on community noise levels can be found at http://www.who.int/peh/.

2. Discuss the historical significance of the Public Health Service Act of 1912. Does that early twentieth-century act have any relevance to you today? If so, how. Be specific.

3. What in your personal opinion is the practical importance of the National Environmental Policy Act (NEPA)? Review NEPA's policy statement that begins, "The Congress, recognizing the profound impact" [...] and rewrite the statement in terms of sustainable development, as discussed in chapter 2.

4. Using Internet resources, ascertain the details of how the Public Health Security and Bioterrorism Preparedness and Response Act of 2002 has improved your state's preparedness to respond to bioterrorism.

5. The Consumer Product Safety Act of 1972 requires the Consumer Product Safety Commission (CPSC) to first rely upon voluntary consumer product standards developed in concert with industry in lieu of CPSC mandated standards. Discuss the benefits and disadvantages of this policy.

6. The Noise Control Act of 1972 is a federal statute without an annual Congressional appropriation, which makes the statute inoperative. Are there benefits to keeping the Act "on the books," rather than its outright repeal by Congress? Be specific.

7. The Occupational Safety and Health Act of 1970 provides the authority for

federal government agencies to develop standards and regulations for purpose of controlling workplace hazards. Discuss the pros and cons of having federal government intervention in workplaces.

8. What are the differences between emission standards and quality standards? Why are both needed for control of environmental pollution?

9. The Information Quality Act came into law without going through the process of developing authorizing legislation. Rather, it was added as a rider to an appropriations bill. Discuss the pros and cons of this route of legislation.

10. Choose any statute from Table 4.1 and discuss its impetus for enactment into a statute.

NOTES

1. Named after Congressman James Delaney (D-NY), who in 1958 authored an amendment to §409 of the Food, Drug and Cosmetic Act, which stated "the Secretary shall not approve for use in food any chemical additive found to induce cancer in man, or, after tests, found to induce cancer in animals."

2. Originally called the Data Quality Act (e.g., [6]). The name was apparently changed by the White House's Office of Management and Budget (OMB).

3. This statement can be construed as an early commitment to what is now called the *precautionary principle*.

4. This statement can be construed as an early commitment to what is now called *sustainable development*.

5. Asbestosis and silicosis are lung diseases caused by inhalation of asbestos fibers and silica particles, respectively.

6. Overlap refers to agents and toxins that are common to the lists developed by DHHS and the USDA.

REFERENCES

1. CPSA (Consumer Product Safety Act), U.S. Code Collection. Available at http://www.thecre.com/fedlaw/legal5c/uscode15-2051.htm, 2003.

2. CPSC (Consumer Product Safety Commission), 2003 Budget and Performance Plan (Operating Plan). Available at http://www.cpsc.gov, 2003.

3. CPSC (Consumer Product Safety Commission), Frequently Asked Questions. Available at http://www.cpsc.gov/about/faq.html, 2003.

4. Lee, M., Environmental Research, Development & Demonstration Authorization Act, Congressional Research Service, Washington, D.C. Available at http://www.cnie.org/nle/leg-8/n.html, 1999.

5. NRC (National Research Council), *Drinking Water and Health*, vol 4, National Academy Press, Washington, D.C., 1982, 489.

6. Weiss, R., "Data quality" law is nemesis of regulation, the *Washington Post*, August 16, 2004.

7. Baba, A., et al., 2005. Legislating "sound science": The role of the tobacco industry, *American Journal of Public Health*, 95, S20, 2005.

8.   OMB (White House Office of Management and Budget), Guidelines for ensuring and maximizing the quality, objectivity, utility, and integrity of information disseminated by federal agencies, *Federal Register*, 67(5365), 8452, February 5, 2002.

9.   USCHAMBER (U.S. Chamber of Commerce), The Data Quality Act. Available at http://www.uschamnber.com/isr/dqa.htm, 2003.

10. Nielsen, M., et al., USEPA announces guidelines implementing the Data Quality Act of 2001, ENVIRON International Corporation, New York, February 20, 2003.

11. OMB (White House Office of Management and Budget), Final Information Quality Bulletin for Peer Review, Office of Information and Regulatory Affairs, Washington, D.C., 2004.

12. Revkin, A.C., Law revises standards for scientific study, National Tribal Environmental Council, March 21. Available at http://www.ntec.org/air/dataquality.html, 2002.

13. OMB Watch, Second lawsuit filed under the Data Quality Act, Washington, D.C. Available at http://ombwatch.org/articleprint/1979/-1/170, 2003.

14. EWG (Environmental Working Group), Data Quality Act challenge. Letter to Dr. David Acheson, U.S. Food and Drug Administration, Washington, D.C., December 22, 2003.

15. EPA Press Advisory, Idaho man sentenced in paint waste case, Washington, D.C., September 30, 2004.

16. Miller, J., Environmental law and the science of ecology, in *Environmental and Occupational Medicine*, (2nd ed.), Rom, W., ed., Little, Brown, Boston, 1992, 1307.

17. Caldwell, L.K., *The National Environmental Policy Act: An Agenda for the Future*, Indiana University Press, Bloomington, 1998.

18. Carson, R., *Silent Spring*, Houghton Mifflin, New York, 1962.

19. Udall, S.L., *The Quiet Crisis*, Holt, Rinehart and Winston, New York, 1963.

20. ELI (Environmental Law Institute), *Environmental Law Deskbook*, Environmental Law Institute, Washington, D.C., 1989, 166, 203, 221, 245.

21. EPA (U.S. Environmental Protection Agency), EPA to launch noise control program. Available at http://www.epa.gov/history/topics/nca/02.htm, 1972.

22. Fong, S. and Johnston, M., Health Effects of Noise, City of Toronto, Toronto Public Health Department, Toronto, 2000.

23. Staples, S.L., Public policy and environmental noise: Modelling exposure or understanding effects?, *American Journal of Public Health* 87, 2063, 1997.

24. Berglund, B. and Lindvall, T., eds., *Community Noise*, World Health Organization, Geneva, 1995.

25. Moran, R.D., Occupational Safety and Health Act, in *Environmental Law Handbook*, Arbuckle, J.G., et al., eds., Government Institutes, Inc., Rockville, MD, 1991, 370.

26. Bingham, E., The Occupational Safety and Health Act, in *Environmental and Occupational Medicine*, (2nd ed.), Little, Brown and Co., Boston, 1992, 1325.

27. OSHA (Occupational Safety and Health Administration), Hazard Communication. Available at www.osha.gov/SLTC/hazardcommunications/index.html, 2002.

28. Blosser, F., Regulatory Programs Beyond OSHA: Primer on Occupational Safety and Health, Bureau of National Affairs, Washington, D.C., 1992.

29. Leigh, J.P., et al., Occupational injury and illness in the United States: Estimates of costs, morbidity, and mortality, *Archives of Internal Medicine*, 157, 1557, 1997.

30. Herbert, R. and Landrigan, P.J., Work-related death: A continuing epidemic, *American Journal of Public Health*, 90, 541, 2000.

31. BLS (Bureau of Labor Statistics), Occupational Injury & Illness Incidence Rates per 100 Full-Time Workers 1973-98. Available at http://www.osha.gov/oshstats/bltable.html.

32. National Commission, Final Report of the National Commission on Terrorism Attacks Upon the United States. Executive Summary, Government Printing Office, Washington, D.C., 2004.
33. Mullan, F., *Plagues and Politics*, Basic Books, New York, 1989.
34. SGPHS (Surgeon General of the Public Health Service), Annual Report of the Public Health Service of the United States, Washington, D.C., 1913.
35. Weise, J., Historic Drinking Water Facts. Drinking Water and Wastewater Program, ADEC Division of Environmental Health, Anchorage, AL, 2003.
36. Snyder, L.P., A new mandate for public health—50th anniversary of the Public Health Service Act, *Public Health Reports*, July-August, 1994.

5 Air and Water Statutes

5.1 INTRODUCTION

Among the most important environmental statutes are those for control of air and water pollution. These statutes are central to the public health necessity of safe air and water in which we come into daily contact. Described in this chapter are the federal Clean Air Act, the Clean Water Act, and the Safe Drinking Water Act. Each act is presented in regard to its history, key public health provisions, and public health significance. Of note will be each statute's embedded policies of regulatory approach to pollution control. Both emission and quality standards are found throughout these three statutes. How the statutes involve interplay and shared responsibilities between federal and state government will be discussed. But it is particularly important to understand these statutes for their public health importance, as discussed throughout the chapter.

5.2 THE CLEAN AIR ACT

5.2.1 History

Pollution of the air we breathe for life's very existence is a problem likely dating from antiquity, perhaps from the time when humans first came into contact with smoke from fires used for warmth and to ward off predators. One source cites an action in the year 1306 when citizens of London petitioned their government to take action to reduce levels of smoke in ambient air. In response, King Edward I issued a royal proclamation to prohibit artificers (i.e., craftsmen) from burning sea coal, as distinguished from charcoal, in their furnaces [1]. This is an example of government taking action against the effects of air pollution, which can be defined as the contamination of the atmosphere by gaseous, liquid, or solid wastes. Given this fourteenth-century example of one government's attempts to improve citizens' air quality, it is not surprising to learn that in the twentieth century the U.S. public's concern about air pollution also led to legislative action.

It is ironic that U.S. federal air pollution control legislation was influenced by a "killer smog"[1] that occurred in London during the winter of 1952, an event in which it was first reported that more than four thousand people died from breathing polluted air caused by a temperature inversion.[2] However, a reassessment of mortality data for December 1952–February 1953 found that more than 12,000 excess deaths occurred due to acute exposure to heavily contaminated ambient air. Pollution levels during the period were 5 to 19 times greater than current air quality standards or guidelines, but similar to those currently found in some developing countries [2]. The primary constituent in the polluted air was smoke from home heating coal-burning stoves and fireplaces.

The state of California provided early and sustained leadership on controlling air pollution. Many of the state's concerns were focused on air pollution in Los Angeles. In 1943, the first recognized episodes of smog occurred in Los Angeles, resulting in limited visibility of approximately three blocks and reports of eye irritation, respiratory discomfort, nausea, and vomiting. The source of the pollution was unknown, but speculated to be an industrial facility. In 1947, California Governor Earl Warren signed into law the Air Pollution Control Act, which authorized the establishment of an air pollution control district (APCD) in every California county, leading to creation of the Los Angeles County APCD, the first of its kind in the U.S. [4]. This is an example of a state taking action to control an environmental hazard before similar action was taken by the federal government. In 1952, Dr. Arie Haagen-Smit, a professor of chemistry at the California Institute of Technology, discovered the nature and causes of photochemical smog. He determined that nitrogen dioxide and hydrocarbons in the presence of ultraviolet radiation from the sun form smog, a key component of which is ozone.

As described by Fromson [1], the first serious congressional recognition of the need for federal air pollution control occurred with the Air Pollution Control Act of 1955. This act provided research and technical assistance for the control of air pollution. The tragic events of London's killer smog in 1952 also raised awareness of the need to address the growing air pollution problem in the U.S. A similar episode of fatal air pollution had occurred during October 23–30, 1948, in Donora, Pennsylvania, where 20 people died and half the city's 12,000 residents became ill from breathing industrial contaminants trapped under a layer of temperature-inverted air [4].

The Clean Air Act requires EPA to set mobile source limits, ambient air quality standards, standards for new pollution sources, and significant deterioration requirements, and to focus on areas that do not attain standards [5].

The Air Pollution Control Act of 1955 declared that states had the primary responsibility for air pollution control. The federal government's role was advisory, providing technical services and financial support to state and local governments. The U.S. Department of Health, Education and Welfare (DHEW) was vested with these responsibilities under the act. In particular, the surgeon general was authorized to conduct a research program, in cooperation with state and local programs, to determine the causes and effects of air contaminants, another example of the policy of federalism at work. This policy approach of conducting research to establish the presence of a hazard to human health and its causal factors was in the tradition of public health.

The next major federal air pollution legislation occurred with the enactment of the Clean Air Act Amendments of 1963, after which the act was called the Clean Air Act (CAA). Whereas the Air Pollution Control Act of 1955 was limited primarily to research and technical and financial assistance to state and local governments, the Clean Air Act

Amendments enhanced federal responsibility for controlling air pollution. At the same time, the act continued Congress's intent that "[t]he prevention and control of air pollution at the source is the primary responsibility of state and local governments" [1].

The Clean Air Act Amendments of 1963 also strengthened the responsibilities of the secretary, DHEW, in several ways. The secretary was authorized to make an investigation involving interstate air pollution, whereas the act of 1955 restricted such investigations to those requested by state or local governments, a congressional policy stance in support of federalism. However, congressional caution was evident in the act of 1963. All DHEW investigations were to be advisory and any recommendations could be ignored by states or local authorities.

The Clean Air Act Amendments of 1963 further expanded DHEW's research and technical assistance programs by authorizing grants for "developing, establishing, or improving" state and local air pollution control programs. These grants took federal air pollution concerns into the arena of source control of air contaminants. Of note, DHEW was directed to develop air quality criteria that were to be "an expression of the scientific knowledge of the effect of various concentrations of pollutants depending on the intended use of a particular [m]ass of air" [1]. However, these criteria were merely advisory and air pollution control agencies were not mandated to adopt the air quality criteria.

Following passage of the Clean Air Act Amendments of 1963, the attention of Congress and environmental groups turned to air pollution caused by automobile emissions [1]. Given the passage of time, it may be difficult for some persons to comprehend the incredulity that accompanied the discovery in California of automobiles' contribution to air pollution, and, more specifically, smog. The discovery of vehicle emissions as the primary constituents of Los Angeles smog prompted federal laboratory research that found increased cancer rates in cancer-resistant mice. These findings were the subject of a 1962 report to Congress from Surgeon General Luther L. Terry. The report added weight to the need for further congressional action to control air pollution.

In 1965, Congress enacted the Motor Vehicle Air Pollution Control Act. The act required federal standards to be promulgated for controlling pollutants emitted from automobiles. The emission standards were to be established on the basis of "technological feasibility and economic costs" of controlling automobile emissions [1]. Upon promulgation of the emission standards, manufacturers of new motor vehicles or new motor engines were prohibited from selling or importing a nonconforming product into commerce.

The Clean Air Act was amended by Congress in 1970, heavily amended in 1977, and again substantively amended in 1990 (Table 5.1). The act's titles are listed in Table 5.2. The effects of unclean air on the public's health remain key motivations for keeping the act enforced. The act, as amended, is a comprehensive, complex statute that controls air pollution emissions and regulates government, business, and community lifestyles that affect the releases of air contaminants into outdoor ambient air. Air pollutants are emitted into the atmosphere from many sources and are broadly characterized by the EPA as deriving from mobile or stationary sources. Examples of the former include vehicles powered by internal combustion engines. Stationary sources of air pollution include industrial smoke stacks, utility companies, incinerators, industrial boilers, and

TABLE 5.1
Clean Air Act and Amendments [5]

Year	Act	Year	Act
1955	Air Pollution Control Act	1973	Reauthorization
1959	Reauthorization	1974	Energy Supply and Environmental Contamination Act
1960	Motor Vehicle Exhaust Study	1977	Clean Air Act Amendments
1963	Clean Air Act Amendments	1980	Acid Precipitation Act
1965	Motor Vehicle Air Pollution Control Act	1981	Steel Industry Compliance Extension Act
1966	Clean Air Act Amendments	1987	Clean Air Act 8-Month Extension
1967	Air Quality Act	1990	Clean Air Act Amendments
1970	Clean Air Act Amendments	1995–6	Relatively minor technical adjustments

residential furnaces. Specific air pollutants are determined by the nature and quantity of fuel combusted (e.g., gasoline, coal, oil, municipal waste) and the physics of combustion. In sum, the 1970 statute federalized air pollution control regulation, with public health protection the basis for much of that regulation, required automotive and other industries to meet emission standards, and created an extensive regulatory system to effect the act's goals.

The 1970 amendments adopted the policy of developing *National Ambient Air Quality Standards* (NAAQS) for individual air contaminants, then placed most of the responsibility on the states to achieve compliance with the standards. The act established two kinds of national air quality standards. *Primary standards,* are based on protection of human health, including the health of sensitive populations such as children, elderly persons, and persons with infirmities (e.g., asthma). *Secondary air quality standards* set limits to protect public welfare, including protection against decreased visibility, damage to buildings, and deleterious ecological effects.

The 1977 amendments added special provisions for geographic areas with air cleaner than national standards in order to prevent their deterioration in air quality, and special provisions were added pertaining to *nonattainment areas,* that is, geographic areas that had failed to meet national air quality standards [6]. Under the 1977 amend-

TABLE 5.2
Clean Air Act's Titles [5]

Title	Name of Title
I	Air Pollution Prevention and Control
II	Emission Standards for Moving Sources
III	General
IV	Acid Deposition Control
V	Permits
VI	Stratospheric Ozone Protection

ments to the act, states were required to develop *State Implementation Plans* (SIPs) that would meet the air quality standards by 1982, except for ozone, for which the deadline was 1987 [7]. The development of individual SIPs begins by dividing each state into geographic air quality control regions, and ascertaining whether and how much air pollution in each region exceeds the limits allowed by the air quality standards. Control requirements are then imposed to reduce emissions from the various sources. The act and its 1977 amendments placed heavy emphasis on reducing emissions from automobiles and other vehicles (i.e., mobile sources of air pollutants)

The 1990 amendments substantively revised the earlier version of the act. Signed into law on November 15, 1990, by President George H.W. Bush, these amendments added comprehensive provisions to regulate emissions of air toxicants, acid rain, and substances thought to be a threat to the ozone layer. In addition, the 1990 amendments added an elaborate permit program and markedly strengthened enforcement provisions and requirements for geographic areas that fail to meet air quality standards (i.e., nonattainment areas), mobile source emissions, and automobile fuels [6]. These amendments finally banned the sale in the U.S. of gasoline that contained lead additives, ending one of the twentieth century's worst environmental health missteps, the use of tetraethyl lead as a gasoline additive.

The 1990 amendments also changed the way hazardous air pollutants are regulated. The act, as amended, in effect, recognizes two kinds of outdoor ambient air pollutants: the six *Criteria Air Pollutants* and, basically, everything else. The latter category comprises what are called *Hazardous Air Pollutants* (HAPs). Before 1990, regulation of HAPs was a two-step process. EPA had to first establish that a pollutant was likely to be hazardous at ambient levels. Once this determination was made (and survived an elaborate hearing process), the second step was to choose the emission sources to be regulated.

Congress became increasingly impatient with EPA's science and risk-based approach for regulating HAPs, because under the pre-1990 act, only a handful of HAPs had been regulated. Sharply curtailing EPA's discretion on how to regulate HAPs, Congress specified more than 180 HAPs in the Act. Moreover, with respect to these substances, Congress shifted the burden of proof. "Whereas before, EPA had to go through an elaborate process to prove a compound *guilty* (emphasis added) before it could be regulated, now EPA must go through an elaborate process to prove a compound *innocent* before it can avoid regulation. Secondly, Congress required that maximum available control technology (MACT) be installed on all sources, regardless of extent of resulting exposure or toxicity. Risk assessment has been related to a residual risk provision that provides for additional action should MACT controls still leave a risk to the maximally exposed individual beyond a relatively stringent level. This shift of the burden of proof requirement of MACT across the board and downgrading of the importance of risk assessment clearly falls within the Precautionary Principle, as does the use of a stringent risk criterion and of the maximally exposed individual rather than the population as the target of concern" [6].

The 1990 CAA amendments also contained a significant environmental policy now called *cap and trade*, a marketplace incentive that was introduced in concept in chapter 2 and discussed there as Trading Pollution Credits. In these amendments, Congress, concerned that acid rain generation and deposition was causing consequential

TABLE 5.3
Air Quality Goals [9]

Goal	Action
1	Mitigate potentially harmful ambient concentrations of the six criteria pollutants (CO, NO_x, SO_2, O_3, Pb, particulate matter)
2	Limit sources of exposure to hazardous air pollutants (HAPs)
3	Protect and improve visibility in wilderness areas and national parks
4	Reduce emissions of substances that cause acid deposition, specially SO_2 and NO_x
5	Curb use of chemicals that have the potential to deplete the stratospheric O_3 layer

detrimental environmental effects on ecosystems, directed the EPA to implement a marketplace approach to reducing sulfur dioxide emissions, the main ingredient of acid rain. The legislation capped national emissions of SO_2 at ten million tons annually, then to decrease thereafter [8]. The amendments provided for emission "allowances" to electric utilities, a primary source of SO_2 emissions, based on their historic emission levels and other factors. Utilities that are able to surpass their EPA emission levels by using new technologies, cleaner fuels and such can sell their unused allowances to other utilities [ibid.]. The five primary air quality goals for the Clean Air Act, as amended, are shown in Table 5.3 [9].

As an outgrowth of the 1990 amendment's cap and trade policy, the EPA annually conducts an acid rain allowance auction. The act established an annual national cap on SO_2 emissions. Each year, EPA issues allowances to existing sources within that cap. In addition, a limited number of those allowances are withheld and auctioned. The auction gives private citizens, brokers and power plants an opportunity to buy and sell SO_2 allowances. According to EPA, "The auctions help ensure that new electric generating plants have a source of allowances beyond those allocated initially to existing units" [10]. The 2005 auction, held at the Chicago Board of Trade, sold 125,011 allowances for a total of approximately $34 million. The average allowance sold for $272. The American Electric Power trade association purchased the largest number of allowances, 75,000, and paid approximately $21 million [ibid.]. Proceeds from the auctions are returned to sources in proportion to the allowances held.

On March 10, 2005, EPA announced the Clean Air Interstate Rule (CAIR), a rule that EPA asserts will achieve the largest reduction in air pollution in more than a decade. This action, called the Interstate Air Quality Rule when it was proposed in January 2004, according to EPA, offers steep and sustained reductions by 2015 in air pollution levels as well as dramatic health benefits at more than 25 times greater than the cost [11]. Through the use of the cap and trade approach, the CAIR targets substantial reductions in levels of sulfur dioxide (SO_2) and nitrogen oxides (NO_x) emissions in more than 450 counties in the eastern United States and will help the counties meet EPA's protective air quality standards for ozone and fine particles. The CAIR covers 28 eastern states and the District of Columbia. States must achieve the required emission reductions using one of two compliance options: (1) meet the state's emission budget by requiring power plants to participate in an EPA-administered interstate cap and trade system that caps emissions in two stages, or (2) meet an individual state emissions budget through measures of the state's choosing.

The CAIR provides a federal framework requiring states to reduce emissions of SO_2 and NO_x. EPA asserts that states will achieve this primarily by reducing emissions from the power generation sector. According to EPA, in many areas, the reductions are large enough to meet air quality standards. The Clean Air Act, as amended, requires that states meet the new national, health-based air quality standards for ozone and $PM_{2.5}$ standards by requiring reductions from many types of sources. Some areas may need to take additional local actions. The CAIR reductions will lessen the need for additional local controls. In August 2005, EPA released a proposed federal implementation plan (FIP) under the CAIR that requires power plants in CAIR states to participate in one or more of three separate cap and trade programs [12].

According to EPA's cost-benefit analysis, by the year 2015, the Clean Air Interstate Rule will result in: (1) $85 to $100 billion in annual health benefits, annually preventing 17,000 premature deaths, millions of lost work and school days, and tens of thousands of nonfatal heart attacks and hospital admissions, (2) nearly $2 billion in annual visibility benefits in southeastern national parks, such as Great Smoky and Shenandoah, and (3) significant regional reductions in sulfur and nitrogen deposition, reducing the number of acidic lakes and streams in the eastern United States.

Some observations about the CAIR are in order. First, the CAIR makes clear the Bush administration's preference for cap and trade mechanisms for reducing air pollution emissions, especially from electric power plants. This approach's primary strength lies with its emulation of the Acid Rain cap and trade program, which has been successful in markedly reducing acid rain across the eastern United States. Further, the CAIR establishes a federal-state partnership that has clear emission reduction goals. Flexibility within CAIR permits states to implement their own cap and trade programs that could meet CAIR goals, and to tailor the programs to a state's specific sources of emissions of SO_2 and NO_x. Less impressive are the 10 years allocated to reach emission goals. Ten years translates to a decade of continued exposure of the affected public to air contaminants of health concern, SO_2 and NO_x.

Opinion about the utility of pollution trading credits (cap and trade) varies. According to EPA Region I, "In no small part, we have market forces to thank for the quick reduction in the pollutants that cause acid rain. EPA set the standard, telling power plants that they needed to reduce by one-half the emissions of sulfur dioxide, and established a trading program, through which companies within the industry could buy and sell pollution trading credits" [13]. These are strong words from EPA in support of a policy for trading pollution credits. EPA's comments indicate that trading pollution credits can accelerate the reduction of pollution emissions. Of note, the White House Office of Management and Budget asserts that the acid rain program under the CAA amendments has accounted for the largest quantified human health benefits of any federal regulatory program implemented since 1995, with annual benefits exceeding costs by more than 40 to 1 [14]. Moreover, industry favors pollution trading credits (PTCs) for revenue generation purposes and the ability to have greater control over how to meet pollution emission regulations.

However, some persons have expressed less enthusiasm for the concept of PTCs, arguing that such schemes do not make moral sense [15]. Such critics assert that by turning PTCs into a commodity, the moral stigma of pollution is removed. In their view there is harm done when pollution becomes a matter for enrichment, rather than being something that is harmful to society. They note that PTCs exist only if pollution exists. Another argument against the use of PTCs is that they can provide a safety net for polluters who buy them, not having met their own emissions limits, thereby prolonging pollution from their facility. On balance, it remains to be seen whether PTCs are a panacea for improving environmental quality or a plague because of prolonging pollution emissions.

◊ ◊ ◊

The 1977 amendments contained a provision (Title I, Part A, §111), called the *New Source Review* (NSR), that gained relatively little attention until the 1990s. The NSR applies when electric power companies make renovations to their facilities and operations. Under the provision, such plants can be considered as newly built plants and therefore must meet more stringent emission standards. The amount of pollution and the amount of electricity produced by older electricity generating units compound the problem of how to adequately implement the Clean Air Act's NSR provisions. Under provisions in the act, EPA requires electricity generating units built or modified after August 17, 1971, to meet uniform national emissions standards for regulated substances emitted from the units. The General Accounting Office (GAO)[3] found that 1,396 older electricity generating units, those built or modified after August 17, 1971, still operate. These older electricity generating plants emitted 59 percent of the sulfur dioxide, 47 percent of the carbon dioxide, and 42 percent of the carbon dioxide from fossil-fuel units in year 2000, while generating 42 percent of all electricity produced by fossil-fuel units [17].

Further, the GAO found that these "older" units emitted sulfur dioxide at levels above the new source standards applicable to new electricity generating units. Further, these "additional" emissions accounted for 34 percent of the sulfur dioxide and 60 percent of the nitrogen oxides produced by older generating units. How to force—through regulatory action or some market-based approach, or some combination of alternatives—old electricity generating units to come into compliance with current emission standards will be a difficult challenge, given the political influence that can be mobilized by the energy industry. Moreover, these older plants produce a considerable amount of the nation's electricity, making their closure for any length of time a difficult proposition. Nevertheless, not to reduce this major source of air pollution would be a serious setback for protecting the public's health.

For approximately twenty years, EPA interpreted the CAA, as amended, as permitting electric utilities to undertake routine maintenance, repair, and replacement activities. In 1996, the Clinton-era EPA proposed rules to reform the NSR, followed in 1999 by litigation against seven utilities companies, alleging that they had engaged in modifications of electric generation units without first obtaining NSR permits. The litigated utilities rebutted EPA's claims by asserting that their renovations did not meet NSR requirements [18]. Later, in 2003, the EPA announced it would drop investigations into seventy power plants for past violations of the Clean Air Act, deciding, rather, to judge the power plants on the basis of new, less stringent air pollution rules set under the NSR regulations [19].

On June 13, 2002, EPA announced changes to the NSR provision; its Administrator stating, "EPA is taking actions now to improve the NSR and thereby encourage emissions reductions" [20]. EPA asserts that the proposed changes would make it easier for companies to make changes in plant operations and maintenance, but without triggering the NSR provision. The companies would be permitted to operate as long as air emissions were not increased. On the same day as the EPA announcement, the American Lung Association (ALA) declared EPA's NSR changes as "[a]ccounting gimmicks that will increase air pollution and threaten public health" [21]. The ALA statement alludes to EPA's recommendation that electric power facilities "[w]ill be allowed to use any consecutive 20-month period in the previous *decade* (emphasis added) as a baseline, as long as current control requirements are taken into account" [20]. It seems noteworthy that the EPA announcement contains no language suggesting that older utility plants would be expected to bring themselves into compliance with emission standards expected of new plants.

In support of the administration's proposed changes to the NSR process, the National Association of Manufacturers (NAM), in its comments on EPA's proposed changes in the NSR, stated: "The NSR program is in need of substantial reform. The NAM recommends that EPA reform efforts target the core NSR program to increase flexibility, establish certainty and simplify the program. With the manufacturing community in an economic downturn, it is important for EPA to quickly and thoroughly consider administrative solutions to eliminate the fundamental problems with the NSR program" [22]. The NAM's comments are clearly focused on economic implications of the NSR; environmental quality and human effects of air contaminants were not addressed in their response to EPA.

On August 24, 2003, the Bush administration announced its decision to release new NSR regulations that would allow coal-fired power plants and refineries to upgrade their facilities without installing costly antipollution equipment, as was the case under the Clinton-era NSR regulations [23]. Under the new regulation, older plants could avoid installing pollution controls when they replace equipment provided the cost does not exceed 20 percent of the cost of replacing a plant's essential production equipment and the new parts are the "functional equivalent" of the worn-out equipment (ibid.). For example, if the replacement cost of a large electric generator was $200 million, then a power utility could spend as much as $40 million on plant upgrades without triggering NSR regulations. Industry officials lauded the new regulation, while environmental groups claimed that older plants would be operated at a higher generating capacity, producing even more pollution. The public health implications of the new regulation seem to have been ignored, given that fine particulate pollution, a major pollutant from older power plants, has been associated with serious lung and cardiovascular disease. In 2003, several states and a coalition of environmental organizations sued EPA over the administration's NSR regulations. In a ruling released in March 2006, the U.S. Court of Appeals for the District of Columbia overturned EPA's NSR regulation. The court ruled that EPA had misinterpreted the original NSR language by exempting new equipment from regulatory review by EPA [24]. Further litigation is expected. This kind of federal judicial intercession in a matter of federal environmental policy illustrates the check and balance concept envisioned by the framers of the U.S. Constitution.

In another area of litigation, several states have taken actions independent of EPA for purpose of limiting emissions from two major sources of CO_2, the primary greenhouse gas. These states' actions were taken because of their concerns that EPA's applicable regulations were inadequate. In one action, twelve states, three cities, and one U.S. territory[4] sued EPA for the agency's alleged failure to regulate CO_2 emissions from new cars and trucks [25]. The suit reached the U.S. Court of Appeals for the District of Columbia, where a three-judge panel found in 2005 that EPA had the administrative discretion to decide not to order reductions in CO_2 emissions from new vehicles [ibid.]. Further appeals of the panel's split decision could occur. This suit illustrates how the judiciary, as discussed in chapter 3, can play a crucial role in deciding environmental issues between federal and state governments.

A nonlitigious approach was taken in 2005 by some northeastern states dissatisfied with EPA's decision not to regulate CO_2 emissions. Nine states[5] announced plans to enact state legislation that would freeze power plant emissions within their borders at their current emission levels. Emissions from more than 600 electricity generating plants would be capped at a total of 150 million tons of CO_2 annually for the nine states. Each state would have its own cap. The 150 million ton cap, starting in 2009, would be sustained through 2015, when reductions would be required, reaching 10 percent in 2020 [26]. This is an example of states' rights in action. In general, states can enact their own environmental standards as long as the standards are not weaker than corresponding federal standards.

5.2.2   Cost and Benefits of Air Pollution Control

The 1990 CAA amendments (§812) required EPA to estimate the costs and benefits of the CAA. In response, EPA has estimated that the total direct compliance costs of the CAA from 1970–1990 were $523 billion in 1990 dollars, while the total monetized benefits exceeded $22,000 billion [27]. In 1999, EPA released a prospective study on the anticipated costs and benefits of the 1990 Clean Air Act Amendments from 1990–2010 [28]. The EPA study assumed a significant decrease in air pollutants over this period, estimating a net benefit as being $510 billion, with an expectation that benefits will again exceed direct compliance costs by approximately four to one [29]. In 2003, the White House's Office of Management and Budget [30] estimated that over the period 1992–2002 federal air pollution rules resulted in an annual benefit of $117–177 billion, with costs estimated at $17–120 billion, a benefits/cost ratio of about 6:1, a significant ratio, given that the benefits were primarily in terms of human health gains.

The costs of air pollution control are essentially apportioned across the economic sectors of the U.S. public. Businesses increase the price they charge for their products, government authorities charge motorists for the cost of vehicle emissions inspections, and taxes are increased to pay for government inspectors and allied personnel who are charged with enforcing air pollution regulations. Some will argue that these kinds of costs" are somehow unfair and without merit. Such arguments find a hearing in the court of cost-benefit analysis, where analysts attempt to associate the costs of regulatory impacts (e.g., more stringent air pollution regulations) against the benefits to society (e.g., improvements in the public's health). This kind of analysis is a most difficult calculus because of the many uncertainties in economic models used in the analysis and limited data on health benefits. As a matter of environmental health policy, current cost-benefit analysis must be improved by enriching the databases on associations between environmental hazards and their consequences to the public's health. Having this type of data benefits policy making, in general, and advances the possibility of using the precautionary principle more effectively.

5.2.3 KEY PROVISIONS OF THE CAA, AS AMENDED, RELEVANT TO PUBLIC HEALTH

Title I—Air Pollution Prevention and Control
Part A—Air Quality and Emissions Limitations
§109: EPA must promulgate primary National Ambient Air Quality Standards (NAAQS) necessary to protect the public health, allowing for an adequate margin of safety and promulgate secondary NAAQS to protect the public welfare, which includes "effects on soils, water, crops, animals, weather, visibility, economic values, and personal comfort and well-being" (§302(h)). §110: Each state must submit State Implementation Plans (SIPs) to EPA for the implementation, maintenance, and enforcement of primary and secondary NAAQS. §110(c): EPA must promulgate a federal implementation plan if a state fails to submit a SIP or revise a SIP that EPA deems inadequate. §111(a)(2): The term "new source" means any stationary source, the construction or modification of which is commenced after the publication of regulations prescribing a standard of performance under this section which will be applicable to such source. (4) The term "modification" means any physical change in, or change in the method of operation of, a stationary source which increases the amount of any air pollutant emitted by such source or which results in the emission of any air pollutant not previously emitted. §112: National Emission Standards for Hazardous Air Pollutants—§112(b): The CAA lists 189 Hazardous Air Pollutants (HAPs) and directs EPA to periodically revise the list. EPA must publish and periodically modify a list of categories and subcategories of major and area sources of HAPs, which are defined in §112(a). §§112(d)(1): EPA must promulgate emission standards for categories and subcategories of major and area sources of HAPs. §112(d)(2): The standards must require the maximum degree of reduction in HAP emissions achievable for new or existing sources in the category or subcategory. §112(d)(5): EPA may promulgate area source standards that provide for using generally available control technologies or management practices in lieu of meeting the §112(d)(2) requirements. §112(e), (I): Strict deadlines are set for promulgation of, and compliance with, the emission standards. §112(f): EPA must report to Congress on residual risks to public health remaining after application of the emission standards. If Congress fails to act, EPA must promulgate additional standards. EPA must promulgate residual risk standards for pollutants classified as known, probable, or possible human carcinogens if the §112(f) emission standards fail to reduce the lifetime cancer risk of the "most exposed" individual to less than one-in-one million. §112(g)(2): Source

modifications must comply with maximum achievable control technology. §113: EPA is authorized to issue administrative compliance and penalty orders, and seek injunctions and civil and criminal penalties. §179(b) authorizes the EPA Administrator to "[i]mpose a prohibition, applicable to a nonattainment area, on the approval by the Secretary of Transportation of any projects or the awarding by the Secretary of any grants, under Title 23, United States Code, other than projects or grants for safety." §182(C)(c)(6): De Minimis rule.[6] The new source review provisions under this part shall ensure that increased emissions of volatile organic compounds resulting from any physical change in, or change in the method of operation of, a stationary source located in the area shall not be considered de minimis for purposes of determining the applicability of the permit requirements established by the act [.]. §211(k): EPA is required to promulgate regulations that establish requirements for reformulated gasoline to be used in gasoline fueled vehicles in specified nonattainment areas.

Part B—Ozone Protection. The 1990 Amendments replaced Part B with Title VI.
Part C—Prevention of Significant Deterioration of Air Quality
    §161: SIPs must contain requirements to prevent significant deterioration of air quality in regions designated as attainment or unclassifiable. §§162, 164(a): A three-tiered classification system is established. Class I areas, which are subject to the greatest emission limitations, include national parks exceeding 6,000 acres and national wilderness and memorial parks exceeding 5,000 acres. All other areas are classified as Class II areas, except that such areas may be redesignated as Class III areas in certain limited circumstances. §165: Preconstruction permits are required for the construction in Prevention of Significant Deterioration (PSD) areas of "major emitting facilities" on which construction began after August 7, 1977. §165(a)(4) Permits must require facilities to employ best available control technology (BACT) for regulated pollutants. §166(a): EPA must promulgate regulations to prevent the significant deterioration of air quality resulting from hydrocarbon, carbon monoxide, photochemical oxidant, and nitrogen oxide pollution emissions. §167: Allows EPA to enforce Prevention of Significant Deterioration (PSD), including prohibition of construction of facilities. §169A(a): EPA must promulgate regulations to address the impairment of visibility in Class I areas resulting from man-made air pollution.

Part D—Plan Requirements for Nonattainment Areas
    §107(d): States are divided into areas; areas are designated as attainment, nonattainment, or unclassifiable. §172(a): Nonattainment areas are further classified based on severity of nonattainment and the availability and feasibility of pollution control measures necessary for attainment. §172(a)(2): Nonattainment areas for primary NAAQS must achieve attainment as expeditiously as practicable, but not later than 5 years after designation. Nonattainment areas for secondary NAAQS must achieve attainment as expeditiously as practicable. EPA may extend the attainment deadlines in certain cases. §172(c): Requirements for the content of nonattainment-area SIPs are set forth. §173(c): Before a new major stationary source, or a modification to an existing source, may be constructed in a nonattainment area, offsetting emission reductions must be obtained from the same source or other sources in the same nonattainment area. §181–192:

Special provisions exist for areas that are nonattainment for ozone, carbon monoxide, particulate matter, sulfur oxides, nitrogen dioxide, and lead.

Title II—Emission Standards for Moving Sources

§§202(a)(b): EPA must establish emission standards for new motor vehicles and engines, subject to specified limitations for hydrocarbon, carbon monoxide, and NO_x emissions by "light-duty" vehicles. EPA may set standards for heavy-duty vehicles after model year 1983, reflecting the greatest degree of emission reduction achievable for that model year. §211: EPA may require motor vehicle fuels to be registered and tested. §211(a): No manufacturer or processor may sell any EPA-designated fuel or additive that has not been registered by EPA. §219(k)(2)(D): Heavy Metals—The gasoline shall have no heavy metals, including lead and manganese. §246: SIPs for states in certain ozone and carbon monoxide nonattainment areas must require a specific percentage of fleet vehicles to be "clean fuel vehicles," beginning with model year 1998.

Enforcement Example: U.S. Announces $94 Million Clean Air Act Settlement with Chrysler Over Emission Control Defects on 1.5 Million Jeep and Dodge Vehicles

In 2005, EPA announced that the U.S. Government had reached a settlement with DaimlerChrysler Corporation (Chrysler) to repair defective emission controls on nearly 1.5 million Jeep and Dodge vehicles from model years 1996 through 2001. The agreement settles allegations that the company violated the CAAct by failing to properly disclose defective catalytic converters installed on the affected vehicles. As part of the settlement, Chrysler agreed to extend the warranty on the catalytic converters installed on approximately 700,000 of the vehicles involved and recall approximately 500,000 of the vehicles to fix a separate defect in the on-board diagnostic system. The total estimated cost to Chrysler to implement the settlement is $90 million. In addition, Chrysler will pay penalties of $1 million and will spend at least $3 million to implement a supplemental environmental project to reduce emissions from diesel engines currently in use [31].

Title III—General

§304: Except as provided in this subsection, any person may commence a civil action on his own behalf against any person (including the United States or any other governmental instrumentality or agency) who is alleged to have violated or to be in violation of (A) an emission standard or limitation under the CAAct or (B) an order issued by the Administrator or a state with respect to such a standard or limitation. §312: The Administrator, in consultation with specified other federal agents, must conduct a comprehensive analysis of the impact of this act on the public health, economy, and environment of the United States. §312(a): No grant which the Administrator is authorized to make to any applicant for construction of sewage treatment works in any area in any state may be withheld, conditioned, or restricted by the Administrator on the basis of any requirement of this Act, except as provided in §312(b). §318A(b): Before publication of notice of proposed rulemaking with respect to any standard or regulation to which this section applies, the Administrator must prepare an economic impact assessment respecting such standard or regulation. §319: Not later than one year after the date of enactment of the Clean Air Act Amendments of 1977 and after notice and opportunity for public hearing, the Administrator must promulgate regulations establishing an air quality monitoring system throughout the United States.

Title IV—Acid Deposition Control

§401(b): The stated goal is to reduce annual sulfur dioxide emissions from fossil fuel-fired electric power plants by 10 million tons below 1980 levels and annual NO_x emissions by 2 million tons below 1980 levels. §404(a): In Phase I, beginning January 1, 2000, 110 plants will receive allowances to emit SO_2 based on 1985–87 fuel consumption. §403(b): In Phase II, beginning January 1, 2000, utilities will receive reduced SO_2 allowances, totaling 8.9 million tons. Allowances may be used, sold, or carried forward. §407: EPA must establish NO_x emission limits for certain types of boilers and issue revised New Source Performance Standards (NSPS) for NO_x emissions from fossil fuel-fired steam-generating units (§111).

Title V—Permits

§172(c): Sources required to obtain permits include "major sources," "affected sources," sources subject to CAAct §111, air toxic sources regulated under CAAct §112, sources required to have new source or modification permits under Title I Parts C or D, and other sources designated by EPA. §§ 502(a),(b),(d): EPA must promulgate standards for a state-administered permit program. States must submit permit programs to EPA for approval.

Title VI—Stratospheric Ozone Protection

§602(a), (b): EPA must publish and revise lists of ozone-depleting substances, designating them Class I or Class II. §604, §605: The Class I list must include specified chlorofluorocarbons, halons, carbon tetrachloride, and methyl chloroform. The Class II list must initially include specified hydrochloroflurocarbons. Class I substances are to be phased out by January 1, 2000 (January 1, 2002, for methyl chloroform) and Class II substances by January 1, 2030 (subject to certain exceptions).

5.2.3.1   Perspective

The Clean Air Act, as amended, is the most complex and comprehensive of the federal environmental statutes. It contains a strong commitment to federalism, requiring the states to enforce many of the act's provisions such as issuing permits to facilities that emit air contaminants into ambient outdoor air. There is also a strong framework of quality standards that are linked to the emission standards. For geographic areas that do not meet air quality standards, the act authorizes such penalties as an area's potential loss of highway transportation funds. Regarding the impact of poor air quality on the public's health, a considerable and impressive body of health data have accrued over time. These data associate specific air contaminants with adverse health effects on the heart, lungs, and cardiovascular system.

5.2.4 PUBLIC HEALTH IMPLICATIONS OF THE CAA

As previously noted, air quality is regulated under the federal Clean Air Act. The act covers both *Criteria Air Pollutants* and *Hazardous Air Pollutants*. The former comprise ozone, particulate matter, carbon monoxide, sulfur dioxide (SO_2), oxides of nitrogen (NO_x), and lead. Hazardous Air Pollutants are toxicants released into ambient air primarily from industrial and business operations. The public health implications of individual air contaminants will be discussed in this section. A subsequent section will discuss the associations between air pollution and specific adverse health effects on the public's health.

National Ambient Air Quality Standards (NAAQS) must be developed by EPA for purpose of protecting human health. Enforcement of these standards over the three decades that began in 1970 has led to a general reduction in emissions of all criteria air pollutants. An examination of Figure 5.1 shows decreases from 1970 to 2003 in emis-

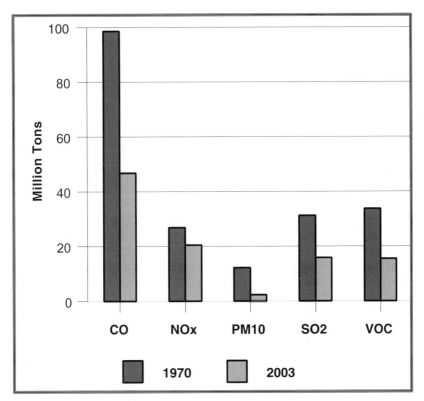

FIGURE 5.1   Air emissions in millions of tons for years 1970 and 2003 [114]. *Note*: CO values are shown as half scale.

sions of particulate matter (PM_{10}[7], 81%), volatile organic compounds (VOC, 54%), carbon monoxide (CO, 53%), sulfur dioxide (SO_2, 49%), and nitrous oxide (NO_x, 24%). These decreases are significant and remarkable. They are significant because all five pollutants are hazardous to human health. The decreases are remarkable because during this 30-year period the U.S. Gross Domestic Product increased by 158 percent, energy consumption increased by 45 percent, and vehicle miles traveled increased by 143 percent [33]. Not shown in Figure 5.1 is a dramatic decrease in emissions of lead, falling from approximately 221,000 tons released in 1970 to approximately 3,000 tons in 2003, a 99 percent reduction, which can be attributed to removal of lead compounds from gasoline used in the United States.

However, as shown in Figure 5.1, during the period 1970–2000, NO_x emissions decreased less than the other criteria air pollutants. EPA attributed this smaller decrease "[t]o growth in emissions from non-road engines (like construction and recreation equipment), diesel vehicles, and power plants" [33]. Because ozone occurs when sunlight interacts with NO_x and hydrocarbons in the air, increased NO_x emissions have contributed to an undesirable and unhealthy trend in increased ground level ozone levels in the southern and north-central regions of the U.S. [ibid.].

The human health effects of ozone and the other Criteria Air Pollutants are summarized in the following sections. Current quality standards for the criteria air pollutants are listed in Table 5.4.

Ozone (O_3) is a highly reactive gas that results primarily from the action of sunlight on nitrogen oxides and hydrocarbons emitted in combustion of fuels. Ozone exposure can produce significant decreases in lung function, inflammation of the lungs' lining, respiratory discomfort, and impair the body's immune system, making people more susceptible to respiratory illness, including pneumonia and bronchitis [34,35]. At sufficient high levels, repeated exposure to ozone for several months can cause permanent structural damage to the lungs [34]. Hospital admissions and emergency room visits increase on days of high ozone pollution in outdoor [e.g., 36–38]. The primary and secondary standard for O_3 is 0.08 ppm, averaged over an eight-hour period. Other conditions of the O_3 standard are given in Table 5.4.

Particulate matter (PM) is a general term that refers to very small, carbon-based, solid particles; dust; and acid aerosols. PM pollution is harmful to people with lung disease such as asthma and chronic obstructive pulmonary disease, which includes chronic bronchitis and emphysema, as well as to people with heart disease. PM air pollution has also been associated with increased premature death from heart failure [34]. The NAAQS for particulate matter are shown in Table 5.4. The primary and secondary standard for PM_{10} is 50 μg/m³, based on an annual arithmetic mean. For $PM_{2.5}$, the primary and secondary standard is 15 μg/m³, based on an annual arithmetic mean and 65 μg/m³ when based on a twenty-four-hour mean. Other conditions of the PM standards are given in Table 5.4.

Carbon monoxide (CO) is a colorless, odorless gas that forms when carbon-containing materials are incompletely combusted. CO is a byproduct of motor vehicle exhaust formed when gasoline is not fully combusted. Vehicle exhaust accounts for about 60 percent of all CO emissions in the U.S. [34]. At very high exposure levels CO is a fatal poison, but these levels far exceed those found in outdoor ambient air. When CO is

TABLE 5.4
Primary National Ambient Air Quality Standards for Criteria Pollutants [89]

Pollutant	Primary Standard	Averaging Times	Secondary Standard
Carbon monoxide	9 ppm (10 mg/m³)	8-hr[1]	None
	35 ppm (40 mg/m³)	1-hr[1]	None
Lead	1.5 µg/m³	quarterly average	SAP[6]
Nitrogen dioxide	0.053 (100 µg/m³)	annual (arith. mean)	SAP
Ozone	0.08 ppm	8-hr[5]	SAP
PM$_{2.5}$	15 µg/m³	annual[3] (arith. mean)	SAP
	65 µg/m³	24-hr[4]	SAP
PM$_{10}$	50 µg/m³	annual[2] (arith. mean)	SAP
	150 µg/m³	24-hr[1]	None
Sulfur dioxide	0.03 ppm (80 µg/m³)	annual (arith. mean)	None
	0.14 ppm (364 µg/m³)	24-hr[1]	None
		3-hr[1]	0.5 ppm (1300 µg/m³)

[1]Not to be exceeded more than once per year.
[2]To attain this standard, the three-year average of the weighted annual mean PM$_{10}$ concentration at each monitor within an area that must not exceed 50 µg/m³.
[3]To attain this standard, the three-year average of the weighted annual mean PM$_{2.5}$ concentrations from single or multiple community-oriented monitors must not exceed 15.0 µg/m³.
[4]To attain this standard, the three-year average of the 98th percentile of twenty-four-hour concentrations at each population-oriented monitor within an area must not exceed 65 µg/m³.
[5]To attain this standard, the three-year average of the fourth-highest daily maximum 8-hour average ozone concentrations measured at each monitor within an area over each year must not exceed 0.08 ppm. [6]Same as primary standard.

inhaled, the molecule binds to hemoglobin in red blood cells, reducing their oxygen-carrying capacity, which results in a reduction of oxygen available to the body's cells. As an air contaminant, CO is a hazard to persons afflicted with angina (severe chest pain) or peripheral vascular disease. The National Ambient Air Quality Standard for CO is 9 ppm, measured as an eight-hour nonoverlapping average not to be exceeded more than once per year. Values that equal or exceed 9.5 ppm are counted as exceeding the CO standard. According to EPA, an area meets the CO NAAQS if no more than one eight-hour value per year exceeds the threshold [ibid.].

Sulfur dioxide emissions occur when sulfur-containing fuels are combusted. Exposure to SO$_2$ at high levels is associated with breathing difficulties, respiratory illness, reduced pulmonary resistance to infectious agents, and aggravation of existing cardiovascular disease. The major source of SO$_2$ emissions are electric utilities. There are three NAAQS for SO$_2$: an annual arithmetic mean of 0.03 ppm, a twenty-four-hour level of 0.14 ppm, and a three-hour level of 0.50 ppm. The first two standards are primary (i.e., health-related) standards, while the three-hour NAAQS is a secondary (welfare-related) standard. The annual mean is not to be exceeded, while the short-term standards are not to be exceeded more than once per year [34].

Oxides of nitrogen (NO_x) is the general term for a group of highly reactive gases, all of which contain nitrogen and oxygen in varying amounts. The NO_x are created when fuel is burned at high temperatures, including internal combustion engines. Fossil fueled electric utilities, motor vehicles, and industrial operations are the primary sources of NO_x. Nitrogen dioxide (NO_2) can irritate the lungs and reduce resistance to respiratory infections such as influenza. The threshold value for both primary and secondary NAAQS for NO_2 is 0.053 ppm, measured as an annual arithmetic mean concentration in ambient air [34].

Lead and lead compounds are air contaminants of previous public health importance in the United States. Emissions from vehicles fueled by lead-containing gasoline were the major source of the public's exposure to lead in the environment. Elimination of lead as a gasoline additive, as the consequence of the 1990 CAA amendments, was a major public health success. As a result, measured levels of lead in blood have declined significantly, decreasing from a national average of approximately 7 µg/dl in the 1980s to about 2 µg/dl in year 2000. This reduction in baseline blood lead concentration lowers the body content of lead from all sources of exposure: air, food, water. The elimination of lead in gasoline significantly reduced the impact of lead on children's health, specifically, hearing impairment and cognitive development. Primary and secondary NAAQS for lead are 1.5 µg/m^3, maximum arithmetic mean averaged over a calendar month [34].

Of policy and public health note, the George W. Bush administration implemented major changes, under the provisions of the Clean Air Act, to regulate emissions from diesel engines. The two major emission components of greatest relevance to human health are particulate matter and sulfur. The Clean Air Nonroad Diesel Rule, adopted on May 11, 2004, requires off-road equipment powered by diesel engines to meet stringent new air pollution regulations [39]. Such equipment is found in construction, agricultural, and industrial equipment. EPA asserts that by the year 2010 the new rule will remove about 99 percent of the sulfur in current sources of diesel fuel, which in turn will effect a lower emission of particulate matter from diesel engines. The Clean Air Nonroad Diesel Rule complements EPA's Clean Diesel Truck and Bus Rule of December 21, 2000. The latter rule, which went into effect on December 21, 2000, requires diesel-powered trucks and buses to dramatically lower emissions from diesel engines. On-highway compliance takes effect in year 2007. EPA estimates that full implementation of the Clean Air Nonroad Diesel Rule will annually prevent up to 12,000 premature deaths, 15 heart attacks, and 6,000 children's asthma-related visits to hospital emergency rooms. The agency also estimates that the overall benefits of the nonroad Diesel program will outweigh the costs by a ratio of 40 to 1 [ibid.]. The benefits to public health of reduced diesel engine pollution will be significant and beneficial, particularly to individuals with lung disease or respiratory impairment.

◊ ◊ ◊

A good illustration of an intersection between environmental policy and public health practice is EPA's Air Quality Index (AQI) [40]. As developed by the EPA, the AQI is a scale of 0 to 500, divided into several color-coded categories. A region's AQI score at any time is based on the highest of five criteria air pollutants: particulate matter (PM), sulfur dioxide (SO_2), carbon monoxide (CO), nitrogen dioxide (NO_2) and ozone (O_3).

TABLE 5.5
Air Quality Index (AQI) and Associated Health Advisories [40]

Air Quality	AQI	Health Advisory
Good	0–50	None
Moderate	51–100	Unusually sensitive people should consider reducing prolonged or heavy exertion
Unhealthy for Sensitive Groups	101–150	People with heart or lung disease, older adults, and children should reduce prolonged or heavy exertion
Unhealthy	151–200	People with heart or lung disease, older adults, and children should avoid prolonged or heavy exertion. Everyone else should reduce prolonged or heavy exertion.
Very Unhealthy	201–300	People with heart or lung disease, older adults, and children should avoid all physical activity outdoors. Everyone else should avoid prolonged or heavy exertion.

The intervals and the terms describing the AQI air quality levels are as shown in Table 5.5. Using the state of Georgia as an example, the Ambient Monitoring Program at the Georgia Environmental Protection Division (EPD), Air Protection Branch, is responsible for measuring air pollutant levels throughout the state [41]. When these levels are reported, the EPD utilizes the Air Quality Index (AQI) to gauge their public health importance and make adjustments to applicable air quality programs.

AQI figures inform the public about whether air pollution levels in a particular location are Good, Moderate, Unhealthy for Sensitive Groups, Unhealthy, or Very Unhealthy. In addition, the AQI can inform the public about the general health effects associated with different pollution levels and describe possible precautionary steps to take if air pollution rises into the unhealthy ranges. Local news media provide alerts when AQI levels are unhealthful, which helps individuals make health-based decisions in support of daily activities.

Air pollution continues to be a widespread public health and environmental problem in the U.S., causing premature death, cancer, and long-term damage to respiratory and cardiovascular systems. Air pollution also reduces visibility, damages crops and buildings, and deposits pollutants on the soil and in bodies of water where they affect the chemistry of the water and the organisms living there. Approximately 113 million people live in U.S. areas designated as nonattainment areas by EPA for one or more of the six Criteria Air Pollutants, for which the federal government has established health-based standards [42]. Unhealthful air is expensive. The estimated annual health costs of human exposure to all outdoor air pollutants from all sources range from $40 billion to $50 billion, with an associated 50,000 premature deaths [43]. Most of the U.S. population lives in expanding urban areas where air pollution crosses local and state lines and, in some cases, crosses U.S. borders with Canada and Mexico [ibid.].

In the United States, the presence of unacceptable levels of ground-level ozone is the largest air contaminant problem, as determined by the number of people affected and the number of areas not meeting federal standards. Motor vehicles account for approximately one-fourth of emissions that produce ozone and one-third of nitrogen oxide emissions. Particulate and sulfur dioxide emissions from motor vehicles represent approximately 20 percent and 4 percent, respectively. Some 76.6 percent of carbon monoxide emissions are produced each year by transportation sources (e.g., motor vehicles) [43].

The problem of air pollution is international in scope. Deaths from air pollution, including indoor and outdoor sources, have been ranked by the World Health Organization (WHO) as one of the leading ten causes of disability. In 1997, WHO joined with others to estimate that approximately 700,000 deaths worldwide occur annually from particulate matter pollution and that 8 million avoidable deaths will occur worldwide by year 2020 [cited in 44]. Although some progress toward reducing unhealthful air emissions has been made, a substantial air pollution problem remains, with millions of tons of toxic air pollutants released globally each year. Reduction of global air pollution levels would mitigate or prevent much of the excess loss of life reported by WHO.

5.2.4.1 Associations between Air Pollution and Human Disease

The effects of polluted air on human health are numerous and significant. The health impacts are greatest for elderly persons, children, cigarette smokers, and persons with lung or heart disease. The health effects of contaminated air are well known to the U.S. public from news reports and the continuous release of new scientific information. What's best known to the public are the deleterious effects of air pollutants on the lungs. These effects are generally well known because of news media reports on lung disease related to air pollution and, more importantly, from weather reports that advise the public when air pollutants have reached hazardous levels. When such conditions occur, persons are advised to remain indoors and reduce activity levels when outdoors. Also, government agencies promote vehicle use reductions on days when pollution levels are hazardous. The sum of these news media and governmental acts is a general awareness among the U.S. public of the health hazards of air pollution, usually focused on the effects on the lungs.

Scientific evidence accrued over many years strongly associates specific air contaminants as contributing to lung disease, including respiratory tract infections, asthma, and lung cancer. As stated by the American Lung Association, "Lung disease claims close to 350,000 lives in America every year and is the third-leading cause of death in the United States. Over the last decade, the death rate for lung disease has risen faster than that of any of the top five causes of death" [35]. Air pollution is therefore a leading contributor to the respiratory health burden on the U.S. population.

The effects of air pollution on children's health is a particularly important subject, as any disease or disability in children reduces their quality of life and can lead to expensive health care costs. Knowing the effects of environmental hazards on children's health is important because they are preventable. In regard to outdoor air pollution, one major study has reported serious consequences to children who resided in areas in California where measured levels of air pollutants exist. In 1992, the California Air Resources Board (CARB) commenced a large-scale, long-term study of the health effects of children's chronic exposures to air pollutants in southern California [45]. Approximately 5,500

TABLE 5.6
Initial Findings from the California Children's Health Study [45]

Correlation was found between lower lung function and more intense air pollution

Slower lung growth was associated with high levels of NO_2, PM_{10}, and $PM_{2.5}$

Breathing capacity was lower for girls living in the most polluted communities

Wheezing was more evident in boys exposed to higher levels of NO_2 and acid vapor

---

children in twelve communities were enrolled in the study. Children's health status was assessed through questionnaires, pulmonary function testing, and monitoring of school absences. The study's major findings to date are summarized in Table 5.6. Two of the initial findings pertain to lung function and lung growth. Gauderman and colleagues [46] elaborated the CARB study's initial findings by reporting the effect of air pollution on lung development of children 10 to 18 years of age. Children (n = 1759) recruited from schools in 12 southern California communities served as the study population [47]. Linear regression was used to examine the relation between outdoor ambient air pollution levels and spirometric measures of lung development. Results showed that over the eight-year period of study, deficits in the growth of FEV(1) (forced expiratory volume in 1 second) were statistically significant with exposure to NO_2, acid vapor, PM$_{2.5}$, and elemental carbon. Of policy note, the results suggest that current levels of air pollution have chronic, adverse consequence to lung development in children.

In another study, the effect of air pollution on the occurrence of birth defects was reported by Ritz and colleagues. [48]. The investigators reviewed data from the California Birth Defects Monitoring Program on neonates and fetuses delivered in Southern California during the period 1987–1993. Monthly exposures to air pollutants were estimated from existing ambient air monitoring stations. Findings showed that odds ratios for cardiac ventricular defects increased in dose-response with increasing prenatal second-month carbon monoxide exposure. Also, second-month ozone exposure was associated with increased risk of aortic artery and valve defects, pulmonary artery and valve anomalies, and spinal defects. This study raises the troubling possibility that ozone and carbon monoxide may be related to birth defects of the heart and cardiopulmonary system.

In another study, children born in California during the period 1975–1987 birth outcomes were evaluated in regard to prenatal exposure to ozone, carbon monoxide, and particulate matter. Investigators reported that ozone exposure during the second and third trimesters of pregnancy and carbon monoxide exposure during the first trimester were associated with reduced birth weights. Specifically, a 12-ppb increase in twenty-four-hour ozone averaged over the entire pregnancy was associated with a 47.2 g lower birth weight. A 1.4-ppm difference in first-trimester CO exposure was associated with a 21.7g lower birth weight [49].

Further troubling findings regarding the adverse health effects of ambient air ozone were published by Bell et al. [50]. Using data from a national air pollution database, investigators estimated a national average relative rate of mortality associated with short-term exposure to ambient ozone for 95 large U.S. urban communities for the period

1987–2000. Findings showed that a 10-ppb increase in the previous week's ozone was associated with a 0.52 percent increase in daily mortality and a 0.64 perfect increase in cardiovascular and respiratory mortality. These findings extend the known association between air pollutants and human health impacts, and suggest that current ambient air quality standards should be further lowered in the interest of public health.

◊ ◊ ◊

Associations between environmental hazards and asthma have become a matter of great public health importance, indeed, a subject of urgency. The well-documented increase in asthma prevalence, particularly in children, has occurred globally, including countries as geographically and culturally diverse as the United States, Mexico, Denmark, and Australia. It is likely that epidemiological investigations in other countries would yield the same result (i.e., asthma prevalence increasing as the result of gene-environment interactions). Asthma now afflicts 14–15 million people in the United States and has become, almost silently to the U.S. public, the number one U.S. childhood illness [51]. According to the American Lung Association, the overall prevalence of asthma (cases per thousand people) rose from 34.8 in 1982 to 56.1 in 1994, an increase of 61 percent; the rise in pediatric asthma incidence rates (under eighteen years) increased from 40.1 to 69.1, an increase of 72 percent over the same years [cited in 51].

Further, the Centers for Disease Control and Prevention (CDC) estimated that 9 million children under 18 years of age have been diagnosed with asthma at some point in their lives, and more than 4 million have had an asthma attach in the past twelve months [52]. The CDC also estimated that approximately 7.5 percent of the U.S. population in year 2002 reported having asthma [53]. In 2002, self-reported asthma prevalence among racial/ethnic minority populations ranged from 3.1 percent to 14.5 percent, compared with 7.6 percent among whites [ibid.]. In this study, there was considerable variation between asthma prevalence according to race/ethnic populations. Current asthma was highest among non-Hispanic respondents of multiple races (15.6%), and lowest in non-Hispanic Native Hawaiians/Pacific Islanders (1.3%); other rates included non-Hispanic Blacks (9.3%), non-Hispanic whites (7.6%), and non-Hispanic Asians (2.9%).

Asthma rates are highest in urban areas. In one report, the incidence of asthma in children residing in central Harlem, New York City, was 25 percent, a strikingly high rate of this severe disease [54]. Moreover, the economic cost of increased prevalence of asthma is estimated to have risen from $6 billion in 1990 to $14 billion in 2000 [cited in 51]. To what extent can this dramatic rise in asthma rates and associated increased economic burden be attributed to air pollution?

A considerable body of published research now exists on environmental factors associated with new cases of asthma (i.e., induction) and exacerbation of an individual's existing asthmatic symptoms. One team of reviewers concluded, "Asthma is associated with production of IgE [Immunoglobulin E] to common environmental allergens, including feces of cockroaches and dust mites, animal dander, fungal spores, and pollens" [55]. "Studies illustrating causal effects between outdoor air pollution and asthma prevalence are scant" [ibid.]. These and other reviewers [e.g., 56] have postulated that asthma induction is more attributable to indoor environmental hazards, including indoor air quality, than to outdoor ambient air pollution, although the latter is important for the health of asthmatics. Some persons with asthma will experience more severe asthmatic symptoms (e.g., shortness of breath, coughing) during periods of elevated air pollution. In one review of the medical literature, a rise in ambient air ozone levels was found associated with an increase in emergency room visits for asthma attacks [57]. In another study, 4,000 school-aged children in twelve southern California communities were investigated to ascertain associations between medically-diagnosed asthma rates in children with putative exposure to environmental hazards [58]. Results showed that asthma diagnosis before five years of age was associated with exposures in the first year of life to wood or soil smoke, soot, or vehicle exhaust, cockroaches, herbicides, pesticides, and farm crops, farm dust, or farm animals.

Risk factors for asthma in children have been investigated by McConnell and associates. In one study, they investigated the association between newly-diagnosed asthma (n = 265 cases) and participation in team sports in a cohort of 3,535 young children [59]. Risk of asthma was investigated in children playing team sports in six communities in Southern California with high daytime ozone levels, six communities with lower ozone levels, and in communities with high or low concentrations of other air pollutants. Findings showed, in high ozone communities, the relative risk of developing asthma in children playing three or more team sports was 3.3, compared with children who played no sports. Further, time spent by children outdoors was associated with a higher incidence of asthma in high ozone areas, but not in areas of low ozone. The investigators opine that outdoor exercise and high ozone concentrations could increase the risk of children developing asthma.

Researchers at the University of Southern California investigated the pollution-asthma link in 208 children who resided since 1993 in ten Southern California cities [60]. Air samplers were placed outside the home of each student in order to measure NO_2 levels. Further, the distance of each child's home from local freeways, as well as how many vehicles traveled within 150 meters of the child's home were determined. Aerodynamic models were used to estimate traffic-related air pollution levels at each child's home. Results showed a link between asthma prevalence in the children and NO_2 levels at their homes. For each increase of 5.7 ppb in average NO_2, the risk of asthma increased by 83 percent. Further, the closer the students lived to a freeway, the higher the students' asthma prevalence. Asthma risk increased by 89 percent for every 1.2 kilometers (about three-quarters of a mile) the closer the students lived to a freeway.

Current knowledge about air pollution and its association with increased asthma prevalence and causing more severe asthmatic symptoms presents important policy questions. For example, if indoor air allergens are a major contributor to increased asthma prevalence, what would be an appropriate role for environmental protection and public health authorities? For instance, some might argue that a person's residence should not be subject to environmental law and regulatory concern, whereas outdoor air pollution belongs to all. However, it is in the tradition of public health to prevent disease and disability wherever possible, and policies on asthma prevention or mitigation must account for both indoor and outdoor air quality contributions.

◊ ◊ ◊

While the effects of air pollutants on lungs are, and will remain, significant in terms of the public's health, scientific evidence is emerging that air pollutants may exert an even greater public health burden as a contributor to cardiovascular and heart disease.

Particularly alarming is the association between particulate matter in air and their contribution to sudden heart failure. Research now implicates moderate levels of air pollution as triggers of fatal heart attacks. It is possible that sudden death, not lung disease, may be the most serious medical threat posed by ambient air pollution, given findings from recent research. For example, Rossi and colleagues [61] examined air pollution levels for the years 1980–1989 in Milan, Italy, for association with deaths on days of elevated pollution. Among the findings, a significant association was found for heart-failure deaths (7% increase/100 $\mu g/m^3$ increase in total suspended particulate [TSP]). Similarly, Neas and colleagues [62] analyzed daily mortality rates among Philadelphia, Pennsylvania, residents from 1973–1980. Investigators found that a 100 $\mu g/m^3$ increase in the forty-eight-hour mean level of TSP was associated with deaths due to cardiovascular disease. In another study, investigators examined air pollution levels in Seoul, Korea, and stroke mortality data over a four-year period [63]. They reported "[t]hat PM(10) and gaseous pollutants are significant risk factors for acute stroke death and that the elderly and women are more susceptible to the effect of particulate pollutants."

Other investigators suggest that exposure to fine (i.e., 2.5 μm in diameter or less) particulate matter ($PM_{2.5}$) decreases heart rate variability, possibly contributing to myocardial infarction (heart attack). For example, Gold and colleagues [64] measured heart rate variability (HRV) in twenty-one Boston, Massachusetts, residents from June to September 1997. Significantly less HRV was associated with elevated $PM_{2.5}$. In another study, Peters and colleagues. [65] compared defibrillator discharge interventions among 100 patients with implanted defibrillators. Patients with ten or more interventions experienced increased cardiac arrhythmias during periods of elevated air pollution.

Investigators at the University of Southern California investigated a large database in regard to chronic health effects of air pollution [66]. Researchers examined data from 22,906 residents of Los Angeles and adjacent areas. They determined air pollution exposure in 267 different zip codes where participants lived, and compiled causes of death for the 5,856 participants who died by year 2000. The effects of exposure to $PM_{2.5}$ was examined across the study areas. Among participants, for each increase of 10 micrograms per cubic meter ($\mu g/m^3$) of fine particles in the neighborhood's air, the risk of death from any cause rose by 11 percent to 17 percent. Ischemic heart disease mortality risks rose by 25 percent to 39 percent for the 10 $\mu g/m^3$ increase in air pollution. The investigators believed particulate matter may promote inflammatory processes, including atherosclerosis, in key tissues. In another study, investigators found an increase in overall mortality associated with each 10$\mu g/m^3$ increase in $PM_{2.5}$ modeled as the overall mean or as exposure in the year of death [67]. $PM_{2.5}$ was associated with increased lung cancer and cardiovascular deaths. Of note, the investigators' database included $PM_{2.5}$ levels that had decreased because of environmental controls. Findings showed improved overall mortality was associated with decreased $PM_{2.5}$. Although further research is needed to clarify the association between air pollution and fatal heart attacks, there are already sufficient data to move forward with public health prevention actions, such as public awareness and physician education campaigns.

◊ ◊ ◊

Regarding the health risk of individual air pollutants, EPA has developed a database to assist local, state, and federal governments involved in air pollution decision making. The

National-Scale Air Toxics Assessment (NATA) is a screening tool that estimates cancer and other health risks from exposure to toxic air pollutants, called *air toxics* by EPA [68]. Air toxics are those air pollutants known or suspected to be carcinogens or known to cause other health effects, such as birth defects or respiratory problems. Risk assessment methods (see chapter 11) are used by EPA to estimate human health risks from lifetime exposure to air pollutants at year 1999 levels as the baseline for their assessment.

EPA released its first NATA, based on year 1996 air emissions data, in year 2002. The second release occurred in 2006, using 1999 national emissions data. The 2006 NATA covers 177 of the Clean Air Act's list of 187 air toxics as well as diesel particulate matter (PM). The assessment includes estimates of cancer or non-cancer health effects for 133 air toxics and diesel PM for which EPA concluded that sufficient data existed.

Major findings revealed by the 2006 NATA assessment include both national estimates of health risk as well as health risks associated with individual air pollutants. On a national scale, the 2006 NATA estimates that people in most of the country have a lifetime cancer risk between 1 and 25 in a million due to inhalation exposure to air toxics. EPA notes that persons residing in transportation corridors can have a risk greater than 50 in a million [68]. For comparison, EPA estimates the national risk of contracting cancer from exposure to radon is approximately 2,000 in a million. Concerning individual air toxics, from a national perspective, benzene is the most significant air toxic for which cancer risk could be estimated, contributing 25 percent of the average individual cancer risk.

For most of the non-cancer health effects, estimated exposure levels to the air toxics covered by the 2006 NATA were generally below those of health concern to EPA. However, more than 92 percent of the U.S. population has hazard index (HI) values for respiratory toxicity greater than 1.0 and more than 17 percent have hazard index values (see chapter 11) greater than 10. EPA observes that because these exposures exceed the no-effect level (HI = 1.0), this result suggests that some people may experience an increased risk of respiratory irritation or other adverse respiratory effects from exposure to some air toxics. Of note, acrolein is the most significant air toxic that causes respiratory problems, contributing about 90 percent of the nationwide average cancer hazard in the year 2006 NATA assessment [68].

The NATA database and findings from research investigations like those cited in this chapter provide public health and environmental protection authorities with essential data from which to set federal, state, and local air pollution control policies.

5.2.4.2 Perspective

The Clean Air Act, as amended, is the most complex and comprehensive of the federal environmental statutes. Its provisions affect daily the whole of the U.S. population. For instance, the act controls the emissions of air contaminants from sources that range from internal combustion engines to emissions from electricity generating plants. This means that the provisions of the Clean Air Act affect anyone who uses an automobile or relies upon electricity for personal or business purposes. No other federal environmental statute has such a broad sweep of societal impacts.

5.3   THE CLEAN WATER ACT

5.3.1   HISTORY

Water suitable for human consumption and other uses has historically been of public health importance. Indeed, Hippocrates, the father of medicine, emphasized (circa 400 BC) the importance of boiling and straining water for health purposes [69]. Modern programs of water quality protection can be dated from the late nineteenth century, when chlorine was found to be an effective water disinfectant when added in low concentrations to drinking water supplies. In 1902, Belgium became the first country to make continuous use of chlorine as an additive to drinking water supplies. Chlorination of public drinking water supplies in the United States dates to 1908, when the Boonton reservoir supply, Jersey City, New Jersey, was chlorinated, triggering a series of lawsuits, which were ultimately decided by courts in favor of water chlorination as a means for water purification [ibid.]

The Clean Water Act established a program of grants to construct sewage treatment plants and a regulatory and standards program to control discharges of chemical and microbial contaminants into U.S. waters [70].

Chlorination of public drinking water supplies must be considered as a "modern" public health triumph. The notion of adding a human poison, chlorine—even at very low concentrations—to a vital resource, drinking water, must have seemed foolhardy to many persons in the early twentieth century. However, as time passed, the marked reduction in waterborne diseases such as cholera, typhus, and dysentery demonstrated the public health benefits of water chlorination and overcame residual public opposition.

Prior to the enactment of federal water quality statutes, states bore the responsibility for dealing with water quality problems, including issues of sanitation and drinking water quality. According to one source, many of the states' water pollution control policies from the late nineteenth century through the first half of the twentieth century comprised two steps:

1.   "First, common-law cases involving adverse effects of pollution upon public health or fish and wildlife resources were brought to court.
2.   Second, statutory regulatory authority was then given to state health or fish and wildlife agencies. Sometimes these two authorities were combined and extended by a water pollution control board" [71].

This approach by states led to considering each case of water pollution as an individual matter, subject to informal negotiations between polluters and state officials and attendant negotiations, all of which took considerable time in general, resulting in litigation if the parties could not agree on a pollution control strategy [71]. Little of this kind of informal approach to pollution control was apparent to the general public unless a particular court action attracted news media attention. Other problems with a state-by-state approach to water quality control included different water quality standards between states and the migration of some polluting industries to states with less stringent water quality standards and controls. Problems with state-based pollution controls contributed to pressure to develop federal water quality standards and regulations in mid-20th century.

The federal government's involvement with water pollution control dates to the turn of the twentieth century, when water pollution control regulations were included in the Rivers and Harbors Act of 1899. This act authorized the regulation of industrial discharges of pollution into waters that might cause navigation problems [71]. Later, the Public Health Service Act of 1912 expressed the first federal policy on the disposal of human wastes, which authorized the Public Health Service to provide technical advice and assistance to communities and for federal research on sanitary waste disposal methods. Over time, more comprehensive, focused federal water quality legislation was enacted by Congress, as described in this chapter.

The principal federal law now governing pollution of the nation's waterways is the Federal Water Pollution Control Act, more commonly called the Clean Water Act (CWA), whose titles are given in Table 5.7. The original purpose of the act was to establish a federal program to award grants to states for construction of sewage treatment plants. Although originally enacted in 1948, the act was completely revised by amendments in 1972, giving the CWA most of its current shape. The 1972 legislation declared as its objective the restoration and maintenance of the chemical, physical, and biological integrity of U.S. waters. Two goals were established: zero discharge of pollutants by 1985 and, as an interim goal and where possible, water quality that is both "fishable" and "swimmable" by mid-1983. While those dates have passed, the goals remain, and efforts to attain them are continuing [72].

The CWA contains a number of complex elements of overall water quality management. Foremost is the requirement in §303 that states must establish ambient water quality standards for water bodies, consisting of the designated use or uses of a water body (e.g., recreational, public water supply, or industrial water supply) and the water quality criteria that are necessary to protect the use or uses. Through permitting, states or EPA impose wastewater discharge limits on individual industrial and municipal facilities in order to ensure that water quality standards are attained. However, Congress

TABLE 5.7
Clean Water Act's Titles [72]

Title	Name of Title
I	Research and Related Programs
II	Grants for Construction of Treatment Works
III	Standards and Enforcement
IV	Permits and Licenses
V	General Provisions
VI	State Water Pollution Control Revolving Funds

recognized in the act that in many cases pollution controls implemented by industry and municipalities would be insufficient, due to pollutant contributions from other unregulated sources [72].

At the heart of the CWA is a system of permits, called the National Pollutant Discharge Elimination System (NPDES), which determines how much pollution can be released into surface (e.g., rivers) and underground water supplies. Each source of pollution must comply with permits specific to the source. Permits are therefore tailored to the size of the pollution source, the toxicity or hazard of individual pollutants, the technology available to reduce pollution levels, and the quality and size of the waterway receiving the pollution discharges [74]. Unfortunately, according to one environmental organization that studied the status of 6,700 permits for major facilities included in the NPDES, about 25 percent of the permits were not current [ibid.]. That is, more than 1,690 polluting facilities were operating without current discharge permits. Given the purpose of pollution discharge permits, it is important to the public's health that they be kept up to date.

In 2005, EPA's Inspector General reported that there remains a large backlog of NPDES permits requiring renewal [75]. According to the report, 1,120 major permit facilities, 9,386 individual minor, and 6,512 general minor permit facilities need permit renewals.

In a third analysis of CWA permits, in 2006 the U.S. Public Interest Research Group reported findings similar to those from the EPA Inspector General [76]. The group's research showed more than 62 percent of industrial and municipal facilities in the United States discharged more pollution into U.S. waterways than the CWA permits allowed. The investigation covered the period between July 2003 and December 2004. The average facility discharged pollution in excess of its permit limit by more than 275 percent, or almost four times the legal limit.

Reflection on these three reports of problems with CWA permits indicates a significant weakness in the permitting policy. Permits to discharge pollution into environmental media are ineffective without a commitment to enforce them.

The Clean Water Act has forced the development and use of technologies to reduce the quantities of pollutants released into waterways. The CWA gave industries until 1977 to install *best practicable control technology* (BPT) to clean up waste discharges. Later amendments to the CWA (Table 5.8) required a greater level of pollutant cleanups, generally requiring that by 1989 industry utilize the *best available technology* (BAT) that is economically feasible. Failure to meet statutory deadlines can lead to enforcement action, although compliance extensions of as long as two years are available for industrial sources utilizing innovative or alternative technology.

Control of pollution discharges has been the key focus of water quality programs. In addition to the BPT and BAT national standards, states are required to implement control strategies for waters expected to remain polluted by toxic chemicals even after industrial dischargers have installed the best available cleanup technologies required under the act, as amended. Development of management programs for these post-BAT pollutant problems was a prominent element in the 1987 CWA amendments and is a key continuing aspect of the act's implementation [77].

The process of deriving treatment requirements to attain specified water quality

TABLE 5.8
Clean Water Act and Major Amendments [72]

Year	Act
1948	Federal Water Pollution Control Act
1956	Water Pollution Control Act
1961	Federal Water Pollution Control Act Amendments
1965	Water Quality Act
1966	Clean Water Restoration Act
1970	Water Quality Improvement Act
1972	Federal Water Pollution Control Amendments
1977	Clean Water Act
1981	Municipal Wastewater Treatment Construction Grants Amendments
1987	Water Quality Act
2000	BEACH Act

is a complicated four-step process: (1) the state first establishes the desired highest use for a surface water, (2) then adopts scientific criteria (made available by EPA) for water quality to support the designated use, (3) determines how much of a pollutant may be discharged into a body of water without violating the criteria, and (4) determines how much of the pollutant may be discharged by a given point source [78]. Some states have chosen to have EPA administer these requirements rather than assume the responsibilities themselves.

Under §303(d) of the CWA, states must identify lakes, rivers, and streams for which wastewater discharge limits are not stringent enough to achieve established water quality standards, after implementation of technology-based controls by industrial and municipal dischargers. For each of these waterbodies, a state is required to set a total *maximum daily load* (TMDL) of pollutants at a level which ensures that applicable water quality standards can be attained and maintained [77]. A TMDL sets the maximum amount of pollution a waterbed can receive without violating water quality standards, including a margin of safety. If a state fails to do this, the EPA is required to develop a priority list for the state and make its own TMDL determination. A TMDL is both a planning process for attaining water quality standards and a quantitative assessment of problems, pollution sources, and pollutant reductions needed to restore and protect a river, stream, or lake. TMDLs may address all pollution sources, including point sources such as municipal sewage or industrial plant discharges; nonpoint sources, such as runoff from roads, farm fields, and forests; and naturally occurring sources, such as runoff from undisturbed lands. The TMDL itself does not establish new regulatory controls on sources of pollution. However, when TMDLs are established, municipal and industrial wastewater treatment plants may be required to install new pollution control technology.

The TMDL program has become controversial in part because of requirements and costs now facing states to implement this 30-year-old provision of the law. In 1999, EPA proposed regulatory changes to strengthen the TMDL program. Industries, cities, farmers, and others may be required to use new pollution controls to meet TMDL requirements. EPA's proposal was widely criticized, and congressional interest has been high. In July 2000, EPA issued final rules to revise the program, stimulating more controversy, although the effective date of the changes was delayed until October 2001. The Bush administration has decided to delay the effective date of the rule to allow for additional review and changes [77].

Enforcement Example: Connecticut Company to Pay $10 Million for Clean Water Act Violations

On August 17, 2004, the Tyco Printed Circuit Group (TPCG) of Stafford, Connecticut, was sentenced on 12 counts of violating the Clean Water Act. The plea agreement calls for TPCG to pay a total of $10 million in fines. Between 1999 and June 2001, TPCG managers at the company's Stafford, Staffordville, and Manchester facilities engaged in a variety of practices that caused the facilities to discharge wastewater with higher than permitted levels of pollutants into municipal sewage treatment systems. The case was prosecuted by the U.S. Attorney's Office in Hartford [73].

5.3.2 AMENDMENTS TO THE CWA

In addition to the 1972 amendments, several other important amendments to the CWA have occurred, as listed in Table 5.8. Amendments enacted in 1977, 1987, and 2000 are particularly relevant for public health purposes.

The 1977 amendments to the act focused on toxic pollutants. The Clean Water Act of 1977 established the basic structure for regulating discharges of pollutants into U.S. waters. Further, §404 established a program to regulate the discharge of dredged and fill material into U.S. waters, including wetlands. The basic premise of §404 is that no discharge of dredged or fill material can be permitted if a practicable alternative exists that is less damaging to the aquatic environment or if the nation's waters would be significantly degraded. Regulated activities are controlled by a permit review process. An *individual permit*, which is the responsibility of EPA, is usually required for potentially significant impacts. However, for discharges thought to have minimal impact, the U.S. Army Corps of Engineers can grant *general permits*, which are issued for particular categories of activities (e.g., minor road crossings, utility line backfill) as an expedited means for regulating discharges [79]. What are called *wetlands* constitute a vital natural resource in the United States and elsewhere. "Wetlands are areas where the frequent and prolonged presence of water at or near the soil surface drives a natural ecosystem, i.e., the kind of soils that form, the plants that grow, and the fish and wildlife that find habitat [80]. Swamps, marshes, and bogs are common types of wetlands. The Everglades in Florida are perhaps the best know U.S. wetland.

Wetlands serve an important environmental health purpose, one in addition to serving as a habitat for great numbers of birds, fish, mammals, plants, and trees. Wetlands are one of nature's water purifiers. Turbid surface waters that flow into wetlands drain off as freshwater. Regrettably, there has been a steady loss of wetlands acreage. For instance, the U.S. Fish and Wildlife Service estimates that between 1986 and 1997, a net of 644,000 acres of wetlands was lost, with 58,400 acres lost annually [81]. The principal causes of loss of wetlands were urban development, agriculture, silviculture (i.e., the growing and culture of trees), and rural development [ibid.]. As a matter of environmental policy, finding the best balance between protection of wetlands and the need for land development is, and will remain in the future, a difficult calculus for policy makers.

In 1987, the CWA was reauthorized and again focused on toxic substances. The amendments authorized citizen suit provisions, and funded sewage treatment plants under a construction grants program. Prior to 1987, the act only regulated pollutants discharged to surface waters from *point sources* (i.e., pipes, ditches, and similar conveyances of pollutants), unless a permit was obtained under provisions in the act. *Non-point sources* (e.g., stormwater runoff from agricultural lands) were covered by the Water Quality Act of 1987, which amended the CWA. Pollution sources are required by the act to treat their wastes to meet the more stringent of two sets of requirements, based on either technologic feasibility or attainment of desired levels of water quality [78]. Since 1972, the nation has invested more than $300 billion (in constant dollars) to build and upgrade wastewater treatment systems [82].

Over the years, the sewage collection system in the United States has grown to more than a million miles of collection pipes. This system carries about 50 trillion gallons of raw sewage daily, delivered to approximately 20,000 sewage treatment plants [83]. This enormous system of sewage collection and treatment represents a vital public health resource to the U.S. public, preventing human exposure to the pathogens found in raw sewage. It also represents an indispensable global environmental contribution by reducing the pollution load deposited in the planet's oceans and seas.

In 2000, the CWA was amended when Congress enacted the Beaches Environmental Assessment and Coastal Health (BEACH) Act, which amended §304 of the act. The act provided EPA with additional authority to regulate the quality of water used for recreation at beaches. The act applies to those 35 states and U.S. territories that have coastal water or border the Great Lakes [84]. The Beaches Act of 2000 contains eight sections [85]. Those sections with the greatest potential for protecting the public's health are as follows:

1. §2(i)(1)(A) "[e]ach State having coastal recreation waters shall adopt and submit to the Administrator water quality criteria and standards for the coastal recreation waters of the State for those pathogens and pathogen indicators for which the Administrator has published criteria under §(a)." If a state fails to adopt water quality criteria and standards or submits water quality criteria and standards less protective than EPA's, EPA is authorized to propose regulations for such states. Note: According to EPA, as of December 2004, 14 states have adopted adequate water quality standards, 21 states or territories have adopted some to none [84].

2.§3(a)(v)(b)(9) requires EPA to publish new or revised water quality criteria for pathogens and pathogen indicators for the purpose of protecting human health in coastal recreation waters. States are required to promptly notify the public, local governments, and EPA of any exceedance or likely exceedance of applicable water quality standards for coastal recreation waters. Note: In November 2004, EPA issued a final regulation on the assessment and monitoring of pathogens in recreation water [ibid.].

3.§4(a)(1)(A) requires EPA to publish performance criteria for monitoring and assessment of coastal recreation waters adjacent to beaches or similar points of access. Under this section, EPA is authorized to provide grants to states and local governments for implementation of monitoring and assessment programs. Under this section, federal agencies with jurisdiction over coastal recreation waters adjacent to beaches must develop and implement a monitoring and notification program for the coastal recreation waters.

The Beaches Act of 2000 is important for several public health and economic purposes. For the 2004 swimming season, of the 3,574 beaches monitored by twenty-eight coastal states and Puerto Rico, 942 had at least one health advisory or closing during the swimming season [85]. A total of 4,906 beach notification actions were taken at the 3,574 beaches. EPA has estimated that Americans annually take a total of 910 million visits to coastal areas and spend about $44 billion at beach locations [83]. Given the large number of beachgoers, exposure to water pathogens would be a considerable public health problem. Further, economic losses due to polluted beach water could be quite large, posing a financial hardship to cities and businesses that benefit from beach recreation. The spirit of the act is in the prevention of disease and disability, the core principle of public health.

5.5.3   KEY PROVISIONS OF THE CWA, AS AMENDED, RELEVANT TO PUBLIC HEALTH

Given the Clean Water Act's stated goal and policies, the current act could be said to consist of three major parts. The first major part consists of the provisions in Title I, which establishes research, investigations, training, and information authorities [72,87]. The provisions of Title II and Title VI constitute the second major part. They authorize federal financial assistance for municipal sewage treatment plant construction. The third major part consists of the regulatory requirements, found throughout the act, that apply to industrial and municipal discharges of pollutants.

Title I contains several important authorities delegated to EPA, including: (1) conduct and promote the coordination and acceleration of research, investigation, experiments, training, demonstration, surveys, and studies relating to the causes, effects, extent, prevention, reduction, and elimination of pollution; (2) encourage, cooperate with, and render technical services to pollution control agencies, and public and private agencies; (3) conduct, in cooperation with state water pollution control agencies and other interested agencies, public investigations concerning the pollution of any navigable waters; (4) establish advisory committees of recognized experts; (5) establish, equip, and maintain a water quality surveillance system; and (6) initiate and promote research for measuring the most effective practicable tools and techniques for measuring the social and economic costs and benefits of activities that are subject to regulation under the CWA [87].

Under **Titles II and VI**, Congress has authorized grants for planning, design, and construction of municipal sewage treatment facilities since 1956 through 1998. Since that time, more than $57 billion has been appropriated for grants to aid wastewater treatment plant construction [72]. Federal grants are made for several kinds of water treatment projects, based on a priority list established by the states. Grants are generally available for up to 55 percent of total project costs [ibid.]. Federal grants awarded to states are another example of environmental health federalism.

5.3.4　PUBLIC HEALTH IMPLICATIONS OF THE CWA

Contamination of water can come from both point (e.g., industrial sites) and nonpoint (e.g., agricultural runoff) sources. Biological and chemical contamination significantly reduces the value of surface waters (streams, lakes, and estuaries) for fishing, swimming, and other recreational activities, and can cause disease in humans. For example, during the summer of 1997, blooms of *Pfiesteria piscicida* were implicated as the likely cause of fish kills in North Carolina and Maryland. The development of intensive animal feeding operations (e.g., large scale swine farms) has worsened the discharge of improperly or inadequately treated wastes, which presents an increased health threat in waters used either for recreation or for producing fish and shellfish [43].

Two surveillance systems have provided relevant data on U.S. water quality. One system is a disease surveillance system operated by the Centers for Disease Control and Prevention (CDC). The other system, which is maintained by EPA, provides data on water quality measurements. Turning first to the CDC system, since 1971, CDC, EPA, and the Council of State and Territorial Epidemiologists have maintained a collaborative surveillance system for collecting and periodically voluntarily reporting data on occurrences and causes of waterborne-disease outbreaks (WBDOs) related to drinking water supplies. Tabulation of recreational water-associated outbreaks was added to the surveillance system in 1978. This surveillance system is the primary source of data concerning the scope and effects of waterborne disease outbreaks in the United States [88].

During the period 2001–2002, a total of sixty-five WBDOs associated with recreational water were reported by 23 states. The outbreaks caused illness among an estimated 2,536 persons; 61 persons were hospitalized, 8 of whom died. According to CDC [88], this is the largest number of recreational water-associated outbreaks to occur since reporting began in 1978. The numbers of recreational water-associated outbreaks have increased significantly during this period ($p < 0.01$). Of these 65 outbreaks, thirty involved gastroenteritis. The etiologic agent was identified in 23 of the 30 outbreaks; 18 of the 30 were associated with swimming or wading pools. Four of the 65 outbreaks involved acute respiratory illness associated with chemical exposure at pools.

In addition to the CDC waterborne disease surveillance system, EPA is required under §305(b) of the Clean Water Act, as amended, to report to Congress on the nation's water quality conditions. Under §305(b), states, territories, and interstate commissions must assess their water quality biennially and report those findings to the EPA. These entities must compare their monitoring results to the water quality standards they have set for themselves. In the year 2000 report to Congress, EPA found that 39 percent of rivers, 45 percent of lakes, and 51 percent of estuaries did not meet water quality standards [89]. The leading causes of inadequate water quality included bacteria, nutrients, metals (primarily mercury), and siltation. EPA cites the following conditions as the primary sources of water degradation: runoff from agricultural lands, sewage treatment plants, and hydrological modifications such as dredging of channels. As a matter of environmental health policy, these statistics indicate that the states, territories, and other governmental entities have a substantial problem if water quality is to be improved.

The problem of inadequate sewage treatment is particularly important, given the huge volume of sewage that must be treated in order to prevent waterborne diseases. As stated previously in this chapter, approximately 50 trillion gallons of raw sewage must be treated every day in the United States. Unfortunately, many of the sewage systems in the country are old and inadequately designed. To be more specific, some sewage-carrying pipes are almost 200 years old, with 100-year-old pipes not uncommon. Moreover, many older municipalities, primarily located in the northeastern and the Great Lakes regions, have sewage collection systems designed to carry both sewage and stormwater runoff. Such combined systems can overflow during heavy rainfall, resulting in raw sewage becoming mixed with stormwater, which can bypass sewage treatment plants. EPA has estimated that 1.3 trillion gallons of raw sewage are dumped annually due to combined sewer overflows. The agency also estimates that 1.8–3.5 million persons in the United States become ill annually from swimming in waters contaminated by sanitary sewage overflows [cited in 83]. To prevent this kind of public health problem will require repairing and upgrading the sewage collection and treatment systems. There is government and private sector consensus that there is a funding gap of $1 trillion for water infrastructure [ibid.]. Regrettably, gathering political support for repair and upgrading of municipal infrastructures can be difficult, sewage systems in particular. There is the tendency to pass infrastructure repairs to succeeding governments. Only when emergencies occur, such as the aftermath of hurricanes or release of large amounts of pollutants in an area or under a court order, do political bodies become energized.

5.5.5 Cost and Benefits of Water Pollution Control

In 2003, the White House's Office of Management and Budget (OMB) estimated that over the period 1992–2002 federal water pollution rules resulted in an annual benefit of $0.89–8.07 billion, with costs estimated at $2.4–2.9 billion. These figures have considerable uncertainty in the monetary benefits of water pollution control, with less uncertainty in the costs. The cost-benefit ratio therefore ranges from <1 to approximately 2.7 [90].

5.6 THE SAFE DRINKING WATER ACT

5.6.1 HISTORY

The Safe Drinking Water Act of 1974 (SDWA)[8] was enacted to protect the quality of drinking water throughout the country. This law focuses on all waters actually or potentially designed for drinking use, whether from surface or underground sources. Congress acted after a nationwide study of community water systems revealed widespread water quality and health risk problems resulting from poor operating procedures, inadequate facilities, and poor management of public water supplies in communities of all sizes. Further, the 1974 act was in response to congressional findings that chlorinated organic chemicals were contaminating major surface and underground water supplies, that widespread underground injection operations were a threat to aquifers, and that the infrastructures of public water supply systems were increasingly inadequate to protect the public health [92].

The SDWA, as amended in 1986 and 1996, gives EPA the authority to set drinking water standards. *Drinking water standards* are regulations that EPA sets to control the level of contaminants in the nation's drinking water. These standards are part of the act's "multiple barrier" approach to drinking water protection, which includes assessing and protecting drinking water sources; protecting wells and collection systems; making sure water is treated by qualified operators; ensuring the integrity of distribution systems; and making information available to the public on the quality of their drinking water. According to EPA, with the involvement of EPA, states, tribal nations, drinking water utilities, communities and citizens, these multiple barriers ensure that tap water in the United States and territories is safe to drink. In most cases, EPA delegates responsibility for implementing drinking water standards to states and tribal nations.

There are two categories of drinking water standards [93]:

4.1. A *National Primary Drinking Water Regulation* (NPDWR or primary standard) is a legally-enforceable standard that applies to public water systems. Primary standards protect drinking water quality by limiting the levels of specific contaminants that can adversely affect public health and are known or anticipated to occur in water. They take the form of Maximum Contaminant Levels or Treatment Techniques, which are described below.

4.2. A *National Secondary Drinking Water Regulation* (NSDWR or secondary standard) is a nonenforceable guideline regarding contaminants that may cause cosmetic effects (such as skin or tooth discoloration) or aesthetic effects (such as taste, odor, or color) in drinking water. EPA recommends secondary standards to water systems but does not require systems to comply. However, states may choose to adopt them as enforceable standards.

Drinking water standards apply to those public water systems that provide water for human consumption through at least fifteen service connections, or regularly serve at least twenty-five individuals. Public water systems include municipal water companies, homeowner associations, schools, businesses, campgrounds, and shopping malls. EPA estimates there are approximately 170,000 public water systems in the country. They are classified by EPA according to the number of people the systems serve, the source of their water, and whether they serve the same customers year-round or on an occasional basis. The three classifications (and the number of people each served in 1999) are: Community Water System (53,923 systems serving 253 million), Non-Transient Non-Community Water System (20,082 systems serving 6.3 million), and

---

The Safe Drinking Water Act establishes primary drinking water standards, regulates underground injection disposal practices, and establishes a groundwater control program [94].

Transient Non-Community Water System (93,729 systems serving 16.8 million) [95]. Community Water Systems, which are public water systems that supply water to the same population all year, constitute the vast majority of systems that supply water to the nation's population. Of the number of Community Water Systems, approximately eleven thousand relied on surface water as the source and served 167 million people, and the remainder relied on ground water, serving approximately 86 million people [ibid.]. Because of the large numbers of people serviced by these water systems, it is sound environmental health policy to protect them from contamination in order to prevent waterborne illnesses.

5.6.2   SAFE DRINKING WATER ACT AMENDMENTS

As indicated in Table 5.9, the SDWA has been amended several times since the original act of 1974. "The first major amendments, enacted in 1986, were largely intended to increase the pace at which EPA regulated contaminants. These amendments required

TABLE 5.9
Safe Drinking Water Act and Major Amendments [94]

Year	Act
1974	Safe Drinking Water Act
1977	Amendments
1979	Amendments
1980	Amendments
1986	Amendments
1988	Lead Contamination Control Act
1996	Amendments

EPA to (1) issue regulations for eighty-three specified contaminants by June 1989 and for twenty-five more contaminants every three years thereafter, (2) promulgate requirements for disinfection and filtration of public water supplies, (3) ban the use of lead pipes and lead solder in new drinking water systems, (4) establish an elective wellhead protection program around public wells, (5) establish a demonstration grant program for state and local authorities having designated sole-source aquifers to develop groundwater protection programs, and (6) issue rules for monitoring injection wells that inject wastes below a drinking water source. The amendments also increased EPA's enforcement authority" [94].

The Lead Contamination Control Act of 1988 added a new part F to the SDWA. It was intended to reduce exposure to lead in drinking water by requiring the recall of lead-lined water coolers, and required EPA to issue a guidance document and testing protocol to help schools and day care centers to identify and correct lead contamination in their drinking water [94]. The primary impetus for the act was a report to Congress that identified water coolers in schools as a potential source of children's exposure to lead in drinking water [96].

In 1996, Congress again made sweeping changes to the SDWA. Originally, the act focused primarily on treatment as the means of providing safe drinking water at the tap. The 1996 amendments modified the existing law by recognizing source water protection, operator training, funding for water system improvements, and public information as important components of safe drinking water programs [97]. Implementation of the 1986 provisions had brought to the fore widespread dissatisfaction among states and communities. These concerns included inadequate regulatory flexibility and unfunded mandates. "As over-arching themes, the 1996 Amendments target resources to address the greatest health risks, increase regulatory and compliance flexibility under the Act, and provide funding for federal drinking water mandates. Specific provisions revoked the requirement that EPA regulate 25 contaminants every 5 years, increased EPA's authority to consider costs when setting standards, authorized EPA to consider overall risk reduction, established a state revolving loan program to help communities meet compliance costs, and expanded the Act's focus on pollution prevention through a new source water protection program" [ibid.]. A cost-benefit analysis and a risk assessment are required before a standard can be set. The standards are initially based on health protection and the availability of technology. They are called *Maximum Contaminant Levels*. The amendments required EPA to promulgate standards that maximize health risk reduction benefits at costs that are justified by the benefits.

The 1996 SDWA amendments required that EPA establish criteria for a program to monitor unregulated contaminates found in drinking water supplies. Further, EPA must publish every five years a list of contaminants to be monitored in public drinking water supplies. One way to approach the requirement to regularly update a regulatory action is to establish a regulatory platform that first establishes criteria for updating—in this case a list of substances—then applying the criteria at specified intervals—in this instance, every five years. To develop such a platform is a policy decision. EPA released its Unregulated Contamination Monitoring Rule (UCMR) in September 1999, which covered twenty-five chemicals and one microorganism [98]. This was in response to the 1996 SDWA amendments that provided for monitoring of no more than thirty contaminants over a five-year period, monitoring only a representative sample of public water systems that serve fewer than 10,000 people, and storing analytical results in a National Contaminated Occurrence Database. The second list, UCMR 1, was proposed in August 2005 and contained 26 unregulated drinking water contaminants that must be monitored by U.S. water suppliers that exceed EPA designated minimum number of customers [99]. The data collected will help EPA determine whether to regulate the contaminants, their occurrence in drinking water, the potential population exposed to each contaminant, and the levels of exposure.

5.6.3 KEY PROVISIONS OF THE SDWA, AS AMENDED, RELEVANT TO PUBLIC HEALTH

There are five parts to the SDWA, as amended. The parts establish various responsibilities and authorities for EPA and the states. Parts A, B, C, and F are the most germane for public health concerns.

Part A—Definitions, §1401(1) "The term *'primary drinking water regulation'* means a regulation which—A) applies to public water systems; B) specifies contaminants which, in the judgment of the Administrator, may have any adverse effect on the health of persons; C) specifies for each such contaminant, either— (i) a maximum contaminant level, if, in the judgment of the Administrator, it is economically and technologically feasible to ascertain the level of such contaminants in water in public water systems, or (ii) if, in the judgment of the Administrator, it is not economically or technologically feasible to so ascertain the level of such contaminant, each treatment technique known to the Administrator which leads to a reduction in the level of such contaminant sufficient to satisfy the requirements of §1412; and D) contains criteria and procedures to assure a supply of drinking water which dependably complies with such maximum contaminant levels, including quality control and testing procedures to insure compliance with such levels and to insure proper operation and maintenance of the system [.]. (2) The term *'secondary drinking water regulation'* means a regulation which applies to public water systems and which specifies the maximum contaminant levels which, in the judgment of the Administrator, are required to protect the public welfare. Such regulations may apply to any contaminant in drinking water A) which may adversely affect the odor or appearance of such water and consequently may cause a substantial number of the persons served by the public water system providing such water to discontinue its use, or B) which may otherwise adversely affect the public welfare. Such regulations may vary according to geographic and other considerations." §1412(b)(4) "Each maximum contaminant level goal established under this subsection shall be set at the level at which no known or anticipated adverse effects on the health of persons occur and which allows an adequate margin of safety. Each national primary drinking water regulation for a contaminant for which a maximum contaminant level goal is established under this subsection shall specify a maximum level for such contaminant which is as close to the maximum contaminant level goal as feasible." §1412(b)(5) "For the purpose of this subjection, the term 'feasible' means feasible with the use of the best technology, treatment techniques and other means which the Administrator finds, after examination for efficacy under field conditions and not solely under laboratory conditions, are available (taking cost into consideration). [...]"

The preceding definitions are important not only for forming a foundation for regulatory actions under the SDWA, but also for their richness in policy implications. Noteworthy among the policy implications are: (1) water quality regulations apply only to public water systems (how to provide any services to private entities must be determined by health providers); (2) maximum contaminant levels goals are based on no known or anticipated human health effects (regulators must therefore decide what constitutes an adverse health effect for each contaminant); (3) maximum contaminant levels must include considerations of economic and technology feasibility (this policy is contentious because national water quality standards will lead to more costly technologies in some states, particularly in some western states, where there are large numbers of small water suppliers who lack financial resources to implement new water treatment technologies); (4) secondary drinking water regulations are based on "public welfare," such as odor or aesthetic properties (this policy of secondary water regulations provides state regulators with authority to consider water properties in addition to toxicological properties).

Part B covers public water systems. It required EPA to set minimum national standards to protect public health from drinking water contaminants. EPA was directed to develop, publish, and promulgate national primary drinking water regulations on eighty-three substances within thirty-six months following enactment of the 1986 amendments. In addition, the 1986 amendments required EPA to devise a priority list of additional contaminants for regulation and, every three years thereafter, issue Maximum Contaminant Levels (MCLs) and Maximum Contaminant Level Goals (MCLGs) for another twenty-five contaminants. The MCL is the National Primary Drinking Water Standard; the MCLG is the desired standard if technology permits its achievement. The MCLG is zero for microbial contaminants. For chemical carcinogens, the MCLG is zero unless a safe level can be established by EPA. Part B authorizes EPA to grant a state primary enforcement responsibility for public water systems under conditions specified in the statute and requires states to develop their plans to protect wellhead areas from contaminants.

In addition to establishing MCLs and MCLGs for specific contaminants in drinking water, EPA has established National Secondary Drinking Water Regulations that set nonmandatory water quality standards for fifteen contaminants [100]. EPA does not enforce these secondary maximum contaminant levels (SMCLs); they serve only as guidelines to assist public water systems in managing their drinking water quality. The SMCLs are based on cosmetic effects, technical effects, and aesthetic considerations, such as taste, color, and odor. EPA considers these contaminants not to present a risk to human health at the SMCL. The National Secondary Drinking Water Regulations were adopted in July 1979 and have been amended several times since then. Two states, California and Florida, have adopted the Secondary Standards as mandatory or required [101].

EPA notes that a variety of problems are addressed by their SMCLs. These include aesthetic effects (undesirable taste or odor), cosmetic effects (effects that do not damage the body but are still undesirable, e.g., skin discoloration), and technical effects (damage to water equipment or reduced effectiveness of treatment for other contaminants). EPA guidance implies that SMCLs were established as an aid to drinking water providers who were experiencing problems with customers' decreased use of unaesthetic drinking water. EPA guidance suggests methods to treat contaminants that exceed SMCLs.

In December 2005, EPA announced the finalization of two new drinking water protection rules. Both rules represent the last phase of a congressionally required rulemaking strategy required by the 1996 amendments to the SDWA [102]. The Long Term 2 Enhanced Surface Water Treatment Rule requires that public water systems that are supplied by surface water sources must monitor for *Cryptosporidia*. Those water systems that measure higher levels of *Cryptosporidia* or do not filter their water must provide additional protection by using EPA approved options for microbial control. The second rule, the Stage 2 Disinfection Byproducts Rule, requires public water systems that have high risks of disinfection byproducts (DBPs) to take corrective action when DBPs exceed drinking water standards.

Part C governs protection of underground sources of drinking water. It directs EPA to regulate underground injection, which is the subsurface emplacement of fluid through a well or dug-hole [92]. Emplacements of fluid could occur, for instance, through a septic tank, a cesspool, a dry well, or a fissure. EPA is directed by the SDWA to regulate state programs that in turn, were to regulate underground injection.

Part F is concerned with prohibiting the use of lead solder, pipes, or flux in drinking water systems and removal from schools of drinking fountains fabricated with lead-lined tanks and lead-containing solder. It required EPA to identify such coolers by brand name and provide this information to states, which in turn, are to develop programs of public education, water testing, and removal actions to replace lead-lined water coolers in schools.

The SDWA and its amendments are important not only for their establishment of standards for contaminants in drinking water systems, but for the use of these standards in other federal regulatory contexts. For instance, drinking water standards are used for some CERCLA sites to establish groundwater cleanup levels. Regarding hazardous waste issues, the 1984 amendments to the Resource Conservation and Recovery Act (RCRA), described later in chapter 8, contain several provisions directly applicable to hazardous waste injection wells. In particular, under these provisions, EPA must review all RCRA-listed hazardous wastes to determine whether injection or other land disposal of those wastes may continue [92].

◊ ◊ ◊

It is important to understand how EPA sets drinking water standards. "The 1996 Amendments to Safe Drinking Water Act require EPA to go through several steps to determine, first, whether setting a standard is appropriate for a particular contaminant, and if so, what the standard should be. Peer-reviewed science and data support an intensive technological evaluation, which includes many factors: occurrence in the environment; human exposure and risks of adverse health effects in the general population and sensitive subpopulations; analytical methods of detection; technical feasibility; and impacts of regulation on water systems, the economy and public health. Considering public input throughout the process, EPA must (1) identify drinking water problems; (2) establish priorities; and (3) set standards" [93].

1. Identify drinking water problems—EPA makes these determinations based on health risks and the likelihood that the contaminant occurs in public water systems at levels of concern. The National Drinking Water Contaminant List (CCL) lists contaminants that (a) are not already regulated under the SDWA, (b) may have adverse health effects, (c) are known or anticipated to occur in public water systems, and d) may require regulations under the SDWA.

2. Establish priorities—According to EPA, contaminants on the CCL are divided into priorities for regulation, health research and occurrence data collection. To support any regulatory decisions, EPA must determine that regulating the contaminants would present a meaningful opportunity to reduce human health risk.

3. Propose and finalize a National Primary Drinking Water Regulation—After reviewing health effects studies, EPA sets a Maximum Contaminant Level Goal (MCLG), the maximum level of a contaminant in drinking water at which no known or anticipated adverse effect on the health of persons would occur, and which allows an adequate margin of safety. MCLGs are nonenforceable public health goals. Since MCLGs consider only public health and not the limits of detection and treatment technology, sometimes they are set at a level that water systems cannot meet. When determining an MCLG, EPA considers the risk to sensitive subpopulations (infants, children, the elderly, and those with compromised immune systems) of experiencing a variety of adverse health effects. Three broad categories of contaminants are considered:

- Non-Carcinogens (not including microbial contaminants): For chemicals that can cause adverse non-cancer health effects, the MCLG is based on the reference dose. A reference dose (RfD) is an estimate of the amount of a chemical that a person can be exposed to on a daily basis that is not anticipated to cause adverse health effects over a person's lifetime. In RfD calculations, sensitive subgroups are included, and uncertainty may span an order of magnitude.
- Chemical Contaminants—Carcinogens: If there is evidence that a chemical may cause cancer, and there is no dose below which the chemical is considered safe, the MCLG is set at zero. If a chemical is carcinogenic and a safe dose can be determined, the MCLG is set at a level above zero that is considered safe.
- Microbial Contaminants: For microbial contaminants that may present public health risk, the MCLG is set at zero because ingesting one protozoa, virus, or bacterium may cause adverse health effects. EPA is conducting studies to determine whether there is a safe level above zero for some microbial contaminants. So far, however, this has not been established.

Once the MCLG is determined, EPA sets an enforceable standard. In most cases, the standard is a *Maximum Contaminant Level* (MCL), the maximum permissible level of a contaminant in water which is delivered to any user of a public water system. The MCL is set as close to the MCLG as feasible, which the Safe Drinking Water Act defines as the level that may be achieved with the use of the best available technology, treatment techniques, and other means which EPA finds are available (after examination for efficiency under field conditions and not solely under laboratory conditions), taking cost into consideration [ibid.].

When there is no reliable method that is economically and technically feasible to measure a contaminant at particularly low concentrations, a *Treatment Technique* (TT) is set rather than an MCL. A treatment technique (TT) is an enforceable procedure or level of technological performance which public water systems must follow to ensure control of a contaminant. Examples of Treatment Technique rules are the Surface Water Treatment Rule (disinfection and filtration) and the Lead and Copper Rule (optimized corrosion control).

After determining a MCL or TT based on affordable technology for large systems, EPA must complete an economic analysis to determine whether the benefits of that standard justify the costs. If not, EPA may adjust the MCL for a particular class or group of systems to a level that "maximizes health risk reduction benefits at a cost that is justified by the benefits" [93]. EPA may not adjust the MCL if the benefits justify the costs to large systems, and small systems unlikely to receive variances.

States are authorized to grant variances from standards for systems serving up to 3,300 people if the system cannot afford to comply with a rule (through treatment, an alternative source of water, or other restructuring) and the system installs EPA-approved variance technology. With EPA approval, states can grant variances to systems serving 3,301–10,000 people. The SDWA does not allow small systems to have variances for microbial contaminants [93].

Under certain circumstances, exemptions from standards may be granted to allow extra time to seek other compliance options or financial assistance. After the exemption period expires, the public water system (PWS) must be in compliance. The terms of variances and exemptions must ensure no unreasonable risk to public health.

Primary standards go into effect three years after they are finalized. If capital improvements are required, EPA's Administrator or a state may allow this period to be extended up to two additional years.

Small systems receive special consideration from EPA and states. According to EPA, more than 90 percent of all PWSs are small, and these systems face the greatest challenge in providing safe water at affordable rates. The 1996 SDWA amendments provide states with tools to comply with standards affordable for small systems. When setting new primary standards, EPA must identify technologies that achieve compliance and are affordable for systems serving fewer than 10,000 people. These may include packaged or modular systems and point-of-entry/point-of-use treatment devices under the control of the water supplier. When such technologies cannot be identified, EPA must identify affordable technologies that maximize contaminant reduction and protect public health. Small systems are considered in three categories: serving 10,000–3,301 people; 3,300–501 people; and 500–25 people.

As of January 2006, EPA has promulgated MCLs or Treatment Technologies for eighty-seven water contaminants [104]. Of this number, eleven are specific to micro-organisms (e.g., *Legionella*), three for disinfectants (e.g., chlorine dioxide), sixteen for inorganic chemicals (e.g., arsenic), fifty-three for organic chemicals (e.g., benzene) and four for radionuclides (e.g., ^{226}radium). As previously stated, this list is updated periodically based on requirements in the 1996 amendments to the SDWA.

Enforcement Example: New York Man Pleads Guilty to Drinking Water Monitoring Violation

A former employee of the New York City Department of Environmental Protection (DEP) pleaded guilty on August 4, 2005, in U.S. District Court for the Southern District of New York to charges that he falsified a log book required by the Safe Drinking Water Act while he was a DEP employee. The defendant admitted that on the date in question, he did not conduct water quality testing for turbidity and that he had made a false entry in the log book, which indicated he had done testing. When sentenced, the defendant faces a maximum sentence up to five years in prison and/or a fine of up to $250,000 [103].

5.6.3.1 Bottled Drinking Water

In the United States, tap water and bottled water are regulated by two different federal agencies, EPA and FDA, respectively. As described, EPA regulates tap water under its Safe Drinking Water Act authorities. Bottled drinking water is regulated as a food product, and as such, has been regulated since 1938 by FDA under the Federal Food, Drug and Cosmetic Act (FDCA). FDA has established specific regulations for bottled water, including standard of identity regulations that define different types of bottled water, such as spring water and mineral water. The agency has also established standard of quality regulations that establish allowable levels for contaminants (chemical, physical, microbial, radiological) in bottled water.

Relevant to this chapter, §305 of the SDWA amendments of 1996 includes language that amends §410 of the FDCA as follows:

"(b)(1) Not later than 180 days before the effective date of a national primary drinking water regulation promulgated by the Administrator of the Environmental Protection Agency for a contaminant under §1412 of the Safe Drinking Water Act , [t]he Secretary [of DHHS] shall promulgate a standard for that contaminant in bottled water or make a finding that such a regulation is not necessary to protect the public health because the contaminant is contained in water in public water systems [b]ut not in water used for bottled drinking water." "(4)(A) If the Secretary does not promulgate a regulation under this subsection within the period described in paragraph (1), the national primary drinking water regulation referred to in paragraph (1) shall be considered [a]s the regulation applicable under this subsection to bottled water." Stated more simply, FDA must adopt EPA's MCLs if the contaminants appear in bottled water.

In addition to FDA, state and local governments also regulate bottled water. FDA relies on state and local government agencies to approve water sources for safety and sanitary quality, as specified in §129.3 of the FDCAct. Additionally, states also regulate the bottled water industry as well as the industry itself [105].

It is clear from the preceding language that congressional intent was to yoke bottled drinking water quality with that of tap water as a means to ensure bottled water quality. Has the intent been realized? According to one national environmental group, the Natural Resources Defense Council (NRDC), "[b]ottled water sold in the U.S. is not necessarily cleaner or safer than most tap water" [106]. This conclusion was predicated on findings from their study of 103 brands of bottled water. Approximately one-third of the waters tested contained levels of contamination in at least one sample that exceeded allowable limits under either state or bottled water industry standards or guidelines. Moreover, the NRDC concluded that bottled water regulations are inadequate because the FDA's rules exempt waters that are packaged and sold within the same state, which is 60–70 percent of all bottled water sales, and approximately 20 percent of states don't regulate bottled water.

5.6.4   PUBLIC HEALTH IMPLICATIONS OF THE SDWA, AS AMENDED

5.6.4.1   U.S. Water Quality

Providing drinking water free of disease-causing agents, whether biological or chemical, is the primary goal of all water supply systems. During the first half of the twentieth century, the causes for most waterborne disease outbreaks were bacteria, whereas beginning in the 1970s, protozoa and chemicals became the dominant causes [107]. Most outbreaks involve only a few individuals [43]. However, failures in water treatment systems have occasionally led to instances of widespread waterborne disease. For example, more than 400,000 people were affected in 1993 when the Milwaukee, Wisconsin, water supply became contaminated with *Cryptosporidia*. A public water supply in Canada was contaminated by a bacterium, *E. coli,* May of 2000. *E. coli* is an intestinal bacterium that causes muscle cramps, fever, nausea, and severe diarrhea, and can cause kidney failure in extreme cases. The outbreak may have caused more than 2,000 cases of illness, including seven deaths [108].

Drinking water contaminated with microbial or chemical contaminants can cause human disease. Sources of drinking water (and percentage of waterborne disease outbreaks) comprise wells (70.5%), springs (5.9%), surface water (11.8%), and a combination of wells and springs (11.8%) [109]. It has been estimated that up to 900,000 people, with 900 deaths, occur annually from waterborne infectious illnesses [110]. Since 1971, CDC, EPA, and the Council of State and Territorial Epidemiologists have maintained a collaborative surveillance system for collecting and periodically reporting data related to occurrences and causes of waterborne-disease outbreaks (WBDOs). This surveillance system is the primary source of data concerning the scope and effects of waterborne disease outbreaks on persons in the United States [111]. During 2001–2002, a total of 31 WBDOs associated with drinking water were reported by 19 states. These 31 outbreaks caused illness among an estimated 1,020 persons and were linked to seven deaths. Of the 24 identified outbreaks, nineteen were associated with pathogens, and five were associated with acute chemical poisonings. Five outbreaks were caused by noroviruses, five by parasites, and three by non-*Legionella* bacteria. All seven outbreaks involving acute gastrointestinal illness of unknown etiology were suspected of having an infectious cause [ibid.]. Shown in Figure 5.2 are the primary etiologic agents associated with WBDOs that were reported to CDC during the reporting period. Approximately 60 percent of disease outbreaks were associated with the presence of microorganisms, 16 percent were associated with chemical contaminants, figures that can be useful for prioritizing resources for use in preventing waterborne disease outbreaks.

The number of drinking water-associated outbreaks decreased from thirty-nine during 1999–2000 to thirty-one during 2001–2002 [111]. Two outbreaks associated with surface water occurred during 2001–2002; neither was associated with consumption of untreated water. The number of outbreaks associated with groundwater sources decreased from twenty-eight during 1999–2000 to twenty-three during 2001–2002; however, the proportion of such outbreaks increased from 73.7 to 92.0 percent. The number of outbreaks associated with untreated groundwater decreased from 17 (44.7%) during 1999–2000 to ten (40.0%) during 2001–2002. Outbreaks associated with private, unregulated wells remained relatively stable, although more outbreaks involving private, treated wells were reported during 2001–2002. Because the only groundwater systems that are required to disinfect their water supplies are public systems under the influence of surface water, these findings support EPA's development of a groundwater rule that specifies when corrective action (including disinfection) is required.

Drinking water supplies can also become contaminated by chemical contaminants (Figure 5.2). The classes of contaminants include disinfection byproducts (DBPs) (from chlorination of water), metals (lead, arsenic), nitrates (from fertilizers, septic tanks), radon (naturally occurring), and pesticides/synthetic organic chemicals. The human

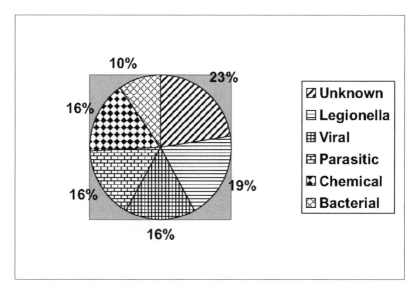

FIGURE 5.2 Etiologic agents and U.S. waterborne disease outbreaks from drinking water, 2001–2002, thirty-one events [111].

health effects caused by specific contaminants will vary according to each substance's toxicity, population at risk, and dose. For example, DBPs are formed when chlorine is added to water supplies for disinfection purposes and then combines with organic materials. EPA has estimated that an upper bound estimate of 2–17 percent of human bladder cancer cases is attributable to DBP exposure [110]. Young children are at elevated health risk from exposure to lead, pesticides, and nitrates. Lead exposure *in utero* is associated with developmental effects in infants; low-level, chronic exposure to pesticides has been associated with neurotoxic and behavioral effects in children; and ingestion of high concentrations of nitrates in infants under four months of age can result in methemoglobinemia, a condition that is fatal in 7–8 percent of cases [ibid.].

In addition to knowing the nature of etiologic agents associated with waterborne disease outbreaks, there is information available from CDC that identifies deficiencies in water quality treatment systems. Water treatment deficiencies included inadequate, interrupted, or no disinfection of groundwater; inadequate or interrupted filtration of surface water; and inadequate disinfection of unfiltered surface water. Water systems are classified by EPA as community systems, noncommunity systems, and individual water systems. During the period 1991–1998, 68 incidents (groundwater and surface water combined) of water system deficiencies occurred. Of this number, 58 (85%) were groundwater sources. Such data can be used by water quality managers to more effectively identify potential problems in their water treatment technologies and operations.

Private wells are generally exempt from meeting water quality standards, unless local authorities require otherwise, which is rarely the case. Given that about 20 percent of drinking water in the United States comes from private wells, a potential public health problem exists. Local health departments generally have information available about the potential health hazard of uninspected private wells and make that available to local residents.

5.6.4.2   Global Water Quality

While the quality of drinking water is generally good in the United States and other developed countries, that is not the case globally. In particular, developing countries struggle to build the environmental health infrastructure necessary to drastically reduce the horrible toll of waterborne diseases. The diseases are caused by ingestion of water contaminated by human or animal feces or urine containing pathogenic bacteria or viruses. Waterborne diseases include cholera, typhoid, dysentery, and other diarrheal diseases. These diseases can be prevented through disinfection of drinking water supplies and sanitary disposal of animal and human bodily wastes.

The United Nations collects data from individual countries in order to estimate the access of human populations to potable water and adequate sanitation. According to their data, which currently cover 89 percent of the world's population, 1.1 billion people lacked access to "improved water supply" and more than 2.4 billion lacked access to "improved sanitation," primarily in developing countries [111]. Given the world's population of more than six billion people in year 2000, approximately 20 percent lacked access to potable water and 40 percent lacked access to proper sanitation. Using the WHO data, Gleick [112] estimated that as many as 135 million people will die from "unmet human needs for water" by year 2020, assuming the lack of sustained public health interventions. To put this figure into perspective, 135 million people exceeds the combined current populations of France (60 million), Spain (40 million), and Canada (31 million) [113]. Imagine the hue and cry should the population of these three countries disappear over the next twenty years. The same anxiety and fear should be mobilized to prevent the same number of deaths from occurring in the developing countries.

As a matter of international environmental health policy, should the developed countries provide the resources necessary to prevent the global loss of life and illnesses attributable to waterborne diseases? The answer is obviously, yes. Altruistic reasons include children's welfare, improved quality of life, decreased disabilities, and increased longevity. Prevention of waterborne diseases would also improve national economic development by increased workforces, lower health care costs, improved social stability, and larger tax bases from employed workers and from the products they would produce.

5.6.4.3 Graywater

The reuse of graywater is an emerging environmental policy issue. *Graywater* is generally defined as all wastewater generated from household activities except that produced from toilets, which is called *blackwater*. Graywater includes water from dish washers, clothes washers, household wash basins, showers, and bathtubs. Such "waste" can be collected and used for outdoor watering of plants, trees, lawns, and irrigation of crops. The average amount of graywater produced in the U.S. is 40 gallons per day per person, which equates to about 65 percent of a household's daily water consumption [115]. This is a significant amount of water that is potentially available for recycling. Graywater is important because several U.S. states are considering its use as a component of water conservation programs. These programs are largely nascent and are in response to shortages of water supplies needed to meet the needs of households, industry, and agriculture. The causes of the shortages vary, but factors include increased human populations, fragile groundwater supplies, greater water demand by industry, and drought conditions. Making maximum use of existing water supplies is an environmental policy that will become increasingly important in many countries, including the United States, as climate changes due to greenhouse gas emissions continue to appear.

Can graywater be used so as to avoid public health consequences? Some microorganisms inevitably contaminate any water that comes into contact with human activities. Some treatment of graywater is therefore advisable before its reuse [112]. Technology for treating graywater contamination is under development both for households and other users. Both on-site and central treatment systems are being considered. In the future, the gradual introduction of graywater collection, treatment, and reuse will become environmental health policy, most likely by action of individual states in arid regions of the United States.

5.7 SUMMARY

This chapter describes what are, arguably, the three most significant environmental statutes in terms of public health benefits, because they impact large numbers of the U.S. population. The Clean Air Act (CAA), as amended, has as purpose the control of contaminants released into outdoor ambient air. The act, as amended, is the most complex and comprehensive of the suite of U.S. environmental statutes. It is the statute that surely has impact on the greatest part of the U.S. population, since air quality is relevant to everyone. As discussed, the CAA also has the greatest amount of health data. The act contains several policies of relevance to public health practice. The CAA adopted the policy of developing *National Ambient Air Quality Standards* (NAAQS) for individual air contaminants, then placed most of the responsibility on the states to achieve compliance with the standards, which is an example of the federalism policy. The act also established two kinds of national air quality standards. *Primary standards* are based on protection of human health and are enforceable under law. *Secondary air quality standards* set limits to protect public welfare, including protection against decreased visibility, damage to buildings, and deleterious ecological effects, and are advisory, although some states have adopted them as legally binding. As policy, having two sets of standards provides policy makers with guidance on priorities and resource management.

The 1990 CAA amendments added an elaborate permit program and markedly strengthened enforcement provisions and requirements for geographic areas that fail to meet air quality standards. The policy of issuing permits to generators of air pollution is a strategy of using emission controls as a means toward achieving the NAAQS. The 1990 amendments finally banned the sale in the U.S. of gasoline containing lead additives, ending one of the twentieth century's worst environmental health missteps, the use of tetraethyl lead as a gasoline additive.

Just as the Clean Air Act controls the emissions of contaminants released into outdoor air in the United States, the Clean Water Act (CWA), as amended, controls the emissions of contaminants into U.S. bodies of water. As with the CAA, the CWA contains a number of policies of importance to public health practice. One of the earliest provisions, continuing today, is the awarding of grants to states for construction of sewage treatment plants, an example of federalism. The public health benefits of treating raw sewage in order to achieve sanitary and healthful conditions are obvious. As another important policy, the CWA requires states or the EPA to issue permits that limit the amount of contaminants discharged into bodies of water. These emission standards are for purpose of meeting water quality standards established under the Safe Drinking Water Act (SDWA). The CWA also adopts the policy that the regulated community (i.e., those entities that release contaminants into water) must use the Best Available Technology in their waste management operations. This policy leads to updates and improvements in waste management as technology changes. The policy also moves the regulated community toward a uniform technology.

The Safe Drinking Water Act, as amended, is the complement of the Clean Water Act in that the act establishes water quality standards that forge emission standards under the CWA. The SDWA contains policies of import to public health practice. The act creates the policy of dual drinking water regulations. Primary regulations are intended to protect human health and are enforceable under law. Secondary regulations, which are voluntary unless adopted as law by individual states, pertain to welfare considerations such as odor, appearance, and taste of water. The SDWA also contains a second set of dual standards. Specifically, the EPA must establish Maximum Contaminant Levels (MCLs) for individual water contaminants, which are legally enforceable by states, unless individual states have promulgated more stringent standards. MCLs, as policy, must consider the availability of technology necessary to achieve desired contaminant levels. Where technology is lacking, Maximum Contaminant Level Goals (MCLGs) are established. The policy of having both MCLs and MCLGs marries the present to the future. Public health benefits when water quality standards shift when new or improved water treatment technologies are adopted.

5.8 POLICY QUESTIONS

1. Visit your school library and obtain three journal papers on asthma in children. Using what you consider to be the key public health issues gleaned from the articles, apply the issues to Appleton's four elements of the precautionary principle, as found in Table 2.4. In particular, discuss the prevention of childhood asthma in terms of the precautionary principle elements and attendant actions. Use critical thinking to the extent possible.

2. Congratulations! You've just been hired as the newest member of the local health department. The director has asked you to become a member of the policy team. You're expected to develop a policy initiative on childhood asthma that will gain the support of the department and, eventually, the county commissioners. Using as background the articles on childhood asthma that were obtained from the local library, describe how you will develop this policy and get it implemented, using the PACM model.

3. Under § 202, Title II, of the Clean Air Act, EPA must establish emission standards for new motor vehicles and engines. (A) Do you agree that EPA should have this authority? Why or why not? Be specific. (B) Using Internet and EPA resources, determine the emissions standards for your personal vehicle.

4. Ethanol is used as a gasoline additive in some states for the purpose of lowering vehicle emissions. Using Internet resources, discuss the benefits and disadvantages of using ethanol as a gasoline additive.

5. Assume that you are in charge of communications for the local health department. Develop a one-page health alert that can be released to the public and medical community during periods of high ambient air levels of ozone.

6. The Clean Water Act requires states to establish ambient water quality standards for bodies of water. Select a lake within your state's borders and determine the applicable water quality standards. Discuss the public health implications of the lake's water quality standards.

7. In the context of public health, discuss the significant differences between a primary drinking water regulation and a secondary drinking water regulation, as found in the Safe Drinking Water Act.

8. If climate-change models are correct, changes in global rainfall patterns are likely, making water conservation a necessary environmental policy. Discuss 10 ways that you can conserve water, now and in the future. List the ways in descending order of effectiveness.

9. What are the primary effects of the criteria air pollutants on human health? *Note*: The criteria air pollutants comprise carbon monoxide, lead, ozone, particulate matter, oxides of sulfur, and oxides of nitrogen. The text describes human health effects as including respiratory disease, cardiovascular disease, acute morbidity, neurotoxicologic effects, and eye irritation. The text stresses the important of effects caused by ozone (pulmonary irritation), particulate matter (cardiovascular effects, premature death), and carbon monoxide (angina of heart).

10. For pollution control purposes, several federal environmental statutes include the embedded policy of granting permits to those who pollute, a kind of command and control policy. Select one such statute and discuss two alternative to a permit policy. Using critical thinking, discuss the likely effectiveness of each alternative.

NOTES

1. The word *smog* was first recorded in 1905 in a newspaper report of a meeting of the Public Health Congress, where Dr. H.A. des Vœux gave a paper entitled "Fog and Smoke" in which he coined the word smog [3].
2. A temperature inversion, which occurs when a cold layer of air settles under a warmer layer, can slow atmospheric mixing and allow pollutants to accumulate hazardously near ground level.
3. The GAO Human Capital Reform Act of 2004 changed the agency's name to Government Accountability Office, effective July 7, 2004 [16].
4. California, Connecticut, Illinois, Massachusetts, Maine, New Jersey, New Mexico, New York, Oregon, Rhode Island, Vermont, Washington, Baltimore, New York City, Washington, D.C., and American Samoa.
5. Conecticut, Deleware, Massachusetts, ME, NH, NJ, NY, RI, VT
6. Short for *de minimis non curat lex*: the law takes no account of trifles. The phrase "de minimus" literally means "of minimum impact." A de minimis standard exempts producers of environmental pollution if the amounts are below some level thought by regulatory agencies to be without public health or environmental consequence [32].
7. PM_{10}, the mass of particles less than 10 micrometers in diameter. Similarly, $PM_{2.5}$ refers to the mass of particles less than 2.5 micrometers in diameter.
8. The "Safe Drinking Water Act" consists of Title XIV of the Public Health Service Act (42 U.S.C. 300f-300j-D) as added by Public Law 93-523 (December 16, 1974) and subsequent amendments [91].

REFERENCES

1. Fromson, J., A history of federal air pollution control, in *Environmental Law Review–1970*, Sage Hill Publishers, Albany, 1970, 214.

2. Bell, M.L. and David, D.L., Reassessment of the lethal London fog of 1952: Novel indicators of acute and chronic consequences of acute exposure to air pollution, *Environmental Health Perspectives*, 109 (supp. 3), 389, 2001.

3. *American Heritage®Dictionary of the English Language*, 3rd ed., 1996, Houghton Mifflin Co., Boston, 1996.

4. CARB (California Air Resources Board), California's Air Quality History—Key Events, Air Resources Board, Sacramento, 2000.

5. McCarthy, J.E., et al., Clean Air Act. Summaries of Environmental Laws Administered by the EPA, Congressional Research Service, Washington, D.C., 1999.

6. Randle, R.V. and Bosco, M.E., Air pollution control, in *Environmental Law Handbook*, Arbuckle, J.G., et al., eds., Government Institutes, Inc., Rockville, 1991, 524.

7. Quarles, J. and Lewis, W. Jr., Mopping up after the Clean Air Act; navigating the rules and fog, *Legal Times*, February 11, 1991.

8. Fiorino, D.J., *Making Environmental Policy*, University of California Press, Berkeley, 1995.

9. Chameides, W., et al., Air quality management in the United States, *Journal Air Polution Management Association*, July, 22, 2005.

10. EPA (U.S. Environmental Protection Agency), EPA announces results of annual sulfur dioxide auction, news brief, Washington, D.C., March 29, 2005.

11. EPA (U.S. Environmental Protection Agency), Clean Air Interstate Rule. Basic Information, Office of Air and Radiation, Washington, D.C., 2005.

12. EPA (U.S. Environmental Protection Agency), Federal plan proposed to ensure power plant pollution cuts, press advisory, Washington, D.C. August 1, 2005.

13. EPA New England, Charting our progress, setting our course for the future: Protecting New England's air quality at the millennium, press release. Boston, December 17, 1999.

14. EPA (U.S. Environmental Protection Agency), EPA celebrates 15th anniversary of Clean Air Act Amendments, news release, Washington, D.C., 2005.

15. Sandel, M.J., It's immoral to buy the right to pollute, *New York Times*, December 15, 1997.

16. GAO (Government Accountability Office), GAO Human Capital Reform Act of 2004. Available at http://www.gao.gov, 2005.

17. GAO (General Accounting Office), Air Pollution. Emissions from Older Electricity Generating Units, GAO-02-709, Washington, D.C., 2002.

18. Edison Electric Institute, New Source Review, Washington, D.C., 2001.

19. Pianin, E., White House to end power plant probes, *Washington Post*, November 6, 2003, A31.

20. EPA (U.S. Environmental Protection Agency), EPA announces steps to increase energy efficiency, encourage emissions reductions. Available at http://www.epa.gov/air/nsr-review/release.html, 2002.

21. ALA (American Lung Association), *State of the Air 2002*, Statement of John L. Kirkwood. Available at http://www.lungusa.org/press/envir/air_061402.html, 2002.

22. NAM (National Association of Manufacturers), Comments on new source review (NSR) 90-day review and report to the President (Docket Number A-2001-19), Washington, D.C., 2001.

23. Pianin, E., Clean air rules to be relaxed, *Washington Post*, August 23, 2003, A01.

24. Janofsky, M., Judges overturn Bush bid to ease pollution rules, *New York Times*, March 16, 2006.

25. DePalma, A., Court says E.P.A. can limit its regulation of emissions, *New York Times*, July 16, 2005.

26. DePalma, A., 9 states in plan to cut emissions by power plants, *New York Times*, August 24, 2005.

27. EPA (U.S. Environmental Protection Agency), Benefits and Costs of the Clean Air Act. Available at http://www.epa.gov/oar/sect812/copy.html, 1999.

28. EPA (U.S. Environmental Protection Agency), Benefits and Costs of the Clean Air Act 1990-2010. Available at http://www.epa.gov/oar/sect812/1990-2010/fullrept.pdf, 1999.

29. Ryan, H.S. and Thompson, K.M., 2003. When domestic environmental policy meets international trade policy: Venezuela's challenge to the U.S. gasoline rule, *Human and Ecological Risk Assess*ment 9, 811, 2003.

30. OMB (White House Office of Management and Budget), Informing Regulatory Decisions: 2003 Report to Congress on the Costs and Benefits of Federal Regulations and Unfunded Mandates on State, Local, and Tribal Entities, Office of Information and Regulatory Affairs, Washington, D.C., 2003.

31. EPA Press Advisory, U.S. announces $94 million Clean Air Act settlement with Chrysler over emission control defects on 1.5 million Jeep and Dodge vehicles, Washington, D.C., December 21, 2005.

32. TNRCC (Texas Natural Resource Conservation Commission). De minimis for air emissions. Available at http://www.tnrcc.state.tx.us/exec/shea/sblga/deminimis.html, 2003.

33. EPA (U.S. Environmental Protection Agency), Latest Findings on National Air Quality: 2000 Status and Trends, Office of Air and Radiation, Washington, D.C. Available at http://www.epa.gov/oar/aqtmd00/, 2001.

34. EPA (U.S. Environmental Protection Agency), 1990 Amendments to the Clean Air Act. Available at http://www.epa.gov/oar/section812/copy.html, 1998.

35. ALA (American Lung Association), EPA Changes to New Source Review 'Accounting Gimmicks that Increase Pollution,' Statement of John L. Kirkwood, President and CEO, American Lung Association. Available at http://www.lungusa.org/press/envir/air_061402.html, 2000.

36. Gwynn, R.C. and Thurston, G.D., The burden of air pollution: Impacts among racial minorities, *Environmental Health Perspectives*, 109(suppl 4), 501, 2001.

37. Burnett, R.T., et al., Association between ozone and hospitalization for acute respiratory diseases in children less than 2 years of age, *American Journal of Epidemiology*, 153(5), 444, 2001.

38. Daggett, D.A., Myers, J.D., and Anderson, H.A., Ozone and particulate matter air pollution in Wisconsin: Trends and estimates of health effects, *Wisconsin Medical Journal*, 99(8), 47, 2000.

39. EPA (U.S. Environmental Protection Agency), New clean diesel rule major step in a decade of progress, news release, Washington, D.C., May 11, 2004.

40. EPA (U.S. Environmental Protection Agency), Air Quality Index—A Guide to Air Quality and Your Health, Office of Air and Radiation, Washington, D.C., 2006.

41. Cleanaircampaign, Health Effects. Available at http://www.cleanaircampaign.com, 2004.

42. EPA (U.S. Environmental Protection Agency), National Air Quality and Trends Report, Office of Air and Radiation, Washington, D.C., 1997.

43. DHHS (U.S. Department of Health and Human Services), Healthy People 2010, vol 1, Washington, D.C., 2000, 8–1.

44. Cifuentes, L., et al, Hidden health benefits of greenhouse gas mitigation, *Science*, 293, 1257, 2001.

45. CARB (California Air Resources Board), The Children's Health Study, Air Resources Board, Sacramento, 2002.

46. Gauderman, W.J., et al., The effect of air pollution on lung development from 10 to 18 years of age, *New England Journal of Medicine*, 351(11), 1057, 2004.
47. Gauderman, W.J., et al., Childhood asthma and exposure to traffic and nitrogen dioxide, *Epidemiology*, 16, November, 2005.
48. Ritz, B., et al., Ambient air pollution and risk of birth defects in Southern California, *American Journal of Epidemiology*, 155, 17, 2002.
49. Salam, M.T., et al., Birth outcomes and prenatal exposure to ozone, carbon monoxide, and particulate matter: Results from the children's health study, *Environmental Health Perspectives*, 113, 1638, 2005.
50. Bell, M.L., et al., Ozone and short-term mortality in 95 US urban communities, 1987–2000, *Journal of the American Medical Association,* 292(19), 2372, 2004.
51. PSR (Physicians for Social Responsibility), Asthma and the Role of Air Pollution, Washington, D.C., 1997.
52. CDC (Centers for Disease Control and Prevention), Summary health statistics for U.S. children: National health interview survey, 2002. Available at http://www.cdc.gov/nchs/data/series/sr_ 10/sr10_223.pdf, 2004.
53. CDC (Centers for Disease Control and Prevention), Asthma prevalence and control characteristics by race/ethnicity—United States, 2002, *Mortality & Morbitity Weekly Report*, 53(07), 145, 2004.
54. Pérez Pérez-Peña, R., Study finds asthma in 25% of children in central Harlem, *New York Times*, April 19, 2003.
55. Clark, N.M., et al., Childhood asthma, *Environmental Health Perspectives*, 107(suppl 3), 421, 1996.
56. Bukowski, J.R., Range-finding study of risk factors for childhood asthma development and national asthma prevalence. *Human and Ecological Risk Assesment*, 8, 735, 2002.
57. Bascom, R., Environmental factors and respiratory hypersensitivity: The Americas, *Toxicology. Letters* (Shannon), 86(2–3), 115, 1996.
58. Salam MT, et al., Early-life environmental risk factors for asthma: Findings from the Children's Health Study, *Environmental Health Perspectives*, 112, 760 2004.
59. McConnell, R., et al., Asthma in exercising children exposed to ozone: A cohort study, *Lancet*, 359, 2002.
60. Gauderman, E., et al., Childhood asthma and exposure to traffic and nitrogen dioxide, *Epidemiology*, 16, 737, 2005.
61. Rossi, G., et al., Air pollution and cause-specific mortality in Milan, Italy, 1980–1989, *Archives of Environmental Health*, 54(3), 158, 1999.
62. Neas, L.M., Schwartz, J., and Dockery, D., A case-crossover analysis of air pollution and mortality in Philadelphia, *Environmental Health Perspectives*, 107(8), 629, 1999.
63. Hong, Y.C., et al. Effects of air pollutants on acute stroke mortality, *Environmental Health Perspectives*, 110(2), 187, 2002.
64. Gold, D.R., et al., Ambient pollution and heart rate variability, *Circulation*, 101(11), 1267, 2000.
65. Peters, A., et al., Air pollution and incidence of cardiac arrhythmia, *Epidemiology*, 11(1), 11, 2000.
66. Jerrett, M., et al., Spatial analysis of air pollution and mortality in Los Angeles, *Epidemiology*, 16 6), 2004.
67. Laden, F., et al., Reduction in fine particulate air pollution and mortality, *American Journal of Respiratory Critical Care Medicine*, 173(6), 667, 2006.
68. EPA (U.S. Environmental Protection Agency), National-Scale Air Toxics Assessment for 1999: Estimated Emissions, Concentrations and Risk, Office of Air and Radiation, Washington, D.C., 2006.

69. Weise, J., Historic Drinking Water Facts. Drinking Water and Wastewater Program, ADEC Division of Environmental Health, Anchorage, 2003.

70. CRS (Congressional Research Service), Summaries of Environmental Laws Administered by the Environmental Protection Agency, The Library of Congress, Washington, D.C., 1991.

71. Ohio DEP, Understanding Ohio's Surface Water Quality Standards, Ohio Department of Environmental Protection, Columbus. Available at http://rol.freenet.columbus.oh.us/PDF_Chap2.pdf, 2003.

72. Copeland, C., Clean Water Act: Summaries of Environmental Laws Administered by EPA, Congressional Research Service, Washington, D.C., 1999.

73. EPA Press Advisory, Connecticut company to pay $16 million for Clean Water Act, Washington, D.C., August 24, 2004.

74. EWG (Environmental Working Group), Clean Water Report Card: How the Regulators are Keeping up with Keeping Our Water Clean, Washington, D.C., 2000.

75. EPA (U.S. Environmental Protection Agency), Efforts to Manage Backlog of Water Discharge Permits Need to be Accomplished by Greater Program Integration, Report 2005-P-00018, Inspector General, Washington, D.C., 2005.

76. USPIG (U.S. Public Interest Research Group), Troubled Waters: An Analysis of Clean Water Act Compliance, Washington, D.C., 2006.

77. Copeland, C., Clean Water Act and Total Maximum Daily Loads (TMDLs) of Pollutants, Congressional Research Service, Washington, D.C., 2002.

78. Miller, J., Environmental law and the science of ecology, in *Environmental and Occupational Medicine*, 2nd ed, Rom, W., ed, Little, Brown and Co, Boston, 1992, 1307.

79. EPA (U.S. Environmental Protection Agency), Section 404 of the Clean Water Act: An overview. Available at http://www.epa.gov/owow/wetlands/facts/fact10.html,2003.

80. EPA (U.S. Environmental Protection Agency), Section 404 of the Clean Water Act: How wetlands are defined and identified. Available at http://www.epa.gov/owow/wetlands/facts/fact11.html, 2003.

81. Dahl, T.E., National Wetlands Inventory. Summary findings. Available at http://wetlands.fws.gov/bha/SandT/SandTSummaryFindings.html, 2000.

82. EPA (U.S. Environmental Protection Agency), 2002–2003: The Year of Clean Water, Office of Water, Washington, D.C., 2003.

83. NRDC (Natural Resources Defense Council), Swimming in Sewage. Washington, D.C., February, 2004.

84. EPA (U.S. Environmental Protection Agency), EPA acts to make beaches cleaner and safer, press release, Office of Public Affairs, Washington, D.C., November 8, 2004.

85. EPA (U.S. Environmental Protection Agency), Beach Act, Office of Water, Washington, D.C., 2004.

86. EPA (U.S. Environmental Protection Agency), EPA's BEACH program: 2004. Swimming season update, Office of Water, Washington, D.C., 2005.

87. ELI (Environmental Law Institute), *Environmental Law Deskbook*, Environmental Law Institute, Washington, D.C., 166, 203, 221, 245, 1989.

88. Yoder, J.S., et al., Surveillance for waterborne-disease outbreaks associated with recreational water—United States, 2001–2002, *Mortality & Morbidity Weekly Report,* 63/SS-8, 2004.

89. EPA (U.S. Environmental Protection Agency), National Air Quality Standards (NAAQS). Available at http://www.epa.gov/airs/criteria.html, 2002.

90. OMB (White House Office of Management and Budget), Informing Regulatory Decisions:

2003 Report to Congress on the Costs and Benefits of Federal Regulations and Unfunded Mandates on State, Local, and Tribal Entities, Office of Information and Regulatory Affairs, Washington, D.C., 2003.

91. NRC (National Research Council), Drinking Water and Health, vol. 4, National Academy Press, Washington, D.C., 1982, 489.

92. Randle, R.V., Safe Drinking Water Act, in *Environmental Law Handbook*, Arbuckle, J.G., et al., eds, Government Institutes, Inc., Rockville, 1991, 149.

93. EPA (U.S. Environmental Protection Agency), Setting standards for Safe Drinking Water. Available at http://www.epa.gov/OGWD/standard/setting.html, 2005.

94. Tiemann, M., Safe Drinking Water Act. Summaries of Environmental Laws Administered by the EPA, Congressional Research Service, Washington, D.C., 1999.

95. Craun, G.F., et al. Outbreaks in drinking-water systems, 1991–1998, *Journal of Environmental Health*, 65, 16, 2002.

96. ATSDR (Agency for Toxic Substances and Disease Registry), The Nature and Extent of Lead Poisoning in Children in the United States: A Report to Congress, Public Health Service, Atlanta, 1988.

97. EPA (U.S. Environmental Protection Agency), Understanding the Safe Drinking Water Act, EPA 810-F-99-008, Office of Public Information, Washington, D.C., December, 1999.

98. EPA (U.S. Environmental Protection Agency), Unregulated Contaminant Monitoring Program, Office of Air and Radiation, Washington, D.C., 2005.

99. EPA (U.S. Environmental Protection Agency), News release: Data sought for 26 drinking water contaminants, Office of Water, Washington, D.C., 2005.

100. EPA (U.S. Environmental Protection Agency), Secondary Drinking Water Regulations: Guidance for Nuisance Chemicals, EPA 810/K-92-001, Office of Drinking Water, Washington, D.C., 1992.

101. Donohue, J.M. 203. Personal communication.

102. EPA Press Advisory, EPA announces new rules that will further improve and protect drinking water, Washington, D.C., December 15, 2005.

103. EPA Press Advisory, New York man pleads guilty to drinking water monitoring violation, Washington, D.C., August 23, 2005.

104. EPA (U.S. Environmental Protection Agency), List of Drinking Water Contaminants & MCLs. Available at http://www.epa.gov/safewater/mcl.html, 2005.

105. FDA (Food and Drug Administration), Bottled Water Regulation and the FDA, Center for Food Safety and Applied Nutrition, Washington, D.C., 2002.

106. NRDC (Natural Resources Defense Council), Bottled water: Pure drink or pure hype? Available at http://www.nrdc.org/water/drinking/nbw.asp, 2000.

107. Craun, G.F., Statistics of waterborne outbreaks in the U.S. (1920-1980), in *Waterborne Disease Outbreaks in the United States*, Craun, G.F., ed, CRC Press, Boca Raton, 1986.

108. AP (The Associated Press), Swedish capital sees less silver pollution thanks to digital photos, November 25, 2005.

109. Barwick, R.S., et al., 2000. *Mortality & Morbidity Weekly Report*, 49(SS04), 1, 2000.

110. PSR (Physicians for Social Responsibility), Water Contamination, PSR Reports 22(2), Washington, D.C., 2000.

111. WHO (World Health Organization), Global Water Supply and Sanitation Assessment 2000 Report. Available at http://www.who.int/water_sanitation_health/Globassessment/Gloval-TOC.htm, 2002.

112. Gleick, P.H., Dirty Water: Estimated Deaths from Water-Related Diseases 2000-2020, Pacific Institute for Studies in Development, Environment, and Security, Oakland, CA, 2002.

113. U.S. Census Bureau, World Population. Available at http://www.census.gov/ipc/www/world.html, 2002.

114. EPA (U.S. Environmental Protection Agency), Air Emissions Trends—Continued Progress Through 2003, Office of Air and Radiation, Washington, D.C., September 22, 2004.

115. Noah, M., Graywater use still a gray area, *Journal of Environmental Health*, 64(10), :22, 2002.

6 Food Safety Statutes

6.1 INTRODUCTION

We humans have evolved from ancestors who hunted feral animals for food, later learning to cultivate grains, vegetables, and fruits. Food grown and prepared before the invention of refrigeration became commonplace in the United States was generally consumed soon after its acquisition. This was in order to avoid contact with pathogens found in deteriorating food. Decay of food is caused by microorganisms that parasitize dead plant or animal tissue. To preserve food, meat was salted or exposed to dense smoke from hardwood fires. Vegetables and fruits were canned or stored in root cellars. With refrigeration came the ability to chill or freeze food and store it safely for short to long periods of time. In modern times, irradiation has been used to preserve food, but this method has not gained widespread acceptance due to concerns that irradiation might cause harmful changes in food, which then could affect consumers of irradiated food. The problem of pathogens in deteriorating food was theoretically overcome though refrigeration. However, the equally important health consequence of how food is prepared remained a problem. Fresh, or properly preserved, food that is contaminated with pathogens from human contact during growing, transporting, preparing, or serving food has the potential to cause human illnesses.

In the United States, an estimated 76 million foodborne illnesses occur annually [1].

Government's involvement in protecting the public against adulterated or impure food is a relatively recent occurrence in the United States. Until the early twentieth century, food safety was deemed to be the responsibility of individual consumers and therefore not a matter for government intervention. As will be described in the following sections, states had the primary authority over issues of food safety—an arrangement that still prevails. However, federal government gradually assumed a strong role in food safety as a matter of interstate commerce. Federal involvement in food safety brought together the public health triad of federal, state, and local governments directed to a common purpose—in this case, prevention of foodborne illnesses.

Foodborne illnesses are a serious environmental health problem, although the U.S. food supply is relatively safe overall. However, how food is transported, prepared, and served can introduce pathogens of potential harm to human health. Mead and colleagues [1] estimate that foodborne diseases cause approximately 76 million illnesses, 325,000 hospitalizations, and 5,000 deaths annually in the United States. Three pathogens, *Salmonella*, *Listeria*, and *Toxoplasma*, are responsible for more than 75 percent of the foodborne illnesses that have known causes. The following sections describe federal laws and state and local government responsibilities that are intended to prevent foodborne illnesses. Two federal statutes to be described are the Federal Meat Inspection Act (and analogous laws for poultry and eggs), and the Food, Drug, and Cosmetic Act.

6.2 FEDERAL MEAT INSPECTION ACT

6.2.1 History

Consumption of meat and meat products has long been part of the human diet, although debate continues about the ethics of raising animals as a food source. Our ancestors, whether indigenous people or colonists, hunted the forests, plains, and bodies of water for birds, mammals, fish, and shellfish as food sources. With the passage of time, rural Americans grew their own food in gardens and processed domesticated animals into meat and meat products. Farmers slaughtered animals in the fall and winter, when temperatures were cool so as to diminish the deterioration of meat products. Salt rubbed into the meat and smoke from wood fires were used to preserve meats so that consumption of the meat could occur during warmer seasons. A family's meat quality and personal health protection were therefore at the mercy of a farmer's skill and resources in food preservation. Government had no role to play in what were essentially personal matters of family diet and health.

As villages and cities grew in numbers and population, meat was supplied by butchers and sold in butcher shops. In the nineteenth century, cities such as Chicago and Cincinnati became renowned as centers of the meatpacking industry. As matters of public health, how animals were slaughtered and under what conditions were matters of indifference to the U.S. public. During this period, until 1906, states had the primary responsibility for protecting the public against impure food, including meat products. Needless to say, food inspection programs varied considerably between the states.

The public's ignorance of the conditions in the meatpacking industry began to change in the early years of the twentieth century. In particular, a major influence on public opinion was Upton Sinclair's 1906 book, *The Jungle*, which graphically described unsanitary conditions and inhumane slaughter of animals in the Chicago meatpacking industry. Sinclair's book, much like Rachael Carson's book *Silent Spring* 56 years later, served to turn a spotlight on a major environmental health problem. On June 30, 1906, Congress enacted the Federal Meat Inspection Act of 1906 (FMIA). The act was substantially amended by the Wholesome Meat Act of 1967. The act was the first federal government involvement in the food safety of meat and meat products.

6.2.2 Key Provisions of the FMIA, as Amended, Relevant to Public Health

The primary goals of the FMIA, as amended, are to prevent adulterated or misbranded livestock and products from being sold as food and to ensure that meat and meat products are slaughtered and processed under humane and sanitary conditions. These requirements apply to animals and their products produced and sold within states as well as to imports, which must be inspected under equivalent foreign standards [2]. The key provisions of the Meat Inspection Act are as follows [3]:

§602. Congressional Statement of Findings: "Meat and meat food products are an important source of the Nation's total supply of food. They are consumed throughout the Nation and the major portion thereof moves in interstate or foreign commerce. It is essential in the public interest that the health and welfare of consumers be protected by assuring that meat and meat food products distributed to them are wholesome, not adulterated, and properly marked, labeled, and packaged. [. . .] [I]t is hereby found that all articles and animals which are regulated under this chapter are either in interstate or foreign commerce or substantially affect such commerce, and that regulation by the Secretary[1] and cooperation by the States and other jurisdictions as contemplated by this chapter are appropriate to prevent and eliminate burdens upon such commerce, to effectively regulate such commerce, and to protect the health and welfare of consumers."

§603. Inspection of Meat and Meat Food Products: *(a) Examination of animals before slaughtering; diseased animals slaughtered separately and carcasses examined.* "For the purpose of preventing the use in commerce of meat and meat food products which are adulterated, the Secretary shall cause to be made, by inspectors appointed for that purpose, an examination and inspection of all cattle, sheep, swine, goats, horses, mules, and other equines before they shall be allowed to enter into any slaughtering, packing, meat-canning, rendering, or similar establishment, in which they are to be slaughtered and the meat and meat food products thereof are to be used in commerce; and all cattle, sheep, swine, goats, horses, mules, and other equines found on such inspection to show symptoms of disease shall be set apart and slaughtered separately from all other cattle, sheep, swine, goats, horses, mules, or other equines, and when so slaughtered the carcasses of said cattle, sheep, swine, goats, horses, mules, or other equines shall be subject to a careful examination and inspection, all as provided by the rules and regulations to be prescribed by the Secretary, as provided for in this subchapter."

(b) Humane Methods of Slaughter: "For the purpose of preventing the inhumane slaughtering of livestock, the Secretary shall cause to be made, by inspectors appointed for that purpose, an examination and inspection of the method by which cattle, sheep, swine, goats, horses, mules, and other equines are slaughtered and handled in connection with slaughter in the slaughtering establishments inspected under this chapter. The Secretary may refuse to provide inspection to a new slaughtering establishment or may cause inspection to be temporarily suspended at a slaughtering establishment if the Secretary finds that any cattle, sheep, swine, goats, horses, mules, or other equines have been slaughtered or handled in connection with slaughter at such establishment by any method not in accordance with the Act of August 27, 1958 (72 Stat. 862; 7 U.S.C. 1901–1906) until the establishment furnishes assurances satisfactory to the Secretary that all slaughtering and handling in connection with slaughter of livestock shall be in accordance with such a method."

§604. Post Mortem Examination of Carcasses and Marking or Labeling; Destruction of Carcasses Condemned; Reinspection: "For the purposes hereinbefore set forth the Secretary shall cause to be made by inspectors appointed for that purpose a post mortem examination and inspection of the carcasses and parts thereof of all cattle, sheep, swine, goats, horses, mules, and other equines to be prepared at any slaughtering, meat-canning, salting, packing, rendering, or similar establishment in any State, Territory, or the District of Columbia as articles of commerce which are capable of use as human food; and the carcasses and parts thereof of all such animals found to be not adulterated shall be marked, stamped, tagged, or labeled as 'Inspected and passed'; and said inspectors shall label, mark, stamp, or tag as 'Inspected and condemned' all carcasses and parts thereof of animals found to be adulterated; and all carcasses and parts thereof thus inspected and condemned shall be destroyed for food purposes by the said establishment in the presence of an inspector. [...]"

§605. Examination of Carcasses Brought into Slaughtering or Packing Establishments, and of Meat Food Products Issued from and Returned Thereto; Conditions for Entry: "The foregoing provisions shall apply to all carcasses or parts of carcasses of cattle, sheep, swine, goats, horses, mules, and other equines or the meat or meat products thereof which may be brought into any slaughtering, meat-canning, salting, packing, rendering, or similar establishment, and such examination and inspection shall be had before the said carcasses or parts thereof shall be allowed to enter into any department wherein the same are to be treated and prepared for meat food products; and the foregoing provisions shall also apply to all such products, which, after having been issued from any slaughtering, meat-canning, salting, packing, rendering, or similar establishment, shall be returned to the same or to any similar establishment where such inspection is maintained. [...]"

§606. Inspectors of Meat Food Products; Marks of Inspection; Destruction of Condemned Products; Products for Export: "For the purposes hereinbefore set forth the Secretary shall cause to be made, by inspectors appointed for that purpose, an examination and inspection of all meat food products prepared for commerce in any slaughtering, meat-canning, salting, packing, rendering, or similar establishment, and for the purposes of any examination and inspection and inspectors shall have access at all times, by day or night, whether the establishment be operated or not, to every part of said establishment; and said inspectors shall mark, stamp, tag, or label as 'Inspected and passed' all such products found to be not adulterated; and said inspectors shall label, mark, stamp, or tag as 'Inspected and condemned' all such products found adulterated, and all such condemned meat food products shall be destroyed for food purposes, as hereinbefore provided, and the Secretary may remove inspectors from any establishment which fails to so destroy such condemned meat food products. [...]"

In summary, the FMIA, as amended, provides a framework to inspect, label, and enforce standards of meat and meat products. Inspections include both visual examination of carcasses as well as tests for microbacterial contamination. Federal meat inspectors bear the public health responsibility for approving or condemning meat or meat products intended for human consumption. As an issue of environmental health policy, some meatpackers have pressured the federal government to transfer to themselves part of the government's meat inspection authorities. Companies assert that federal meat inspectors slow the production of meat products, and they, the meatpacking industry, can do the inspection task more efficiently and at lesser cost. As a matter of public policy, redelegation of government public health authorities into the hands of food producers should generally be avoided due to potential conflicts of interests.

6.2.3 Public Health Implications of the FMIA, as Amended

Pathogens in meat and meat products have the potential to cause human illnesses. There, however, are few data on the incidence of foodborne illnesses specific to meat and meat products. That does not mean that foodborne illnesses from meat and other foods are inconsequential. Quite the contrary.

"[o]ne in four Americans is estimated to have a significant foodborne illness each year" [4].

Since 1996, estimates of the incidence of foodborne illnesses have been compiled by the Centers for Disease Control and Prevention (CDC) through its Foodborne Diseases Active Surveillance Network (FoodNet). Nine foodborne diseases are monitored in selected U.S. cities. Data from this network indicate that one in four Americans are estimated to have a significant foodborne illness each year [4]. This is a considerable burden of illness afflicting the U.S. population.

The incidence of foodborne illness attributable to pathogens in meat and meat products is unknown, given limitations in health surveillance systems. One source examined 3,500 foodborne food outbreaks, representing 115,700 individual illnesses between the years 1990 and 2003, and found that beef and beef dishes were associated with 338 outbreaks and 10,795 cases, which represented about 9 percent of the total cases [5]. Moreover, it is known that contaminated meat has been associated with individual outbreaks of illness. For example, in 1996 CDC investigated an outbreak of *Salmonella* serotype Thompson infections that were associated with a restaurant in Sioux Falls, South Dakota. Fifty-two infections were found in persons who had eaten food prepared by the restaurant. Results of the investigation revealed that cooking times and storage temperatures for roast beef were inadequate to prevent *Salmonella* proliferation [6].

Episodes of meat-associated foodborne disease can occur because of failures in meat production, preparation, or delivery to consumers. A significant provision of the Federal Meat Inspection Act gives the U.S. Department of Agriculture (USDA) authority to take action against producers of meat or meat products are found to be unsafe for human consumption. Specifically, §673 of the act states, "…[any carcass, part of a carcass, meat or meat product…is capable of use as human food and is adulterated or misbranded…shall be liable to be proceeded against and seized and condemned, at any time, on a libel of information in any United States district court or other proper court.…" States with their own food safety statutes can also suspend operations or close facilities found to be producing meat or meat products contaminated with pathogens. Preventing contaminated meat from reaching consumers is, in public health terms, an act of primary prevention, i.e., hazard interdiction. Two examples will suffice to illustrate the interdiction of contaminated ground beef from reaching consumers. In October 2002, USDA recalled 27.4 million pounds of poultry found contaminated with *Listeria*, made into delicatessen products, produced at a Pennsylvania processing plant, the largest recall in U.S. history. In July 2002, the department recalled 19 million pounds of *E. coli* contaminated ground beef produced by a Colorado beef products plant, following illnesses in nineteen persons who had consumed ground beef produced by the plant [7].

Under the FMIA, USDA must regulate the operations of meatpackers for purposes of preventing contaminants in meat from reaching consumers. However, how USDA develops and enforces its meat inspection authorities has historically been subject to policy challenges. The meatpacking industry has argued that federal meat inspectors should have lesser authority to inspect meat and meat products, asserting that such inspections impede a plant's productivity. The industry preferred an inspection system whereby meat inspections would be conducted by a plant's personnel, but overseen by federal meat inspectors. To date, the industry's proposal has not been fully implemented.

In 2002, USDA announced more stringent regulations that are intended to reduce *E. coli* contamination of meat and meat products, particularly ground beef. In support of the revised regulations, USDA noted that 43 percent of animal carcasses were contaminated with *E. coli* [8]. Moreover, the department referenced CDC data showing foodborne transmission of *E. coli* annually causes more than 62,000 illnesses and fifty-two deaths. Under the proposed regulations, no slaughter plants would be exempt from random *E. coli* testing (some small production meat processors were previously exempt). Further, the new regulations would require meat processors to add microbiological testing to actions required of them.

In addition to the FMIA, analogous acts pertain to poultry and egg products [23]. The Poultry Products Inspection Act and the Egg Products Inspection Act mandate inspections of producers of those products and authorize the USDA to take actions similar to those in the FMIA in order to prevent contaminated poultry and eggs from causing foodborne ill nesses. The Food Safety and Inspection Service of the USDA is the administrative unit that bears the responsibility for enforcing the provisions of the FMIA, the Poultry Products Inspection Act, and the Egg Products Inspection Act. All three acts require states to cooperate with the USDA and require the states to establish their own statutes that comply with the three federal statutes. This, of course, in another example of federalism in action, a characteristic of the main body of federal environmental health legislation.

◊ ◊ ◊

Of particular public health importance are foodborne illnesses in school children, because they are particularly susceptible to some pathogens (e.g., *Escherichia coli*) [9]. While relatively few instances of foodborne illnesses in school meals occur annually, those that do occur are preventable. Daniels and colleagues [10] examined the CDC Foodborne Outbreak Surveillance System for foodborne disease outbreaks in U.S. schools for the years 1973 through 1997. Over this span, 604 outbreaks were found; the median annual number was 22 outbreaks (range 9–44). *Salmonella* was the most often identified pathogen, but in 60 percent of outbreaks, no etiologic agent had been reported. Because the CDC surveillance system comprises data voluntarily reported by state epidemiologists, the system likely contains fewer actual number of foodborne diseases in schools than what actually occur.

A follow-up study by the General Accounting Office [9], using the same CDC data as Daniels and colleagues, concluded that "[w]hile school foodborne outbreaks from all schools constituted less than 4 percent of total U.S. foodborne outbreaks reported to CDC from 1973 through 1999, they were responsible for about 10 percent of all outbreak illnesses during this period." Stated in different words, a single outbreak can involve many children.

Public health surveillance data on foodborne illnesses in schools convey both good and bad news. It is good news that outbreaks of foodborne illnesses in schools are relative few, perhaps 4 percent of annual foodborne illnesses. The bad news is that these outbreaks occur at all, since they are preventable through better training of food service personnel in schools, improved food preparation, and adjustments to surveillance systems so as to more readily capture foodborne illness in schools. As a matter of environmental health policy, should states be required to report foodborne illnesses linked to schools to a federal health surveillance system?

6.3   FOOD, DRUG, AND COSMETIC ACT

6.3.1   History

Americans of the twenty-first century expect not to be harmed by the food and medicinal drugs and therapeutic devices with which they come into contact. The expectation is the product of personal experience (e.g., few of us have had protracted illnesses from eating impure food) and there is general trust in public health systems (e.g., restaurant inspections). While episodes of illness occur as the result of impure food (e.g., undercooked meat in hamburgers), the current situation is vastly different from that of our ancestors.

In the nineteenth and early twentieth centuries, any government control of food and drugs was the responsibility of states. State laws, if existing, varied greatly between states. In that era, use of chemical preservatives and toxic colors added to food was virtually uncontrolled [11]. Instances of morbidity surely occurred, given current bacteriological and toxicological knowledge, but no health reporting system was in place then to record the extent of morbidity. As public concern grew in the late nineteenth century about unsanitary conditions in the meatpacking industry, a similar concern arose about the harm caused by drugs, medications, and concoctions sold for alleged medicinal purposes. "Medicines" containing opium, morphine, heroin, and cocaine were sold without any restriction [ibid.]. Moreover, labels gave no indication about the ingredients of "over-the-counter" drugs and medications. The policy of "buyer beware" prevailed during this period.

During the 1870s, the grassroots Pure Food Movement arose and soon became the principal source of political support for federal food and drugs legislation [ibid.]. In 1903, Dr. Harvey W. Wiley became the director of the U.S. Department of Agriculture's Division of Chemistry and soon thereafter aroused public opinion against impure consumer products that his staff had identified. In a sense, Dr. Wiley was serving as a surrogate surgeon general, informing the public and advocating for public health legislation. Strenuous opposition to Wiley's campaign for a federal food and drug law came from whiskey distillers and the patent medicine firms, many of which thought they would be put out of business by federal authorities and regulation of their industries. Supporting the need for federal legislation were agricultural organizations, some food processors, public health professionals, and state food and drug officials. The political scale was tipped toward legislative action through the intercession of club women who rallied behind the pure food cause [ibid.]. Remarkably, Congress enacted on the same day, June 30, 1906, both the Pure Food and Drugs Act and the Federal Meat Inspection Act.

The Food and Drugs Act of 1906 prohibited the manufacture and interstate shipment of adulterated and mislabeled foods and drugs. The law enabled the federal government to initiate litigation against alleged illegal products, but lacked affirmative requirements to guide compliance with the law. The 1906 law also lacked key provisions necessary to make it effective in identifying harmful food and drug products. For example, food adulteration continued to flourish because judges could find no authority in the law for any standards of purity and content established by FDA [13]. In time, the 1906 law became obsolete because FDA lacked enforcement authorities and due to technological changes in how food and drugs were produced. The provisions of the 1906 Food and Drugs Act simply were not sufficiently robust to keep up with technology changes in the food and drug industries. Thirty-two years were to pass before the act was updated.

6.3.2 KEY PROVISIONS OF THE FDCA, AS AMENDED, RELEVANT TO PUBLIC HEALTH

In June 1938, President Franklin D. Roosevelt signed into law the Federal Food, Drug, and Cosmetic Act (FDCA), which replaced the Food and Drugs Act of 1906. The 1938 law contained many significant changes, including those shown in Table 6.1 [13]. Even the 1938 law was found to be in need of further improvements. For instance, the 1938 law prohibited poisonous substances, but required no evidence that food ingredients were safe for human consumption.

In 1949, Congress began lengthy hearings on the Food, Drug, and Cosmetic Act, resulting in three substantive amendments: the Pesticide Amendment of 1954, the Food Additives Amendment of 1958, and the Color Additive Amendments of 1960 [13]. These amendments effectuated an environmental health policy that no substance can legally be introduced into the U.S. food supply unless there has been a prior determination that it is safe. Moreover, these amendments required manufacturers to conduct the research necessary to establish their products' safety. FDA became a reviewer of manufacturers' data, with the authority to reject products or to request more data from manufacturers.

6.3.3 PUBLIC HEALTH IMPLICATIONS OF THE FDCA, AS AMENDED

The FDCA, with its amendments, has resulted in removal of unsafe food additives from the U.S. food supply, tighter pesticide regulations on levels of these substances in food sources, and review of drugs and medical devices intended for commerce. These actions benefit the public's health, and represent primary prevention measures. However, at least one public interest organization, the Center for Science in the Public Interest

TABLE 6.1
Major Changes Made in 1938 to the Federal Food and Drug Act of 1906 [13]

Drug manufacturers were required to provide scientific proof that new products could be safely used before putting them on the market	Addition of poisonous substances to foods was prohibited except where unavoidable or required in production
Proof of fraud was no longer required to stop false claims for drugs	Safe tolerances were authorized for residues of such substances, e.g., pesticides
Specific authority was provided for factory inspections	Federal court injunctions against violations were added to the previous legal remedies of product seizures and criminal prosecutions.
Food standards were required to be established when needed	Cosmetics and therapeutic devices were regulated for the first time

(CSPI), has recommended that additional measures should be taken to protect the U.S. food supply. Their policy recommendations stem from their study of foodborne illness reports. The CSPI examined records from 3,500 food illness outbreaks[2] in the United States during the period 1990 to 2003. This represented 115,700 individual cases of foodborne illness. Their food illness database was established by gathering data from CDC, contacts with state agencies, and by searching newspapers and other publications that report foodborne illness. The CSPI reported the top five causes of foodborne illness during the thirteen-year span were as follows:

- Seafood and seafood dishes (720 outbreaks; 8,044 cases of illness)
- Produce and produce dishes (428 outbreaks; 23,857 cases)
- Poultry and poultry dishes (355 outbreaks; 11,898 cases)
- Beef and beef dishes (338 outbreaks; 10,795 cases)
- Eggs and egg dishes (306 outbreaks; 10,449 cases)

The CSPI observed that "Foods regulated by the Food and Drug Administration (FDA) were the vehicles in two-thirds of the outbreaks in CSPI's database, while foods (meat, poultry) regulated by the U.S. Department of Agriculture (USDA) were the vehicles in one-fourth of the outbreaks" [5]. As a means to set policy, data such as the CSPI's can be quite useful to identify weak links in the system of food and food products protections. For example, the large number of foodborne illnesses associated with produce and produce products suggest that improved inspection and preparation procedures should be implemented at sources of produce production (i.e., farms and produce suppliers), transportation, and delivery.

Of note for prevention of foodborne illness, is FDA's *Food Code*, which is a set of guidelines that represent best practices for the retail and food service industries [14]. The *Food Code* was first issued by FDA in 1993 and is currently updated every four years. According to FDA, more than 1 million retail and food service establishments use the *Food Code's* provisions as a model to develop or update their own food safety rules. While the *Food Code* is not mandated of states, a survey found that 48 of the 56

U.S. states and territories, which cover 79 percent of the U.S. population, have adopted food safety codes modeled after the *Food Code* [15]. As environmental health policy, the widespread voluntary adoption of the *Food Code* by states, territories, and the food industry is an example of how prevention of foodborne illness can be reduced by adoption of a common set of food safety practices.

Enforcement Example: FDA Initiates Seizure of Ginseng

On December 16, 2004, at the request of the FDA, the U.S. District Court for the District of New Jersey issued a warrant for the seizure of imported ginseng held for sale at FCC Products, Inc., Livingston, NJ. U.S. Marshals, accompanied by an FDA investigator, later seized the ginseng. The FDA had determined that the ginseng was adulterated under the Federal Food, Drug, and Cosmetic Act, containing unsafe pesticide chemical residues. The pesticide chemical residues, procymidone and quintozene, had been deemed unsafe because FDA had not established tolerance levels for the pesticides. [12].

6.4   STATE AND LOCAL HEALTH DEPARTMENTS' FOOD SAFETY AUTHORITIES

State and local governments have the primary responsibility for enforcing food safety regulations. The authorities of states vary, as do the degrees of local government involvement. States typically establish standards for the transportation, storage, preparation, and serving of food by food service establishments. Local health departments conduct inspections of restaurants and commercial food processors, typically under authorities in state laws. Issuance of permits to food service establishments is at the heart of state food safety laws. Without approved permits, food service establishments cannot legally operate. As an illustration of one state's approach to food safety, consider the state of Georgia. In Georgia, two state agencies have the primary responsibilities for protecting the public against foodborne illness. One agency, the Division of Public Health, has the state's primary authority for illness attributable to food services. The other state agency, the Georgia Department of Agriculture, has authority to regulate and inspect food supplies.

The key public health food safety provisions administered by the Georgia Division of Public Health include the following:[3]

Section II 290-5-14-02: "Selected Provisions: (1)(a): It shall be unlawful for any person to operate a food service establishment, or mobile food unit, a temporary food service operation or a restricted food service operation without having first obtained a valid food service permit from the health authority pursuant to this Chapter... (d): The permit shall be prominently displayed at all times, as near the main entrance as practicable. (e) The permit shall be the property of the health authority and shall be returned within seven days to the local health authority when the food service establishment ceases to operate or is moved to another location or when the permit is revoked."

Section III 290-5-14-03: "Selected Provisions: Food Care: (1) Food Supplies: (a): Food shall be in sound condition, free from spoilage, filth, or other contamination and shall be safe for human consumption. (b) Food shall be obtained from approved sources that comply with all laws relating to food processing and shall have no information on the label that is false or misleading.... (d) Fluid milk and fluid milk products used or served shall be pasteurized and shall meet the Grade A quality standards as established by law. Dry milk and dry milk products shall be made from pasteurized milk and milk products.... (g) Only clean whole eggs, with shell intact and without cracks or checks, or pasteurized liquid, frozen, or dry eggs or pasteurized dry egg products shall be used, except that hard-boiled, peeled eggs, commercially prepared and packaged, may be used. (2) Food Protection: (a) At all times, including while being stored, prepared, displayed, served, or transported, food shall be protected from potential contamination, including toxic materials, dust, insects, rodents, unclean equipment and utensils, unnecessary handling, cross contamination, coughs and sneezes, flooding, drainage, and overhead leakage or overhead drippage from condensation. (3) Food Storage: (a) Food, whether raw or prepared, if removed from the container or package in which it was obtained, shall be stored in an approved, clean, and covered container except during necessary periods of preparation of service... (g) Enough conveniently located refrigeration facilities or effectively insulated facilities shall be provided to assure the maintenance of perishable and potentially hazardous food at required temperatures during storage. (4) Food Preparation: (a) Food shall be prepared with the least possible manual contact with suitable utensils, and on surfaces that prior to use have been cleaned, rinsed and sanitized to prevent cross-contamination. (b) Raw fruits and vegetables shall be thoroughly washed with potable water under pressure before being cooked or served. A separate sink shall be provided for this purpose. (5) Food Display and Service:... (g) Food on display shall be protected from consumer contamination by the use of packaging or by the use of easily cleanable counter, serving line or salad bar protective devices, display cases, or by other effective means. (6) Food Transportation: (a) During transportation, food and food utensils shall be kept in covered containers or completely wrapped or packaged so as to be protected from contamination and spoilage."

Section XI 290-5-14-11: "Selected Provisions: Compliance Procedures: (1) Permits: (a) Issuance: Permits shall be issued by the health authority. Such permits shall be valid until suspended or revoked. (2) Inspections: (a) Inspection Frequency: An inspection of a food service establishment shall be performed at least twice annually. Additional inspections of the food service establishment shall be performed as often as necessary for the enforcement of this Chapter.[4] (b) Access: Representatives of the health authority, after proper identification, shall be permitted to enter any food service establishment or operation at any reasonable time for the purpose of making inspections to determine compliance with this Chapter."

A moment of reflection on Georgia's food safety law and regulations shows a program centered on permits issued to food service establishments. Without a permit from the state's public health department (or county health department, if delegated by the state), no food service operations are allowed to operate. Moreover, the state can

revoke a permit if sufficient unsanitary conditions are found by local health department inspectors. Of particular note are *critical violations* found by health inspectors,
as distinguished from *minor violations*. Critical violations are those findings that have
direct implications for the public's health (e.g., service personnel not wearing protective
gloves or food stored at temperatures that permit the growth of bacteria).

The state's regulations provide detailed specifications on food transportation,
storage, preparation, and service. While the regulations and public health systems of
inspections and reporting are generally impressive, they are only as effective as available budgets and personnel permit.

◊ ◊ ◊

The Georgia Department of Agriculture's food safety authorities complement
those of the Georgia Division of Public Health. The department's primary food safety
authorities derive from several Georgia State laws, and include the following [17]:

- Enforce state laws, rules, and regulations by conducting sanitation inspection of retail food stores, salvage food operations, mobile meat trucks, and rolling stores.
- Inspect food storage warehouses, wholesale bakeries, bottled water, and flavored
  drink processors, seafood processors, and wholesale fish dealers, and sanitation
  in establishments where food is handled and manufactured.
- Enforce federally mandated programs of inspection and sampling of dairy farms
  and dairy processing plants. This authority extends to the inspection of out-of-state
  milk products shipped to Georgia, along with authority to inspect tanker trucks,
  route trucks, and warehouses that are used to transport or store dairy products.
- Respond to consumers' inquiries about sanitary conditions relative to food and
  foodborne illness.

◊ ◊ ◊

A comparison of the food safety authorities administered by the two Georgia state agencies shows both similarities and differences. Regarding similarities, both the Division
of Public Health and the Department of Agriculture derive their food safety authorities
from state laws. Without authorizing statutes, the agencies would have no specific food
safety authority. Also, the prevention of foodborne illness is at the heart of both agencies' authorities and programs. This prevention focus is achieved primarily by requiring
food supplies and food service establishments to be registered under state control and,
second, to conduct inspections of food producers, transporters of food products, storage
facilities where food products are stored, and food service establishments.

Regarding differences between the two agencies' food safety authorities, the Division of Public Health focuses on the registration and inspection of food service establishments; whereas the Department of Agriculture focuses on registration and inspection of
food producers, transporters, and those who store food, such as warehouse operators. As
a policy observation, this kind of sharing of public health responsibility for food safety
is much like the duality of responsibility found throughout environmental health. For
example, on matters of toxic substances, the EPA has primacy in controlling the release
into the environment of substances that can harm human and ecological health; whereas
the U.S. Public Health Service agencies conduct research on the toxicity and human
health implications of toxic substances and work with states to collect surveillance

health data and exposure data that can be used to help determine regulatory standards developed by EPA or other regulatory agencies.

In addition to Georgia's responsibilities in food protection, city and county health departments play a critical role as well. For example, the DeKalb County Department of Health's food protection unit reviews and approves plans for new food service establishments, issues permits, and conducts ongoing inspections. Approximately 1,800 food establishments and services are inspected by the county each year. The results from restaurant inspections are made available to the public by: (1) posting a copy of the inspection report in a prominent place in each restaurant inspected, and (2) placing the inspection reports on the county's web site. As environmental health policy, providing the public with information with which to make personal health decisions is a matter of right-to-know.

In addition, the department evaluates and issues temporary event food service permits for festivals, carnivals, and fairs. Hotels and motels are evaluated and inspected for food safety. The unit also investigates all foodborne illness complaints and refers for follow-up any significant findings to disease surveillance programs operated by the state of Georgia's Division of Public Health.

◊ ◊ ◊

There are other important contemporary public health issues associated with food products and services in addition to food safety. These include food labeling, unhealthy food, and tobacco use in restaurants facilities. Each of these issues engages important environmental health policies. The policies involve federal, state, and local levels of government; the food industry; and, especially, individuals.

Food labeling refers here to the federal requirements under the FDCA, as amended, and covers most prepared foods, such as breads, cereals, canned and frozen foods, snacks, desserts, beverages, and such [18]. Nutrition labeling for fruits, vegetables, and fish is voluntary. Food labels give consumers information upon which to base food purchasing decisions. The labels must state the number of calories in the product, total fat, cholesterol, sodium, protein, carbohydrates, vitamin A, vitamin C, calcium, and iron content. Further, the label must identify the product's ingredients. Product labels therefore contain information of relevance to an individual's personal health. For example, persons who must restrict their sodium intake can avoid food products high in sodium. Similarly, those seeking to control their caloric intake can use product labels to estimate daily calorie consumption. While food labeling can serve beneficial public health purposes, the benefits are only as good as the quality of the data on the label, as placed there by the food producer. There is little verification by federal agencies of the accuracy of food labels; however, some states do conduct laboratory evaluations of food products and have state regulatory authority to prevent the sale of mislabeled products. Local government agencies are rarely involved with food labeling verification.

Unhealthy food has become a public health issue in large measure because of obesity as a major public health problem in the United States. While obesity is a consequence of many factors, poor dietary choices are one significant contributing factor. A diet of "fast food" has been accused of contributing to obesity in the U.S. population. Such food, often impregnated with high levels of fat, cholesterol, and excess calories has been the subject of personal injury litigation by persons who allege that their obesity was caused by "addictive" fast food. While these lawsuits have yet to run their course in the courts, the food industry has brought pressure on some state legislatures to enact legislation to protect it from "frivolous" lawsuits. As policy, such laws will likely eventually find their way into federal courts for examination of constitutionality.

Prohibition of tobacco smoking in public places became a public health issue, commencing in the 1990s. Generally, the battleground over this issue has been with local government, although five states[5] have enacted laws as of January 2004 to restrict tobacco sales, distribution, and use in public places. Prohibition of tobacco smoking in restaurants and other food service establishments has occurred in 72 U.S. communities by way of local ordinances [19]. The environmental health policy of reducing exposure to secondhand tobacco smoke has generally prevailed over the counter argument that private businesses (e.g., restaurants) should be spared the regulatory hand of government. In particular, the allegation of loss of customers and revenue due to no-smoking regulations has characterized the opposition from food service operators. However, in a study by CDC [ibid.] of revenues reported by restaurants in El Paso, Texas, before and following the enactment of no-smoking ordinances, it was found that no statistically significant difference in income had occurred.

6.5   FOOD SAFETY POLICY ISSUES

Several environmental health policy issues pertain to food safety. Federalism is one such issue. The entry in 1906 of the federal government into the areas of meat inspection, food, drugs, and cosmetics, somewhat diminished the food safety role of free enterprise in the food industry. Heretofore, food safety was largely a matter of "let the buyer beware," supported by state food safety laws. Outbreaks of foodborne illnesses were considered then as a matter of personal health and consumer choice. While individual consumer choice and an informed public remain essential for preventing foodborne illnesses, stronger federal and state laws, girded with local health departments' inspections, are essential for food safety.

Another policy issue is how to inform the public about food service establishments that fail to meet standards of food safety. Some local health departments place on their Web sites the results, current and past, of restaurant inspections. How these are presented to the public is a challenge. The inspection report must be factually accurate, but should not create unrealistic fears in the public. This difficult balance in health communication has led some food safety authorities to suggest that Internet posting of individual food service scores is inappropriate, possibly raising unreasonable fears in the public.

There are several arguments against posting food establishment's inspection reports on the Internet or given to local news media. Some inspectors have expressed concern that the public could be misled by unabridged inspection scores, citing problems in inspection procedures that do not clearly distinguish between critical and noncritical findings [20]. They note, depending on the kind of inspection system used, that a restaurant with a score of 95, based on a critical health finding like prepared food left unrefrigerated, would be seen as preferable to a restaurant with a score of 88, based primarily on administrative failures, such as inappropriate placement of the food inspection score within the food service establishment. Further, some food inspectors have expressed their concern that their professional relationship with food service managers can be hindered when inspection scores are made available to the public [ibid.].

On the other hand, in support of communicating food service inspection reports to the public is the acceptance by many health departments that posted reports help improve food safety. In a study of foodborne-disease hospitalizations in Los Angeles County, California, it was found that restaurant hygiene grading with public posting of results was an effective means for reducing the incidence of foodborne disease [21]. Investigators reported a decrease of 13 percent in the number of foodborne-diseases in the year following implementation of a public posting program for restaurant inspections. As this study suggests, public perception can be a powerful motivator for change. Much like how the Toxics Release Inventory data have led to voluntary reductions of emission from industrial facilities, food establishments fear a poor rating of their services. Some therefore argue that public availability of inspection scores help reinforce food quality standards and practices [22].

Regardless of which side one takes on the argument about the public's access to food inspection reports, the trend seems clear. The U.S. public will continue to want access to government information that has health and safety relevance to them. This trend has been accelerated because of the rapid growth of the Internet and the public's access to it. Moreover, the well-publicized news media reports of occasional food poisonings have compounded the public's concerns and personal interests. The challenge is therefore not whether to report food establishment ratings, but how to do it in a responsible manner.

6.6 SUMMARY

Described in this chapter are two major federal environmental health statutes that are intended to enhance food safety in the United States. As public health policy, the Federal Meat Inspection Act, which dates from 1906, requires that meat and meat products are subject to federal inspection before entering the human food supply. Similarly, the Food, Drug, and Cosmetic Act, also dating from 1906, as public health policy prohibits the distribution of adulterated and mislabeled foods and drugs. Both these federal statutes therefore adopt a policy of limited federal government involvement in inspection of food quality prior to the release of food products into commerce and for human consumption. By this process, adulterated or impure food is interdicted before entering the food chain. This, of course, is an example of the core principle of public health, prevention of disease and disability.

A policy of note is found in the FDCA, the matter of burden of proof. Under this policy, the information upon which the FDA bases its drug safety decisions is provided to them by the producers of the drugs under review. Whether this is good public policy is often the subject of sharp and thoughtful debate.

In distinction to other environmental hazards, federal involvement in food safety is rather limited and involves multiple partners in the public health effort to prevent foodborne illnesses. To be more specific, food safety requires the active participation of government, private sector entities, and individual food consumers to a degree not found in issues of air pollution, water contamination, toxics control, and waste management. Indeed, states have food quality responsibilities that exceed those of the federal government, as illustrated in this chapter by Georgia's food quality statute. Moreover, private sector entities such as food producers, transporters, and food servers (e.g., restaurants) have quite significant roles and responsibilities for protecting against foodborne illness. However, in distinction to other environmental hazards, individuals play the critical role in protecting themselves against foodborne illness. For public health purposes, how individuals prepare food in the home is critical. After all, even the most wholesome food, if prepared under unsanitary conditions, has the portent to cause human illness.

6.7   POLICY QUESTIONS

1. Let's consider the matter of food safety. Should food safety be a concern of local health departments through inspections of restaurants and other places of commercial food service? If so, why? If not, why?

2. Assume you were recently hired by an urban municipal health department. Your first assignment is to design a public health program to improve food safety in public establishments. (A) Discuss the nature and impact of foodborne illness that would be of concern to your health department. (B) Using this material, design a public health program to prevent foodborne illness, choosing any four elements of the eight elements shown in Figure 1.1. Use critical thinking, as described in chapter 1, to the extent possible.

3. Summarize the public health benefits of the Federal Meat Inspection Act. Discuss the ethical implications, if any, of the Act.

4. The Food, Drug, and Cosmetic Act, as amended, gives the FDA the authority to approve drugs to be placed into commerce. Assume that the act did not exist, leaving the manufacturer solely with the responsibilities for the safety of their products. Discuss the public health implications of this kind of market-driven arrangement.

5. Visit a local restaurant and look for a posted food inspection report. Describe the impact, if any, on your patronage of the selected restaurant. (A) What aspects of the food inspection report were of greatest importance to your decision? In your opinion, should food inspection reports be available to the public? If so, how? (B) Discuss the pros and cons of making restaurant inspection scores available to the public. (C) Some county health departments post restaurant scores on the internet. Using such a Web site, select a restaurant known to you and access its restaurant score and other background information. Critique the adequacy of the restaurant inspection information made available to you.

6. As discussed in chapter 6, states have a major responsibility for protecting the public against foodborne illnesses. Discuss your state's responsibilities for food safety. Be specific in regard to which state agencies have specific responsibilities.

7. How do FDA's and EPA's regulatory responsibilities differ with regard to bottled

watter?

8. Discuss the assertion "...with refrigeration came the ability to chill or freeze food and store it safely for long periods of time" in the context of reducing foodborne illnesses.

9. As discussed in the chapter, foodborne illnesses will affect annually about one in four Americans. A substantial but unknown amount of illnesses occur because of poor food preparation practices in the home. Discuss some practical means of preventing foodborne illnesses caused by home food preparation.

10. Discuss the pros and cons of giving meat industry inspectors the authority to supplant government meat inspectors.

NOTES

1. This refers to the Secretary, U.S. Department of Agriculture.
2. Outbreaks of food poisonings are clusters of illness that result from ingestion of food contaminated by a common substance [5].
3. The cited provisions are only a small part of a larger set of provisions [16].
4. Chapter refers to material in the state of Georgia's Division of Public Health materials [16].
5. California, Connecticut, Delaware, Maine, and New York [19].

REFERENCES

1. Mead, P.S., et al., Food-related illness and death in the United States, *Emerging Infectious Diseases*, 5, 607, 2000.
2. House Agriculture Committee, Federal Meat Inspection Act of 1906 Available at http://agriculture. House.gov/glossary/federal_meat_inspection_act_of 1906.htm, 2002.
3. WSDA (Washington State Department of Agriculture), Federal Meat Inspection Act. Available at http://www.wa.gov/agr/IBP/Federal%20meat%20inspection%20act.htm, 2002.
4. Tauxe, R.V., Emerging foodborne pathogens, *International Journal of Food Microbiology.*, 78(1-2), 31, 2002.
5. CSPI (Center for Science in the Public Interest), Outbreak Alert!, 6th ed., Washington, D.C., March, 2004.
6. Shapiro, R., et al., Salmonella Thompson associated with improper handling of roast beef at a restaurant in Sioux Falls, South Dakota, *Journal of Food Production* 62(2), 118, 1999.
7. *New York Times*, Random testing for *E. coli* is set for meatpacking sites, Sept. 26, 2002.
8. USDA (U.S. Department of Agriculture), New Measures to Address E. Coli 0157:H7 Contamination. Backgrounder, Washington, D.C., September, 2002.
9. GAO (General Accounting Office), School Meal Programs: Few Instances of Foodborne Outbreaks Reported, But Opportunities Exist to Enhance Outbreak Data and Food Safety Practices, Washington, D.C., 2003.
10. Daniels, N.A., et al., Foodborne disease outbreaks in United States schools, *Pediatric Infectious Diseases Journal,* 21(7), 623, 2003.
11. CFSAN (Center for Food Safety and Nutrition, FDA), The Story of the Laws Behind the Labels. Part I. 1906—The Federal Food, Drug, and Cosmetic Act, Food and Drug

Administration, Washington, D.C., 1981.

12. FDA (Food and Drug Administration), FDA initiates seizure of ginseng because of potentially risky pesticide residues, FDA Talk Paper No. T04-59, Washington, D.C., December 16, 2004.

13. CFSAN (Center for Food Safety and Nutrition, FDA), The Story of the Laws Behind the Labels. Part II. 1938—The Federal Food, Drug, and Cosmetic Act, Food and Drug Administration, Washington, D.C., 1981.

14. FDA (Food and Drug Administration). FDA food code. Available at http://www.cfsan.fda. gov/~dms/foodcode.html, 2005.

15. FDA (Food and Drug Administration), Real progress in food code adoptions, Center for Food Safety and Applied Nutrition, Washington, D.C. Available at http://www.cfsan.fda. gov/~ear/fcadopt.html, 2005.

16. State of Georgia, Rules and Regulations: Food Service. Available at http://www.ph.dhr. state.ga.us/publications/foodservice, 2002.

17. GDA (Georgia Department of Agriculture), Homepage. Available at http://www.agr.state. ga.us/html/food_safety_inquiries.html, 2004.

18. CFSAN (Center for Food Safety and Nutrition, FDA), Food Labeling. Available at http:// vm.cfsan.fda.gov/label.html, 2004.

19. CDC (Centers for Disease Control and Prevention), Impact of a smoking ban on restaurant and bar revenues—El Paso, Texas, 2002, *Mort & Morb Weekly Report*, 53, February 27, 2004.

20. *Journal of Environmental Health*, Should restaurant inspection reports be published? *Journal of Environmental Health*, April, 27, 2000.

21. Simon, P.A., et al., Impact of restaurant hygiene grade cards on foodborne-disease hospitalizations in Los Angeles County, *Journal of Environmental Health*, March, 32, 2005.

22. Almanza, B.A., Nelson, D.C., and Lee, M.-L., Food service health inspectors' opinions on the reporting of inspection in the media, *Journal of Environmental Health*, June, 9, 2003.

23. FSIS (Food Safety and Inspection Service), Acts & Authorizing Statutes, U.S. Department of Agriculture, Washington, D.C., 2006.

7 Pesticides and Toxic Substances Statutes

7.1 INTRODUCTION

Humankind has known since antiquity that some substances possessed harmful properties. For instance, ancient peoples gradually learned which noxious plants to avoid eating; in effect, practicing the core principle of public health, prevention of disease and disability. Similarly, humankind learned to avoid venomous creatures whose bites could cause harmful health effects. The common factor between noxious plants and venomous creatures would over time become revealed to be chemical substances that possess toxic properties. In time, the study of chemical substances' harmful properties would be called *toxicology*.

The Industrial Revolution led to the manufacture of machines and products that involved especially the use of metals. In the process, metals had to be mined, smelted, forged, and fabricated into machinery for uses in agriculture, industrialization, transportation, and consumer commerce. From the nineteenth century through to the mid-twentieth century, industrial processes often exposed workers to metal fumes and other harmful substances, and if exposure levels were sufficiently great, adverse health consequences occurred. While acute exposures to high levels of toxic substances certainly occurred, there was also a gradual shift to exposure to substances that manifested their toxicity over long periods of time. For example, lead poisoning and metal fume fever were occupational health outcomes for many workers. As workplace conditions gradually improved in the industrialized countries, workers' exposure to metals lessened, but did not disappear. The toxicity of metals had not changed, but exposure levels had decreased, lessening the adverse health effects in workers.

In the mid-twentieth century, the manufacture of synthetic chemicals became a significant economic force and commercial reality, in part, due to the resource demands of World War II. The chemical industry had arrived, generating products such as therapeutic drugs, pesticides, herbicides, plastics, synthetic rubber, and consumer goods. The production and use of these products brought exposure to new, synthesized substances for which toxicology information was lacking. Moreover, the exposures were experienced by persons in the general environment, not solely confined to workplace environments. Exposure occurred at lower levels through contamination of environmental media such as outdoor ambient air and community drinking water supplies. The toxicological implications had changed from those of dealing with the consequence of short-term, high to medium levels of chemical substances, to the condition of long-term exposure to low concentrations of substances found in essential environmental media (i.e., air, water, and food).

One source observes that approximately 10 million chemical compounds have been synthesized in laboratories since the beginning of the twentieth century, but only about 1 percent are produced commercially and can possibly come into contact with living organisms [1]. Although many substances found in commerce lack adequate toxicity data, there already exist ample data to characterize a large number of substances as being deleterious to human health. Shown in Table 7.1 are the major human organ systems[1] know to be affected by toxic substances, illustrated by specific substances shown in the table. Standard references in toxicology contain more comprehensive listings of substances hazardous to human health (e.g., NIOSH's *Registry of Toxic Effects of Chemical Substances* [2], which contains detailed toxicological and industrial hygiene information on a large number of chemicals).

In recognition of the need to control the release of hazardous substances into the environment, Congress has enacted three major statutes: the Federal Insecticide, Fungicide and Rodenticide Act, the Toxic Substances Control Act, and the Food Quality Protection Act. Each of these statutes is discussed in the following sections.

7.2 FEDERAL INSECTICIDE, FUNGICIDE, AND RODENTICIDE ACT

7.2.1 HISTORY

Although federal pesticide legislation was first enacted in 1910, its aim was to reduce economic exploitation of farmers by manufacturers and distributors of adulterated or ineffective pesticides. Congress did not address the potential risks to human health posed by pesticide products until it enacted the 1947 version of the Federal Insecticide, Fungicide, and Rodenticide Act (FIFRA). The U.S. Department of Agriculture (USDA) became responsible for administering the pesticide statutes during this period. However, responsibility was shifted to EPA when that agency was created in 1970. Broader congressional concerns about long- and short-term toxic effects of pesticide exposure

---

TABLE 7.1
Toxicity Endpoints and Illustrative Toxic Substances

Endpoint	Example Substances
Cancer	arsenic, asbestos, beryllium, cadmium, chromium, PAHs
Cardiovascular diseases	carbon monoxide, lead, ozone
Developmental disorders	cadmium, endocrine disruptors, lead, mercury
Immune dysfunction	formaldehyde
Liver diseases	ethyl alcohol, carbontetrachloride
Nervous system disorders	lead, manganese, methyl mercury, organophosphates, PCBs
Reproductive disorders	cadmium, endocrine disruptors, DDT, PCBs, phthalates
Respiratory diseases	nitrogen dioxide, particulate matter, sulfur dioxide
Skin diseases	dioxins, nickel, pentachlorophenol

TABLE 7.2
FIFRA Amendments [3]

Year	Act
1947	Federal Insecticide, Fungicide, and Rodenticide Act (FIFRA)
1964	Amendments
1972	Federal Environmental Pesticide Control Act
1975	FIFRA Extension
1978	Federal Pesticide Act
1980	Amendments
1988	Amendments
1990	Food, Agriculture, Conservation, and Trade Act
1991	Food, Agriculture, Conservation, and Trade Act Amendments
1996	Food Quality Protection Act

on pesticide applicators, wildlife, nontarget insects and birds, and on food consumers subsequently led to a complete revision of the FIFRA in 1972 (Table 7.2). The 1972 law, as amended, is the basis of current federal policy. Substantial changes were made to the FIFRA in 1988 in order to accelerate the process of reregistering pesticides, and again in 1996. The 1996 amendments facilitated registration of pesticides for special (so-called minor) uses, reauthorization of collection of fees to support reregistration, and a requirement to coordinate regulations between the FIFRA and the Federal Food, Drug, and Cosmetic Act (FDCA).

The Federal Insecticide, Fungicide, and Rodenticide Act governs pesticide products and their use in the United States. [3]

As detailed by Schierow [3], the FIFRA, as amended, requires EPA to regulate the sale and use of pesticides in the United States through registration and labeling of the estimated 21,000 pesticide products currently in use [ibid.]. The act directs EPA to restrict the use of pesticides as necessary in order to prevent unreasonable adverse effects on humans and the environment, taking into account the costs and benefits of various pesticide uses. The FIFRA, as amended, prohibits sale of any pesticide in the United States unless it is registered and labeled indicating approved uses and restrictions. It is a violation of the law to use a pesticide in a manner that is inconsistent with the label instructions. EPA registers each pesticide for each approved use (e.g., to control boll weevils on cotton). In addition, the act requires EPA to reregister older pesticides based on new data that meet current regulatory and scientific standards. Establishments that manufacture or sell pesticide products must be registered by EPA. Facility managers are required to keep certain records and to allow inspections by EPA or state regulatory representatives.

Pesticides are broadly defined in the FIFRA §2(u) as chemicals and other products used to kill, repel, or control pests. Familiar examples include pesticides used to kill insects and weeds that can reduce the yield and harm the quality of agricultural commodities, ornamental plantings, forests, wooden structures, and pastures. But the broad definition of *pesticide* contained in the act also applies to products with less familiar "pesticidal uses." For example, substances used to control mold, mildew, algae, and other nuisance growths on equipment, in surface water, or on stored grains are considered to be pesticides. Disinfectants and sterilants, insect repellents and fumigants, rat poison, mothballs, and many other substances are also included.

When pesticide manufacturers apply to EPA to register a pesticide's active ingredient, pesticide product, or a new use of a registered pesticide under the FIFRA §3, EPA requires them to submit scientific data on pesticide toxicity and behavior in the environment. EPA may require data from any combination of more than one hundred different tests, depending on the toxicity and degree of exposure. To register a pesticide's use on food, EPA also requires applicants to identify analytical methods that can be used to test food for pesticide residues and to determine the amount of pesticide residue that could remain on crops, as well as on (or in) food products, assuming that the pesticide is applied according to the manufacturer's recommended rates and methods [3].

Based on the data submitted, EPA determines whether and under what conditions the proposed pesticide's use presents an unreasonable risk to human health or the environment. If the pesticide is proposed for use on a food crop, EPA also determines whether a *safe* level of pesticide residue, called a *tolerance*, can be established under the act. A tolerance must be established before a pesticide registration may be granted for use on food. If any registration is granted, EPA specifies the approved uses and conditions of use, including safe methods of pesticide storage and disposal, which the registrant must explain on the product label. The FIFRA, as amended, requires that federal regulations for pesticide labels preempt state, local, and tribal regulations. Use of a pesticide product in a manner inconsistent with its label is prohibited [ibid.].

EPA may classify and register a pesticide product for general or restricted use. Products known as *restricted-use pesticides* are those judged to be more dangerous to the applicator or to the environment. Such pesticides can be applied only by people who have been trained and certified. Individual states and Indian tribes generally are responsible for training and certifying pesticide applicators [ibid.].

The FIFRA §3 also allows conditional, temporary registrations if: (1) the proposed pesticide ingredients and uses are substantially similar to currently registered products and will not create additional significant environmental risks, (2) an amendment is proposed for additional uses of a registered pesticide and sufficient data are submitted indicating that there is no significant additional risk, or (3) data requirements for a new active ingredient require more time to generate than normally allowed, and use of the pesticide during the period will not cause any unreasonable adverse effect on the environment and will be in the public interest.

The FIFRA §3 directs EPA to make the data submitted by the applicant publicly available within thirty days after a registration is granted. However, applicants may claim certain data are protected as trade secrets under §10. If EPA agrees that the data are protected, the agency must withhold the data from the public, unless the data pertain to the health effects or environmental fate or effects of the pesticide's ingredients. Information may be protected if it qualifies as a trade secret and reveals: (1) manufacturing processes; (2) details of methods for testing, detecting, or measuring amounts of inert ingredients; or (3) the identity or percentage quantity of inert ingredients [3].

Companies sometimes seek to register a product based upon the registration of similar products, relying upon the data provided by the original registrant that is publicly released. This is allowed. However, §3 of the act provides for a ten-year period of *exclusive use* by the registrant of data submitted in support of an original registration or a new use. In addition, an applicant who submits any new data in support of a registration is entitled to compensation for the cost of data development by any subsequent applicant who supports an application with that data within fifteen years of its submission. If compensation is not jointly agreed upon by the registrant and applicant, binding arbitration can be invoked [ibid.].

Most pesticides currently registered in the United States are older pesticides and were not subject to modern safety reviews. Amendments to the act in 1972 directed EPA to reregister approximately 35,000 older products, thereby assessing their safety in light of current knowledge. The task of reregistering older pesticides has been streamlined by reviewing groupings of products having the same active ingredients, on a generic instead of an individual product basis. Nevertheless, the task for registrants and EPA remains immense and costly. In 1988, in order to accelerate the process of reregistration, Congress imposed a 10-year reregistration schedule. To help pay for the additional costs of the accelerated process, Congress directed EPA to require registrants to pay reregistration and annual registration maintenance fees on pesticide ingredients and products. Many of the 35,000 products will not be reviewed and their registrations will be canceled because registrants do not wish to support reregistration. The 1996 amendments to the FIFRA extended EPA's authority to collect maintenance fees through fiscal year 2001. Exemptions from fees or reductions are allowed for minor-use pesticides, public health pesticides, and small business registrants [3].

7.2.2 KEY PROVISIONS OF THE FIFRA, AS AMENDED, RELEVANT TO PUBLIC HEALTH

In its current construction, the FIFRA, as amended, has the following major functions [3]:

(1) *Pesticide Registration*—All new pesticide products used in the United States must first be registered with EPA. To register a new pesticide requires the submission to EPA of the product's complete chemical formula, a proposed label, and a full description of the tests made of the product and the results upon which the claims are based. Manufacturers can ask for trade secret protection to protect information claimed to be vital to commercial propriety.

(2) *Control Over Pesticide Usage*—EPA has authority to restrict use of pesticides. The act permits the classification of pesticides into general and restricted categories, with the latter category available only to certified applicators. Certification standards are developed by EPA to regulate how certified applicators apply restricted pesticides.

(3) *Removal of Pesticides from the Market*—The act mandates EPA to take action against those pesticide products considered a risk to public health and/or the environment. EPA's actions can include a *cancellation* order (which is used to initiate review of the substance, during which the product can continue to be manufactured and placed in commerce), or a *suspension* order (which is an immediate ban on the production and distribution of a pesticide product). There also are different administrative procedures attending a cancellation order or a suspension order that would determine how quickly EPA's action would take effect.

(4) *Imports and Exports*—§17 of the FIFRA directs that imports of pesticide products will be subject to the same requirements of testing and registration as domestic products. However, the act excludes U.S. exports from coverage other than for certain record keeping provisions.

The FIFRA has several implications for hazardous waste generation and management, primarily through linkage to other federal statutes. The Resource Conservation and Recovery Act of 1976 gives EPA the authority to regulate the disposal of generated hazardous wastes, including the disposal of pesticides from manufacturers. The federal Waste Pollution Control Act of 1972, under §301, requires all industrial enterprises, including pesticide manufacturers and formulators, to apply to EPA for discharge permits if they release effluent into any body of water. The same statute, §307 permits pesticides to be controlled as toxic substances, thereby leading to the development of special discharge standards. The CERCLA, as amended, directs ATSDR, in consultation with EPA and the National Toxicology Program, to initiate a program of research to fill gaps in scientific knowledge for prioritized CERCLA hazardous substances. The program of research is, by statute, to be coordinated with EPA's authorities under the FIFRA and the Toxic Substances Control Act, in both instances possibly leading to EPA rulemaking requiring manufacturers of a particular hazardous substance to fill the research gaps identified by ATSDR.

The FIFRA was amended substantially by the Food Quality Protection Act of 1996, which is discussed in a subsequent section of this chapter.

7.2.3 PUBLIC HEALTH IMPLICATIONS OF THE FIFRA, AS AMENDED

Pesticides are chemical substances evolved by nature or synthetically produced to be biologically active. As such, pesticides are intentionally harmful to living organisms, often with biological specificity. Given the mortal purpose of pesticides, their public health implications might seem obvious. However, the implications are a complicated proposition. For example, it can be argued that pesticides have benefited the public's health by reducing mosquito infestation, thereby reducing the number of persons at risk of contracting malaria or West Nile disease. However, some pesticides used to control mosquitoes are environmentally persistent and can cause serious harm to ecological systems. An example is the use of DDT in the tropics for malaria control, even though it causes ecological degradation. DDT and other chemicals are called *Persistent Organic Pollutants* and their use and management are the subject of an international treaty, which is discussed in chapter 9.

The FIFRA provides some human and ecological health protection by requiring EPA to register pesticides, control their uses, and remove those found harmful from the U.S. market. In this regard, the act serves as a gatekeeper over which pesticides get into the general environment. But this gatekeeping does not provide complete prohibition of pesticides and similar chemicals from migrating into the U.S. environment. This is because many pesticides are approved for use in the United States because of their desirable properties of pest eradication, which can increase crop yields and improve food quality. Are the pesticides in the environment potentially harmful to human and ecological health? And if harmful, does this necessitate further effort to reduce pesticide levels and to increase public health action?

The presence of pesticides, herbicides, and rodenticides in the U.S. environment raises questions about the potential impact on human and ecological health. The U.S. Geological Survey (USGS) [5] observes that about one billion pounds of conventional pesticides are used each year in the United States. In 2006 the USGS reported the findings from a 10-year program of surveillance of pesticide levels in U.S. rivers, fish, and private wells. The report is based on data from 51 major river systems from Florida to the Pacific Northwest, Hawaii, and Alaska, and a regional study conducted in the High Plains aquifer system. The USGS study, which covers the years 1992–2001, found that pesticides seldom occurred alone but almost always as complex mixtures. Most stream samples and about half the well samples contained two or more pesticides, and frequently more [ibid.].

Findings showed pesticides were present throughout the year in most streams in urban and agricultural areas of the United States. When the USGS measurements were compared with EPA drinking water standards and guidelines, the pesticides were seldom found at concentrations likely to affect humans. Concentrations of individual pesticides were almost always lower than the standards and guidelines, representing fewer than 10 percent of the sampled stream sites and about 1 percent of domestic and public supply wells. As concerns fish tissues, organochlorine pesticides and their degradates were found in greater than 90 percent of fish in streams that drained agricultural, urban, and mixed-land-use settings. Pesticides were less common in ground water. More than 80 percent of urban streams and more than 50 percent of agricultural streams had concentrations in water of at least one pesticide that exceeded a water quality benchmark for aquatic life, which suggests the need for further control of pesticide releases into the environment.

Concerning the general toxicity of pesticides, one source examined the scientific literature for evidence of pesticides' carcinogenicity and reproductive toxicity [6]. The investigators used EPA data on carcinogenicity of chemicals. They found that of the 250 pesticides evaluated by EPA, 12 of the 26 of greatest annual use in the United States had been classified as carcinogens in one of EPA's carcinogenesis categories.[2] Chronic exposure at lower levels has been associated with adverse neurological and behavioral conditions in young children [7]. More recently, chronic exposure of adults to pesticides has produced features of Parkinson's disease (cited in [8]).

A study conducted by Columbia University investigators found that insecticide exposures were widespread among minority women in New York City during pregnancy [9]. The study consisted of 314 mother-newborn pairs and insecticide measurements in maternal ambient air during pregnancy as well as in umbilical cord plasma at delivery. For each log unit increase in cord plasma chlorpyrifos levels, birth weight decreased by 42.6 g and birth length decreased by 0.24 cm. Combined measures of cord plasma chlorpyrifos and diazinon (adjusted for relative potency) were also inversely associated with birth weight and length. Birth weight averaged 186.3 g less among newborns possessing the highest compared with lowest 26 percent of exposure levels. Further, the associations between birth weight and length and cord plasma chlorpyrifos and diazinon were highly statistically significant among newborns born before the years 2000–2001 when EPA phased out residential use of these insecticides. Among newborns born after January 2001, exposure levels were substantially lower, and no association with fetal growth was apparent. This investigation affirms the toxicological adage, "The dose makes the poison."

In another study with dose-dependent results, investigators from the National Cancer Institute (chapter 3) examined cancer rates in a large cohort of pesticide applicators [10]. The study involved a total of 54,383 pesticide applicators in Iowa and North Carolina. Exposure to the widely used pesticide chlorpyrifos was found to be associated with increased rate of lung cancer. The incidence of lung cancer was found statistically significantly associated with chlorpyrifos lifetime exposure-days, suggesting a dose-dependent effect. This study and the one from Columbia University imply that environmental health policies regarding pesticide use and application should be further strengthened to mitigate or decrease exposure to pesticides.

In summary, the implications of pesticides and similar chemicals in community environments are of continuing concern to environmental and public health authorities, given the purpose of the chemicals. The FIFRA provides the main federal framework for managing the hazard of pesticides. For EPA-approved pesticides, more than one billion pounds are used annually in various agricultural and other commercial applications in the United States. Given the commercial value of pesticides, there will be continued releases of them into environmental media. This reality emphasizes the importance of policies that are committed to monitoring of pesticide levels in water, food, and human tissues and for conducting research on potential human and ecological impacts.

7.3   THE FOOD QUALITY PROTECTION ACT

7.3.1   HISTORY

In 1996, the Congress enacted major legislation that changed how pesticides are regulated. The Food Quality Protection Act of 1996 (FQPA) revises the Federal Insecticide, Fungicide, and Rodenticide Act (FIFRA) and the federal Food, Drug, and Cosmetic Act (FDCA). The FQPA resulted from the confluence of efforts by special interest groups. Children's health advocates desired greater protections under the FIFRA for children exposed to pesticides. The pharmaceutical and food industries desired a repeal of the Delaney Clause in the FDCA. The FQPA legislation constituted the first major revision in decades in U.S. pesticides laws. This dramatically altered how pesticides are registered, used, and monitored in the food chain. The legislation was passed without

TABLE 7.3
Food Quality Protection Act's Titles [11]

Title	Name of Title
I	Suspension—Applicators
II	Minor Use Crop Protection, Antimicrobial Pesticide Registration Reform, and Public Health Pesticides
III	Data Collection Activities to Assure the Health of Infants and Children and Other Measures
IV	Amendments to the Federal Food, Drug, and Cosmetic Act
V	Fees

a dissenting vote in either the House of Representatives or Senate and signed into law by President Clinton on August 3, 1996.

The overall purpose of the FQPA is to protect the public from pesticide residues found in the processed and unprocessed foods they eat. Essentially, the FQPA amended the FIFRA and the FDCA so that a single health-based standard would be issued to alleviate problems concerning the inconsistencies between the statutes. The health-based standard would be based on a "reasonable certainty of no harm."

The FQPA's titles are given in Table 7.3. The act provides a standard for pesticide residues in both raw and processed foods. The standard is "reasonable certainty of no harm." The law requires EPA to review all pesticide tolerances within 10 years, giving particular attention to exposure of young children to pesticide residues. Furthermore, EPA must consider a substance's potential to disrupt endocrine function when setting tolerances. The statute requires EPA to give consideration to effects of pesticides on the public's health, requiring the Secretary of DHHS to provide information to EPA on pesticides that protect the public's health [11].

It is worth noting that the Delaney Clause in the FDCA was replaced by a risk-based approach (chapter 11). The Delaney Clause had required the FDA to ban *any* food additive that caused cancer in laboratory animals or humans, leading to bans some thought were not always pertinent to human health. This was a zero risk policy; total elimination of a substance leads to no risk, at least in theory. Moreover, the Delaney Clause was enacted in 1958, when analytical technology was, by today's standards, relatively crude. As technology became ever more precise, it became possible to measure very minute levels of some carcinogens in food. Under the Delaney Clause, such substances had to be eliminated from the food chain, whether they posed an actual health risk or not. The FQPA gives government the authority to apply a de minimis standard, rather than a zero risk standard.

The most publicized incident pertaining to the Delaney Clause concerned the artificial sweetener saccharin. The noncaloric sweetener has been used for more than 100 years to sweeten beverages and food, replacing calories that would have come from use of natural sweeteners. In 1977, acting under the Delaney provisions, FDA proposed to ban the use of saccharin as a food additive. The agency's proposal was driven by the findings from a toxicology study that showed an excess frequency of urinary bladder tumors in rats fed large amount of sodium saccharin [12].

Given the rat data, under the Delaney Clause, FDA had no alternative but to initiate action to ban the dietary uses of saccharin. However, consumer advocates and public health officials expressed great concern that the loss of saccharin would lead to use of natural sweeteners (e.g., sugar), which would increase calories in food, lessening the effectiveness of body weight reduction programs, and also complicate the dietary needs of diabetics. Moreover, a considerable number of scientists questioned the relevance of the rat data for its application to humans. The hue and cry against FDA's proposed ban of saccharin led Congress in 1977 to enact a moratorium to prevent FDA's proposed action. FDA withdrew its proposed ban of saccharin in 1991.

7.3.2   Key Provisions of the FQPA Relevant to Public Health

Title I—Suspension–Applicators

§102–Suspension: Allows EPA to suspend a pesticide registration in an emergency situation without simultaneously issuing a notice of intent to cancel. §103–Tolerance: Reevaluation as Part of Reregistration: Specifies that tolerances and exemptions from tolerances must be reassessed as part of reregistration to determine whether they meet the requirements of the FDCA §106–Periodic Registration Review: Allows continued sale and use of existing stocks of suspended or canceled pesticides under conditions determined by the EPA Administrator to be consistent with the FIFRA. Required EPA to periodically review registration of pesticides and to establish by regulation a procedure for periodic review. The stated goal is to accomplish a periodic review of a pesticide's registration every fifteen years. §120–Training for Maintenance Applicators and Service Technicians: Creates two new types of pesticide applicators: maintenance applicators and service technicians. Authorizes states to establish minimum training requirements for these applicators. EPA's role on these minimum training requirements is to ensure that states understand these provisions.

Title II—Minor Use Crop Protection, Antimicrobial Pesticide Registration Reform, and Public Health Pesticides

§210–Defines minor use. Allows EPA to waive data requirements for a minor use as long as the EPA Administrator can determine the minor use's incremental risk and that the incremental risk would not present an unreasonable adverse effect. Required EPA to establish a minor use program to coordinate minor use activities and consult with growers. Required USDA to coordinate its minor use responsibilities with EPA. §224–Registration Requirements for Antimicrobial Pesticides: EPA must identify re-forms to the registration process, consistent with risk assessment with an antimicrobial pesticide and the type of review appropriate to evaluate the risks. §230–Public Health Pesticide Definitions: Amends the definition of unreasonable adverse effects on the environment by specifying that the risks and benefits of public health pesticides are considered separate from the risks and benefits of other pesticides. §232-§234–Reregistration: Allows EPA to exempt public health pesticides from reregistration. Instructs DHHS to provide benefits and use information if a public health use pesticide is subject to a cancellation notice.

Title III—Data Collection Activities to Assure the Health of Infants and Children and Other Measures

This title contains provisions on data collection activities to assure the health of infants and children, and integrated pest management.

Title IV—Amendments to the Federal Food, Drug, and Cosmetic Act

Key amendments relevant to public health include the following:

- Requires surveys to document dietary exposure to pesticides among infants and children.
- An additional tenfold margin of safety must be applied to risk assessments involving infants and children.
- Requires that pesticide residues be allowed in foods only if long-term exposure does not jeopardize human health and use of the original pesticide does not threaten domestic food production.
- Establishes that the EPA Administrator consider with higher priority a petition for allowing in foods pesticide chemical residues that pose less human health risk than residues of other pesticides of similar use.
- Limits the sharing of information and data on pesticides permitted in food, except, e.g., when nonconfidentiality is necessary to protect public health.
- Allows a high, 30-day-turnover-time priority for a state to petition the EPA Administrator for permission to regulate pesticide chemical residues in food that present a significant public health threat.
- Requires the EPA Administrator, in consultation with the Secretaries of USDA and DHHS, to annually publish and display in large grocery stores information for the general public on pesticides in food.
- Requires the EPA Administrator to take steps necessary to protect public health if any substances such as pesticides are found to stimulate hormones' effects in the human body.
- Requires the EPA Administrator to review current permits in place for pesticide chemical residues in food, giving highest priority to permits that may present the most significant public health risk.

7.3.3 PUBLIC HEALTH IMPLICATIONS OF THE FQPA

In theory, the public health benefits of the FQPA could be quite consequential, particularly in terms of protecting children from the harmful effects of pesticides. Because children lack fully developed organ systems that are necessary for detoxifying hazardous substances, and their rates of absorption of toxic substances are greater than for adults, they are at greater risk of adverse health effects from exposure to pesticides than are adults. Therefore, prevention of exposure to pesticides is consistent with improved public health. The FQPA contributes to this kind of primary prevention by requiring EPA to develop more protective risk assessments of hazardous substances. In particular, the FQPA directs EPA to incorporate an additional safety factor of 10 for risk assessments specific to children. Specifically, the law focused on making sure that food was safe for children, requiring that permissible exposures to pesticides be reduced tenfold to protect infants and children unless EPA was presented with "reliable data" showing that so great a reduction was unnecessary.

The Food Quality Protection Act of 1996 amends the FIFRA and the Food, Drug, and Cosmetic Act in order to provide a risk-based standard for pesticides and residues.

Although Congress enacted the FQPA of 1996 without a dissenting vote in either the House or Senate, legislation has already been introduced, which in the opinion of many public health officials, would drastically weaken the protective provisions of the act [13]. The legislation, which is attributed to lobbying efforts by chemical and agribusiness interests, would reverse the current burden of proof, requiring EPA to provide detailed justification before it sought to apply any additional safety margin for children. Moreover, EPA would face new obligations to justify the use of computer models or statistical assumptions used in their risk assessments of pesticides "in the absence of data that could be obtained." The proposed legislation would make it more difficult for EPA's development of pesticide regulations for which children might be at risk of exposure.

7.4   TOXIC SUBSTANCES CONTROL ACT

7.4.1   HISTORY

Federal legislation to control toxic substances was originally proposed in 1971 by the President's Council on Environmental Quality during the Nixon administration. Its report, *Toxic Substances*, defined a need for comprehensive legislation to identify and control chemicals whose manufacture, processing, distribution, use, or disposal was potentially dangerous and not adequately regulated under other environmental statutes. The enactment of the Toxic Substances Control Act of 1976 (TSCA) was influenced by episodes of environmental contamination such as the contamination of the Hudson River and other waterways by PCBs, the threat of stratospheric ozone depletion from chlorofluorocarbon (CFC) emissions, and contamination of agricultural produce by polybrominated biphenyls (PBBs) in the state of Michigan. The episodes, together with more exact estimates of the costs of imposing toxic substances controls, opened the way for final passage of the legislation. President Ford signed the TSCA into law on October 11, 1976 [14].

The TSCA authorizes EPA to screen existing and new chemicals used in manufacturing and commerce to identify potentially dangerous products or uses that should be subject to federal control [14].

The TSCA directs EPA to execute the following key actions [ibid.]:

- Require manufacturers and processors to conduct tests for existing chemicals,
- Prevent future risks through premarket screening and regulatory tracking of new chemical products,
- Control unreasonable risks already known or as they are discovered for existing chemicals,
- Gather and disseminate information about chemical production, use, and possible adverse effects to human health and the environment.

At the time of enactment, the law allowed continued production of the 62,000 chemicals already in commercial use, which were called *existing chemicals*. Another 18,000 chemicals have been introduced into commerce since 1976, known as *new chemicals*. In sum, approximately 80,000 chemicals potentially fall under the regulatory provisions of the TSCA. However, the chemical industry asserts that only about 15,000 chemicals are actively made, which would reduce their testing burden [15].

The act authorizes EPA to screen existing and new chemicals used in manufacturing and commerce in order to identify potentially dangerous products or uses that should be subject to federal control. As enacted, the TSCA also included a provision requiring EPA to take specific measures to control the risks from polychlorinated biphenyls (PCBs). Subsequently, three titles have been added to address concerns about other specific toxic substances: asbestos in 1986, radon in 1988, and lead in 1992. The amendments to the TSCA are listed in Table 4.4.

EPA may require manufacturers and processors of chemicals to conduct and report the results of tests to determine the effects of potentially dangerous chemicals on living organisms. Based on test results and other information, EPA may regulate the manufacture, importation, processing, distribution, use, or disposal of any chemical that presents an unreasonable risk of injury to human health or the environment. A variety of regulatory tools are available to EPA under the act, ranging in severity from a total ban on production, import, and use to a requirement that a product must bear a warning label at the point of sale.

TABLE 7.4
Toxic Substances Control Act and Major Amendments [14]

Year	Act
1976	Toxic Substances Control Act
1986	Asbestos Hazard Emergency Response Act
1988	Radon Program Demonstration Act
1989	Asbestos School Hazard Abatement Reauthorization Act
1990	Radon Measurement Act
1992	Residential Lead-Based Paint Hazard Reduction Act

7.4.2   Key Provisions of the TSCA, as Amended, Relevant to Public Health

The TSCA is a statute intended to protect the public's health from exposure to toxic substances. As described in the following sections (adapted from [14]), the TSCA provides EPA with sweeping authorities to regulate chemical substances.

Testing of Chemicals. §4 of the TSCA directs EPA to require the development of test data on existing chemicals when certain conditions prevail: (1) the manufacture, processing, distribution, use, or disposal of the chemical "may present an unreasonable risk," or (2) the chemical is produced in very large volume and there is a potential for a substantial quantity to be released into the environment or for substantial or significant human exposure. Under either condition, EPA must issue a rule requiring tests if: (a) existing data are insufficient to resolve the question of safety, and (b) testing is necessary to develop the data.

Premanufacture Notification. §5 requires manufacturers, importers, and processors to notify EPA at least 90 days prior to producing or otherwise introducing a new chemical product into the United States. Any information or test data that is known to, reasonably ascertainable by, or in possession of the notifier, and that might be useful to EPA in evaluating the chemical's potential adverse effects on human health or the environment, must be submitted to EPA at the same time. The act also requires EPA to be notified when there are plans to produce, process, or use an existing chemical in a way that differs significantly from previously permitted uses so that EPA may determine whether the new use poses a greater risk of human or environmental exposure or effects than the former use.

Each year EPA receives between 1,500 and 3,000 premanufacture notices (PMNs); most of these chemicals never go into commercial distribution [16]. EPA has 45 days after notification (or up to 90 days if it extends the period for good cause) to evaluate the potential risk posed by the chemical. If EPA determines that there is a reasonable basis to conclude that the substance presents or will present an unreasonable risk, the administrator must promulgate requirements to adequately protect against such risk. Alternatively, EPA may determine that the proposed activity related to a chemical does not present an unreasonable risk. This decision may be based on the available data, or, when no data exist to document the effects of exposure, on what is known about the effects of chemicals in commerce with similar chemical structures and used in similar ways.

The TSCA notification does not require chemical manufacturers to report how their compounds are used or monitor where their products end up in the environment. Neither do companies have to conduct health and safety testing of their products either before or after they are entered into commerce. According to one source, 80 percent of all applications to produce a new chemical are approved by EPA with no health and safety data submitted. Eighty percent are approved in three weeks [17]. As policy, the lack of health and safety data is inconsistent with prudent public health practice because it goes counter to the prevention core of public health practice.

Regulatory Controls. The alternative means available to EPA for controlling chemical hazards that present unreasonable risks are specified in §6 of the TSCA. EPA has the authority to: prohibit or limit the amount of production or distribution of a substance in commerce; prohibit or limit the production or distribution of a substance for a particular use; limit the volume or concentration of the chemical produced; prohibit or regulate the manner or method of commercial use; require warning labels or instructions on containers or products; require notification of the risk of injury to distributors and, to the extent possible, consumers; require record-keeping by producers; specify disposal methods; and require replacement or repurchase of products already distributed.

Information Gathering. §8 of the TSCA requires EPA to develop and maintain an inventory of all chemicals, or categories of chemicals, manufactured or processed in the United States. All chemicals not on the inventory are, by definition, new and subject to the notification provisions of §5. These chemicals must be added to the inventory if they enter commerce. Chemicals need not be listed if they are only produced in very small quantities for purposes of experimentation or research.

To aid EPA in its duties under the TSCA, it was granted considerable authority to collect information from manufacturers. EPA may require maintenance of records and reporting of: chemical identities, names, and molecular structures; categories of use; amounts manufactured and processed for each category of use; descriptions of byproducts resulting from manufacture, processing, use, and disposal; environmental and health effects; number of individuals exposed; number of employees exposed and the duration of exposure; and manner or method of chemical disposal. In addition, manufacturers, processors, and distributors of chemicals must maintain records of significant adverse reactions to health or the environment alleged to have been caused by the substance or mixture. Industry also must submit lists and copies of health and safety studies. Studies showing adverse effects previously unknown must be submitted to EPA as soon as they are completed or discovered.

Imminent Hazards. §7 provides EPA with authority to take emergency action through federal district courts to control a chemical substance or mixture that presents an imminent and unreasonable risk of serious widespread injury to health or the environment.

Relation to Other Laws. §9 allows EPA to refer cases of chemical risk to other federal agencies (e.g., OSHA, FDA) with the authority to prevent or reduce the risk. For statutes under EPA's jurisdiction, the TSCA gives the administrator discretion to decide if a risk can best be handled under the authority of the TSCA.

Enforcement and Judicial Review. §11 authorizes EPA to inspect any facility subject to the TSCA requirements and to issue subpoenas requiring attendance and testimony of witnesses, production of reports and documents, answers to questions and other necessary information. §16 authorizes civil penalties, not to exceed $25,000 per violation per day, and affords the defendant an opportunity to request a hearing before an order is issued and to petition for judicial review of an order after it is issued. Criminal penalties also are authorized for willful violations. §17 provides jurisdiction to U.S. district courts in civil actions to enforce the TSCA §15 by restraining or compelling actions that violate or comply with it, respectively. Chemicals may be seized and condemned if their manufacture, processing, or distribution violated the act. §20 authorizes civil suits by any person against any person in violation of the act. It also authorizes suits against EPA to compel performance of nondiscretionary actions under TSCA. §21 provides the public with the right to petition for the issuance, amendment or repeal of a rule requiring toxicity testing of a chemical, regulation of the chemical, or reporting.

Confidential Business Information. §14 provides broad protection of proprietary confidential information about chemicals in commerce. Disclosure by EPA employees of such information generally is not permitted except to other federal employees or when necessary to protect health or the environment. Data from health and safety studies of chemicals are not protected unless their disclosure would reveal a chemical process or chemical proportion in a mixture. Wrongful disclosure of confidential data by federal employees is prohibited and may result in criminal penalties.

Chemical Categories. §26 allows EPA to impose regulatory controls on categories of chemicals, rather than on a case-by-case basis. Examples of chemical categories regulated by EPA under §26 include polychlorinatedbiphenyls (PCBs) and chlorofluorocarbons.

Other Provisions. §10 directs EPA to conduct and coordinate among federal agencies research, development, and monitoring that is necessary to the purposes of the act. §22 waives compliance when in the interest of national defense. §23 provides protection of employees who assist in carrying out the provisions of the act (i.e., whistleblowers). §27 authorizes research and development of test methods for chemicals by the Public Health Service in cooperation with EPA. §28 Grants to states are authorized to establish and operate programs to prevent or eliminate unreasonable risks to health or the environment.

It is apparent that the TSCA gives EPA broad authority to: (1) induce testing of existing chemicals, currently in widespread commercial production or use, (2) prevent future chemical risks through premarket screening and regulatory tracking of new chemicals, (3) control unreasonable risk of chemicals, and (4) gather and disseminate information about chemical production, use, and possible adverse effects to human health and the environment [14].

Enforcement Example: EPA Settles PFOA Case Against DuPont for Largest Environmental Administrative Penalty In Agency

On December 14, 2005, EPA announced that DuPont will pay $10.25 million to settle violations alleged by EPA over the company's failure to comply with federal law. Under the settlement, DuPont is also committing to $6.25 million for Supplemental Environmental Projects (SEPs). The settlement would resolve DuPont's violations related to the synthetic chemical Perfluorooctanoic Acid (PFOA) under provisions of both the Toxic Substances Control Act (TSCA) and the Resource Conservation and Recovery Act (RCRA). PFOA is used in the manufacturing process of fluoropolymers, including some Teflon® products. Seven of the eight counts involve violations of the TSCA's Section 8(e)—the requirement that companies report to EPA substantial risk information about chemicals they manufacture, process, or distribute in commerce. For the other SEP, DuPont will spend $1.25 million to implement the Microscale and Green Chemistry Project at schools in Wood County, West Virginia, where PFOA is produced [18].

7.4.3 Amendments to the TSCA

Starting in 1986, several important amendments to the act provide important public health authorizations to EPA and other federal agencies in order to undertake programs on asbestos, radon, and lead. Two amendments are specific to reducing the hazard of asbestos in schools. The Asbestos Hazard Emergency Response Act of 1986 amends the TSCA to direct the EPA administrator to promulgate regulations for asbestos hazard abatement in schools and set standards for ambient interior concentrations of asbestos after completion of response actions in schools. Other key provisions include: informing and protecting the public during the phases of asbestos abatement, authorizes each state governor to establish administrative procedures for reviewing school asbestos management plans, directs the EPA Administrator to make grants to local educational agencies, and makes local educational agencies liable for civil penalties. The Asbestos School Hazard Abatement Reauthorization Act of 1989 amended the 1986 act by deleting certain reporting requirements of states, directed state governors to maintain records on asbestos in schools, and made accreditation requirements of schools' asbestos removal workers applicable to persons working with asbestos in public or commercial buildings [14].

The TSCA has been amended twice for the purpose of reducing the risk of radon gas in the ambient air of residential buildings. The Radon Program Demonstration Act of 1988 established the national goal of making the air within buildings as free of radon as the outside ambient air. The act contains several significant provisions. EPA is directed to: make available to the public information about radon's hazards, develop model construction standards for buildings, assist state radon programs, provide technical assistance to states, make grants to states on an annual basis for radon assessment and mitigation, and to establish regional radon training centers in at least three institutions of higher learning. The Omnibus Budget Reconciliation Act of 1990 authorized EPA to conduct research on radon and radon progeny measurement methods and mandated an EPA study on the feasibility of establishing a mandatory radon proficiency testing program [ibid.].

Of particular importance to public health, given the toxicity of lead in the environment, Title X of the Housing and Community Development Act of 1992 amended several federal statutes, including TSCA, for the purpose of reducing the health hazard of lead in community and workplace environments. The act directs the Department of Housing and Urban Development to assess lead-based paint hazards in federally assisted housing, and requires housing agencies to take action on evaluating and reducing lead-based hazards. The act amends the TSCA by requiring that contractors and laboratories be federally certified. EPA is directed to conduct a comprehensive program to promote safe, effective, and affordable monitoring, detection, and abatement of lead-based paint and other lead exposure hazards. Also, the Secretary of Labor was directed to issue an interim final regulation for workers' exposure to lead in the construction industry.

7.4.4 Public Health Implications of the TSCA, as Amended

Adverse effects on health can be caused by many chemical substances in the environment. The nature and effects depend on such factors as the potency of the substance, the route and extend of exposure, and an individual's personal characteristics such as genetics, age, and health status. As shown in Table 7.1, all of the body's major organs and organ systems can be potentially affected by exposure to chemicals that can be toxic under the appropriate circumstances. The public health implications of toxic substances can be especially great when a toxic substance is pervasive or widely spread within an environmental medium. Consider the example of lead. As was discussed in chapter 5, lead is one of the six criteria air pollutants. Until removed in the United States as an additive in gasoline, ambient air lead was a significant source of lead exposure to children and adults, raising blood lead levels. Given the known association between prenatal exposure to lead and the adverse effects on children's cognitive development, it was a public health success when lead was removed from gasoline.

Another pervasive source of lead exposure comes from the legacy of lead-based paint, used in the United States for decades, until banned as an additive to paint. Lead-based paint used in older housing became a public health problem when young children ate paint chips and were additionally exposed to lead-laden dust. Some lead exposures were lethal, depending on the amount of paint ingested. Cities and states found themselves having to respond to an epidemic of childhood lead poisonings. For some states, removing lead-based paint and conducting health surveillance of children with potential or actual exposure to household lead sources became a pressing financial obligation. In 2006, the state of Rhode Island successfully litigated three paint companies known to have produced lead-based paint in past years [19]. This sent a shock wave throughout the paint industry, since costs to them could run in the billions of dollars nationwide as other states pursue their own litigation. Given the public health gravity of these two examples from the U.S. experience with lead, one would expect the potential benefits of the TSCA would be substantive in regard to preventing the adverse effects of toxic substances.

Unfortunately, the potential consequential benefits to the public's health of the TSCA have not materialized. Of the major environmental health laws, the TSCA stands out as the major disappointment in public health performance. While there have been some positive impacts, particularly due to the act's amendments, the larger promise of the TSCA has not been realized. At its core, the TSCA provides EPA with the authority to assess and control chemicals in commerce (i.e., existing chemicals) and new chemicals proposed for manufacture. The intent is to protect the public from "unreasonable risk" to human health and the environment. Given these laudable purposes, why hasn't the TSCA lived up to its potential as an environmental health force?

One reason the act has failed is because of the large number of chemicals, 80,000, that fall under regulatory coverage. In theory, EPA could require producers of these chemicals to conduct toxicity testing under the TSCA's authorities. However, under TSCA, EPA must find that a chemical presents an "unreasonable risk" before the agency can mandate toxicity testing. Moreover, EPA must determine that any risks are not outweighed by a chemical's economic and societal benefits for each way in which the substance might be used [14]. These risk and benefits determinations pose a significant challenge to EPA, owing to deficiencies in toxicological data for many substances and uncertainties in substances' benefits.

The shortcomings of the act have been described by former EPA Assistant Administrator Lynn Goldman, "TSCA has not proven to be a successful tool for managing existing chemicals; indeed, it has created a situation in which new chemicals, which may be more benign, are subject to substantially more risk management activities and reviews than older and possibly more risky ones (which are not managed at all). Likewise, the TSCA procedure of referring chemicals to other EPA programs or agencies for risk management has not been effective." [16] Regarding existing chemicals, only five[3] have been regulated under the TSCA. As perspective, more than 60,000 chemicals comprise the EPA inventory of existing chemicals. A major reason for EPA's failure to regulate more existing chemicals is the TSCA's unreasonable risk provision, which sets a hurdle too high for the routine regulation of chemicals [ibid.].

New chemicals are also regulated under the act's provisions. Imposition of these provisions is meant to serve as primary prevention measures to keep hazardous substances out of commerce. As Goldman observes, "EPA's process of premanufacture approval is the *only* (emphasis added) safeguard used by the federal government to guard against such risks. Since 1992, very little progress has been made by EPA in addressing the impacts of new chemicals" [ibid.].

In 2004, the Government Accountability Office (GAO)[4] released a comprehensive study of EPA's TSCA authorities and programs [20]. The shortcomings of the act as an effective public health instrument were the salient findings. The GAO stated that they reviewed EPA's TSCA's efforts "[t]o control the risks of new chemicals not yet in commerce, (2) assess the risks of existing chemicals used in commerce, and (3) publicly disclose information provided by chemical companies under TSCA."

The GAO's primary findings, in order of the study's three purposes, were as follows. Regarding new chemicals, since 1979, when EPA began reviewing chemicals for potential placement on the TSCA's inventory, the GAO found that, on average, about 700 new chemicals are introduced into commerce each year. Of the 32,000 new chemicals submitted to EPA by chemical companies only about 570 were designated for chemical companies to submit premanufacture notices for any significant new uses of the chemical, thereby providing EPA with the data to assess risks to human health or the environment from new uses of the chemical. More disturbing, EPA estimated that most premanusfacture notices do not include test data of any type, and only about 15 percent include health or safety test data. EPA reported to the GAO that they had taken actions to reduce the risks of more than 3,500 of the 32,000 new chemicals they had reviewed. Of public health significance, GAO concluded, "EPA's reviews of new chemicals provide limited assurance that health and environmental risks are identified before the chemicals enter commerce" (ibid., p. 2)

In regard to existing chemicals, GAO found that while EPA has authority under the TSCA to require chemical companies to develop test data after an EPA finding of need, this authority has been used for fewer than 200 of the 62,000 chemicals in commerce since 1979 (ibid., p. 7). GAO concluded that "EPA does not routinely assess the risks of all existing chemicals and EPA faces challenges in obtaining the information necessary to do so" (ibid., p. 7). As noted by GAO, in the late 1990s, in cooperation with chemical companies and national environmental groups, EPA implemented its High Production Volume (HPV) Challenge Program [20]. Under this program, chemical companies voluntarily provide test data on about 2,800 chemicals produced or imported in amounts of one million pounds or more annually. While this testing program seems quite positive in terms of potential new chemical data, there has been no assessment to date of the program's quality and utility for EPA's chemical regulatory purposes.

As to the third part of GAO's study, according to EPA officials, about 95 percent of premanufacturing notices for new chemicals submitted by chemical companies contain some information that is claimed by companies as being confidential business information (ibid., p. 7). GAO opined that this limits EPA's ability to share health relevant information with the public, including state environmental and health agencies.

GAO recommended that Congress provide EPA with additional authorities under the TSCA to improve its assessment of chemical risks. It was also recommended that the EPA Administrator take specific actions to improve EPA's management of its chemicals programs. But given the fact that Congress has failed over almost thirty years to improve TSCA, any acceptance of GAO's recommendation will be problematic.

If the TSCA's authorities, as administered by EPA, have led to regulating only five existing chemicals over the life of the statute and regulatory actions on only about 10 percent of new chemicals, one can ask why the TSCA has not been changed for the better? In other words, why hasn't such an important law been fixed? The answer lies in part to the legislative challenges and uncertainties when amending any major federal statute. Bringing any existing statute back before Congress or a state legislature always runs the risk of changes for the worst. As policy, it is sometimes better to deal with the "devil we know" than with an unknown one!

7.5 SUMMARY

Three federal statutes on the control of pesticides and toxic substances are described in this chapter. Each statute has the public health objective of preventing or reducing human contact with chemical substances that could exert toxic effects. Each statute has some interesting policies of relevance to public health. Of the three statutes, the Federal Insecticide, Fungicide and Rodenticide Act (FIFRA) is the oldest, dating to 1910 when federal pesticides legislation was first enacted by Congress. The core purpose of the act is to control the release into the environment of pesticides and other chemical substances expressly designed to kill specific life forms.

There are several FIFRA policies of importance to public health. The act requires that pesticides must be registered with EPA and used only under prescribed conditions of application. This can be considered as the permit policy, without calling it as such in the act. It is also a kind of command and control policy, in that manufacturers are commanded to register their products with EPA, which has authority to control how the products are used. Two other policies of relevance to public health practice include: (1) public disclosure of pesticides information, unless it is classified as a trade secret; and (2) holding pesticides imported into the United States to the same requirements as domestically produced pesticides. The former policy is a statement of the public's right-to-know; the latter policy closes a potential gap in the distribution and application of pesticides in the United States. It is an extra measure of prevention that is consistent with hazard elimination practices by public health officials.

The Food Quality Protection Act (FQPA) is somewhat misnamed. As described in this chapter, the FQPA is primarily about updating and strengthening the regulation and control of pesticides. In particular, the act targets the need for extra protection of children potentially exposed to pesticides. EPA is directed to apply an additional safety factor of 10 in risk assessments where infants and children may at risk of exposure. This policy, extra protection for children, is consistent with the public health practice of special attention given to vulnerable populations.

The third statute discussed in this chapter, the Toxic Substances Control Act (TSCA), is intended to regulate chemical substances. In particular, substances that have toxic properties are to be banned, or given restricted use, from commerce in the United States. In a sense, this is a kind of quarantine for toxic substances. The act also adopts the policy of requiring chemical producers, importers, and processors to give premanufacture notification (PMN) to EPA. This information is to be used by EPA for evaluating chemicals' potential adverse effects on human health and the environment. This policy places responsibility on the chemical industry to test their products and furnish the information the EPA as a component of the PMN. This is an example of accountability as public policy in action.

7.6 POLICY QUESTIONS

1. Discuss the practical significance of the Food Quality Protection Act. (A) Using EPA resources, ascertain the Act's impact on that agency's children's health program. (B) In your opinion, should the act have repealed the Delaney Clause, formerly a component (and embedded policy) of the Food, Drug, and Cosmetic Act, as amended? Why?

2. The Federal Insecticide, Fungicide, and Rodenticide Act requires EPA to regulate the sale and use of pesticides in the United States of products known as "restricted-use pesticides," which are those assessed by EPA to be dangerous to the applicator or to the environment. Using EPA resources, identify such a pesticide and discuss why it was classified for restricted use. What special precautions were developed for the pesticide's use and application?

3. Using Internet resources of the responsible federal agencies, develop a summary of the programs, policies, and progress that comply with Title X of the Housing and Community Development Act of 1992.

4. The Toxic Substances Control Act, as amended, divides chemicals into two broad categories: existing and new. Discuss EPA's regulatory responsibilities and regulatory policies for each category.

5. The Toxic Substances Control Act, § 7, provides EPA with the authority to take emergency action through the federal district courts in order to control a chemical substance or mixture that presents an imminent and unreasonable risk of serious widespread injury to human health or the environment. Discuss why EPA must work through a court in order to interdict an "imminent" hazard.

6. Assume that you work for a county health department. The county has become infested with mosquitoes, raising anxiety in the public that mosquito-borne

diseases could result. Your department decides to use malathion, a pesticide, to periodically spray those areas known to have high concentrations of mosquitoes. You are assigned the task of informing the public of the department's plans. What do you say to the public?

7. Why should the government require that pesticides be registered?
8. What are trade secrets and how do they relate to the FIFRA?
9. Examine the warning label on a commercially available pesticide. Discuss its content in the context of personal and public health.
10. Section 26 of the TSCA permits EPA to impose regulatory controls on categories of chemicals, not just individual chemicals. Discuss the advantages to public health of regulating categories of chemicals.

NOTES

1. The endpoints and associated toxic substances were obtained from ATSDR Toxicological Profiles [4].
2. *Atrazine, metolachlor,* 2-4-D, *metan sodium,* methyl bromide, glyphosate, *dichloropropene,* chlorpyrifos, *cyanazine, pendimethalin, trifluralin, acetochlor,* alachlor, dicamba, EPTC, *chlorothaloni,* copper hydroxide, propanil, terbfos, *mancozeb, fluometuron,* MSMA, ben-tazone, diazinon, *parathion,* sodium chlorate. The twelve pesticides shown in italic have been classified by EPA as carcinogenic in one of EPA's carcinogenesis categories. (See chapter 9).
3. PCBs, chloroflurocarbons, dioxin, asbestos, and hexavalent chromium.
4. Formerly named the General Accounting Office.

REFERENCES

1. Yassi, A., et al., *Basic Environmental Health,* Oxford University Press, Oxford, 2001, 61.
2. RTECS (Registry of Toxic Effects of Chemical Substances). Available at http://www.cdc.gov/niosh/rtecs/default.html, 2004.
3. Schierow, L., Federal Insecticide, Fungicide, and Rodenticide Act. Summaries of Environmental Laws Administered by the EPA, Congressional Research Service. Available at http://www.NLE/CRSreports/BriefingBooks/Laws/l.cfm, 1999.
4. ATSDR (Agency for Toxic Substances and Disease Registry), *ATSDR ToxProfiles 2004,* Department of Health and Human Services, Public Health Service, Agency for Toxic Substances and Disease Registry, Division of Toxicology, Atlanta, GA, 2004.
5. USGS (U.S. Geological Survey), Pesticides in the Nation's Streams and Ground Water, 1992–2001, Washington, D.C., 2006.
6. NCAP (Northwest Coalition for Alternatives to Pesticides), Are pesticides hazardous to our health?, *Journal of Pesticide Reform,* 19(2), 2, 1999.
7. NRC (National Research Council), Committee on Pesticides in the Diets of Infants and Children, National Academy Press, Washington, D.C., 1993.
8. Betarbet, R., Sherer, T.B., and Greenamyre, J.T., Animal models of Parkinson's disease, *Bioessays,* 24(4), 308, 2002.

9. Whyatt, R.M., et al., Prenatal insecticide exposures and birth weight and length among an urban minority cohort, *Environmental Health Perspectives*, 112, 1125, 2004.

10. Lee, J.L., et al., Cancer incidence among pesticide applicators exposed to chlorpyrifos in the agricultural health study, *Journal of the National Cancer Institute*, 96, 1781, 2004.

11. EPA (U.S. Environmental Protection Agency), Summary of FQPA Amendments to FIFRA and FFDCA. Available at http://www.epa.gov/oppfead1/fqpa/fqpa-iss.htm, 2003.

12. CCC (Calorie Control Council), Saccharin, Atlanta. Available at http://www.saccharin.org/backgrounder.html, 2003.

13. *Washington Post*, Pesticide coalition tries to blunt regulation, May 21, 2000, A01.

14. Schierow, L., Toxic Substances Control Act. Summaries of Environmental Laws Administered by the EPA, National Library for the Environment. Available at http://www.cnie.org/nl3/leg-8/k.html, 1999.

15. Avril, T., U.S. chemical regulation testing leaves much unknown, *Philadelphia Inquirer*, November 4, 2003.

16. Goldman, L.R., Preventing pollution? U.S. toxic chemicals and pesticides policies and sustainable development, *ELR News & Analysis*, Environmental Law Institute, 32 ELR 11018, September, 2002.

17. EWG (Environmental Working Group), Body burden: The pollution in people, Washington, D.C., 2003.

18. EPA Press Advisory, EPA settles PFOA case against DuPont for largest environmental administrative penalty in agency history, Washington, D.C., December 14, 2005.

19. Creswell, J., The nuisance that may cost billions, *New York Times*, April 2, 2006.

20. GAO (Government Accountability Office), Chemical Regulation: Options Exist to Improve EPA's Ability to Assess Health Risks and Manage Its Chemical Review Program, Report GAO-05-458, Washington, D.C., 2005.

8 Solid and Hazardous Waste Statutes

8.1 INTRODUCTION

Involvement of the federal government in regulation of solid and hazardous wastes was not part of the environmental movement of the early 1960s [1]. Environmentalists had given priority to supporting legislation that would improve air and water quality. Moreover, states and municipalities had long had the responsibility for managing municipal waste collection, waste dumps, and sanitary landfills. In colonial America and the agrarian period that followed, farmers disposed of their own solid wastes, much of which was recycled as fertilizer for soil and crop enrichment. Towns and cites during this period continued the longstanding practice of creating open waste dumps, usually located at a distance from occupied areas. Human wastes were disposed of in privies and some cities established rudimentary sewage management facilities. These were local responsibilities; the federal government simply was not involved until early in the twentieth century.

Perhaps the earliest federal involvement in solid waste management is found in the Public Health Service Act of 1912, which states, "The Public Health Service may study and investigate the diseases of man and conditions influencing the propagation and spread thereof, including sanitation and sewage. [...]" [2]. While this authority led to research on waste disposal, almost two decades were to pass before there appeared federal legislation specific to management of solid and hazardous wastes. In 1965, Congress enacted the Solid Waste Disposal Act (SWDA). It was the first federal statute focused solely on waste management. Congress had found "[t}hat the problem presented by solid waste disposal was national in scope and necessitated federal action in assistance and leadership" [3]. However, the act also stated that the collection and disposal of solid waste should continue to be primarily the function of state, regional, and local agencies. Under the SWDA, funds were made available for research on solid waste disposal. The act, in effect, continued, but more directly focused, research on waste disposal already authorized in the Public Health Service Act of 1912.

The proper disposal of hazardous wastes became a concern of Congress commencing in the 1970s. Described in this chapter are the Resource Conservation and Recovery Act, which deals with the permitted disposal of solid and hazardous wastes and the Comprehensive Environmental Response, Compensation, and Liability Act, enacted by Congress to address the environmental and human health problems caused by uncontrolled hazardous waste, particularly abandoned hazardous waste sites. Also described are federal statutes for controlling dumping of waste into oceans and for preventing environmental pollution from oil spills. The chapter concludes with a discussion of the

Pollution Prevention Act, a statute that focuses expressly on the prevention of pollution through means of recycling and improved waste management.

8.2 RESOURCE CONSERVATION AND RECOVERY ACT

8.2.1 HISTORY

The Resource Conservation and Recovery Act of 1976 (RCRA) established the federal program that regulates solid and hazardous waste management. The RCRA amends earlier legislation, the Solid Waste Disposal Act of 1965, but the amendments were so comprehensive that the act is commonly called RCRA rather than by its official title [4]. As overview, the RCRA defines solid and hazardous waste, authorizes EPA to set standards for facilities that generate or manage hazardous waste, and establishes a permit program for hazardous waste treatment, storage, and disposal facilities. As policy, controlling waste releases through a permitting system for individual waste generators emulates the permitting system in the Clean Water Act. Amendments to the RCRA have set deadlines for permit issuance, prohibited the land disposal of many types of hazardous waste without prior treatment, required the use of specific technologies at land disposal facilities, and established a new program regulating underground storage tanks. EPA is also given authority to inspect hazardous waste facilities coverable under the RCRA and is given enforcement powers to ensure compliance with federal RCRA requirements [ibid.].

$$\Diamond \ \Diamond \ \Diamond$$

The amounts of waste generated in the United States are huge and as such bring challenges to waste managers. Three categories of waste identified by the National Research Council (NRC) are municipal solid waste, medical wastes, and hazardous waste [5]. An appreciation of the generated amounts and composition of these wastes is useful for public health considerations.

The Solid Waste Disposal Act and the Resource Conservation and Recovery Act provide regulation of solid and hazardous wastes [4].

Municipal Solid Waste—This category of waste is defined as "[t]he solid portion of the waste (not classified as hazardous or toxic) generated by households, commercial establishments, public and private institutions, government agencies, and other sources" [ibid.]. Shown in Figure 8.1 are the top eight generators in the United States of municipal solid waste in 1997, with a total volume of 217 million tons. In 2004, EPA estimated that approximately 230 million tons of municipal solid waste or garbage are generated each year in the United States, which equates to about 4.6 pounds of solid waste produced per person per day[1] [6]. Noteworthy from the NRC data is that about 45 percent of this total represents paper and paperboard products. In the European Union, the average amount of municipal waste generated in 2003 ranged from 2.1 to 3.5 lbs/person/day across the 25 member states, of which two-thirds came from households [7].The environmental

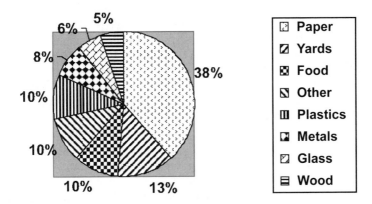

FIGURE 8.1 U.S. municipal solid waste composition percentages, by weight, 1997 [24].

health policy implication is that recycling of paper and paperboard could significantly reduce the amount of municipal solid waste that is taken to permitted landfills, thereby increasing the landfills' volume and area.

Medical Wastes—These wastes constitute a particularly important hazard to human health. Medical wastes are generated throughout the United States health care system. Hospitals, in particular, produce the greatest volume of wastes, generating, according to one source, about 26 lbs. of waste per bed per day [5]. An estimate of the annual volume of generated medical wastes is unknown. A crude division of medical wastes consists of that part which contains infectious pathogens (e.g., HIV) and wastes that do not pose an infectious hazard. Under the provisions of the RCRA, infectious medical wastes must be incinerated or otherwise handled by permitted waste disposal facilities. Noninfectious medical wastes under the RCRA can be handled as municipal solid waste and taken to permitted landfills.

Hazardous Waste—Under the provisions of the RCRA, hazardous waste is a waste material that can be categorized as potentially dangerous to human health or ecosystems. Shown in Figure 8.2 are the five greatest generators of hazardous waste. As noted in

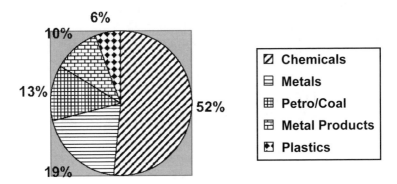

FIGURE 8.2 Top five generators of hazardous waste by percentage of total hazardous waste generated in the U.S. [24]

the figure, the chemical and allied industries are the top generators. According to EPA, more than 12,000 hazardous waste producers annually produce more than 40 million tons of hazardous waste regulated under the RCRA [6]. Of this amount, about 4 percent is hazardous waste produced in households (i.e., more than 1.6 million tons). Leftover household products that contain corrosive, toxic, ignitable, or reactive ingredients are considered by EPA to be household hazardous waste [8]. Products such as paints, cleaners, oils, batteries, pest poisons, and pesticides contain potentially hazardous substances that require special care for proper disposal.

As previously described, the amount of municipal, medical, and hazardous wastes produced in the United States is enormous. If one accepts the proposition that waste generation is fundamentally wasteful, having important consequences such as environmental quality (e.g., air pollution, landfills), economic burdens (e.g., cost of waste disposal), health impacts (e.g., effects of air pollution on children's health), and social disruption (e.g., disputes on where to site landfills), what can be done to lessen these impacts? Some waste will always be inevitable, but a policy of waste reduction and minimization comports with good public health practice.

As shown in Table 8.1, EPA promotes the three Rs of waste reduction: Reduce, Reuse, and Recycle [9]. As elaborated by EPA:

- Reduce—"Source reduction, often called waste prevention, means consuming and throwing away less. Source reduction includes purchasing durable, long-lasting goods and seeking products and packaging that are as free of toxics as possible. It can be as complex as redesigning a product to use less raw material in production, have a longer life, or be used again after its original use is completed. Because source reduction actually prevents the generation of waste in the first place. It is the most preferable method of waste management and goes a long way toward protecting the environment" [ibid.].
- Reuse—"Reusing items by repairing them, donating them to charity and community groups, or selling them also reduces waste. Use a product more than once, either for the same purpose or for a different purpose. Reusing, when possible, is preferable to recycling because the item does not need to be reprocessed before it can be used again" [ibid.].
- Recycle—"Recycling turns material that would otherwise become waste into valuable resources and generates a host of environmental, financial, and social benefits. After collection, materials (e.g., glass, metals, plastics, and paper) are separated and sent to facilities that can process them into new materials or products" [ibid.].

TABLE 8.1
The Three Rs of Waste Reduction [9]

R	Action
Reduce	Consume & discard less
Reuse	Do not discard if reuse is possible
Recycle	Convert waste into usable products

According to EPA, recycling is one of the best environmental success stories of the late twentieth century. By the agency's assessment, recycling, including composting, diverted 68 million tons of material away from landfills and incinerators in 2001, an increase of 34 million tons from that in 1990 [ibid.]. Further, EPA credited curbside waste recycling programs with producing a diversion of about 30 percent of the U.S.'s solid waste in 2001.

◊ ◊ ◊

The RCRA contains a statement of national environmental health policy, "The Congress hereby declares it to be the national policy of the United States that, wherever feasible, the generation of hazardous waste is to be reduced or eliminated as expeditiously as possible. Waste that is nevertheless generated should be treated, stored, or disposed of so as to minimize the present and future threat to human health and the environment" [10]. This pollution prevention policy set a course for subsequent federal and state regulatory action.

The RCRA is a regulatory statute designed to provide "cradle-to-grave" control of hazardous waste by imposing management requirements on generators of hazardous waste and transporters, and upon owners and operators of treatment, storage, and disposal facilities. The statute applies principally to operating waste management facilities, whereas the Comprehensive Environmental Response, Compensation, and Liability Act (CERCLA) applies mainly to uncontrolled hazardous waste sites. The RCRA deals both with hazardous waste and nonhazardous waste, although the main emphasis in the act is on the former. More than 500,000 companies and individuals in the United States who generate more than 172 million metric tons of hazardous waste each year are covered under RCRA regulatory programs [11,12]. The RCRA, as amended, represents a significant challenge to the regulated community. In particular, industry is challenged to find new ways to minimize, treat, and dispose of hazardous waste. The use of innovative technologies, like bioremediation, to reduce waste is the subject of active development.

Wastes covered under the RCRA are defined in the statute. The RCRA, Subtitle A, defines *solid waste* as being "any garbage; refuse; sludge from a waste treatment plant, water supply treatment plant, or air pollution control facility; and other discarded material including solid, liquid, semisolid, or contained gaseous material resulting from industrial, commercial, mining, and agricultural operations, and from community activities; but does not include solid or dissolved material in domestic sewage, or solid or dissolved materials in irrigation return flows or industrial discharges which are point sources subject to permits under §402 of the Federal Water Pollution Control Act, as amended, or source, special nuclear, or byproduct material as defined by the Atomic Energy Act of 1954, as amended" [10].

Under the RCRA, Subtitle A, *hazardous waste* "means a solid waste, or combination of solid wastes, which because of its quantity, concentration, or physical, chemical, or infectious characteristics may either cause, or significantly contributed to an increase in mortality or an increase in serious irreversible or incapacitating reversible illness; or pose a substantial present or potential hazard to human health or the environment when improperly treated, stored, transported, or disposed of, or otherwise managed" [ibid.].

The RCRA therefore applies to almost any waste regardless of its physical form. The EPA regulations have further clarified the definition of waste coverable under the RCRA. It is a complex and comprehensive statute, implemented by EPA through a set of extensive regulations.

8.2.2 AMENDMENTS TO THE RCRA

Commencing in 1980, the Congress has enacted six amendments to the RCRA (Table 8.2). Two amendments were substantive and made major changes in how solid waste is managed in the United States The other amendments were largely technical adjustments to existing legislation. The amendments address specific areas of solid and hazardous waste management, but have not changed the basic thrust of the RCRA's basic principles.

The Solid Waste Disposal Act Amendments of 1980 banned open waste dumps, thereby eliminating a public health hazard that had existed since antiquity.

The Used Oil Recycling Act of 1980 amended the Solid Waste Disposal Act by defining the terms *used oil*, *recycled oil*, *lubricating oil*, and *re-refined oil*. The EPA Administrator was directed to promulgate regulations to establish performance standards and other requirements necessary to protect the public health and environment from hazards of recycled oil. Moreover, EPA was authorized to provide grants to states with approved solid waste plans that: (1) encourages the use of recycled oil, (2) discourages uses hazardous to the public's health and environment, (3) calls for informing the public of the uses of recycled oil, and (4) establishes a program for the collection and disposal of oil in a safe manner [4].

TABLE 8.2
Solid Waste Disposal Act and Amendments [4]

Year	Act
1965	Solid Waste Disposal Act
1970	Resource Recovery Act
1976	Resource Conservation and Recovery Act
1980	Used Oil Recycling Act
1980	Solid Waste Disposal Act Amendments
1984	Hazardous and Solid Waste Amendments
1988	Medical Waste Tracking Act
1992	Federal Facility Compliance Act
1996	Land Disposal Program Flexibility Act

The Solid Waste Disposal Act Amendments of 1980 were substantive and gave EPA broader powers to deal with illegal disposal of hazardous waste. Two provisions were of special import. One important provision prohibited open dumping of solid waste and hazardous waste. This prohibition brought to close in the U.S. a human practice that dates to antiquity. Moreover, banning of open dumps was a major public health contribution. Gone were the open dumps that were rife with disease-carrying rodents and other creatures and provided human access to areas that contained decomposing food, hazardous chemicals, and dangerous vermin. Further, standards were developed for the sanitary disposal of solid waste in dump sites that are designed to prevent releases of hazardous substances into ambient air and underground aquifers [ibid.].

The other important provision authorized the EPA administrator to issue orders requiring individual facility operators to do monitoring, testing, analysis, and reporting necessary to abate hazards to human health or the environment. Other key changes included: (1) transference from EPA to the Department of the Interior all responsibilities for managing coal mining wastes, (2) expanded EPA's standards applicable to generators of hazardous waste and their responsibility for the arrival of wastes at waste management facilities, (3) set forth criminal and civil penalties for failures to comply with waste management permits, and (4) directed each state to submit to EPA an inventory of hazardous waste storage and disposal sites.

An even more significant set of amendments to the SWDA were the Hazardous and Solid Waste Amendments of 1984, comprising six titles and accompanying subtitles. Title I revised findings and objectives of the act to include minimizing the generation and the land disposal of hazardous waste, and declared it to be the national policy that, wherever feasible, the generation of hazardous waste is to be reduced or eliminated as expeditiously as possible, but without stating how. This title states, "Waste that is nevertheless generated should be treated, stored, or disposed of so as to minimize the present and future threat to human health and the environment."

One of the most significant provisions of the 1984 amendments to the RCRA is the prohibition of land disposal of hazardous wastes [11]. In a phased approach, the act bans the disposal in landfills of bulk or noncontainerized liquid hazardous wastes, and hazardous wastes containing free liquids. The act then required EPA to determine whether to ban, in whole or in part, the disposal of all RCRA hazardous wastes in land disposal facilities. At the same time, EPA must establish treatment standards for each restricted waste based on the Best Demonstrated Available Technologies (BDAT). If the restricted waste is first treated to BDAT levels, the treated waste or residue can then be placed in land disposal facilities.

The Medical Waste Tracking Act of 1988 had the purpose "To amend the Solid Waste Disposal Act to require the Administrator of the Environmental Protection Agency to promulgate regulations on the management of infectious waste" [13]. The act had been precipitated by the discovery of medical waste that had washed ashore along the coasts of some northeastern states, particularly in New Jersey. Two provisions were at the heart of the Act. One provision required the EPA, in cooperation with five states, to establish a two-year demonstration program to track listed medical wastes, and segregate, contain, and label such wastes to protect waste handlers and the public. The act also directed the ATSDR administrator to report to the Congress within two years of this act's enactment on the health effects of medical waste. The outcomes of these two provisions are discussed in a subsequent section.

The Federal Facility Compliance Act of 1992 amended the Solid Waste Disposal Act to: (1) waive the sovereign immunity of the United States for purposes of enforcing federal, state, interstate, and local requirements with respect to solid and hazardous waste management; (2) make federal employees subject to criminal sanctions under such laws; (3) prohibit federal agencies from being subject to such sanctions; (4) require the secretary of energy to develop a treatment capacity and technology plan for each facility at which DOE generates or stores mixed wastes; (5) direct the administrator to promulgate regulations identifying when military munitions become hazardous waste and providing for the safe transportation and storage of such waste; (6) exclude from the definition of "solid waste" solid or dissolved material in domestic sewage; and (7) require the administrator to establish a program to assist small communities in planning and financing environmental facilities and compliance activities [4].

Enforcement Example: Idaho Man Sentenced in Paint Waste Case

The former director and corporation secretary for Ponderosa Paint Company in Boise, Idaho, was sentenced on September 16, 2004, to pay a $50,000 fine, pay an additional $40,000 in restitution for cleanup costs incurred by the EPA. He will also spend 30 days in home confinement and serve six months supervised release. The former director pleaded guilty to a charge of being an accessory after-the-fact to transportation of hazardous waste without a manifest in violation of the Resource Conservation and Recovery Act. He was responsible for disposing of approximately 20,000 gallons of waste oil-based paint that had accumulated at the Ponderosa facility between 1995 and 2000. Some of the paint waste was illegally transported to private property in Wilder, Idaho, and burned in a pit. The case was investigated by the Portland Area Office of EPA's Criminal Investigation Division, the FBI, and Idaho State and local officials. The case was prosecuted by the U.S. Attorney's Office in Boise [14].

The Land Disposal Program Flexibility Act of 1996 exempted small landfills located in arid or remote areas from groundwater monitoring requirements, provided there is no evidence of groundwater contamination. This act also exempts hazardous waste from RCRA regulation if it is treated to a point where it no longer exhibits the characteristics that made it hazardous, and is subsequently disposed in a facility regulated under the Clean Water Act or Safe Drinking Water Act

8.2.3 Key Provisions of the RCRA, as Amended, Relevant to Public Health

The subtitles of the RCRA, as amended, are listed in Table 8.3. Those with particular relevance for public health policies and practices are discussed in the following sections.

TABLE 8.3
RCRA, as Amended, Subtitles [10]

Subtitle	Name of Subtitle
A	General Provisions
B	Office of Solid Waste; Authorities of the Administrator
C	Hazardous Waste Management
D	State or Regional Solid Waste Plans
E	Duties of Secretary of Commerce in Resource and Recovery
F	Federal Responsibilities
G	Miscellaneous Provision
H	Research, Development, Demonstration and Information
I	Regulation of Underground Storage Tanks
J	Demonstration Medical Waste Tracking Program

Subtitle B—*Office of Solid Waste; Authorities of the EPA Administrator*—Establishes the EPA Office of Solid Waste and the authorities of the EPA Administrator in carrying out the provisions of the Act.

Subtitle C—*Hazardous Waste Management*—This part of the act is specific to the management of hazardous waste. §3001: requires that the Administrator develop and promulgate criteria for identifying the characteristics of hazardous waste. §§3002,3001(d),3003, 3004: require EPA to compile listings of hazardous wastes and to develop standards applicable to: generators of hazardous waste; transporters of hazardous waste; owners and operators of hazardous waste treatment, storage, and disposal facilities. Permits for treatment, storage, and disposal of hazardous waste are required. EPA is given authority to inspect hazardous waste facilities coverable under the RCRA and is given enforcement powers to ensure compliance with federal RCRA requirements. Each state is required to submit to EPA a continuing inventory that describes the location of each site at which hazardous waste has at any time been stored or disposed of. Similarly, each federal agency must provide the same kind of information to EPA. §3004(u): authorizes EPA or a state to require corrective action for all releases of hazardous waste or constituents from any solid waste management unit at a TSDF seeking a permit under Subtitle C, regardless of the time at which the waste was placed in the unit. Under §3008(h) EPA is authorized to assess a civil penalty to any interim status facility that has released hazardous waste into the environment. §3005: requires that each application for a final determination regarding a permit for a landfill or surface impoundment shall be accompanied by information reasonably ascertainable by the owner or operator on the potential for the public to be exposed to hazardous wastes or hazardous constituents through releases related to the unit. The exposure information is to be provided to EPA, which in turn, shall make it available to ATSDR for public health purposes. When EPA or a state determines that a particular landfill or surface impoundment poses a substantial potential risk to public health, they may request that ATSDR conduct a health assessment of the population at potential risk. However, ATSDR can conduct the requested health assessment "[…] If funds are provided in connection with such request the Administrator of such Agency [i.e., ATSDR] shall conduct such health assessment." §3017: sets forth requirements on the export of hazardous waste to other countries. In general, the exporter must furnish EPA with information about the nature and amount of the waste, the country of destination, the ports of entry, the manner of transport, and the name and address of the ultimate treatment, storage, or disposal facility. Following receipt of the export information, EPA must request the secretary of state to contact the receiving country to obtain that country's written consent to receive the exported hazardous waste.

Subtitle D—*State or Regional Solid Waste Plans*—Regulation of nonhazardous waste, under RCRA, is the responsibility of the states. The federal involvement is limited to establishing minimum criteria that prescribe the best practicable controls and monitoring requirements for solid waste facilities. Disposal of solid waste in open dumps is prohibited, but the RCRA provides EPA with no enforcement authority for banning open dumps. (EPA's enforcement authority under the RCRA covers only hazardous waste.)

Subtitle E—*Federal Responsibilities*—Federal statutes sometimes exempt federal agencies from an act's coverage. This may be for reasons of national security, economic factors, or political reasons. However, the RCRA holds each department, agency, and instrumentality of the executive, legislative, and judicial branches of federal government having jurisdiction over any solid waste management facility or disposal site, or engaged in any activity resulting, or which may result, in the disposal or management of solid waste or hazardous waste to the same expectations and requirements "[a]s any person is subject to such requirements." Only the President can exempt a department's solid waste management facility if it is in the Nation's paramount interest.

Subtitle F—*Research, Development, Demonstration, and Information*—EPA is given authority to conduct research and studies on a range of areas that include: adverse health and welfare effects of solid waste releases; resource conservation systems; production of usable forms of recovered resources; identification of solid waste components; improvements in land disposal practices; methods for disposal of, or recovery of resources; methods of hazardous waste management; and adverse effects on air quality. EPA also is directed to develop, collect, evaluate, and coordinate information on subjects pertaining to solid waste.

Subtitle G—*Regulation of Underground Storage Tanks*—EPA is directed to develop a comprehensive regulatory program for "underground storage tanks" [15]. The RCRA directs EPA to promulgate release, detection, prevention, and correction regulations applicable to all owners and operators of underground storage tanks (UST), as may be necessary to protect human health and the environment. EPA estimates there are 700,000 UST facilities with about two million tanks covered by this regulation.

Subtitle H—*Demonstration Medical Waste Tracking Program*—As previously stated, the Medical Waste Tracking Act of 1988 required EPA to create a demonstration program for tracking the shipment and disposal of medical wastes in a selected number of states. The participating states were Connecticut, New Jersey, New York, Rhode Island, and the commonwealth of Puerto Rico [16]. Apparently no report of the demonstration project's findings was prepared [ibid.], and there was no follow-up by

EPA on the development of federal regulations that would have mandated tracking of infectious medical waste [ibid.], which was the original intent of the act.

However, the act's requirement that ATSDR prepare a report to Congress on the hazard presented by uncontrolled medical waste was accomplished, with the primary finding that such waste was not a national public health hazard, but that medical waste posed a greater than supposed hazard to municipal waste workers and that medical waste from in-home health care was a previously unrecognized health hazard. Given these findings, the ATSDR report contributed to states' enacting more stringent regulations and codes for purpose of controlling medical waste management [17]. This is an example of where a demonstration project, together with a comprehensive public health analysis, dissuaded Congress on the need for comprehensive regulations on an environmental hazard.

8.2.4 PUBLIC HEALTH IMPLICATIONS OF THE RCRA, AS AMENDED

The human health consequences of permitted incinerators and landfills have not been the subject of any sustained program of research. However, one study of adverse birth outcomes in populations residing near landfill sites found small excess risks of congenital anomalies (neural tube defects, hypospadias, abdominal wall defects) and low to very low birth weight babies. The landfills in the study included some hazardous waste sites [18]. The National Research Council reported that few studies have tried to establish a link between an incinerator and illness in the surrounding area, and that most studies found no adverse health effects [5]. In contrast, some studies have shown that municipal incinerator workers have been exposed to high concentrations of dioxins and metals, but any adverse health effects have not been pursued in follow-up studies [19].

Medical waste incinerators are of public health concern because highly toxic dioxins are formed as a byproduct of incinerated plastic materials. EPA has issued standards to reduce emissions from waste incinerators, based on a standard of "maximum achievable control technology" (MACT) [5], and emissions should decrease over time. As public health policy, emissions from incinerators merit scrutiny by state environmental departments in order to assure that harmful emissions are not occurring.

8.2.5 ILLUSTRATIVE STATE SOLID WASTE LAW

As previously noted, the Solid Waste Disposal Act (SWDA), as amended, places solid waste management as primarily the responsibility of the states. This responsibility is effectuated by enactment of state laws, regulations, and solid waste codes. While state laws vary in content according to specific needs and circumstances, all state laws contain provisions that require permits to manage solid waste. This provision reflects requirements found in the SWDA, as amended. Other provisions are illustrated in one state's solid waste law. Following are selected provisions of the state of Georgia's solid waste code [20].

8.2.5.1   Permits[2]

"(a) No person shall engage in solid waste or special solid waste handling in Georgia or construct or operate a solid waste handling facility in Georgia [...] without first obtaining a permit from the director authorizing such activity.

(b)(1) No permit for a biomedical waste thermal treatment technology facility shall be issued by the director unless the applicant for such facility demonstrates to the director that a need exists for the facility for waste generated in Georgia by showing that there is not presently in existence within the state sufficient disposal facilities for biomedical waste being generated or expected to be generated within the state [...]

(c) On or after March 30, 1990, any permit for the transportation of municipal solid waste from a jurisdiction generating solid waste to a municipal solid waste disposal facility located in another county shall be conditioned upon the jurisdiction generating solid waste developing and being actively involved in, by July 1, 1992, a strategy for meeting the state-wide goal of waste reduction by July 1, 1996.

8.2.5.2   Permit Revocation

(e)(1) The director may suspend, modify, or revoke any permit issued pursuant to this Code section if the holder of the permit is found to be in violation of any of the permit conditions or any order of the director or fails to perform solid waste handling in accordance with this part or rules promulgated under this part [...].

8.2.5.3   Site Modification

(2) Prior to the granting of any major modification of an existing solid waste handling permit by the director, a public hearing shall be held by the governing authority of the county or municipality in which the municipal solid waste facility or special solid waste handling facility requesting the modification is located not less than two weeks prior to the issuance of any permit under this Code section and notice of such hearing shall be posted at the site of such facility and advertised in a newspaper of general circulation serving the county or counties in which such facility is located at least 30 days prior to such hearing.

(3) Except as otherwise provided in this part, major modifications shall meet the siting and design standards applicable to new permit applications in effect on the date the modification is approved by the director [...] .

(4) No vertical expansions shall be approved under this subsection unless:

(A) The owner or operator demonstrates compliance with all standards not varied by the director;

(B) The owner or operator has installed a surface and ground-water monitoring system approved by the division under currently promulgated rules and has submitted the initial sampling results to the division;

(C) The owner or operator has implemented or installed a methane gas monitoring program or system approved by the division under currently promulgated rules and has submitted the initial sampling results to the division;

(D) The owner or operator has a closure and postclosure care plan approved by the division under currently promulgated rules;

(E) Where noncompliance with the standards for surface water, ground water, or methane gas has been determined, the owner or operator has a schedule and corrective action plan approved by the division for returning the site to compliance within six months of the director's approval of the corrective action plan. If the owner or operator cannot demonstrate that the site can be returned to compliance within said six-month period, the director shall not issue a permit to expand the site vertically but shall order the facility to prepare a final closure plan, including the cessation of waste receipt within six months of the final effective date of the order […] .

8.2.5.4   Site Inspection

"(j) The director or his designee is authorized to inspect any generator in Georgia to determine whether that generator´s solid waste is acceptable for the intended handling facility [...] ."

A bit of reflection on the Georgia code shows several environmental health policies. The core policy is the permitting of waste managers, with provisions for permit revocation. State and federal permits for management of environmental hazards is a feature of many statutes. Permits provide a legal means for application of the command and control policy that leads to regulations and actions that are intended to ensure accountability of solid waste managers.

Other policies of note in the Georgia code include: (a) provisions for upgrading solid waste facilities when they undergo major modifications, with the intention of keeping such facilities in compliance with current management practices, and (b) provisions for on-site inspections by state inspectors. In a sense, both provisions are an expression of the public health policy of prevention of disease and disability.

8.3   COMPREHENSIVE ENVIRONMENTAL RESPONSE, COMPENSATION, AND LIABILITY ACT *(Superfund)*

8.3.1   History

The CERCLA (aka Superfund) was enacted in 1980 and was reauthorized by the Superfund Amendments and Reauthorization Act of 1986 [12]. The statute was a direct consequence of the discoveries of releases of hazardous substances from abandoned landfills into community residences, in particular, the community of Love Canal, a suburb of Niagara Falls, New York, which was evacuated following the discovery that it overlay an abandoned chemical dump. Love Canal captured the public's attention because of intense news media coverage. Rarely did a day pass without national news media interviewing Love Canal residents, who expressed their concerns about the health of their children and future generations of children. They associated health problems in the community's children with the release of noxious chemicals that had seeped into their homes. As will be subsequently described in this chapter, health investigations confirmed some of the residents' fears that adverse health outcomes had occurred.

The horror of Love Canal occurred during the waning months of President Jimmy Carter's administration. In 1979, the federal government offered to buy the homes of Love Canal residents and assist with their relocation elsewhere. Approximately 950 families were evacuated from the Love Canal area [21]. Over the next 20 years, more than $400 million was spent to remediate the Love Canal area. While this is an impressive expenditure, it is noteworthy that the 21,000 tons of chemical soup that characterized the Love Canal site are still there. To remediate the site, EPA capped it with a thick layer of clay, installed pumps and drains to control runoff of chemicals from the site, and replaced miles of contaminated sewer pipe. The chemicals themselves were left in the contained site and the area was surrounded by a fence [22].

As policy, leaving hazardous substances in place, but interdicting human contact with them has evolved into a risk management decision by EPA, states, and some private sector entities. The theory is that interdiction of contact between hazardous substances and humans and ecological systems will prevent adverse effects. The costs of containment are generally less than for removal of contaminated soil or water. In theory, the cost savings could be used to remediate more sites that require cleanup. On the other side of this argument is the problem that failure to remove hazardous contamination can simply prolong the life of a hazardous waste site and there are no guarantees that future cleanup actions will ever be taken. Moreover, even contained sites, using current best available technology, will in time deteriorate. Therefore the costs of site maintenance and upkeep is passed along to future generations.

In March 2004, EPA removed the Love Canal site from its list of most significant uncontrolled hazardous waste sites, ending 21 years of government and community concern that uncontrolled hazardous waste was a threat to public health. Today the former Love Canal neighborhood is called Black Creek Village, a neighborhood constructed largely of new houses [22].

The Comprehensive Environmental Response, Compensation, and Liability Act, as revised, requires EPA, in cooperation with states, to identify and remediate uncontrolled hazardous waste sites, identify Potentially Responsible Parties (PRPs), collect cost-recovery fees from PRPs, and directs ATSDR to address public health concerns [12].

The CERCLA was therefore the product of great public concern that toxic materials could invade private homes and cause harm to children and future generations. The intent of the law is stated to be, "To provide for liability, compensation, and emergency response for hazardous substances released into the environment and the cleanup of inactive hazardous waste disposal sites." The CERCLA's basic purposes are to provide funding and enforcement authority for remediating (i.e., cleaning up) uncontrolled hazardous waste sites, and for responding to hazardous substance spills. The statute includes provisions for remediating waste sites, responding to public health concerns,

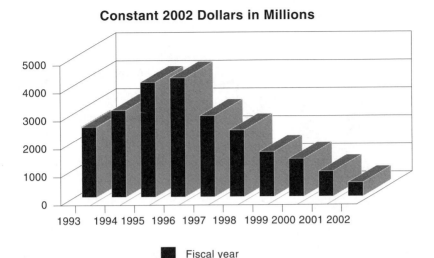

Constant 2002 Dollars in Millions

FIGURE 8.3   The balance of the Superfund Trust Fund available for future appropriations
[23].

enforcement authorities to identify "potential responsible parties," and emergency
removal of chemical spills.

The Hazardous Substance Response Trust Fund was created by the CERCLA,
which was intended by Congress to be a source of funds for site remediation when
other sources were unavailable. Congress appropriated $1.6 million to the trust fund
[29]. News media considered this to be a lot of money, and christened the law as "Superfund," a name that soon became the unofficial name of the CERCLA, at least in the
mind of the public. The 1986 CERCLA amendments changed the name of the trust
fund to Hazardous Substance Superfund [ibid.]. Until 1995 the trust fund was financed
primarily by a tax on crude oil and certain chemicals and an environmental tax on select
corporations [23]. The authority for these taxes expired in December 1995 and has not
been reauthorized by Congress. Neither the Clinton nor George W. Bush administrations
sought reauthorization of the Superfund taxes. The trust fund also receives revenue from
interest accrued on the unexpended balance, recovery of cleanup costs from responsible
parties, and collections of fines and penalties [ibid.]. This trust fund fulfills part of the
act's philosophy of "the polluter pays" for environmental cleanup. (The other part of
this philosophy is the authority given to EPA to identify polluting parties and force them
to bear the cost of site remediation.).

One effect of not imposing the Superfund tax has been a general decrease in the
Hazardous Substance Trust Fund, as shown in Figure 8.3. Less reliance on trust fund
monies means more reliance on funds from general revenue (i.e., taxpayers pay a greater
portion of CERCLA site cleanups). In EPA's fiscal year 2004 budget request for the
CERCLA's program, the general fund appropriation approximated 80 percent of the
program's total appropriation [23].

◊ ◊ ◊

The CERCLA, as amended, places special emphasis on those uncontrolled hazardous
waste sites (HWS) ranked by EPA to pose the greatest hazard to human health and

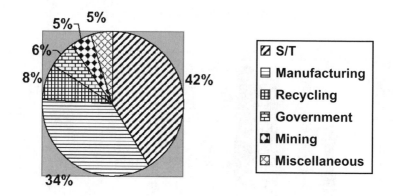

FIGURE 8.4   Categories of NPL sites [24].

natural resources damage. The worst of the HWS comprise what is called the National Priorities List (NPL). As policy, ranking the worst HWS provides decision makers at EPA and states with a means to prioritize sites and remediate first those sites posing the potential greatest risk to human and ecological health. Under the CERCLA, *Potentially Responsible Parties* (PRPs) must be identified by EPA and costs recovered from them to pay for site remediation. Further, the CERCLA stipulates that all NPL sites must receive a health assessment conducted by ATSDR for purpose of ascertaining any adverse effects in persons impacted by the sites. Shown in Figure 8.4 are the broad categories of NPL sites [24]. Uncontrolled waste storage/treatment facilities and former manufacturing facilities constitute about 75 percent of NPL sites. Both categories represent industrial operations that operated in the past and then went out of operation, leaving a legacy of hazardous waste in the environment.

Sites can be placed on the NPL by three mechanisms: (a) EPA's Hazard Ranking System (HRS), (b) states or territories designate one top-priority site regardless of HRS score, or (c) ATSDR has issued a health advisory that recommends moving people away from the site [25]. The HRS uses a structured analysis for scoring and ranking sites. This approach assigns numerical values to factors that relate to risk, based on conditions at the site under consideration. The factors are grouped into three categories: (a) likelihood that a site has released or has the potential to release hazardous substances into the environment, (b) characteristics of the waste (e.g., toxicity and waste quantity), and (c) people or sensitive environments affected by the release of hazardous substances [ibid.]. Sites proposed by EPA for placement on the NPL are published in the *Federal Register* for public comment over a 60-day period. Sites can be deleted from the NPL if EPA determines that no further action is required to protect human health or the environment.

◊ ◊ ◊

At the heart of the CERCLA process is the identification, inspection, remediation, and closure of NPL sites. The processes used by EPA to effectuate this process is complex in details, but fairly straight forward as an overall process, if considered as a step-by-step program. The following steps, when followed in the order given below, comprise what can be called the CERCLA process for remediation of uncontrolled hazardous waste sites [26].

Site Discovery—Potential CERCLA sites are typically identified through state and county inspections and reports from concerned citizens. Federal facilities are required to conduct investigations of past waste management activities, in response to CERCLA §120(d) (Federal Facilities). All potential CERCLA sites are reported to EPA.

Preliminary Assessment and Site Investigation—The first step in the CERCLA process comprises two initial studies known as the Preliminary Assessment (PA) and Site Investigation (SI). Both studies include collecting and reviewing available information to determine the magnitude of the problem posed by the site. At the conclusion of the SI, the site is then scored by the EPA using the HRS [27]. The HRS considers potential relative risks to public health and the environment from release or threatened release of chemicals at the site.

NPL Listing—If the overall potential risks at a site are determined to be significant based on its HRS score, the site will be nominated for placement on the NPL. The NPL is a listing by the EPA of the top-priority sites that are eligible for investigation and remediation under the CERCLA. Typically, sites must receive a score of at least 28.5 out of 100 points in order to be included on the NPL.

Remedial Investigation and Feasibility Study—After a site has been placed on the NPL, two related studies, the Remedial Investigation (RI) and the Feasibility Study (FS), are conducted. An RI/FS may take several years to complete, depending on the size and scope of the site. This phase includes comprehensive sampling and data collection to evaluate the nature, extent, and magnitude of impacts both on- and off-facility. As part of the RI/FS, a risk assessment is performed to identify and quantify the risks that the site poses to public health, welfare (e.g., odors, appearance) and the environment. The risk assessment evaluates current and future risks in the absence of any remediation and helps determine the need for and extent of remediation requirements. During the FS, remedial alternatives are identified and evaluated based on technical feasibility, protectiveness, effectiveness, impacts to the community, institutional concerns, conformance with other applicable relevant or appropriate environmental laws, and costs. A preferred cleanup alternative is proposed as part of the FS.

Removal Actions—A removal action can be conducted at any time during the CERCLA process if the site poses an immediate threat to public health or the environment. A Removal Action is an immediate (short-term) action, such as the implementation of a temporary alternative water supply that is taken to safeguard public health or the environment. In cases where more than a six-month planning period exists before a removal action will begin, an Engineering Evaluation/Cost Assessment (EE/CA) is prepared in order to identify the objectives of the removal action and evaluate various alternatives with respect to cost, effectiveness, and implementability. After the Removal Action is completed, the environmental investigation or remediation process resumes according to the appropriate step in the CERCLA process.

Proposed Plan and Public Comment Period—Upon completion of the FS, a Proposed Plan is published that summarizes the remediation alternatives evaluated in the FS. The Proposed Plan describes the preferred cleanup strategy proposed by the lead agency (e.g., EPA) and the supporting regulatory agencies. The Proposed Plan is then submitted for public comment for a 30-day period, which may be extended an additional thirty days upon timely receipt of a request from a member of the community.

Record of Decision—At the conclusion of the public comment period and following consideration of all community comments, the lead agency with regulatory approval will make the final remedy selection. This final remedy selection is issued in a Record of Decision (ROD), a legal public document that sets forth and explains the remediation alternatives to be used at a CERCLA site. The ROD includes a Responsiveness Summary that contains responses to all public comments received during the public comment period on the Proposed Plan.

Remedy Design and Implementation—After the ROD is signed, the Remedial Design (RD) phase of work is initiated. The RD includes preparation of engineering reports, technical drawings, and specifications to describe implementation of the selected remedy. Upon approval of the RD by the supporting regulatory agencies, the Remedial Action (RA), or the actual construction and implementation of the selected cleanup alternative, is initiated. The final long-term remedial action may take one to two years to construct, although treatment may take several more years. The RA is implemented until cleanup objectives are achieved.

NPL De-Listing—A site can be removed from the NPL upon determination that no further response is required to protect human health or the environment. Under §300.425(e) of the National Contingency Plan, as amended (chapter 4), a site may be de-listed after all appropriate response actions are completed. Partial deletions can also be conducted at NPL sites. For example, soil remediation may be completed and be de-listed prior to de-listing groundwater at the same NPL site.

Long-Term Monitoring/Review—After RD/RA activities have been completed, the site is monitored to ensure the effectiveness of the response. Typically, CERCLA sites undergo reviews every five years after implementation of the remedy in order to evaluate the continued protectiveness of the remedy.

◊ ◊ ◊

The CERCLA's central philosophy is to require parties, called the Potentially Responsible Parties (PRPs), to bear the costs of remediating sites to which the parties had contributed wastes and for costs attending environmental and health problems created by releases of substances from the waste sites. The CERCLA gives EPA broad legal authority to identity PRPs for each NPL site. PRPs include the past and current owners or operators of a site, those who arranged for the transportation of hazardous substances to the site, and those who arranged for the treatment or disposal of the substances, and they are subject to retroactive liability. This is established under the legal concept of *retroactive joint and several liability* for parties whose wastes had contributed to environmental degradation at the waste sites. The concept means that a company or other accountable entities that long ago had disposed of wastes could now be held responsible for all of a site's remediation costs unless other responsible parties can be identified and costs shared. Needless to say, retroactive joint and several liability has often led to litigation as to who pays what portion of a site's remediation.

The CERCLA authorizes EPA to pay for site cleanups out of the Hazardous Substance Superfund and, where possible, out of the liability scheme of the act, that is, from costs recovered from the PRPs. The liabilities of PRPs cover not only the actual costs of remediation, but also the site investigation, feasibility study, design costs, and cost of health studies. The types of parties who may be liable for site-associated costs are specified by the CERCLA to be: (1) current and past "owners or operators" of the site; (2) parties who transported wastes to the site; and (3) parties (usually referred to as "generators") who arranged for wastes to be disposed or treated, either directly with an owner/operator or indirectly with a transporter [29]. It is common to have multiple Potentially Responsible Parties associated with a particular CERCLA site.

In addition to the identification, ranking, remediation, and cost recovery provisions of the act that pertain to uncontrolled releases of hazardous substances from sites, the act contains the following provisions of note [28]:

- Removal actions are conducted by EPA in instances where a short-term, limited response to a manageable environmental release is indicated (e.g., spills of hazardous substances from transportation mishaps), rather than a long-term remedy (i.e., remediation) is indicated, such as for an NPL site.
- As discussed in chapter 3, the CERCLA of 1980 created ATSDR within the U.S. Public Health Service for the purpose of investigating public health implications of hazardous substances in the community environment, with emphasis on those released from NPL sites. ATSDR conducts public health assessments of all NPL sites, develops Toxicological Profiles for priority hazardous substances, conducts epidemiological and other applied research, provides medical education for physicians and other health professionals, responds to emergency releases of hazardous substances (e.g., through transportation spills), and maintains a national registry of persons exposed to specific hazardous substances known to have been released from NPL sites.

Sometimes confusion arises about the differences between the CERCLA and the RCRA. For example, what does one law cover that the other law does not? Is there a difference between the authorities of EPA and the states? Do the laws have different purposes? The key differences between the CERCLA and the RCRA are summarized in Table 8.4. As shown in the table, states have the primary responsibility for RCRA facilities, whereas the federal government (i.e., EPA) has primacy on CERCLA sites. It should be noted that most states also have their own programs to remediate those uncontrolled hazardous waste sites that were not designated as NPL sites by EPA.

TABLE 8.4
Comparison of Key Differences between the RCRA and the CERCLA

RCRA	CERCLA
Focus in on *controlled* facilities that treat, store, and destruct solid and hazardous waste	Focus is on *uncontrolled* hazardous waste sites
States have primacy in setting emission standards and enforcement	Federal government has primacy in setting cleanup standards and enforcement

8.3.2   KEY PROVISIONS OF THE CERCLA, AS AMENDED, RELEVANT TO PUBLIC HEALTH

The CERCLA, as amended in 1986, contains four titles, under which are found the various sections that constitute the statute. Several standard references contain all sections in the statute (e.g., [29]).

Title I—*Provisions Relating Primarily to Response and Liability:*

§101—Definitions—(14) The term *hazardous substance* means (A) any substance designated pursuant to §11(b)(2)(A) of the Federal Water Pollution Control Act [...], (B) any element, compound, mixture, solution, or substance designated pursuant to §9602 of this title, (C) any hazardous waste having the characteristics identified under or listed pursuant to §3001 of the Solid Waste Disposal Act [...], (D) any toxic pollutant listed under §307(a) of the Federal Water Pollution Control Act [...], (E) any hazardous air pollutant listed under §112 of the Clean Air Act [...], and (F) any imminently hazardous chemical substance or mixture with respect to which the Administrator has taken action pursuant to §7 of the Toxic Substances Control Act (15 U.S.C. 2606). The term does not include petroleum, including crude oil or any fraction thereof which is not otherwise specifically listed or designated as a hazardous substance under subparagraphs (A) through (F) of this paragraph, and the term does not include natural gas, natural gas liquids, liquefied natural gas, or synthetic gas usable for fuel (or mixtures of natural gas and such synthetic gas). (16) The term *natural resources* means land, fish, wildlife, biota, air, water, ground water, drinking water supplies [...]. (23) The terms ''remove'' or ''removal'' means the cleanup or removal of released hazardous substances from the environment, [o]r the taking of such other actions as may be necessary to prevent, minimize, or mitigate damage to the public health or welfare or to the environment [...]. (24) The terms *remedy* or *remedial action* means instead of or in addition to removal actions in the event of a release or threatened release of a hazardous substance into the environment, to prevent or minimize the release of hazardous substances so that they do not migrate to cause substantial danger to present or future public health or welfare or the environment. [...] The term includes the costs of permanent relocation of residents and businesses and community facilities where the President determines that, alone or in combination with other measures, such relocation is more cost-effective than and environmentally preferable to the transportation, storage, treatment, destruction, or secure disposition offsite of hazardous substances, or may otherwise be necessary to protect the public health or welfare [...].

§104—(i) Agency for Toxic Substances and Disease Registry; establishment, functions, etc. (1) There is hereby established within the Public Health Service an agency, to be known as the Agency for Toxic Substances and Disease Registry, which shall report directly to the Surgeon General of the United States. health officials, effectuate and implement the health related authorities of this chapter. [...]

§117—Public participation—(a) Proposed plan—Before adoption of any plan for remedial action to be undertaken by the President, by a state, or by any other person, under §9604, 9606, 9620, or 9622 of this title, the President or state, as appropriate, shall take both of the following actions: (1) Publish a notice and brief analysis of the proposed plan and make such plan available to the public. (2) Provide a reasonable opportunity for submission of written and oral comments and an opportunity for a public meeting at or near the facility at issue regarding the proposed plan and regarding any proposed findings under §9621(d)(4) of this title (relating to cleanup standards). The President or the state shall keep a transcript of the meeting and make such transcript available to the public. The notice and analysis published under paragraph (1) shall include sufficient information as may be necessary to provide a reasonable explanation of the proposed plan and alternative proposals considered. (b) Final plan. Notice of the final remedial action plan adopted shall be published and the plan shall be made available to the public before commencement of any remedial action. [...] (d) Publication—For the purposes of this section, publication shall include, at a minimum, publication in a major local newspaper of general circulation. In addition, each item developed, received, published, or made available to the public under this section shall be available for public inspection and copying at or near the facility at issue. (e) Grants for technical assistance—(1) Authority. Subject to such amounts as are provided in appropriations Acts and in accordance with rules promulgated by the President, the President may make grants available to any group of individuals which may be affected by a release or threatened release at any facility which is listed on the National Priorities List under the National Contingency Plan. Such grants may be used to obtain technical assistance in interpreting information with regard to the nature of the hazard, remedial investigation and feasibility study, record of decision, remedial design, selection and construction of remedial action, operation and maintenance, or removal action at such facility.

Enforcement Example: EPA Wins Suit vs. W.R. Grace

On August 28, 2004, a federal district judge ruled that W.R. Grace and Co. owes the federal government every penny sought in a $54.5 million lawsuit aimed at getting the company to pay for Superfund cleanup costs in Libby, Montana. The EPA started work in Libby after a November 1999 news reports detailed how asbestos from a former W.R. Grace and Co. vermiculite mine had sickened hundreds of residents. Asbestos is a contaminant of the local vermiculite ore. The EPA sued Grace in March 2001, seeking repayment of cleanup costs through December 31, 2001. The company filed for bankruptcy just days after the federal agency filed its lawsuit. During a three-day trial in January, EPA argued that Grace should pay for actual cleanup work, health screenings conducted in coordination with the Agency for Toxic Substances and Disease Registry and overhead costs [30]. In December 2005, a federal appeals court upheld the federal judge's ruling [31].

---

§120—Federal facilities—(a) Application of chapter to Federal Government: (1) In general—Each department, agency, and instrumentality of the United States (including the executive, legislative, and judicial branches of government) shall be subject to, and comply with, this chapter in the same manner and to the same extent, both procedurally and substantively, as any nongovernmental entity, including liability under §9607 of this title.

§121—Cleanup standards—(b) General rules—[T]he President shall select a remedial action that is protective of human health and the environment, that is cost effective, and that utilizes permanent solutions and alternative treatment technologies or resource recovery technologies to the maximum extent practicable. [...]

§311—Research, development, and demonstration—(a) Hazardous substance research and training (1) The Secretary of Health and Human Services [s]hall establish and support a basic research and training program (through grants, cooperative agreements, and contracts [...]. (b) Alternative or innovative treatment technology research and demonstration program [...]. (c) Hazardous substance research [...]. (d) University hazardous substance research centers [...].

Title II—*Miscellaneous Provisions*—includes sections on transportation of hazardous materials (§202), leaking underground storage tanks (§205), citizen suits (§206), Indian tribes (§207); research, development, and demonstration (§209), Department of Defense environmental restoration program (§211), oversight and reporting requirements (§212), Love Canal property acquisition (§213).

Title III—*Emergency Planning and Community Right-to-Know*—The Emergency Planning and Community Right-to-Know Act of 1986 was enacted as a freestanding provision of the Superfund Amendments and Reauthorization Act of 1986 [32], constituting Title III of the CERCLA, as amended. Title III is a significant statement from Congress concerning the obligations of government and private industry to protect the public from releases of hazardous substances. Under Title III, state and local governments are required to develop emergency plans for responding to unanticipated environmental releases of acutely toxic materials [33]. (Title III required different kinds of notifications according to groups of chemical substances.) Additionally, businesses covered by Title III must notify state and local emergency planning entities of the presence and amounts in inventory of hazardous materials on their premises and to notify federal, state, and local authorities of planned and uncontrolled environmental releases of those substances. Regulatory agencies, in turn, are required to make available to the public the data on releases of substances in the environment.

Title III includes sections on establishment of state commissions, planning districts, and local committees (§301); substances and facilities covered and notification (§302), comprehensive emergency response plans (§303), emergency notification (§304), emergency training and review of emergency systems (§305), material safety data sheets (§311), emergency and hazardous chemical inventory forms (§312), toxic chemical release forms (§313), trade secrets (§322), providing information to health professionals (§324); public availability of plans, data sheets, and follow-up notes (§324); enforcement (§325), and regulations (§328).

EPA has developed a reporting system for businesses to use when providing information required of them for placement in the Toxics Release Inventory (TRI).[3] The TRI database can be accessed through the EPA web site (http://www.epa.gov/tri-explorer). Citizen and environmental groups have used the TRI data to bring attention to the amounts of hazardous substances released by industry within geographic areas of concern. The impact has been to bring the weight of public opinion to bear on companies' reduction of emissions.

Another use of TRI data is for research purposes. As an example of one organization's use of TRI data, the Greater Boston Physicians for Social Responsibility found that of the 20 TRI chemicals with the greatest total releases into the environment, about 75 percent were known or suspected neurotoxicants [34]. On a more positive note, EPA reported in 2005 that TRI data showed that the amount of toxic substances released into the U.S. environment had declined 42 percent between the years 1999 and 2003 [35]. It is unclear if this decline is a product of relaxed changes in how industry reports emissions data to EPA, or actual reductions due to public pressure, or reductions as a consequence of federal, state, or local emission standards.

Title IV—*Radon Gas and Indoor Air Quality Research*— includes sections on findings (§402); radon gas and indoor air quality research programs (§403), authorizations (§405).

8.3.3 PUBLIC HEALTH IMPLICATIONS OF THE CERCLA, AS AMENDED

The consequences to the health of persons who live near uncontrolled hazardous waste sites (HWS) are both feared and real. Communities located near these kinds of sites often create grassroots groups that actively express their fears that excess cancer rates and reproductive health problems are caused by substances released from HWS in their midst. Although relating a specific community's health problems to a given HWS is investigatively very challenging, there is, nonetheless, a compelling body of epidemiological and toxicological data that associates adverse effects on community health with residential proximity to HWS.

The effects on human reproductive outcomes from exposure to hydrocarbon solvents such as trichloroethylene released from HWS constitute the strongest evidence for a HWS–human health association. According to one source who comprehensively reviewed relevant epidemiological literature, residential proximity to HWS is associated with lower birth weight and an increased risk of congenital malformations that include defects of the heart, neural tube, and oral palate [19]. Particularly compelling were findings from two investigations of two NPL sites: Love Canal, New York, and Lipari, New Jersey [36,37, respectively]. For both sites, average birth weights decreased during the span of time when documented releases of hazardous substances were migrating from the waste sites. For both HWS, when the releases of substances were interdicted, mean birth weights in the geographic areas returned to normal. In another study of congenital anomalies and residence near hazardous waste landfill sites, 245 cases of chromosomal anomalies were compared with 2,412 controls who lived near 23 such landfills in Europe [38]. After adjusting for confounders, a higher risk of chromosomal anomalies was found in people who lived close to sites (0–3 km) than in persons who lived farther away.

Any association between residential proximity to HWS and elevated cancer rates is less well substantiated than for adverse reproductive effects. A review of the epidemiological literature indicates elevated rates of certain cancers, primarily those of the urinary bladder and gastrointestinal tract, in counties that contain HWS and for which groundwater contamination was either documented or assumed (e.g., [39,40]). There also exists published work that associates increased rates of childhood leukemia with the presence of trichloroethylene (TCE) in municipal wells that supplied segments of Woburn, Massachusetts, with residential water (e.g., [41,42]).

There are also toxicological data that have importance for cancer rates in communities impacted by releases of substances from HWS. One source examined the most frequently occurring substances released into groundwater supplies and noted that of these 30 chemicals, 18 were known or reasonably anticipated to be human carcinogens [19]. Given the long latency associated with most cancers, whether these toxicological observations portend any increase in cancer rates will not be known for many years.

In general, there are sufficient published scientific data to designate uncontrolled hazardous waste sites as major threats to the public's health. Given this knowledge, and as a matter of public health, remediation of HWS is an example of primary disease prevention in action. That is, adverse health effects are prevented by elimination of the causal hazard.

8.3.4 Criticisms and Successes of the CERCLA

The CERCLA stands atop the federal environmental laws in regard to criticism. The volume and rancor of criticism exceeds that of any other federal environmental statute. Yet, the CERCLA has been a valuable statute in terms of environmental restoration and reduced public health impacts. Some of the principal criticisms of the CERCLA are shown in Table 8.5. The law is alleged to be unfair because it holds polluters accountable for their past actions, actions that were sometimes in compliance with existing laws pre-1980 when the CERCLA was enacted into law. Critics also allege that the CERCLA program has remediated too few NPL sites, that the costs of cleanups are too great, cleanups take too long, too much litigation accompanies site cleanup, and moreover, there are no health problems in communities located near CERCLA sites. Some of these criticisms have been voiced by industry and trade groups who have historically opposed the CERCLA. Also, some states have expressed some of the same criticisms, preferring to adopt their own state-based waste site remediation programs, rather than having to accept cleanup standards and priorities from the federal government. These criticisms can best be rebutted by examining the successes of the CERCLA.

The criticisms of the act should be viewed in comparison to the successes of the CERCLA program, shown in Table 8.6. Remediation of HWS is a contribution to the

---

TABLE 8.5
Criticisms of the CERCLA Program

Unfair enforcement

Too many sites to remediate

Site remediation is too costly

Takes too long to remediate sites

Little impact on community health

Too litigious

TABLE 8.6
Successes of the CERCLA Program

Many sites remediated

Emergency removal of hazardous substances

Human health effects database

Emergency of environmental justice

Toxics Release Inventory (TRI)

Improved emergency responding

Improved management of hazardous waste

public's health. Removal of a site's hazardous substances and interdiction of human exposure pathways (e.g., well water used for drinking) prevent adverse human health effects. Since 1980, 966 NPL sites have had all cleanup construction completed [43]. As of fiscal year 2004, there were 1,529 NPL sites [44] that were in various stages of site remediation. Of note, over the life of the CERCLA program, EPA has reached settlements with private parties with an estimated value exceeding $30 billion, as of fiscal year 2002 [45]. These funds are used for site remediation and related purposes.

In November 2003, EPA reported to Congress that CERCLA site cleanups had declined for the third consecutive year. For fiscal year 2002, EPA reported 40 sites had been remediated, compared to 47 the previous fiscal year [45]. EPA attributed the decrease in site cleanups to increasing size and remediation complexity of sites currently on the NPL, necessitating greater time to remediate sites. The fact that Congress and the George W. Bush administration had decreased funds for the EPA CERCLA program was not mentioned by EPA. In fiscal year 2003, the administration decreased funding for 33 NPL site cleanups; EPA regional offices had requested $46.7 million for cleanup costs at 54 long-term remediation projects, whereas the administration recommended $33.2 million, which is approximately a 30 percent reduction in requested funds [46]. This trend in underfunding site cleanups continued in fiscal year 2004, which amounted to a shortfall of $250M [47]. As shown in Figure 8.5, an overall decrease in identification and cleanup of NPL sites has occurred since year 2000. Simply stated, fewer funds translates into fewer CERCLA sites being remediated. A decrease in site cleanups is antithetical to environmental health, both human and ecological.

As another success, the act provides EPA with the authority to conduct emergency removals of hazardous substances from sites. Examples of such sites include urban abandoned warehouses that contain barrels of hazardous chemicals and train derailments where industrial chemicals are spilled. EPA has conducted more than 6,400 removal actions through fiscal year 2000 [48]. Emergency responding by local authorities to chemical releases have been substantially improved under Title III of the CERCLA. State and local governments are required to develop emergency plans for responding to unanticipated environmental releases of several acutely toxic materials [33]. Additionally, businesses covered by Title III must notify state and local emergency planning entities of the presence and amounts in inventory of hazardous materials on their

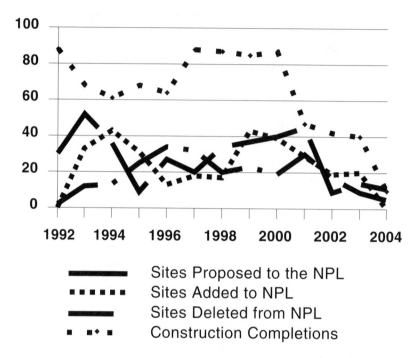

FIGURE 8.5   NPL site statistics [62].

premises and to notify federal, state, and local authorities of planned and uncontrolled environmental releases of those substances. Regulatory agencies, in turn, are required to make available to the public the data on releases of substances in the environment. This is a public health success story in terms of preventing human exposure to hazardous substances.

Further, as the result of the CERCLA, a considerable body of illuminating public health and science findings have accrued. The public health findings in the preceding section resulted from funding from the CERCLA programs at ATSDR and the National Institute of Environmental Health Sciences. Without these funds, waste site health investigations and basic toxicological research would likely not have occurred. Moreover, these research findings have utility for other sources of environmental contamination (e.g., air pollutants) when the contaminants are the same as found released from CERCLA sites.

In another area, environmental justice concerns arose from minority communities' fear that hazardous waste disposal was targeting their communities (chapter 10). In response, both the Clinton and George H.W. Bush administrations generated policies to guard against environmental injustices. Without the CERCLA, it is doubtful that the federal government would have had the resources and resolve to institute environmental justice offices at EPA and other federal agencies.

Because of the CERCLA, the public now has access to Toxics Release Inventory (TRI) data, which are provided to EPA by generators of pollutants released into the environment. TRI data can be used by individuals, community groups, and local government to identify pollution sources of public health concern.

Lastly, management of hazardous waste has undoubtedly improved in the United States due to the act. Generators of hazardous waste know the penalties that come with violating waste management regulations. Not only can violations result in civil penalties, private party litigation can result in large monetary settlements against the generators. The so-called midnight dumpers, who existed pre-1980 and literally dumped liquid hazardous waste along roads in rural America, have largely faded into history.

8.4 OCEAN DUMPING ACT

8.4.1 HISTORY

The world's oceans have historically served as rich sources for life support, both aquatic and terrestrial. Fish and other aquatic life have long been a staple of the human diet. The very vastness of the oceans has led, unfortunately, to their use by humankind as places to dump anthropogenic wastes. As human populations increased in number and industrial pollution appeared due to global industrialization, the need for inexpensive waste disposal methods emerged. Bodies of water were selected for waste dumping in the mistaken belief that dilution of pollution would minimize any threats to human and ecological health. Of course, the flaw in this theory is the failure to acknowledge that any pollution sink has a limit on its capacity to receive waste, even a sink as vast as the oceans.

By the late 1960s and early 1970s, it had become evident that the oceans were in ecological trouble. Centuries of marine pollution had taken their toll. Beaches had become contaminated from pollution washed ashore, marine life had become bearers of chemical pollutants and solid waste, and reefs were dying from the pollution load. Actions to reverse this march of marine pollution began with the adoption of key international protocols to reduce pollution of the oceans and seas.

The Ocean Dumping Act focuses on the regulation of intentional disposal of materials into ocean waters and authorizing related research [49].

In June 1972, the United Nations Stockholm Conference on the Human Environment called for action to protect marine environments, followed by a treaty drafted in London at the Intergovernmental Conference on the Convention on the Dumping of Wastes at Sea [51]. The Convention on the Prevention of Marine Pollution by Dumping of Wastes and Other Matter, also called the London Convention, entered into effect on August 30, 1975, after fifteen countries had ratified the treaty, including the United States. In 1972, the U.S. Congress enacted the Marine Protection, Research, and Sanctuaries Act. Title I of the act is commonly called the Ocean Dumping Act (ODA) and implements the agreements the United States made at the London Convention. This is therefore an example of an U.S. environmental statute that emerged from the need to comply with an international treaty.

Enforcement Example: Ship's Captain and Engineers Arrested on Ocean Dumping Charges

The captain of the *MN Katerina*, the ship's chief engineer; and the ship's second engineer were all arrested in the Los Angeles area on September 21 on charges that they had allegedly been involved in the dumping of oil-contaminated water into the Pacific Ocean. The Coast Guard inspected the ship on September 14 and 15, 2004. During these inspections, they discovered that the ship's oil-water separator was not being used and that a bypass had been constructed around the separator. All three defendants are charged with failing to properly maintain the *Katerina's* Oil Record Book, making false statements to Coast Guard investigators and obstructing justice by falsifying records [50].

The Ocean Dumping Act has two basic aims: to regulate international ocean disposal of materials, and to authorize related research. Title I contains permit and enforcement provisions for ocean dumping [49].

Title I of the act prohibits all ocean dumping, except that allowed by permits, in any ocean waters under U.S. jurisdiction, by any U.S. vessel, or by any vessel sailing from a U.S. port. Banned is any dumping of radiological, chemical, and biologic warfare agents and any high-level radioactive waste. Permits for dumping any other materials can be issued by EPA where it is determined that the dumping will "[n]ot unreasonably degrade or endanger human health, welfare, or amenities, or the marine environment, ecological systems, or economic potentialities." Permits issued under the ODA must specify the kind of material to be disposed, the amount to be transported for dumping, the location of the dumpsite, the length of time the permit is valid, and special provisions for surveillance [ibid.]. Subsequent amendments to the ODA prohibited the dumping of municipal sewage sludge and industrial waste.

Four federal agencies have responsibilities under the ODA [ibid.]. EPA has primary authority for regulating ocean disposal of all substances except dredged soils, for which the Army Corps of Engineers has authority. Provisions concerning general and ocean disposal research are contained in **Title II**, under which the National Oceanic and Atmospheric Administration has authority for researching the effects of human-induced changes to the marine environment, while EPA is authorized to carry out research and demonstration activities pertaining to phasing out sewage sludge and industrial waste dumping. The U.S. Coast Guard is charged with maintaining surveillance of ocean dumping [ibid.].

Title III of the ODA authorizes the establishment of marine sanctuaries. **Title IV**

establishes a regional marine research program, and **Title V** addresses coastal water quality monitoring.

8.4.2 PUBLIC HEALTH IMPLICATIONS OF THE ODA

It is unknown to what extent the ODA and international agreements structured to reduce ocean pollution have been effective. However, it is apparent that large quantities of ocean pollutants exist, as measured by what washes up on beaches. In 1991, the Center for Marine Conservation organized 118,200 volunteers to clean 3,800 miles of U.S. coastline in one day. They picked up 2.8 million pounds of trash [52]. That amount obviously represents only a tiny fraction of what is in the oceans and the creatures that live there. If one assumes that human health is dependent on a healthy ecosystem, then pollution in the seas and oceans of the world has the potential to degrade the public's health, particularly as it pertains to human consumption of fish and other food taken from the ocean.

8.5 ACT TO PREVENT POLLUTION FROM SHIPS

8.5.1 HISTORY

Ocean pollution has doubtless occurred from the time that humans first found ways to traverse great bodies of water. Whether human waste of mariners, or damaged goods being dumped, or oil from leaking vessels, the oceans and its resident creatures have been victims of pollution from ships. As ocean pollution became a concern to the global environmental protection community, international maritime treaties emerged to regulate and control pollution from ships. These treaties are in addition to laws enacted by individual nations, such as the U.S. Ocean Dumping Act.

The Act to Prevent Pollution from Ships implements the international MARPOL Convention. The act restricts pollution from ships, establishes record keeping of materials released from ships, and provides penalties for violations of the act [53].

On an international scale, the International Maritime Organization (IMO), an agency of the United Nations, has led the development and implementation of treaties to control ocean pollution from ships. Created in 1948, the IMO's purposes include, "[t]o provide machinery for cooperation among Governments in the field of governmental regulation and practices relating to technical matters of all kinds affecting shipping engaged in international trade; to encourage and facilitate the general adoption of the highest practicable standards in matters concerning maritime safety, efficiency of navigation and prevention and control of maritime pollution from ships" [54]. In 1973 the IMO issued the International Convention for the Prevention of Pollution from Ships, modified by the Protocol of 1978, both being referred to as MARPOL 73/78. The MARPOL Convention covers pollution by chemicals, goods in packaged form, sewage, garbage, and air pollution, as well as accidental and operational oil pollution (i.e., oil pollution from operation of a ship).

Enforcement Example: Chairman of Shipping Line Sentenced for Illegal Dumping of Waste Oil

The Chairman of Sabine Transportation Company, Cedar Rapids, Iowa, was sentenced by a U.S. District judge in Miami, Florida, to 33 months in prison and fined $60,000 for violations of the Act to Prevent Pollution from Ships. The company was found guilty by a jury for dumping grain contaminated with diesel oil into the ocean during a vessel's trip from Singapore to Portland, Oregon. Crew members had informed the U.S. Coast Guard of the dumping of approximately 440 metric tons of oil-contaminated wheat. The company had previously been fined $2 million for similar dumping violations [55].

As an international treaty, adoption of the MARPOL Convention requires national legislation for its implementation. In the United States, the Act to Prevent Pollution from Ships implements the provisions of the Convention. As stated in the act, it "[a]pplies to all U.S. flag ships anywhere in the world and to all foreign flag vessels operating in the navigable waters of the United States or while at a port or terminal under the jurisdiction of the United States. The oil and noxious liquid substances provisions apply only to seagoing ships. The regulations implementing Annex I and Annex II of MARPOL limit discharges of oil and noxious substances, establish report requirements for discharges, and establish specific requirements for monitoring equipment and record keeping aboard vessels" [53]. The act contains provisions for both criminal and civil penalties for violations.

8.5.2   PUBLIC HEALTH IMPLICATIONS OF THE ACT TO PREVENT POLLUTION FROM SHIPS

As with the Ocean Dumping Act, the impact on human health of pollution dumped from ships is unknown. However, the consequences of marine pollution on ecosystems and marine life are undeniable and consequential. The deleterious effects of ocean pollution on birds and other wildlife dependent on the oceans have been well chronicled in news media reports. Moreover, the presence of toxic substances in the tissues of marine life has also been well reported. In brief, ocean pollution affects the quality of human life and can contribute to adverse human health consequences. An environmental health policy such as that contained in the Act to Prevent Pollution from Ships that attempts to prevent ocean pollution is consistent with pollution prevention goals in much of the body of U.S. environmental statutes.

8.6   OIL POLLUTION ACT

8.6.1   HISTORY

The Oil Pollution Act of 1990 (OPA) was enacted by Congress in recognition of the large number of oil spills occurring annually in the United States, and in particular, the legislation was markedly influenced by the 1989 oil spill in Alaska from the ruptured vessel *Exxon Valdez*. The act is a comprehensive amendment to the Clean Water Act, §311. The act is designed to enhance oil spill prevention, preparedness, and response capabilities and authorities of government agencies [56,57]. The act established a new liability and compensation regime for oil pollution incidents in the aquatic environment and provided the resources necessary for the removal of discharged oil.

The Oil Pollution Act expands oil spill prevention, preparedness, and response capacities of the federal government and industry [58].

The Oil Spill Liability Trust Fund, which is similar in statutory concept to the CERCLA's Hazardous Substance Superfund, created a $1 billion fund to be used to respond to, and provide compensation for damages caused by, discharge of oil. Similar to authorities in the CERCLA, the OPA provides that the responsible party for a vessel or a facility from which oil is discharged, or which poses a substantial threat of a discharge, is liable for certain damages and costs of removal of oil. In addition, the OPA provides new requirements for contingency planning both by government and industry and establishes requirements on construction (e.g., §4115 of the act mandates newly constructed tank vessels must be equipped with double hulls, with the exception of vessels used only to respond to discharges of oil or hazardous substances), manning, and licensing for tank vessels [56].

Several federal agencies are responsible for implementing the OPA. In general, EPA is responsible for oil spill prevention, preparedness, and response activities associated with non-transportation related onshore activities. The U.S. Department of Transportation is generally responsible for oil spill planning and response activities for tank vessels, transportation-related onshore facilities, and deepwater ports. The U.S. Department of Interior is responsible for oil spill planning and response activities for offshore facilities except deepwater ports. States that have laws governing oil spill prevention and responses are covered under the act. For example, §1019 provides states the authority to enforce, on the navigable waters of the state, OPA requirements for evidence of financial responsibility.

When oil spills occur, the authorities and resources of the National Contingency

Plan (chapter 4) are used to quickly respond to emergency conditions that involve the release of oil or hazardous substances. The NCP, because it is based on law, brings together the coordination between federal government agencies and others in order to protect the public's health and the well-being of ecosystems. As policy, yoking government agencies in a legally binding rope of coordination well serves the public, since such cooperation is often difficult to achieve in the absence of law.

8.6.2 PUBLIC HEALTH IMPLICATIONS OF THE OPA

The direct consequences to human health of the OPA are not great. When spills or other releases of oil have occurred, few acute or chronic impacts on the public's health have occurred. For instance, ocean spills of oil released from large tanker ships have certainly fouled the surrounding environment, but had little impact on the health of persons tasked with emergency responding, beach restoration, waterfowl rescue, and ecosystem remediation. This can be attributed to the training of responding personnel, use of protective equipment, and low toxicity of neat oil.

In contrast, the indirect impacts of oil spills on the environment are considerable. Large ocean spills of oil can contaminate fish, mammals, birds, and other living creatures, both in the water and on land. Although both the sea and land can absorb large scale oil pollution, there is always a residual amount that can work its way into the human food chain. Although the health risk is very small, any chronic effects have not been investigated. Of a more acute nature, ocean spills that contaminate focal geographic areas can lead indigenous people to forego customary seafood and turn to less healthful alternative foods.

8.7 POLLUTION PREVENTION ACT

8.7.1 HISTORY

The Pollution Prevention Act of 1990 (PPA) is an example of a federal environmental statute whose genesis derives substantively from EPA initiative. While the executive branch of the U.S. government has the authority, indeed, the responsibility, to propose legislation to the Congress, in fact, the main body of federal environmental law has generally been the outcome of Congressional interest and action. With the election of President George H.W. Bush came the appointment of William Reilly as EPA administrator. Reilly was at the time the president of the World Wildlife Federation, a respected and venerable international environmental organization. He is the only EPA administrator to have served as the leader of a major environmental organization.

One of Reilly's priorities was pollution prevention [1]. Pollution prevention was seen to be preferable to dealing with "after-the-fact" pollution management. This prevention orientation is, of course, identical to the central thesis of public health—prevention of disease and disability. Reilly's pollution prevention priority found favor in Congress, contributing to the enactment of the PPA.

The PPA declares "[i]t to be the national policy of the United States that pollution should be prevented or reduced at the source whenever feasible, pollution that cannot be prevented should be recycled in an environmentally safe manner, whenever feasible; pollution that cannot be prevented or recycled should be treated in an environmentally safe manner whenever feasible; and disposal or other release into the environment should be employed only as a last resort and should be conducted in an environmentally safe manner [59]." This act encourages voluntary reduction of pollution and is, therefore, a complement to the command and control regulatory approach whereby pollution control is mandated of polluting sources.

Under authorities in the PPA, EPA is directed to implement several key actions of relevance to public health, as follows:

The Pollution Prevention Act of 1990 seeks to prevent pollution through reduced generation of pollutants at their point of origin [59].

---

§13103: EPA Activities—EPA is directed to establish an office to promote multi-media (i.e., air, water, soil) approaches to source reduction of pollution. EPA is directed to review its regulations before and after proposal to determine their effects on source reduction, promoting source reduction by other federal agencies, and methods to assure the public's access to data collected under federal environmental statutes, and facilitating use of source reduction methods by business.

§13104: Grants to States for State Technical Assistance Programs—EPA is directed to make matching grants to states for programs that promote business' use of source reduction techniques.

§13105: Source Reduction Clearinghouse—EPA is required to establish a Source Reduction Clearinghouse to serve as a center for source reduction technology transfer, to undertake outreach and educational programs, and to collect and compile information from states that received grants under the provisions of the PPA.

§13106: Source Reduction and Recycling Data Collection—Each owner or operator of a facility required to file an annual toxic chemical release form under the Emergency Planning and Community Right-to-Know Act[4] must also include a toxic chemical source reduction and recycling report for the preceding year. According to one source, "For each reported toxic chemical, the report to EPA shall include the quantity of the chemical entering the waste stream prior to recycling, treatment, or disposal; the amount of the chemical treated; the amount of the chemical recycled and the process used; the source reduction practices used with respect to the chemical; the amount of the chemical expected to be reported during the next two years; the techniques used to identify source reduction techniques; and the amount of any toxic chemical released into the environment as the result of a catastrophic event" [60].

§13107: EPA Report—The act required EPA to report biennially to Congress on actions to implement the strategy to promote source reduction of pollution. The biennial report is required to contain an industry-by-industry analysis of data submitted by facilities to EPA; barriers to source reduction of pollution; recommendations on incentives to encourage research, development, and investment in source reduction; analyses of cost and technical feasibility of source reduction; and evaluation of methods to improve the public's access to the data collected by EPA under the provisions of the PPA.

◊ ◊ ◊

The requirements of the PPA were extended to U.S. federal agencies under Presidential Executive Order 12856, signed by President Clinton on August 3, 1993 [61]. The Executive Order states, "[t]he head of each Federal agency is responsible for ensuring that all necessary actions are taken for the prevention of pollution with respect to that agency's activities and facilities, and for ensuring that agency's compliance with pollution prevention and emergency planning and community right-to-know provisions established pursuant to all implementing regulations issued pursuant to EPCRA (i.e., Emergency Planning and Community Right-to-Know Act, which is Title III of the CERCLA) and the PPA (i.e., Pollution Prevention Act of 1990)."

It should be noted that federal environmental laws generally exempt U.S. federal government agencies from responsibilities required of the private sector (e.g., businesses and corporations), unless specifically specified by statute. Another example of an environmental statute that specifies actions required of federal agencies is found in the CERCLA, where uncontrolled hazardous waste sites under the control of federal agencies must be remediated as is the situation with private sector responsibility. Presidential executive orders, such as the aforementioned one on pollution prevention, is one way to bring federal agencies into compliance with environmental laws directed at the private sector.

8.7.2   PUBLIC HEALTH IMPLICATIONS OF THE PPA

At its heart, the Pollution Prevention Act encourages reduction of pollution at its source. This is a complementary strategy to the command and control approach, which figures so prominently in environmental protection actions taken by federal regulatory agencies. Critics of command and control allege that, in concept, it allows too little flexibility in how to reduce specific sources of pollution and mandates actions not always consistent with current technologies. Supporters of command and control argue that absent strong, mandated, specific regulatory statements from government, pollution reduction would not occur, that is, voluntary actions to reduce pollution would be unreliable and uncoordinated across pollution generators. In fact, both sides of the command and control argument have some truth, and the PPA attempts to bridge these two positions.

Given that air pollution, water contamination, hazardous waste sites, and toxic substances in workplaces and community environments are considered as hazards to human health, it is good practice to eliminate or reduce pollution sources. Toward that end, the PPA contributes to improved public health.

8.8   SUMMARY

The two major statutes discussed in this chapter are the Solid Waste Disposal Act (SWCA), which upon amendment became known as the Resource Conservation and Recovery Act (RCRA), and the Comprehensive Environmental Response, Compensation, and Liability Act (CERCLA). Both statutes contain a number of important environmental health policies of relevance to public health. From the 1980 SWDA amendments, open dumps were prohibited in the United States, thereby eliminating places where disease-carrying vermin prospered, and placed upon states the responsibility for solid waste management, which is an example of federalism at work. The RCRA, as amended, also requires permits for those facilities that conduct hazardous waste management activities and authorizes the EPA to inspect hazardous waste facilities and issue corrective action orders when standards are not being met by facility operators. The policies of establishing national standards, issuance of permits for control of environmental hazards and inspection of facilities, in concept, are the same found in some other environmental statutes (e.g., the Clean Air Act as amended). These policies' purpose is to control releases of potentially hazardous substances into the environment.

The CERCLA also contains several important policies of relevance to public health. The act identifies uncontrolled hazardous waste sites and ranks those that are the most hazardous to human health and the environment (i.e., the NPL sites). The highest ranked sites are remediated preferentially before those of lesser hazard, a policy that comports with the public health practice of addressing first those hazards of greatest threat to human health. The act also contains the policy of "the polluter pays for the costs of their pollution," which is a statement of accountability to the public. The act is unique among federal environmental statutes in its creation of a federal public health agency, ATSDR, for the purpose of addressing a specific environmental hazard—in this case, human exposure to uncontrolled hazardous substances.

Other acts discussed in this chapter consisted of the Oil Pollution Act, the Ocean Dumping Act, the Act to Prevent Pollution from Ships, and the Pollution Prevention Act. All four acts are intended to prevent or reduce the amount of pollution released into the environment. Reduced pollution levels are consistent with improved public health protection, since the opportunity for human exposure is lessened.

8.9 POLICY QUESTIONS

1. Select one of these laws (Clean Air Act, Clean Water Act, Safe Drinking Water Act, Resource Conservation and Recovery Act), as amended, and (A) describe how its purpose and provisions have affected you personally. (B) Then assume the law you selected does not exist. Discuss the ramifications of no law, and discuss any actions you would take to protect your personal health.

2. In this book, we have discussed the following federal environmental laws: NEPA, CAA, CWA, SDWA, RCRA, CERCLA, FMIA, FIFRA, and FQPA. For *each* of the nine statutes: (A) State the law's primary purpose. (B) State the law's key policies. (C) Identify any law(s) that Congress, in your opinion, should update. Discuss the nature of the update. Be specific. (D) Should any of the laws be abolished? If so, which one(s) and why? Be specific.

3. Chapters 4–8 contain a summary and consequences of the major federal environmental health laws. (A) With that as background material, discuss which of these laws apply to the purchase and consumption of a cheeseburger and cup

of coffee purchased at a local fast food establishment. (B) Discuss the state and local health department laws or ordinances that would apply. (C) In a public health context, discuss which, in your opinion, of the federal, state, and local regulations is the most essential in preventing foodborne illnesses from commercial establishments such as the place where the cheeseburger and coffee were purchased? (D) What is the role of private business in protecting the safety of the food supply?

4. The Resource Conservation and Recovery Act, as amended, covers the management of municipal solid waste and other wastes. Discuss five actions that you can take to reduce the volume of your household waste. Estimate in pounds the annual volume of waste reduction that you can achieve.

5. Using Internet resources, ascertain your state's programs and policies that pertain to the management of hazardous waste. Discuss what you consider to be the single most effective policy.

6. The Comprehensive Environmental Response, Compensation, and Liability Act (also called the Superfund Act), as amended, requires EPA to rank uncontrolled hazardous waste sites and place the most hazardous sites on the National Priorities List (NPL). Using EPA resources, ascertain the number of NPL sites in your state. Discuss the status of any site's remediation. If there are no NPL sites in your state, select an adjacent state that contains one or more NPL sites.

7. Many states have state-based CERCLA programs. Ascertain if your state has such a program. If so, how many state-lead uncontrolled hazardous waste sites are being remediated in your state. Select the site nearest to your residence and describe its remediation status.

8. In a public health context, in your opinion, do the successes of the federal CERCLA statute, as amended, outweigh its criticisms (Tables 8.5, 8.6)? Why? Be specific.

9. Discuss the essential public health differences between the RCRA and the CERCLA. Be specific.

10. What U.S. Public Health Service agency was created by the Comprehensive Environmental Response, Compensation, and Liability Act? Why?

NOTES

1. Solid waste refers here to household and industrial wastes, not bodily wastes.
2. These headings were added for purpose of enhancing clarity. They do not appear in the cited Georgia code.
3. As described in this chapter, the Pollution Prevention Act of 1990 requires that industrial facilities also report data on recycling of wastes and other information on pollution prevention.
4. Title III of the Comprehensive Environmental Response, Compensation, and Liability Act, as amended by the Superfund Reauthorization Act of 1986.

REFERENCES

1. Landy, M.K., Roberts, M.J., and Thomas, S.R., *The Environmental Protection Agency: Asking the Wrong Questions from Nixon to Clinton*, Oxford University Press, New York, 1994.
2. SGPHS (Surgeon General of the Public Health Service), Annual Report of the Public Health Service of the United States, Washington, D.C., 1913.
3. Sherrod, H.F., *Environmental Law Review—1971*, Sage Hill Publishers, Albany, NY, 1971, 323.
4. McCarthy, J.E. and Tiemann, M., Solid Waste Disposal Act/Resource Conservation and Recovery Act, Congressional Research Service, Report RL 30022. Available at http://www.ncseonline.org/nle/crsreports/briefingbook/laws/h.cgm, 1999.
5. NRC (National Research Council), *Waste Incineration and Public Health*, National Academy Press, Washington, D.C., 2000.
6. EPA (U.S. Environmental Protection Agency), Basic facts about waste, Office of Solid Waste and Emergency Response, Washington, D.C., 2004.
7. EEA (European Environmental Agency), Household Consumption and the Environment, Report No. 11/2005, Office for Official Publications of the European Communities, Luxembourg, 2005.
8. EPA (U.S. Environmental Protection Agency), Municipal Solid Waste. Household Hazardous Waste, Office of Solid Waste and Emergency Response, Washington, D.C., 2005.
9. EPA (U.S. Environmental Protection Agency), Municipal Solid Waste. Reduce, Reuse, and Recycle, Office of Solid Waste, Washington, D.C., 2005.
10. EPA (U.S. Environmental Protection Agency), RCRA, Superfund & EPCRA Hotline Training Module, EPA530-R-99-063, Office of Solid Waste and Emergency Response, Washington, D.C., February, 2000.
11. Case, D.R., Resource Conservation and Recovery Act (RCRA), in *Environmental Law Handbook*, Arbuckle, J.G., eds., Government Institutes, Rockville, MD, 406, 1991.
12. Reisch, M., Superfund. Summaries of Environmental Laws Administered by the EPA, Congressional Research Service Report RL 30022, Congressional Research Service, Washington, D.C. Available at http://www.ncseonline.org/nle/crsreports/briefingbooks/laws/j.cfm, 1999.
13. EPA (U.S. Environmental Protection Agency), Medical Waste Tracking Act of 1988. Available at http://www.epa.gov/epaoswer/other/medical/download.htm, 1989.
14. EPA Press Advisory, Idaho man sentenced in paint waste case, Washington, D.C., September 30, 2004.
15. Bosco, M.E. and Randle, R.V., Underground storage tanks, in *Environmental Law Handbook*, Arbuckle, J.G., et al., eds, Government Institutes, Rockville, MD, 443-470, 444, 1991.
16. Rodenbeck, S.E. 2005. Personal communication.
17. Lichtveld, M.Y., 2005. Personal communication.
18. Elliott, P. et al., Risk of adverse birth outcomes in populations living near land sites, *British Medical Journal*, 323, 363, 2001.
19. Johnson, B.L., *Impact of Hazardous Waste on Human Health*, CRC Press, Lewis Publishers, Boca Raton, FL, 1999.
20. State of Georgia, Solid Waste Code 12-8-24, Department of Natural Resources, Atlanta, 2003.
21. DePalma, A., Love Canal declared clean, ending toxic horror, *New York Times*, March 18, 2004.

22. DePalma, A., Pollution and the slippery meaning of 'clean,' *New York Times*, March 28, 2004.

23. GAO (General Accounting Office), Superfund Program: Current Status and Future Fiscal Challenges, Washington, D.C., 2003.

24. Skinner, J., Hazardous waste treatment trends in the U.S., *Waste Management & Res.*, 9, 55, 1991.

25. EPA (U.S. Environmental Protection Agency), How Sites are Placed on the NPL. Available at http://www.epa.gov/superfund/programs/npl_hrs/nplon.htm, 2003.

26. NASA (National Aeronautics and Space Administration), CERCLA Overview. Available at http://jplwater.nasa.gov/NMOWeb/Home/new_page_1.htm, 2004.

27. EPA (U.S. Environmental Protection Agency), Introduction to the HRS, Office of Solid Waste, Washington, D.C. Available at http://www.epa.gov/superfund/programs/npl_hrs/hrsint.htm, 2002.

28. Johnson, B.L., Implementation of Superfund's health-related provisions by the Agency for Toxic Substances and Disease Registry, *Environmental Law Reporter*, 20, 10277, 1990.

29. ELI (Environmental Law Institute), *Environmental Law Deskbook*, Environmental Law Institute, Washington, D.C., 166, 203, 221, 245, 1989.

30. Smith, E.S., EPA wins suit vs. W.R. Grace,. *The Missoulian*, August 28, 2004.

31. *Helena Independent Record*, W.R. Grace must pay $54 million for asbestos cleanup, December 1, 2005.

32. Arbuckle, J.G., Randle, R.V., and Wilson, P.A.J., Emergency planning and community right-to-know act, (EPCRA), in *Environmental Law Handbook*, Arbuckle, J.G., et al., eds, Government Institutes,Rockville, MD, 190, 1991.

33. Arbuckle, J.G., et al., eds, *Environmental Law Handbook*, Government Institutes, Rockville, MD, 471, 1991.

34. GBPSR (Boston Chapter Physicians for Social Responsibility), In *Harm's Way: Toxic Threats to Child Development*, Boston, 2000.

35. EPA Press Advisory, 2003 Toxics Release Inventory shows continued decline in chemical releases, Washington, D.C., May 11, 2005.

36. Vianna, W.K. and Polan, A.K., Incidence of low birth weight among Love Canal residents, *Science*, 226, 1217, 1984.

37. Berry, M. and Bove, F., Birth weight reduction associated with residence near a hazardous waste landfill, *Environmental Health Perspectives*, 105, 866, 1997.

38. Vrijheid, M. et al., Chromosomal congenital anomalies and residence near hazardous waste landfill sites, *The Lancet*, 359, 320, 2002.

39. Najem, G.R. et al.. Clusters of cancer mortality in New Jersey municipalities; with special reference to chemical toxic waste disposal sites and per capita income, *International Journal of Epidemiology*, 14, 528, 1985.

40. Griffith, J. et al., Cancer mortality in U.S. counties with hazardous waste sites and ground water pollution, *Archives of Environmental Health*, 44, 69, 1989.

41. Lagakos, S.W., Wessen, B.J., and Zelen, M., An analysis of contaminated well water and health effects in Woburn, Massachusetts, *Journal of the American Statistical Association*, 81, 583, 1986.

42. MDPH (Massachusetts Department of Public Health), Woburn Childhood Leukemia Follow-up Study, Boston, 1996.

43. *Waste News*, $524M spend on Superfund cleanups in fiscal 2005, EPA reports, November 23, 2005.

44. EPA (U.S. Environmental Protection Agency), Superfund National Accomplishments Summary Fiscal Year 2004, Office of Solid Waste, Washington, D.C., 2005.

45. *New York Times*, EPA: Superfund cleanup finishes decline, November 4, 2003.
46. Seelye, K.Q., Bush slashing aid for E.P.A. cleanup at 33 toxic sites, *New York Times*, July 1, 2003.
47. *Washington Post*, Federal toxic waste program's budget is stagnant, Nov. 25, 2004.
48. EPA (U.S. Environmental Protection Agency), Superfund cleanup figures, Office of Solid Waste and Emergency Response, Washington, D.C., 2003.
49. Copeland, C., Ocean Dumping Act: A Summary of the Law. Report RS 20028, Congressional Research Service, Washington, D.C., 1999.
50. EPA Press Advisory, Ship's captain and engineers arrested on ocean dumping charges, Washington, D.C., September 30, 2004.
51. EPA (U.S. Environmental Protection Agency), London Convention. Available at http://www.epa.gov/owow/ocpd/icnetbri.html, 2004.
52. *New York Times Magazine*, World's oceans are sending an S.O.S., May 3, 1992.
53. NOAA (National Oceanographic and Atmospheric Administration), Act to Prevent Pollution from Ships. Available at http://www.csc.noaa.gov/opis/html/summary/apps.htm, 2005.
54. IMO (International Maritime Organization), Introduction to IMO. Available at http://www.imo.org/About/mainframe.asp?topic_id=3, 2005.
55. EPA (U.S. Environmental Protection Agency), News for release. Chairman of shipping line sentenced for illegal dumping of waste oil, Washington, D.C., April 4, 2005.
56. EPA (U.S. Environmental Protection Agency), OPA Q's & A's: Overview of the Oil Pollution Act of 1990, Office of Solid Waste and Emergency Response, Washington, D.C., 1991.
57. USFWS (U.S. Fish and Wildlife Service), Digest of Federal Resource Laws of Interest to the U.S. Fish and Wildlife Service. Available at http://laws.fws.gov/lawsdigest/oilpoll.html, 2005.
58. EPA (U.S. Environmental Protection Agency), Oil Pollution Act Overview. Available at http:www.epa.gov/oilspill/opaover.htm, 2004.
59. Schierow, L., Pollution Prevention Act of 1990. Congressional Research Service Report RL 30022. Available at http://www.ncseonline.org/nle/crsreports/briefingbooks/laws/c.cfm, 1999.
60. O'Grady, M.J., ed, *Environmental Statutes Outline*. Environmental Law Institute, Washington, D.C., 1998.
61. White House, Presidential Documents, Executive Order 12856, Federal compliance with right-to-know laws and pollution prevention requirements, *Federal Register*, 58 (No. 150), August 6, 1993.
62. EPA (U.S. Environmental Protection Agency), Number of NPL site actions and milestones, Available at http://www.epa.gov/superfund/sites/query/queryhtm/nplfy.htm, 2004.

9 International Environmental Health Programs

9.1 INTRODUCTION

Pollution of air, water, soil, and food respects no geographic boundaries. For instance, untreated chemical waste released into a river will eventually add to the pollution load of the ocean receiving the river's outflow. Ocean currents will distribute what began as local pollution throughout the planet. In the course of pollution migration, contaminants can cross international boundaries. How cross-boundary pollution is prevented or otherwise managed can therefore require the involvement of regional (e.g., European Union) and global (e.g., United Nations [UN]) agencies. This involvement often includes technical assistance and support.

Because we live in an increasingly political world and complex environment, knowledge of laws, regulations, and agreements that control or influence the management of environmental hazards is topically important. Global political systems have changed remarkably in recent years because of the restructuring of many national governments, greater regional collaboration in trade (e.g., the North American Free Trade Agreement), and increased international trade competition.

Furthermore, as global geopolitical systems have become more complex and integrated across national borders, the world community has broadened what it considers to be the "environment." For example, the World Commission on Environment and Development (sometimes called the Brundtland Commission) noted the importance of infectious diseases and the link to poverty as a *global environmental* (emphasis added) concern [1]. They stated, "[C]ertain infectious diseases show signs of new gains as a result of increasing poverty and an inability to meet persons's basic needs. Malnutrition remains a serious obstacle to health and to the development of human resources." For others, the "environment" includes the areas stated by the UN's World Health Organization [2]: "Environmental health comprises those aspects of human health, including quality of life, that are determined by physical, chemical, biological, social and psychosocial factors in the environment. It also refers to the theory and practice of assessing, correcting, controlling and preventing those factors in the environment that can potentially affect adversely the health of present and future generations." Regardless of differences in definitions, the world community has indicated through national legislation and international treaties its concern over environmental conditions that can adversely affect the health and well-being of humans and ecological systems.

This chapter will describe international agencies' roles in administering environmental health programs, in particular, the programs of the United Nations, as well as the roles of two non-UN organizations, the World Bank and the World Trade Organization. In addition, one regional organization, the European Union, is discussed in terms of it environmental health activities. The chapter concludes with a presentation of international environmental health rankings.

9.2 UNITED NATIONS

The United Nations is a global organization that has its roots in efforts to prevent war and preserve peace among nations. It is unique as the only global organization banded together for peaceful purposes. That is not to say that the UN has always been successful in preventing conflicts between nations and peoples. The continuing conflicts in several regions of the world illustrate the problems faced by UN peacekeeping goals. A more successful aspect of the UN has been its humanitarian programs, including those for environmental protection and human health enhancement. Specific examples of these accomplishments are discussed in the sections that follow.

Nations have made efforts since the nineteeth century to establish international organizations to peacefully settle disputes. In 1899, the International Peace Conference was convened in The Hague, The Netherlands, to explore ways to settle disputes peacefully, prevent wars, and codify rules of warfare [3]. An outcome of the conference was the adoption of the Convention for the Pacific Settlement of International Disputes and established the Permanent Court of Arbitration. These outcomes failed to prevent World War I, a particularly nasty war amongst several European nations (primarily, Austria, France, Germany, Great Britain and its Commonwealth allies, Italy, Russia). The United States was eventually drawn into the war. In 1919, the Treaty of Versailles brought an end to World War I. The terms of settlement imposed on Germany eventually contributed to economic collapse in that country, a condition that fostered resentment in Germany and contributed to the start of World War II.

The UN's Millennium Declaration establishes measureable goals in seven key areas, one of which is specific to protecting the environment.

The Treaty of Versailles established the League of Nations in 1919. Its purpose was " [T]o promote international cooperation and to achieve peace and security" [3]. It was headquartered in Geneva, Switzerland, chosen because of that country's history of political neutrality in world conflicts. Although President Woodrow Wilson was a principal in the League's establishment, the United States did not become a member, owing to concern in Congress that membership could bring international entanglements not in the best interests of the country. This stance was part of a political policy of isolationism (i.e., a belief that the United States should isolate itself from the problems of Europe and other regions). In time, the League of Nations ceased operations after failing to prevent World War II.

In 1945, following the conclusion of World War II, representatives of fifty countries met in San Francisco to convene the United Nations Conference on International Organization. The name *United Nations* had been coined in 1942 by President Franklin D. Roosevelt, although he did not live long enough to attend the San Francisco meeting. The global representatives to the conference drafted the United Nations Charter. It was signed on June 26, 1945, by representatives of the fifty countries. The UN officially came into existence on October 24, 1945, after the Charter had been ratified by China, France, the Union of Soviet Socialist Republics (USSR), the United Kingdom, the United States, and by a majority of other signatories [3]. New York City was selected as the UN's headquarters. Currently, 191 countries constitute the UN's membership.

In September 2000, the member states of the UN met in New York City to set its international agenda for the start of the twenty-first century. The resulting Millennium Declaration established measurable goals in seven key areas, one of which is specific to protecting the environment [4]. Under this goal is stated the following:

1. "We must spare no effort to free all of humanity, and above all our children and grandchildren, from the threat of living on a planet irredeemably spoilt by human activities, and whose resources would no longer be sufficient for their needs.
2. We reaffirm our support for the principles of sustainable development, including those set out in *Agenda 21*, agreed upon at the United Nations Conference on Environment and Development.
3. We resolve therefore to adopt in all our environmental actions a new ethic of conservation and stewardship and, as first steps, we resolve:
    *   To make every effort to ensure the entry into force of the Kyoto Protocol, preferably by the tenth anniversary of the United Nations Conference on Environment and Development in 2002, and to embark on the required reduction in emissions of greenhouse gases.
    *   To intensify our collective efforts for the management, conservation and sustainable development of all types of forests.
    *   To press for the full implementation of the Convention on Biological Diversity and the Convention to Combat Desertification in those Countries Experiencing Serious Drought and/or Desertification, particularly in Africa.
    *   To stop the unsustainable exploitation of water resources by developing water management strategies at the regional, national and local levels, which promote both equitable access and adequate supplies.
    *   To intensify cooperation to reduce the number and effects of natural and man-made disasters.
    *   To ensure free access to information on the human genome sequence" [4].

This collection of objectives is a powerful commitment to improved global environmental conditions, within an architecture of sustainable development. Specifically, the UN member states commit to actions to reduce greenhouse gases, conserve forests, reduce the creeping expansion of deserts, protect water resources, and reduce the number of environmental disasters. As a matter of environmental health policy, these objectives would make a notable improvement in human and ecological health. However, the objectives must face the reality of global political challenges. For example, as will be described, enforcing the Kyoto Protocol in order to reduce emissions of greenhouse gases has yet to achieve full approval by the U.S. government.

UNEP's programs are directed to encourage sustainable development through sound environmental practices globally.

Component organizations of the UN are located in several countries, including those that specialize in environmental protection, human health, and labor issues. For the purposes of this book, four UN agencies are of particular relevance: United Nations Environment Programme, the World Health Organization, the International Labour Organization, and the Food and Agriculture Organization. These agencies will be described in the following sections.

9.2.1 UNITED NATIONS ENVIRONMENT PROGRAMME

The United Nations Environment Programme (UNEP) was founded in 1972. Its programs are directed to encourage sustainable development through sound environmental practices globally. UNEP's activities span a wide spectrum of environmental issues, including protection of atmospheric and terrestrial ecosystems, promotion of environmental science acquisition and dissemination of information, and systems to respond to natural and anthropogenic emergencies and disasters [5]. The headquarters of UNEP are based in Nairobi, Kenya, with offices in Paris, Geneva, Osaka, The Hague, Washington, DC; New York, Bangkok, Mexico City, Manama, Bahrain; Montreal, and Bonn.

UNEP's priorities include:

1. "Environmental information, assessment and research, including environmental, emergency response capacity and strengthening of early warning and assessment functions,
2. Enhanced coordination of environmental conventions and development of policy documents,
3. Fresh water,
4. Technology transfer and industry,
5. Support to Africa" [ibid.].

UNEP, like other UN agencies, has developed information networks and databases that are available through the Internet. These include: the Global Resource Information Database (GRID), the International Register of Potentially Toxic Chemicals (IRPTC), and the UNEP.Net, a web-based interactive catalogue that provides access to environmentally relevant geographic, textual, and pictorial information. Like other UN resources, these UNEP environmental networks are particularly useful and relevant to the needs of developing countries [5].

UNEP has declared twenty-seven principles for encouraging environmentally responsible development [6]. Two of the principles seem especially appropriate for making environmental health policy. Principle 10 states, "[E]nvironmental issues are best handled with the participation of all concerned citizens, at the relevant level. At the national level, each individual must have appropriate access to information concerning the environment that is held by public authorities, including information on hazardous materials and activities in their communities, and the opportunity to participate in decision-making processes." This principle is a statement of the public's right-to-know policy discussed in chapter 1.

Two actions, one in the United States and the other in the Europe, illustrate the application of Principle 10. In the United States, as described in chapter 8, the Toxics Release Inventory (TRI) provision of the Comprehensive Environmental Response, Compensation, and Liability Act, as amended in 1986, makes information about hazardous substances available to the public. Similarly, in 2004, the European Environment Agency, which is discussed later in this chapter, announced the release of Europe's first industrial pollution register, a database similar in concept and content to the TRI [7]. The European Pollutant Emission Register (EPER) is intended to report about 90 percent of point source pollution released by Europe's major industrial facilities. In this regard, the policy of using emissions data, such as in the EU database and the TRI, to bring pubic pressure on industrial sources of pollution will continue to be effective.

UNEP's Principle 14 asserts, "States should effectively cooperate to discourage or prevent the relocation and transfer to other states of any activities and substances that cause severe environmental degradation or are found to be harmful to human health." International cooperation will be needed to effectively implement this principle and thereby avoid waste generation and environmental degradation that result from industries relocating from industrialized countries into developing countries [8].

Of note, UNEP has had major involvement in the two global environmental summits on the environment. The United Nations Conference on Environment and Development (UNCED) was held in Rio de Janeiro, Brazil, in 1992. The other summit, the World Summit on Sustainable Development, was held in Johannesburg, South Africa, in 2002. Both meetings are discussed in subsequent sections of this chapter. Also to be discussed are several international treaties of importance to national environmental health policies. These include the UN Framework Convention on Climate Change, the Basel Convention for management of hazardous waste, and the Convention on Persistent Organic Pollutants.

9.2.1.1 UN Conference on Environment and Development

The member states of the UN met in Rio de Janeiro, Brazil, from June 3–14, 1992, to conduct the UN Conference on Environment and Development (UNCED), also called the Earth Summit.[1] On June 14, 1992, the conference approved *Agenda 21*, which is a comprehensive statement of concerns, findings, and recommended actions in specific environmental areas [6]. *Agenda 21* is a significant agreement among the nations of the world. It is a statement of global consensus and political commitment at the highest level for economic development and environmental cooperation. *Agenda 21* is organized into forty chapters, commencing with a Preamble, which states, "Humanity stands at a defining moment in history. We are confronted with a perpetuation of disparities between and within nations, a worsening of poverty, hunger, ill health and illiteracy, and the continuing deterioration of the ecosystems on which we depend for our well-being [ibid.]." Of note to global environmental health policy, *Agenda 21's* chapter 9, *Protecting and Promoting Human Health Conditions*, addresses environmental hazards as contributors to adverse human health conditions. Objective 6.40 states the following [9]:

> *Agenda 21* is a significant agreement among the nations of the world. It is a statement of global consensus and political commitment at the highest level for economic development and environmental cooperation.

"The overall objective is to minimize hazards and maintain the environment to a degree that human health and safety is not impaired or endangered and yet encourage development to proceed. Specific programme objectives are:

A. By the year 2000, to incorporate appropriate environmental and health safeguards as part of national development programmes in all countries;
B. By the year 2000, to establish, as appropriate, adequate national infrastructure and programmes for providing environmental injury, hazard surveillance and the basis for abatement in all countries;
C. By the year 2000, to establish, as appropriate, integrated programmes for tackling pollution at the source and at the disposal site, with a focus on abatement actions in all countries;
D. To identify and compile, as appropriate, the necessary statistical information on health effects to support cost/benefit analysis, including environmental health impact assessment for pollution control, prevention and abatement measures" [9].

To implement these four objectives would have cost about $3 billion, according to the UN Secretariat in 1992. To give perspective to this global public health cost, the cost of *one* U.S. B-2 stealth bomber is about $2 billion [10].

Additional objectives in *Agenda 21* and plans for implementation address specific environmental hazards, including protection of the atmosphere and freshwater resources, combating deforestation, managing fragile ecosystems, promoting sustainable agriculture and rural development, conservation of biological diversity, protection of the oceans; and sound management of toxic chemicals, hazardous wastes, and solid wastes [9].

Of relevance to environmental health policy, proper management of hazardous waste was of importance to the preparers of *Agenda 21*. Several chapters relate to hazardous waste and human health issues; two are particularly germane. Chapter 20, *Environmentally Sound Management of Hazardous Wastes Including Prevention of Illegal International Traffic in Hazardous Wastes*, opens by stating, "Effective control of the generation, storage, treatment, recycling and reuse, transport, recovery, and disposal of hazardous wastes is of paramount importance for proper health, environmental protection and natural resource management, and sustainable development. This will require the active cooperation and participation of the international community." Furthermore, "Prevention of the generation of hazardous wastes and the rehabilitation of contaminated sites are the key elements, and both require knowledge, experienced people, facilities, financial resources and technical and scientific capacities."

The four targets established in chapter 20 of *Agenda 21* are: (1) preventing or minimizing the generation of hazardous wastes as part of an overall integrated production approach; eliminating, or reducing to a minimum, transboundary movements of hazardous wastes, consistent with the environmentally sound and efficient management of those wastes; and ensuring that environmentally sound hazardous waste management options are pursued to the maximum extent possible within the country of origin; (2) ratification of the Basel Convention on the Control of Transboundary Movements of Hazardous Wastes and their Disposal and the expeditious elaboration of related protocols; (3) ratification and full implementation of regional agreements bearing on transboundary shipments of hazardous wastes; and (4) elimination of the export of hazardous wastes to countries that prohibit the import of such wastes [6]. Chapter 20 details a series of recommended policies and actions that support each overall target.

In a similar voice of concern, chapter 16, *Environmentally Sound Management of Biotechnology*, contains the following statement: "Despite increasing effort to prevent waste accumulation and to promote recycling, the amount of environmental damage caused by overconsumption, the quantities of waste generated and the degree of unsustainable land use appear likely to continue growing" [6].

An emphasis on biotechnology is relevant to hazardous waste management because biotechnology can be used under certain circumstances to destroy hazardous waste. *Agenda 21* defines biotechnology as "[a] set of enabling techniques for bringing about specific man-made changes in deoxyribonucleic acid (DNA), or genetic material, in plants, animals and microbial systems, leading to useful products and technologies" [6]. Biotechnology is specifically recommended to increase the availability of food, feed, and renewable raw materials; improve human health; and enhance protection of the environment. Bioremediation is also described as a means "[t]o prevent, halt, and reverse environmental degradation through the appropriate use of biotechnology in conjunction with other technologies, while supporting safety procedures as an integral component of the program" [ibid.]. This statement conveys international support for bioremediation as a means to reduce hazardous wastes and as a method for waste site remediation.

These statements from chapters 16 and 20 of *Agenda 21* commit UN member states to a "cradle-to-grave" approach for reducing the volume of hazardous wastes, and the need to take an ecological perspective and sustainable development approach toward reducing the impacts of hazardous wastes.

How the nations of the world implement *Agenda 21* and other agreements from the Rio Conference will shape future impacts of environmental hazards on ecological systems, environmental quality, and human health. However, little will be achieved if national governments do not commit to actions that support the UNCED agreement and effect collaborative actions across national borders.

9.2.1.2   UN Framework Convention on Climate Change

In 1992, concerned about scientific projections of catastrophic change in the global climate, the nations of the world agreed to the terms of the United Nations Framework Convention for Climate Change (FCCC). The text of the convention was signed in 1992 at the Earth Summit in Rio de Janeiro. Currently, more than 180 countries, including the United States have since ratified the agreement. The FCCC adopts as a general objective the "[s]tabilization of greenhouse gas concentrations in the atmosphere at a level that would prevent dangerous anthropogenic interference with the climate system. Such a level should be achieved within a time frame sufficient to allow ecosystems to adapt naturally to climate change, to ensure that food production is not threatened, and to enable economic development to proceed in a sustainable manner" (cited in [11]). It was a historic agreement that targeted the need to protect the global environment against catastrophic change due to global warming caused by increased amounts of carbon dioxide and other gases in the atmosphere. The gases, called *greenhouse gases*, have increased in concentrations in the Earth's atmosphere, beginning in the Industrial Revolution circa 1850. For instance, atmospheric concentrations of carbon dioxide, the principal greenhouse gas, are now 30 percent greater than levels 200 years ago [ibid.].

The UN Framework Convention for Climate Change targeted the need to protect the global environment against catastrophic change due to global warming caused by increased amounts of carbon dioxide and other greenhouse gases in the atmosphere.

The greenhouse gases are relatively inefficient in heat transference properties, thereby contributing to heat buildup on earth. Greenhouse gases contribute to the greenhouse effect, which is the warming of the air because of a lessened ability of the Earth to radiate energy through its atmosphere. According to scientific consensus, global warming of 1 to 3.5°C will occur over the next 100 years unless greenhouse gas concentrations are decreased [12]. A significant increase in global temperature would cause cataclysmic social changes. Polar ice caps would melt, raising ocean levels, and coastal cities and other areas would be flooded. Arid areas would increase in size, causing loss in food production, with consequences for regional famine. National economies could be devastated by climate change, according to various socioeconomic models.

Some greenhouse gases occur naturally in the atmosphere, while others result from anthropogenic activities. Naturally occurring greenhouse gases include water vapor, carbon dioxide, methane, nitrous oxide, and ozone. Anthropogenic activities, however, have increased the atmospheric concentrations of some greenhouse gases. For example [13], carbon dioxide is formed when solid waste, fossil fuels, wood, and wood products are burned. Methane is emitted during the production and transport of coal, natural gas, and petroleum; decomposition of organic waste in landfills, and from livestock flatulence. Remarkably, it was reported in 2006 that terrestrial plants under aerobic conditions produce significant amounts of methane, estimated to be 10 to 30 percent of annual levels of atmospheric methane. This startling finding has significant implications for control of atmospheric levels of the greenhouse gas methane [14]. Nitrous oxide is considered a major greenhouse gas due to its long atmospheric lifetime (approximately 120 years) and radiative forcing effects. In the United States, anthropogenic emissions of N_2O are primarily generated by agriculture soil management, mobile and stationary sources of fossil fuel combustion, adipic acid production, and nitric acid production [ibid.]. Other greenhouse gases that are not naturally occurring include hydroflurocarbons, perflurocarbons, and sulfur hexafluoride. Although many factors have shaped the increase in atmospheric greenhouse gas concentrations, the most significant factors are human population growth, fossil fuel combustion, and deforestation

The social and economic consequences of global warming scenarios, buttressed by measured increases in global temperature and ocean levels over the past century, motivated the United Nations to organize what has become the Framework Convention on Climate Change. As a treaty, the Framework Convention went into force for signatories on March 21, 1994 [15]. The Framework Convention was what its name implies—a structure upon which to build specific action plans and set environmental goals for emissions of greenhouse gases. In 1997, nations met in Kyoto, Japan, to establish specific goals for control of greenhouse gases. The Kyoto Protocol commits developed countries (i.e., those with fully developed national economies) to reduce their collective emissions by more than 5 percent below 1990 levels of six[2] key greenhouse gases by the period 2008–2012, with different targets negotiated for individual countries [16]. The Protocol was to take global effect when ratifying countries accounted for at least 55 percent of all industrialized countries' 1990-level emissions. In 1990, the United States accounted for 36.1 percent and Russia 17.4 percent of greenhouse emissions from industrialized countries [ibid.]. The Clinton administration agreed to an overall reduction of 7 percent in U.S. gas emissions, but as it was a treaty, ratification by the U.S. Senate was required.

In 2000, the Bush administration announced that the United States would not honor the Kyoto Protocol and that the United States would withdraw from its ratification, thereby reversing the Clinton administration's position. The administration's basis for opting out of the Protocol is its stated concern that lower U.S. emissions of greenhouse gases would have dire consequences on the national economy, an economy strongly reliant on carbon-based sources of energy, such as petroleum.

In October 2004, Russia endorsed the Kyoto Protocol, which had already been ratified by 120 countries. Russia's approval of the Protocol pushed the treaty past the 55 percent emissions reduction required for global adoption [17]. Countries that have not ratified the Protocol are not bound to its terms and conditions. Nevertheless, global adoption of the Kyoto Protocol represents a significant step toward reducing the potentially cataclysmic consequences of global climate change.

The U.S. opposition to the Kyoto Protocol raises serious, indeed vital, questions about the country's national environmental health policy on climate change. In particular, are short-term, political gains (where no U.S. politician wants to be seen as harming the nation's economy) worth the potentially long-term, uncertain, catastrophic effects of climate change? There is no more important policy issue facing the United States than what to do about climate change. The political process is likely unable to develop and effectuate the necessary environmental health policies that would support the Kyoto Protocol. Only pressure from an informed public will force the necessary policy debate within the U.S. government.

Failure by the U.S. government to honor the Kyoto Protocol portends serious human health and environmental consequences. As assessed by climate scientists, the body of scientific evidence regarding global warming is convincing and warrants immediate action by national governments and international organizations. A warming trend of about 1°F has been recorded since the late nineteenth century, occurring in both the northern and southern hemispheres, and over the oceans [18]. Moreover, confirmation of a warmer climate is evidenced by melting glaciers, decreased snow cover in the northern hemisphere and warming of the seas. The burning of fossil fuels by vehicles and power plants releases more than twenty-five billion tonnes[3] of CO_2 annually [19]. The United States is the largest emitter of CO_2, releasing about 5.8 billion tonnes in year 2003, followed by China's release of 3.5 billion tonnes [ibid.]. CO_2 constituted about 82 percent of total U.S. greenhouse gas emissions in 1999 [20]. Combustion of fossil fuel was the dominant source of CO_2 emissions, which grew by 13 percent during the decade of the 1990s [ibid.]. While much remains unknown about the future extent of global warming, both the mechanism of global warming and its initial effects are known.

There is no more important policy issue facing the U.S. than what to do about global climate change.

The mechanism of global warming is well established. Scientists long ago agreed that greenhouse gases trap heat in the Earth's atmosphere and help warm the planet—an occurrence that is vital if life on Earth is to be sustainable. About two-thirds of solar energy reaching Earth is absorbed by the Earth's surface. The heat radiates back to the atmosphere, where some of it is trapped by greenhouse gases, including carbon dioxide [21]. However, with increased volumes of greenhouse gases comes increased global warming. Measurements of some greenhouse gases have occurred for a century or more. For example, it is known from such measurements that CO_2 concentrations in the atmosphere have increased more than 30 percent since the Industrial Revolution (i.e., circa 1750) [20]. As greenhouse gases continue to increase due to human activities, further rises in global temperature are projected. For example, the Intergovernmental Panel on Climate Change (IPCC), a group of international scientists associated with the World Meteorological Society and the United Nations Environment Programme, predicts further global warming of 2.2–10°F by the year 2100 [ibid.].

Regarding environmental effects of global warming, in 2001 the IPCC reported the following consequences:

- "Global mean surface temperature will rise by 1.4–5.8°C. Warming will be greatest over land areas, and at higher altitudes;
- The projected rate of warming is greater than anything humans have experienced in the last 10,000 years;

- The frequency of weather extremes is likely to change leading to an increased risks of floods and drought. There will be fewer cold spells but more heat waves;
- The frequency and intensity of El Niño may be affected;
- Global mean sea level is projected to rise by 9–88 cm by the year 2100" (cited in [21]).

In another report, acidification of oceans is occurring according to the British Royal Society, which observes that ocean water is now 8.1pH, a decrease of 0.1 over that of 200 years ago, which translates to a 30 percent increase in hydrogen ions in the water. The Society predicts that the pH of ocean water near the surface will decrease to 7.7–7.9 by year 2100 [22]. The increased acidity could reduce populations of plankton, disrupting the ocean food chain and harming fisheries.

The consequences of these projected changes in global and regional environments will cause serious impacts on public health, national economies, and environmental quality. Public health impacts have been categorized by the IPCC as follows (cited in [21]):

- Increased frequency of heat waves. Heat waves bring heat-related mortalities and morbidities, together with loss of economic productivity;
- Regional variations in precipitation patterns would cause problems with freshwater supplies, producing an increase in waterborne diseases as human populations use impure water supplies in lieu of freshwater;
- Food production would be compromised in regions with lesser precipitation and increased ambient temperature;
- Coastal flooding due to rising ocean levels would cause coastal flooding, placing human populations at risk;
- Vector-borne diseases would increase as vectors such as mosquitoes expand their region of activity.

An attempt to estimate the global human death toll attributable to climate change was undertaken by the WHO, which estimated that approximately 150,000 deaths in year 2000 were attributable to changes in global climate experienced over the preceding decade (cited in [23]). The public health impacts of global warming will be hardest on regions and countries that have the least resources to defend themselves against the consequences of global warming. For example, lack of a public health infrastructure in some of the world's poorest countries will lead to inabilities to prevent heat-related illness and consumption of impure drinking water. In the long term, of course, the primary prevention of global warming health effects must be through marked reductions in the generation and release into the atmosphere of greenhouse gases. And that is what the Kyoto Protocol attempts to address.

In addition to the IPCC and WHO findings on the consequences of global warming, other findings of note include the following:

- Frog populations are decreasing in several areas of the world due to global warming. Warmer temperatures have been associated with outbreaks of a skin fungus that is fatal to frogs. The fungus proliferates with warmer temperatures [24].

- Ancient ice samples show that for at least 650,000 years, levels of CO_2, methane, and nitrous oxide were at stable concentrations until humans began intensively clearing forests and burning coal, oil, and other fossil fuels [25].
- Year 2005 was the second warmest year on record, with record-breaking heat, drought, storms, and flooding worldwide, according to the United Nations' weather service. Warmer-than-average Arctic temperatures brought the extent of sea ice to an all-time low [26].

Failure of the U.S. government to implement the Kyoto Protocol places the U.S. population and regions of the world at risk of the consequences of global warming. Moreover, this failure violates at least three key public policies: precautionary approach, sustainable development, and the public's right-to-know.

First, as discussed in chapter 2, the precautionary approach to environmental hazards states, "[W]here there are threats of serious or irreversible damage, lack of all scientific certainty shall not be used as a reason for postponing cost-effective measures to prevent environmental degradation." It is clear that global warming is already causing serious damage to the environment and human health (glacial melting, CO_2 increases, acidification of the oceans, regional increases in health-related deaths) and measures are warranted to prevent further environmental degradation. The Kyoto Protocol is one such measure. Adoption by the United States and other governments of a policy to forego the use of fossil fuels would be a key prescription to mitigating global warming. Such a policy in the long run would have great economic benefits to those countries that develop the technologies to replace fossil fuels. The historical precedent is the development and commercialization of pollution reduction equipment, which found a ready market in countries lacking such technologies.

Second, the policy of sustainable development is also violated by lack of adherence to the Kyoto Protocol. As noted in chapter 2, sustainable development is "[D]evelopment which meets the needs of the present without compromising the ability of future generations to meet their own needs." Failure to adopt the Kyoto Protocol surely puts in question the well-being of future generations, which will face higher ambient air temperatures, wider spread of vector-borne diseases, and diminished food supplies. Future generations will suffer the consequences of current inactions on prevention of global warming. They will not suffer their plight gently.

Third, failure to act aggressively to prevent global warming also violates one of the U.S. public's key policy expectations, right-to-know. As discussed in chapter 1, the U.S. public expects to be informed about hazards and risks that they may confront. In the United States little effort has been expended by federal government to actively communicate the serious consequences of global warming. Both Congress and the executive branch have been silent on communicating the effects of global warming. Environmental groups have developed health warnings about temperature rises and heat-related illness, but a national communication strategy is not yet evident. On the other hand, those opposed to actions to prevent global warming, presumably on the basis of alleged economic impact, have developed media campaigns to discredit global warming. In particular, some radio talk show commentators argue that global warming is nothing more than a naturally occurring cycle, therefore, why take costly actions to address a problem that doesn't exist. As previously stated, consideration of the precautionary approach and sustainable development argue to the contrary.

9.2.1.3 World Summit on Sustainable Development

As a follow-up to the 1992 Earth Summit, the World Summit on Sustainable Development (WSSD) was held from August 26 to September 4, 2002, in Johannesburg, South Africa, to elaborate on *Agenda 21*. In 1992, *Agenda 21* represented a significant agreement among the nations of the world. It remains a statement of global consensus and political commitment at the highest level for economic development and environmental cooperation. However, in the decade that followed the Earth Summit, progress in meeting the objectives of *Agenda 21* was disappointing, according to UN Secretary-General Kofi Annan. In 2002, prior to the World Summit on Sustainable Development meeting, Secretary-General Annan stated, "Attempts to promote human development and to reverse environmental degradation have not, in general, been effective over the last decade. Too few resources, a lack of political will, a piecemeal and uncoordinated approach and continued wasteful patterns of production and consumption have conspired to thwart efforts to implement sustainable development, or development that is balanced between people's economic and social needs and the ability of the earth's resources and ecosystems to meet present and future needs" [27].

Agenda 21's grand dreams of sustainable development, environmental protection, and elimination of poverty can only be made into international and national policies if there is political resolve, accompanied by monetary funds and other resources. A primary objective of the 2002 summit was to develop detailed steps and to identify quantifiable targets for better implementation of *Agenda 21*. In attendance were heads of state and other senior government officials,[4] national delegates, and leaders from nongovernment organizations, businesses, and other major concerned groups. Two areas of focus at the summit were alleviation of global poverty and protection of the natural environment and human health.

Only time will reveal the true worth of the WSSD. However, as a matter of environmental health policy, having an agenda, stated goals, and targets for global improvement is an important resource for long-range national and international planning. According to the UN [28], highlights of commitments and implementation initiatives from the WSSD Plan of Implementation pertinent to environmental health policy are as follows:[5]

- "Water and Sanitation—Commitment to halve the proportion of people without access to sanitation by 2015; this matches the goal of halving the proportion of people without access to safe drinking water by 2015.
- Energy—Commitments: To increase access to modern energy services increase energy efficiency and to increase the use of renewable energy; To phase out, where appropriate, energy subsidies; To support the NEPAD objective of ensuring access to energy for at least 35 percent of the African population within 20 years.
- Health—Commitments: By 2020, chemicals should be used and produced in ways that do not harm human health and the environment; To enhance cooperation to

reduce air pollution; To improve developing countries' access to environmentally sound alternatives to ozone depleting chemicals by 2010.

- Agriculture—Commitments: The GEF will consider inclusion of the Convention to Combat Desertification as a focal area for funding; In Africa, development of food security strategies by 2005.
- Biodiversity and Ecosystem Management—Commitments: To reduce biodiversity loss by 2010; To reverse the current trend in natural resource degradation; To restore fisheries to their maximum sustainable yields by 2015; To establish a representative network of marine protected areas by 2012; Undertake initiatives by 2004 to implement the Global Programme of Action for the Protection of the Marine Environment from Land Based Sources of Pollution."

Secretary-General Annan stated at the conclusion of the Johannesburg Summit, "This Summit makes sustainable development a reality. This Summit will put us on a path that reduces poverty while protecting the environment, a path that works for all peoples, rich and poor, today and tomorrow. Governments have agreed here on an impressive range of concrete commitments and action that will make a real difference for people in all regions of the world" [29]. The Secretary-General's statement seems to portend hope for the future, but also bears the ring of hyperbole, given the history of broken promises from the Rio Summit. Moreover, what isn't stated is the level of effort and resources that will have to be mobilized in order for the Summit's commitments to become reality in their implementation. Also, it should be noted that the Summit's agreements are not legally binding, which lessens any nation's resolve to act to implement the agreements.

Some environmental organizations have expressed disappointment that the Summit did not result in more definitive commitments and resources. For example, the World Wildlife Federation stated, "The Summit failed to address energy issues, the harmful effects of trade liberalization and subsidies, made lukewarm statements to support the Biodiversity Convention, and compromised on toxic chemicals to the extent that the outcome was weaker than previous international agreements" [30]. While this criticism is harsh, perhaps appropriately so, the Summit's agreements, if implemented, would lead to significant improvements in global water quality, sanitation, and fisheries protection. As a matter of environmental health policy, considerable pressure will need to be brought on national governments in order to make the Summit's Plan of Implementation become a reality. Failure of the nations of the world to implement the agreements from the Rio Summit should serve as a warning about implementing the Johannesburg agreements.

9.2.1.4 The Basel Convention

The Basel Convention on the Control of Transboundary Movements of Hazardous Wastes and Their Disposal (referred to as the Basel Convention) is influencing waste management on an international scale [31,32]. Negotiated under the auspices of UNEP, the Basel Convention was adopted in 1989 [33]. It covers transboundary movements of hazardous waste and household waste ash and prohibits movement between non-parties to the convention, unless the countries have a separate agreement that ensures sound waste management. The convention specifies that government-to-government notice must be given, and consent obtained, before hazardous waste is exported. It sets an "environmentally sound management" standard as the basis for all transboundary movements of hazardous waste. Countries that export waste must not allow a shipment to proceed, even if an importing country has agreed to accept it, if evidence suggests that wastes will not be managed in an environmentally sound manner.

According to UNEP, the Basel Convention is the response of the international community to problems caused by the annual global production of 400 million tonnes of hazardous wastes. The primary principles of the Basel Convention are [ibid.]:

- Transboundary movements of hazardous wastes should be reduced to a minimum consistent with their environmentally sound management.
- Hazardous wastes should be treated and disposed of as close as possible to their source of generation.
- Hazardous waste generation should be reduced and minimized at the source.

The convention entered into force as an international treaty ninety days after the twentieth signatory country, Australia, ratified it on May 5, 1992. As of October 1996, 100 parties (i.e., member states of the UN or political/economic organizations) had ratified the Basel Convention [33]. Ratification of the treaty signals a party's readiness to fully implement the convention, including having the necessary legislative authorities to enforce its terms. On March 21, 1990, the United States signed the Basel Convention, indicating that it will not take any action that would defeat the objective and purpose of the convention [34]. Each signatory to the treaty must ratify it through legislative action. Although the U.S. Senate has ratified the Basel Convention [35], no congressional actions have provided the authorizing legislation necessary to implement the convention's provisions.

The impact of the Basel Convention on the United States and other industrialized countries is unclear, in part because the amount of exported waste is uncertain. According to EPA, less than 1 percent of the hazardous waste generated in the United States and less that 10 percent of that generated in European countries is exported [36]. Legislation in support of implementing the conditions of the Basel Convention, together with further amendments to the Resource Conservation and Recovery Act, would give EPA additional regulatory authorities to control the export of nonhazardous solid waste, limit exports of solid waste based on the management of exported waste in the receiving country, limit imports of certain solid waste, and administer a registration or permit program for waste exports and imports [ibid.].

Some business groups in the United States oppose U.S. ratification of the Basel Convention because, they assert, it would restrict free trade. Environmental organizations argue that free trade agreements lead to the dumping of hazardous waste in poor nations [37]. These trade and environmental issues will influence any future congressional action on ratification of the Basel Convention.

9.2.1.5 Convention on Persistent Organic Pollutants (POPs)

A significant public health and ecological problem is the presence of chemicals that persist in the environment for long periods of time. In response to this problem, UNEP has led the development of a global treaty to ban or severely restrict the production and use of what are called *persistent organic pollutants* (POPs). POPs are a set of chemicals that are toxic, persist in the environment for long periods of time, and biomagnify as they move up through the food chain. They have been linked to adverse effects on human health and animals, such as cancer, damage to the nervous system, reproductive disorders, and disruption of the immune system. Because they circulate globally through the atmosphere, oceans, and other pathways, POPs released in one part of the world can travel to regions far from their source of origin [38].

In 1997, UNEP was asked by several governments to commence negotiations of treaties to reduce and/or eliminate releases of POPs into the environment. At the same time, academic and government scientists and various environmental organizations also suggested similar action by UNEP. On December 10, 2000, in Johannesburg, South Africa, diplomats from 122 countries completed the text of a legally binding treaty for control of POPs, the Convention on Persistent Organic Pollutants, which was adopted on May 22, 2001, in Stockholm, Sweden, covering twelve chemicals.[6] The convention is linked to the 1992 Rio Summit, stating in the Stockholm Convention's Article 1 as its objective, "Mindful of the precautionary approach as set forth in Principle 15 of the Rio Declaration on Environment and Development, the objective of this Convention is to protect human health and the environment from persistent organic pollutants" [39]. The Stockholm Convention represents an international treaty, requiring signatory countries to the treaty to ratify it through treaty approval mechanisms. The treaty requires all parties to the treaty to stop production and new uses of intentionally produced POPs, with limited country-specific and general exceptions. All new manufacture of PCBs is banned, and parties are to take steps to reduce use of existing PCBs. DDT use is restricted to vector control (e.g., to control malaria-bearing mosquitoes), and is slated for ultimate elimination as cost-effective alternatives become available. Parties will also be required to implement rigorous controls on sources of POP byproducts to reduce releases. The treaty also includes requirements for safe handling and disposal of POPs in an environmentally sound manner [38].

In a sense, the POPs treaty continues the work advocated by Rachael Carson in her book *Silent Spring*, published in 1962. Carson's concerns about the persistence of DDT in the environment are echoed in the POPs treaty, together with concerns for the ecological and human health effects of eleven other POPs.

The treaty also includes provisions restricting trade of POPs for which uses or production continue to exist and bans all export of POPs, except for environmentally sound management once there are no longer any uses allowed. In addition, a strong financial and technical assistance provision in the agreement provides support to developing countries and countries in economic transition to assist them in implementing the obligations under the treaty. Finally, the treaty includes a science-based procedure to allow for the addition of other chemicals to the agreement [ibid.].

The POPS treaty took effect on May 17, 2004, after fifty signatory countries gave their ratification [40]. The United States signed the Convention on Persistent Organic Pollutants (POPs) on May 23, 2001. On April 11, 2002, the administration submitted the treaty to the U.S. Senate for ratification [41]. In order to ratify the treaty, Congress would have to first amend U.S. chemicals and pesticides laws, including the Toxic Substabces Control Act and the the Federal Insecticide, Fungicide, and Rodenticide Act , in order to give EPA the authority to ban or restrict domestic production, use, and export of POPs. Senate ratification is necessary to provide resources and authorities for implementation of the POPs treaty. In December 2005, legislation was introduced in the U.S. House of Representatives that would allow the United States to join the POPs treaty and two other international environmental agreements[7] that are intended to reduce the global hazard of POPs [42]. Similar action is required of the U.S. Senate.

9.2.2   WORLD HEALTH ORGANIZATION

The World Health Organization (WHO) is the UN agency that specializes in human health issues. It was created on April 7, 1948, and has its headquarters in Geneva, Switzerland, with regional offices in Brazzaville, Copenhagen, New Delhi, Washington, D.C., Cairo, and Manila. The WHO Director-General heads the organization. WHO states that its objective is "the attainment by all peoples of the highest possible level of health. Health is defined as a state of complete physical, mental, and social well-being, not merely the absence of disease or infirmity" [43]. WHO is governed by the UN's member states (currently 191 countries) through the World Health Assembly, which meets biennially to consider major policy questions about WHO's programs and priorities. WHO has two sources of funds. Its regular budget derives from dues paid by the UN's member states. In 2002, the regular budget was approximately $800 million for a two-year period. The second, and larger, source of funds comes from donor countries. These are voluntary funds (called extrabudgetary) and amount to about $1 billion per budget period [44]. For purpose of comparison, the budget for the U.S. National Institutes of Health was approximately $28.4 billion for fiscal year 2005 [45].

WHO states that its objective is the attainment by all peoples of the highest possible level of health.

WHO is the UN's lead international political body for global, regional, and national responses to major threats to human health. It provides technical assistance, research, and special services on matters of disease and disability. Perhaps the most notable success of WHO was its leadership in global eradication of the scourge of smallpox. This was achieved by mobilizing medical talent and resources from many countries and energetic application of disease prevention principles. While industrialized countries had already conquered smallpox through vaccination of national populations, many developing countries lacked the resources to vaccinate their people. WHO's teams of specialists worked with health professionals in countries with endemic smallpox outbreaks to isolate new cases of the disease, quarantine affected populations, and vaccinate those at risk. In time, the strategy reduced the new cases of smallpox to zero.

WHO maintains a global disease surveillance network, operates regional laboratories and investigates outbreaks, such as the Avian Flu. It compiles and evaluates epidemiological data provided by its 191 member countries. In recent years it has produced ambitious analyses of the global burden of specific diseases, as well as estimates of the contribution that various modifiable "risk factors"—such as unclean water, smoking, obesity, and unsafe sex practices—make to illness and the world's 55 million annual deaths.

In the developing world, WHO offers advice to ministries of health and provides technical services that many nations cannot do on their own, such as establishing standards for exposure to hazardous chemicals. It promulgates treatment guidelines for specific diseases, compiles an "essential medicines" list for governments and public agencies, and promotes optimal—and often underused—disease-fighting strategies, such as the use of insecticide-treated sleeping nets in areas where mosquitoes carry malaria.

WHO operates under three-decade-old regulations that established its responsibilities for combating global health problems. WHO's regulations were established during the era when yellow fever, cholera, and the plague were the infectious diseases of global concern [46]. In combating such diseases, WHO then, and now, works in cooperation with affected member states of the United Nations. In this mode, WHO can request health information from a particular country or propose specific interventions. However, the affected country is under no binding arrangement to provide the requested information or accept the proposed intervention.

To take action against a recalcitrant country would require WHO to refer the problem to the UN Security Council, a politically complicated proposition with an uncertain outcome. Rather, it relies on a "bully pulpit" approach for its global health effectiveness. Much like the U.S. surgeon general, who has relatively little direct control over public health programs, the WHO can bring pressure on a country by making public their concerns about a particular health problem. Whether WHO should have greater authority to intercede in a sovereign nation's affairs is, of course, a serious policy issue. One can argue that a nation must have the authority to reject intercessions from WHO and others. After all, shouldn't a country be in the best position to know what is in its own best interests? On the other hand, modern day diseases can easily and rapidly emigrate from one country to another. Shouldn't a global body, such as the WHO, have the authority to interdict disease outbreaks before they become an epidemic or pandemic? These are questions being debated by WHO and the member nations of the UN.

9.2.2.1   WHO's Global Health Risk Factors

In 2002, WHO made a major contribution to a better understanding of global health risk factors [47]. Following a three-year study of 25 risk factors, WHO published a ranked list, with supporting documentation, of the ten leading global health risk factors. While the factors were found to vary by region of the world, the list shown in Table

TABLE 9.1
WHO's Ten Leading Health Risk Factors (adapted from Table 3.11 in [47])

Risk Factor	DALYs (millions)	Number of Premature Deaths (in millions)
Underweight	138	3.7
Unsafe sex	92	2.9
High blood pressure	64	7.1
Tobacco consumption	59	5.0
Alcohol consumption	58	1.8
Unsafe water, sanitation & hygiene	54	1.7
High cholesterol	40	4.0
Indoor smoke	39	—
Iron deficiency	35	1.0
Obesity	33	0.5

9.1 represents a global integration of regional risk factors. The principal metric used by WHO to compare risk factors was the DALY (disability-adjusted life year). One DALY is equal to the loss of one healthy year of life. The concept of the DALY comes from the World Development Report [48,49], an endeavor of the World Bank. Some have criticized the DALY concept, arguing that its information dataset consists of sex, age, disability status, and time period, but not socioeconomic status. Incorporation of socioeconomic data, critics assert, would give greater weight to the illness of more disadvantaged populations [44].

The following narrative is excerpted from the WHO report [47]. Each of the ten health factors is accompanied there by proposed public health interventions, but which are not repeated here.

Underweight/Under-Nutrition—Childhood and maternal underweight was estimated to cause 3.7 million deaths in 2000, about 1.8 million in Africa. This accounted for about one in fourteen deaths globally. Undernutrition was a contributing factor in more than half of all child deaths in developing countries. Since deaths from undernutrition all occur among young children, the loss of healthy life years is even more substantial: about 138 million DALYs, 9.5 percent of the global total.

Undernutrition is mainly a consequence of inadequate diet and frequent infection, leading to deficiencies in calories, protein, vitamins, and minerals. Underweight remains a pervasive problem in developing countries, where poverty is a strong underlying cause, contributing to household food insecurity, poor childcare, maternal undernutrition, unhealthy environments, and poor health care.

Unsafe Sex—HIV/AIDS caused 2.9 million deaths in 2000, or 5.2 percent of the global total. It also caused the loss of 92 million DALYs (6.3% of all) annually. Life expectancy at birth in sub-Saharan Africa is currently estimated at 47 years; without AIDS it is estimated that it would be around 62 years. Current estimates suggest that 95 percent of the HIV infections prevalent in Africa in 2001 are attributable to unsafe sex. In the rest of the world the estimated percentage of HIV infections prevalent in 2001 that are attributable to unsafe sex ranges from 25 percent in Eastern Europe to 90 percent or more in parts of South America and the developed countries of the Western Pacific.

High Blood Pressure and Cholesterol—Worldwide, high blood pressure is esti-mated to cause 7.1 million deaths, about 13 percent of the global fatality total. Across WHO regions, research indicates that about 62 percent of strokes and 49 percent of heart attacks are caused by high blood pressure.

High blood pressure levels damage the arteries that supply blood to the brain, heart, kidneys and elsewhere. Cholesterol is a fat-like substance found in the bloodstream that is a key component in the development of atherosclerosis, the accumulation of fatty deposits on the inner lining of arteries.

High cholesterol is estimated to cause a loss of 4.4 million deaths (7.9% of global total) and a loss of 40.4 million DALYs (2.8% of total), although its effects often over-lap with high blood pressure. This amounts to 18 percent of strokes and 56 percent of global ischemic heart disease.

Tobacco Consumption—WHO estimates that tobacco caused about 4.9 million deaths worldwide in 2000, or 8.8 percent of the total, and was responsible for 4.1 per-cent of lost DALYs (59.1 million). In 1990, it was estimated that tobacco caused just 3.9 million deaths, demonstrating the rapid evolution of the tobacco epidemic and new evidence of the size of its hazard, with most of the increase occurring in developing countries.

Alcohol Consumption—Alcohol consumption has health and social consequences via intoxication (drunkenness), dependence (habitual, compulsive, long-term heavy drinking), and other biochemical effects. Intoxication is a powerful mediator for acute outcomes, such as vehicle crashes or domestic violence, and can also cause chronic health and social problems. Global alcohol consumption causes 1.8 million deaths an-nually deaths (3.2% of global) and 4.0 percent of DALYs (58 million). Most of all the increase in alcohol consumption is occurring in developing countries.

Unsafe Water and Sanitation—WHO estimates that approximately 1.7 million deaths annually (3.1% of global deaths) and 3.7 percent of lost DALYs (54.2 million) worldwide are attributable to unsafe water, sanitation, and hygiene. Of this burden, about one-third occurred in Africa and one-third in Southeast Asia. Overall, 99.8 percent of deaths associated with these risk factors are in developing countries, and 90 percent are deaths of children. Various forms of infectious diarrhea make up the main burden of disease associated with unsafe water, sanitation, and hygiene.

Indoor Smoke from Solid Fuels—Cooking and heating with solid fuels such as dung, wood, agricultural residues, or coal are likely to be the largest source of indoor air pollution globally. When used in simple cooking stoves, these fuels emit substantial amounts of pollutants, including respirable particulates, carbon monoxide, nitrogen and sulfur oxides, and benzene. According to WHO, nearly half of the world cooks with solid fuels. This includes more than 75 percent of people in India, China, and adjacent countries, and 50–75 percent of people in parts of South America and Africa. In total, 2.7 percent of DALYs globally are attributable to indoor smoke from burning solid fuels.

Iron Deficiency—Iron deficiency is one of the most prevalent nutrient deficiencies in the world, affecting an estimated two billion people with consequences for maternal and perinatal health and child development In total, 800,000 (1.5%) of deaths worldwide are attributable to iron deficiency, 1.3 percent of all male deaths and 1.8 percent of all female deaths. Attributable DALYs are even greater, amounting to the loss of about 25.9 million healthy life years (2.5% of global DALYs) because of nonfatal outcomes like cognitive impairment.

Obesity, Overweight, and High Body Mass—Overweight and obesity lead to adverse metabolic effects on blood pressure, cholesterol, triglycerides, and insulin resistance. Risk of coronary heart disease, ischemic stroke, and type-2 diabetes mellitus increase steadily with increasing body mass index (BMI). In the WHO study, 58 percent of diabetes mellitus globally, 21 percent of ischemic heart disease, and 8–42 percent of some cancers were attributable to BMI greater than 21 kg/m^3.

The WHO health risk factors study is a remarkable contribution to global public health. Its findings, ranked health risk factors, constitute a roadmap for regional and national programs of interventions that would reduce the loss of DALYs and premature loss of life. To follow the roadmap will require international cooperation, national resources, and political resolve. National environmental health policies will need to be articulated and adopted. Absent national resolve and policy infrastructure, the WHO health risk factors will languish much like the recommendations in *Agenda 21*.

There are major political challenges facing the implementation of interventions to reduce the impact of health risk factors identified in the WHO report. For example, some have questioned WHO's management structure. The criticisms include allegations that the WHO headquarters in Geneva and its six regional offices do not effectively coordinate programs, leading to too little impact of disease prevention efforts, particularly in developing countries [44]. Contributing to the alleged lack of cooperation is the fact that WHO regional directors are selected by the member states of their region, not by the WHO Director-General, nor the WHO World Health Assembly. Because of how regional directors are selected, the WHO regions have a degree of autonomy from the Geneva headquarters. Another challenge to WHO is the emergence of other organization, in particular the World Bank, that are receiving funds from donors in support of global health projects.

9.2.2.2 WHO's Environmental Health Programs

WHO's priorities have traditionally focused on preventing infectious and, to a somewhat lesser extent, chronic diseases. Human health consequences of environmental hazards have received lesser attention and have suffered from lack of resources. As will be discussed subsequently, WHO has provided leadership in organizing and promoting the activities of the International Agency for Research on Cancer and the International Programme on Chemical Safety. Further, WHO has taken significant actions to reduce the global devastation caused by tobacco usage.

In support of action on WHO's global health risk factors (Table 9.1), WHO has led the development of an international tobacco control program. The WHO Framework

Convention on Tobacco Control is the first international treaty solely addressed to an environmental health issue. The convention, which is an international treaty, was adopted during WHO's 56th World Health Assembly, on May 28, 2003, in Geneva, Switzerland. The convention was opened for signature by all members of the WHO, or members of the UN, and by regional economic interest organizations. The treaty will go into effect after 40 governments have ratified it.

As noted by WHO [50], tobacco usage is the second major cause of death globally. Approximately five million people die each year from tobacco smoking or other use. As perspective, the loss of 5 million persons approximately equals the depopulation of the metropolitan area of Atlanta, Georgia. WHO estimates that without controls on tobacco production and usage, 10 million deaths will occur annually by year 2025. The Framework Convention on Tobacco Control is intended to diminish the global toll on human health attributable to tobacco.

The treaty provides a set of principles and a framework for action on tobacco issues, which include product advertising standards, excise taxes, and cigarette smuggling [ibid.]. The convention was signed on May 10, 2004, by the U.S. Secretary of Health and Human Affairs, but has yet to be submitted to the U.S. Senate for treaty ratification [51]. Senate ratification and other congressional action would provide authorities and resources to U.S. federal agencies for implementation of the treaty's provisions.

9.2.2.3   International Agency for Research on Cancer

The International Agency for Research on Cancer (IARC) is a component organization of the WHO. It was created in May, 1965, and is based in Lyon, France [52]. IARC's mission "[i]s to coordinate and conduct research on the causes of carcinogenesis, and to develop scientific strategies for cancer control" [53]. IARC is involved in both epidemiological and laboratory research and disseminates scientific information through publications, meetings, courses, and fellowships. IARC's program of work has four main objectives, as listed in Table 9.2. Of the four program areas listed in the table, Identifying the Causes of Cancer has received the greatest public attention, primarily because of the issuance of cancer risk documents on individual chemical and physical agents.

TABLE 9.2
IARC's Program of Work [53]

IARC Program	Illustrative Activity
Monitoring global cancer occurrence	Studying cancer incidence, mortality, and survival in many countries
Identifying the causes of cancer	More than 870 agents and exposures have been examined for evidence of carcinogenicity
Elucidation of mechanisms of carcinogenicity	Laboratory research examining the interaction between carcinogens and DNA
Developing scientific strategies for cancer control	Programs are directed to finding ways to prevent human cancer

Since 1970, IARC has published assessments of the carcinogenic risks to humans from a variety of agents, mixtures of agents, and exposure circumstances. These assessments, known as the IARC Monographs, are prepared by international experts, assisted by IARC staff. Each monograph is prepared by an international working group that is specific to the agent under review. More than 870 agents (chemicals, groups of chemicals, complex mixtures, occupational exposures, cultural habits, biological or physical agents) have been evaluated [53]. Each monograph includes basic information about an agent's physical and chemical properties, methods of analysis, production volumes, toxicological data, and epidemiological findings. Sections of the monographs review the evidence for the agent's carcinogenicity. The monographs are available to an international audience of researchers, public health officials, and regulatory authorities. The monographs are particularly relevant to developing countries, where resources to develop similar documents may be lacking.

A significant feature of IARC Monographs is the classification of a chemical or physical agent's potential to cause cancer in humans. An IARC finding that a particular agent is a human carcinogen has genuine public health importance. Such a statement from IARC can be the impetus for international regulatory actions (e.g., trade bans), public health education programs, and legislative actions.

9.2.2.3.1   IARC Categories of Carcinogenicity

In the course of developing the IARC Monographs, working groups are asked to categorize each agent or exposure circumstance as to its carcinogenicity. Over time, IARC has developed guidelines for use in the categorization process. Although the guidelines provide considerable direction to a monograph's working group, scientists' professional judgment is still required. For example, different scientists may disagree over the quality and implications of the same toxicological study or epidemiological investigation. These disagreements are usually worked out in the course of assigning a category (e.g., Group 2A) of carcinogenicity for a particular agent. Following are IARC's carcinogenicity criteria [53]. Shown in Table 9.3 are IARC's current categories of carcinogens.

The agent, mixture or exposure circumstance is described according to the wording of one of the following categories, and the designated group is given. The categorization of an agent, mixture, or exposure circumstance is a matter of scientific judgment, reflecting the strength of the evidence derived from studies in humans and in experimental animals and from other relevant data.

These guidelines on carcinogenicity classification are in effect a policy statement from IARC, because they specify a course of action to be followed by working groups that develop individual monographs. Without such a policy, each working group would be able to make its own rules for carcinogenicity determination, making it impossible to compare carcinogenicity levels across monographs.

Shown in Table 9.3 is a comparison of IARC's grouping of carcinogens and those of the EPA. There are obvious similarities and some minor differences in wording. Even though the two sets of carcinogen categories have very similar wording, occasionally IARC and EPA will come to different conclusions as to an agent's carcinogenicity. This is because IARC and EPA work groups may differ when reviewing the same scientific

TABLE 9.3

Comparison of IARC [53] and EPA [87] Carcinogen Groups

IARC	EPA
Group 1: The agent (mixture) is carcinogenic to humans. The exposure circumstance entails exposures that are carcinogenic to humans. This category is used where there is sufficient evidence of carcinogenicity in humans.	Group A (human carcinogens): Compounds for which human data are sufficient to demonstrate a cause-and-effect relationship between exposures and cancer incidence in humans.
Group 2A: The agent (mixture) is probably carcinogenic to humans. This category is used when there is limited evidence of carcinogenicity in humans and sufficient evidence of carcinogenicity in experimental animals. Group 2B: The agent (mixture) is possibly carcinogenic to humans. This category is used for agents, mixtures and exposure circumstances for which there is limited evidence of carcinogenicity in humans and less than sufficient evidence of carcinogenicity in experimental animals. It may also be used where there is inadequate evidence of carcinogenicity in humans but there is sufficient evidence of carcinogenicity in experimental animals.	Group B1: Compounds for which limited human data suggest a cause-and-effect relationship between exposure and cancer incidence. Group B2: Compounds for which animal data are sufficient to demonstrate a cause-and-effect relationship between exposure and cancer incidence in animals, and human data are inadequate or absent.
Group 3: The agent (mixture) is not classified as to its carcinogenicity to humans. This category is used most commonly for agents, mixtures and exposure circumstances for which the evidence of carcinogenicity is inadequate in humans and inadequate or limited in experimental animals.	Group C (possible human carcinogen): Compounds for which animal data are suggestive to demonstrate a cause-and-effect relationship between exposure and cancer in animals.
Group 4: The agent (mixture) is probably not carcinogenic in humans. This category is used for agents or mixtures for which there is evidence suggesting lack of carcinogenicity in humans and in experimental animals.	Group D (not classifiable as to human carcinogenicity): Compounds for which human and animal data are inadequate to either suggest or refute a cause-and-effect relationship for human carcinogenicity.
	Group E (evidence of noncarcinogenicity): Compounds for which animal data are sufficient to demonstrate the absence of a cause-and-effect relationship between exposure and cancer incidence in animals.

data as to what is "sufficient" evidence. However, both sets of categories serve their purpose of providing guidance on weight-of-evidence assessment for the carcinogenicity of individual chemical compounds and mixtures.

9.2.2.3.2   IARC Policy Issues

Some controversy has arisen over how IARC constitutes its working groups and how some data on mechanisms of toxicity are accepted by the groups. An example of this kind of criticism follows. The critic is a former IARC staff member.

"[a] new attitude seems to have pervaded the *IARC Monograph* program, resulting in an increasing influence of or partiality for industry and a diminishing dedication to public and occupational health and safety concerns, and for primary prevention. Some of this attitude comes from an apparent misguided scientific zest prematurely to endorse purported or hypothetical mechanisms of chemical carcinogenicity or modes of action of chemicals causing cancer in experimental animals" [54]. Two issues of scientific policy are evident in the preceding quotation. First, what is the proper role of industry representatives on committees like the IARC working groups? Second, how and when should mechanistic data be used in the development of documents like the IARC Monographs? Some elaboration of these two issues follows.

Regarding industry representation on advisory committees, critics of such representation seem to believe that business entities want to bias the process so as to minimize the impact on themselves. Inherent in this argument is that industry scientists are willing to skew scientific debates to positions more favorable to industry. This notion takes dead aim at the traditional view of scientists as objective persons who base their opinions solely on research data and scientific principles.

On the other hand, industry scientists argue that they often possess industry-specific knowledge and experience that is vital to any discussion of an agent associated with their industry. Given the polar extremes of full to no industry representation on advisory committees, what should organizations like IARC do? Often such organizations rely on conflict of interest declarations to determine who should serve on an advisory committee. These statements require a prospective committee member to declare his or her financial holdings, current and past work history, and committee memberships (e.g., National Academy of Science committees). Conflict of interest statements are reviewed by committee managers (e.g., IARC administrators) to ascertain any potential bias or conflicts that might occur. Some persons would argue that conflict of interest statements are inadequate to protect an advisory committee's integrity from scientists with biased, vested interests. Such critics argue that the only genuine protection is to exclude industry scientists from all advisory committees. A more moderate position is to adopt a policy of industry involvement and that of other interested groups, but restrict their numbers on any advisory committee and specify in advance the nature of their participation (e.g., as advisers, not members, of work groups).

The second policy issue, using mechanistic data to establish advisory guidelines or regulatory standards, is a contemporary concern of persons who assert that industry scientists want to use such data in lieu of experimentally-derived data. The issue is a most interesting one. A policy response is necessarily complicated, complicated by the issues of costs of doing experimental studies, the role of basic science, and political motives of the interested parties. In the ideal situation, mechanisms of toxicity data, buttressed by experimental data, would be available upon which to conduct risk assessments and to develop regulatory standards. Having both kinds of data is seldom the situation. Rather, some persons argue that knowing a substance's basic mechanisms of toxicity (e.g., liver toxicity) is sufficient knowledge to forego costly tests on laboratory animals. This argument, by implication, leads to fewer laboratory animals being subjected to toxicity testing, then killed for pathologic examination. However, experimentalists argue that certainty in the knowledge of a substance's basic mechanisms of toxicity cannot be assured unless experimental testing is conducted. On their side of the argument is a long record of animal toxicity testing that has been useful in identifying human carcinogens.

These arguments are likely not going to be resolved until more data are obtained on the predictive power of mechanisms of toxicity data.

9.2.2.4 International Programme on Chemical Safety

The International Programme on Chemical Safety (IPCS) resulted from the UN Conference on the Human Environment, held in Stockholm in 1972. From the conference came the recommendation that programs, to be guided by WHO, should be undertaken for the early warning and prevention of harmful effects of chemicals to which human populations were being exposed [55]. The IPCS functions through the cooperation of the WHO, UNEP, and the International Labor Organization. These three organizations coordinate the development of technical reports, share personnel and other resources, and work together on education programs that address the impacts of chemical hazards on human health.

The two main roles of the IPCS are to establish the scientific health and environmental risk assessment basis for safe use of chemicals and to strengthen national capabilities for chemical safety. The latter role is particularly important for developing countries, which often lack the technical and economic resources to develop national programs in chemical safety. WHO has the overall administrative responsibility for the work of the IPCS, working through a central office that is based in Geneva, Switzerland. IPCS's work is divided into four main areas: risk assessment of specific chemicals, risk assessment of methodologies, risk assessments for food safety, and management of chemical exposures [55]. Much of the IPCS work is conducted in collaboration with regional and national organizations that address chemical safety issues. These organizations include the U.S. Environmental Protection Agency, the U.S. National Institute of Environmental Health Sciences, the U.S. Agency for Toxic Substances and Disease Registry, the European Commission, the International Life Sciences Institute, the International Union of Pure and Applied Chemistry, the International Union of Toxicology, and others.

The IPCS develops and coordinates several products and services of considerable importance to global environmental health. In particular, several information resources—some of which overlap each other—on chemical substances are available to environmental and health officials, as well as the general public. These documents include the following [55]:

> *Environmental Health Criteria* (EHC) documents, which are reasonably comprehensive reports of a substance's toxicity, exposure routes, and human health effects. Recommended Exposure Levels (RELs) are usually contained in each document [56]. Approximately 250 chemicals have been subjects of EHC documents. The primary audience for these documents consists of national policy makers, environmental and health officials, and government and private sector risk assessors.
> *International Chemical Safety Cards* (ICSCs) are cards that summarize essential health and safety information on chemicals. They are intended for use by workers and employers in factories, agriculture, construction, and other workplaces. They provide their users with a quick, credible resource for use in preventing chemical emergen-

cies and responding to them if they occur. ICSCs are similar to Material Safety Data Sheets developed by chemical producers and some national governments.

Concise International Chemical Assessment Documents (CICADs) are summary documents that provide information on the relevant scientific information pertinent to the adverse effects of a specific substance on human health and the environment. As stated by the IPCS, "The primary objective of CICADs is characterization of hazard and dose-response from exposure to a chemical. CICADs are not a summary of all available date on a particular chemical; rather, they include only that information considered critical for characterization of the risk posed by the chemical" [55]. The primary audience appears to be practicing risk assessors, whether in government or industry.

Methodological publications are part of an effort to improve the methodology of chemical risk assessment, developed by expert panels convened by the IPCS [57]. The documents include such documents as *Human Exposure Assessment, Biomarkers in Risk Assessment, Principles for Evaluating Health Risks to Reproduction Associated with Exposure to Chemicals, and Guidelines on Studies in Environmental Epidemiology.* The documents are used by national governments, professional organizations, and individual risk assessors. The IPCS also conducts regional and local training sessions in risk assessment, using their methodological publications as teaching materials.

Chemical incidents and emergencies are a global problems, irrespective of whether they occur in Industrialized or developing countries. Such incidents include spills of oil from tankers, explosions in chemical factories, and mishaps in overland transportation of chemical products and substances. The primary role of IPCS in such episodes is to interact with public health and medical authorities. More specifically, the IPCS provides guidance and training to member states in their planning on how to respond to chemical incidents and emergencies. The IPCS also serves as a source of technical information, advice, and assistance on the health implications of chemical incidents. In particular, WHO keeps a *World Directory of Poisons Centres* for access by first responders and health professions responding to chemical incidents and emergencies.

INCHEM is an IPCS database that offers access to " [t]housands of searchable full-text documents from international bodies on chemical risks and chemical risk management" [58]. The database can be accessed through the Internet and is free of charge. Included in the INCHEM database are the IPCS's EHCs, CICADS, Health and Safety Guides, International Chemical Safety Cards, and documents from non-IPCS sources. This database would seem to have a broad-based audience, ranging from emergency responders to academic researchers.

INTOX [59] is an IPCS database that is primarily directed to poison centers and health care providers who respond to chemical poisonings. Poison centers, in particular, need information on the toxicity of toxins and toxicants when caring for victims of exposure to both natural hazards (e.g., snake venom) as well as anthropogenic chemicals (e.g., industrial solvents). INTOX gives health professionals direct access to a database that will assist them in the diagnosis and treatment of poisonings, complemented by a data management software. The INTOX system is a primary resource for health professionals in developing countries, where local databases on poisonings may not exist.

Regarding global environmental health policy, in 1989, WHO's Regional Office for Europe, located in Copenhagen, issued a significant statement about the environment in their *European Charter on Environment and Health* [60]. The charter was a product of the First European Conference on Environment and Health. The conference was convened by government ministers and other senior representatives from the environment and health administrations of twenty-nine European countries and the Commission of the European Communities. The charter established a series of entitlements and responsibilities for governments and individuals, articulated principles for public policy, set strategic elements in support of public policies, and stated priorities for actions needed to protect human health and the environment [ibid.].

The *European Charter* also contains principles relevant for public policy; three are particularly germane to hazardous waste issues. Principle 8 states, "The entire flow of chemicals, materials, products and waste should be managed in such a way as to achieve optimal use of natural resources and to cause minimal contamination." Principle 9 avers, "Governments, public authorities and private bodies should aim at both preventing and reducing adverse effects caused by potentially hazardous agents and degraded urban and rural environments." And Principle 11 declares, "The principle should be applied whereby every public and private body that causes or may cause damage to the environment is made financially responsible." This latter principle parallels the philosophy and liability provisions in the Comprehensive Environmental Response, Compensation, and Liability Act: polluters must pay for the consequences of their pollution. Statements from the WHO are important because some countries adopt them as principles to help shape national legislation.

9.2.3 International Labour Organization

The International Labour Organization (ILO) was founded in 1919 as a provision of the Treaty of Versailles. It is based in Geneva, Switzerland, with regional offices in other regions of the world. According to ILO history, there were three primary motivations for ILO's establishment [61]. One motivation was humanitarian. Specifically, the unhealthy, unsafe, and economically exploited condition of workers was deemed unacceptable. Second, there was a political motivation for ILO's creation. Industrialists of that era were concerned that workers would create social unrest, perhaps even revolution, given the Russian Revolution of 1917, which was led by V.I. Lenin, and involved large numbers of workers and Russian peasants. Third, there was an economic motive to the creation of ILO. Industrialists and some national leaders were concerned that some nations might adopt social polices of workers' welfare that would put themselves at economic disadvantage if other nations did not adopt similar reforms [ibid.]

The ILO formulates international labor standards in the form of Conventions and Recommendations that set minimum standards of basic labor rights and promotes the rights and welfare of workers.

"The ILO formulates international labour standards in the form of Conventions and Recommendations setting minimum standards of basic labour rights: freedom of association, the right to organize, collective bargaining, abolition of forced labour, equality of opportunity and treatment, and other standards regulation conditions across the entire spectrum of work related issues" [61]. Unique among the UN agencies, the ILO operates as a tripartite structure: government/labor/management. As a matter of policy, the ILO implements a tripartite structure with workers and employers participating as equal partners with governments in the organization's work. This structure forces consensus-seeking on matters of workplace policy and practices. ILO labor standards do not override national standards; they are advisory unless adopted by national governments. For example, ILO workplace standards do not replace the OSHA standards that are in force in the United States

The ILO provides technical assistance in several areas related to labor and workplace conditions, including the areas of employment policy, vocational training, labor administration, working conditions, labor statistics, and occupational safety and health. The technical services offered by the ILO have particular relevance in developing countries, where national labor and work environment resources may not exist.

9.2.4 Food and Agriculture Organization

The UN's Food and Agriculture Organization (FAO) was created in 1943, when forty-four nations, meeting in Hot Springs, Virginia, committed themselves to founding a permanent organization for food and agriculture [62]. FAO is headquartered in Rome, Italy, with regional offices in Accra, Ghana; Bangkok, Thailand; Cairo, Egypt; Rome, Italy; and Santiago de Chile, Chile. The FAO functions as an organization representing 187 member countries and the European Union. FAO's mandate is "[t]o raise levels of nutrition, improve agricultural productivity, better the lives of rural populations and contribute to the growth of the world economy." The budget for 2004–2005 was US$ 749.1 million.

The FAO helps developing countries and countries in transition to modernize and improve agriculture, forestry and fisheries practices and ensure good nutrition for all.

Serving both developed and developing countries, FAO acts as a neutral forum where all nations can meet as equals to negotiate agreements and debate policy. FAO is also a source of knowledge and information, helping developing countries and countries in transition to modernize and improve agriculture, forestry and fisheries practices and to ensure good nutrition for all. The focus of FAO since its creation has been on the needs of developing rural areas, where 70 percent of the world's poor and hungry people reside [ibid.]. FAO "[p]rovides the kind of behind-the-scenes assistance that helps people and nations help themselves. If a community wants to increase crop yields but lacks the technical skills, we introduce simple, sustainable tools and techniques. When a country shifts from state to private land ownership, we provide the legal advice to smooth the way. When a drought pushes already vulnerable groups to the point of famine, we mobilize action. And in a complex world of competing needs, we provide a neutral meeting place and the background knowledge needed to reach consensus" [ibid.].

The FAO is a significant global policy-maker on issues that include food, agriculture, pesticides, and fisheries. The work and responsibilities of the FAO are quite likely to increase in both magnitude and importance as challenges arise in food production, combating hunger, control of pesticides and hazardous substances, and ocean pollution, among others. These problems will be exacerbated due to increased global population and the effects of global climate change.

9.2.5 UNITED NATIONS ECONOMIC COMMISSION FOR EUROPE

The United Nations Economic Commission for Europe (UNECE) was established in 1947 to encourage economic cooperation among its member states. The commission was established to foster economic recovery in Europe after World War II. It is one of five regional commissions under the administrative direction of United Nations headquarters. The UNECE has 55 member states, and reports to the UN Economic and Social Council. The United States is a member. The UNECE's headquarters are located in Geneva, Switzerland. "The UNECE strives to foster sustainable economic growth among its 55 member countries. To that end UNECE provides a forum for communication among States; brokers international legal instruments addressing trade, transport and the environment, and supplies statistics and economic and environmental analysis" [63].

Of environmental health policy note, there is a treaty on water quality that was developed under the auspices of the UNECE. The Convention on the Protection and Use of Transboundary Watercourses and International Lakes was agreed upon at Helsinki, Finland, on March 17, 1992 [64]. The convention's general provisions provide insight into its environmental health polices:

- The parties shall take all appropriate measures to prevent, control and reduce any transboundary impact.
- The parties shall, in particular, take all appropriate measures.
- To prevent, control and reduce pollution of waters causing or likely to cause transboundary impact.
- To ensure that transboundary waters are used with the aim of ecologically sound and rational water management, conservation of water resources and environmental protection.
- To ensure that transboundary waters are used in a reasonable and equitable way, taking into particular account their transboundary character, in the case of activities which cause or are likely to cause transboundary impact.
- To ensure conservation and, where necessary, restoration of ecosystems.
- Measures for the prevention, control and reduction of water pollution shall be taken, where possible, at source.

- These measures shall not directly or indirectly result in a transfer of pollution to other parts of the environment.
- The precautionary principle, by virtue of which action to avoid the potential transboundary impact of the release of hazardous substances shall not be postponed on the ground that scientific research has not fully proved a causal link between those substances, on the one hand, and the potential transboundary impact, on the other hand.
- The polluter-pays principle, by virtue of which costs of pollution prevention, control and reduction measures shall be borne by the polluter.
- Water resources shall be managed so that the needs of the present generation are met without compromising the ability of future generations to meet their own needs.
- The Riparian[8] Parties shall cooperate on the basis of equality and reciprocity, in particular through bilateral and multilateral agreements, in order to develop harmonized policies, programmes and strategies covering the relevant catchment areas, or parts thereof, aimed at the prevention, control and reduction of transboundary impact and aimed at the protection of the environment of transboundary waters or the environment influenced by such waters, including the marine environment.
- The application of this Convention shall not lead to the deterioration of environmental conditions nor lead to increased transboundary impact.
- The provisions of this Convention shall not affect the right of Parties individually or jointly to adopt and implement more stringent measures than those set down in this Convention [64].

A perusal of these provisions reveals several policies that correspond to five of the public's policy expectations that were discussed in chapter 1: (1) prevention of pollution is strongly emphasized in the provisions, (2) rational water management can be interpreted as a matter of accountability, (3) the precautionary principle is advocated as a key policy, (4) the polluter-pays principle is endorsed, and (5) reference to present and future generations is consistent with a policy of sustainable development. It is interesting to observe the incorporation of the public's policy expectations into a major regional environmental treaty.

In order to implement the general provisions of the convention required the development of the Protocol on Water and Health. The protocol entered into force on August 4, 2005, following ratification by the minimum sixteen countries [66]. The protocol calls on the ratifying countries to: strengthen their health systems, improve planning for and management of water resources, improve the quality of water supplies and sanitation services, address future health risks, and ensure safe recreational water environments. Each country has the responsibility for its implementation of the protocol.

According to WHO, the Protocol on Water and Health is the world's first legally binding international agreement that expressly intends to reduce water-related diseases [ibid.].

9.3 WORLD BANK

At the end of World War II, the victorious World War II Allies (United Kingdom, United States, and USSR chief among them) were faced with the reality of rebuilding the nations they had defeated (Germany, Italy, Japan). Other countries in Europe caught up in the war also needed rebuilding of their national economies and physical infrastructure. Only the U.S. mainland had remained relatively free from the wounds of World War II. Protected by the vast Atlantic and Pacific oceans, the United States had been largely immune to the presence of foreign troops and damage to the country's infrastructure.

The World Bank makes loans to national governments for them to make improvements in their economies and infrastructure.

Under the Truman administration, the Marshall Plan[9] poured financial and technical aid into Europe. In 1947, U.S. Secretary of State George C. Marshall had proposed a solution to the widespread hunger, unemployment, and housing shortages in Europe. Marshall proposed that the European nations themselves set up a program for reconstruction, with U.S. help. This proposal led to congressional enactment of the Economic Assistance Act of 1948. Over the four years of the Marshall Plan's life, about $13.3 billion was appropriated by Congress [67]. A component of the plan to rebuild Europe was the creation in 1944 of the International Bank for Reconstruction and Development (IBRD), known more simply then as the World Bank. Following the rebuilding of Europe, the World Bank engaged the needs of the world's poorest countries, which became known as *developing countries*.

During the decade of the 1990s, the World Bank became increasingly important in global programs of health and environmental protection. Based in Washington, D.C., the World Bank Group currently consists of five institutions that are owned by member countries. Two of the institutions, The International Bank for Reconstruction and Development and The International Development Association, constitute what is now called the World Bank [68]. The World Bank functions as a bank in the sense that it raises capital from its investment in global financial markets (e.g., national stock exchanges) and makes loans to national governments. Funds invested by the group come from member countries and individual donors.

Of the two groups constituting the World Bank, The International Development Association (IDA) "[h]elps provide access to better basic services (such as education, health care, and clean water and sanitation) and supports reforms and investments aimed at productivity growth and employment creation" [68]. The IDA was established in 1960 to provide assistance to countries that are too poor to borrow at commercial rates. They issue interest-free loans (called IDA credits), which borrowers must repay in 35 to 40 years. Contributions to the IDA constitute $6–7 billion annually, deriving from approximately 40 countries, including a mix of industrialized countries (e.g., France, Germany, Japan, United Kingdom, and U.S.) and some developing countries (e.g., Botswana, Turkey) [ibid.]. The IDA has become a significant source of funds for public health and environmental projects in developing countries that quality for IDA credits. For example, approximately $300 million was committed in 2001 to projects in the categories of health, nutrition, and population ($96 million); education ($88 million); social protection ($62 million); and water and sanitation ($38 million) [ibid.].

The World Bank's monetary credits to developing countries in support of environmental health projects have grown over time, in part because of donors' increased funding. As a matter of policy, some donor countries view the World Bank as preferable to UN agencies because of its more stringent control over how funds are allocated and spent. Another matter of political importance is whether developing countries should be permitted to forego repayment of World Bank credits. Some environmentalists and social activists have argued that unless debt relief is given to poor countries, they will remain in debt, chilling any prospects of social and economic development. There is merit in this argument, but such debt relief would have to be managed in ways that do not prompt irresponsible borrowing in the future.

9.4   WORLD TRADE ORGANIZATION

Environmental issues and trade in goods and services are intertwined in ways both good and bad. Trade in goods has historically generally brought economic prosperity to individuals, cities, and nations. Indeed, trade was essential for the economic growth of the fledgling United States. Economic prosperity promotes social development and cultural growth. Social development includes such positive benefits as job creation, more capital to invest in business enterprises, and more banking services. Economic prosperity can contribute to cultural growth by establishing educational institutions, libraries, fine arts, civil rights, and access to health care. In a positive environmental context, economic prosperity provides, through taxation, philanthropy, and other means, the resources to help finance the infrastructures needed for sewage treatment, air pollution control, water purification, food safety, waste management, and public health programs. On the negative side, numerous examples exist of abuses accompanying the production of goods to be traded. Of course, the most egregious, vile example was (sadly, still is) the trade in human slaves. Other negative examples include pollution generated by production of goods, disease caused by international transportation of goods, and occupational health and safety problems, including child labor, found in some workplaces.

The World Trade Organization functions as a voluntary body of nations. It seeks consensus on trade policies and has the authority to make binding decisions on trade disputes within its membership.

Following the end of World War II, nations were eager to stabilize economic development and trade. War had ravaged much of Europe, the USSR, Japan, and China. Even the United States, which had been largely spared physical damage to its infrastructure, needed economic recovery to pay for the monetary cost of World War II. The mechanism chosen for trade stabilization was the Global Agreement on Trade and Tariffs (GATT), created in 1947 by 23 trade-dependent nations, including the U.S. GATT provided the forum for resolving trade issues (e.g., tariff disputes over goods traded between the United States and Europe).

GATT was replaced by the World Trade Organization (WTO), which came into existence on January 1, 1995. Like GATT, the WTO, based in Geneva, Switzerland, functions as a voluntary body of nations. The WTO generally operates by seeking consensus among its member countries. Unlike GATT, WTO's decisions are binding and can be enforced by withdrawing trade benefits from a country that has violated WTO rules [69]. WTO members are expected to enact WTO policies and directions and accept findings on both general and specific trade issues. The WTO states, "The WTO is a rules-based, member-driven organization—all decisions are made by the member governments, and the rules are the outcome of negations among members" [70]. The organization's membership, as of April 2003, comprised 146 countries. WTO's functions are stated to be: (1) administering WTO trade agreements, (2) serving as a forum for trade negotiations, (3) handling trade disputes, (4) monitoring national trade policies, (5) providing assistance and training for developing countries, and (6) cooperating with other international organizations [ibid.]. Settling trade disputes is an important and often quite visible function of the WTO.

Dissatisfaction with the WTO has been expressed in words and deeds, e.g., demonstrations by environmental groups and organized labor. Some environmental groups believe that WTO decisions on trade have lessened environmental protections. For example, Friends of the Earth notes that U.S. restrictions on the import of shrimp from countries where fishermen catch shrimp with nets that kill endangered sea turtles were set aside by the WTO, which found the United States in violation of trade rules [71]. The effect of this WTO decision was to negate U.S. protection of an endangered species, sea turtles, in favor of commercial shrimpers with no evident environmental conscience. As a matter of U.S. domestic policy, WTO's authority to negate environmental and public health protections is very troubling. Do the benefits of "free trade" exceed the costs to environmental quality and public health? This is a calculus yet to be performed.

Organized labor in the United States has also expressed reservations about aspects of WTO's operation. One organization asserts, "[t]he push to reduce all trade barriers in all sectors has exacerbated social tensions, frayed social safety nets and highlighted national differences in labor laws and environmental protection. Problems with the WTO arise because its rules are seen as too intrusive by some countries (in overriding legitimate domestic laws) and because of the absence of rules in such crucial areas as labor rights. [B]ecause workers' rights (other than prison labor) are not included in WTO rules, countries may not withdraw trade preferences from WTO members even for egregious violation of workers' rights" [69]. As policy, how will the United States and other industrialized countries with well developed protections for workers' rights work to improve the WTO's record of lowering environmental and labor protections in the guise of "free trade"?

9.5 EUROPEAN UNION

This chapter has to this point focused on UN global agencies or non-UN agencies with global programs. There also exist organizations with regional focus. One of the most significant is the European Union (EU). The history of the EU derives from the need for ways to prevent wars among European nations. For centuries, Europe was the scene of frequent and bloody conflicts. In the twentieth century alone, political instability in Europe led to two world wars, conflicts that spilled out of Europe and engaged the United States and other countries in regional and global warfare. As noted by the EU [72], following World War II, a number of European leaders sought for ways to secure a lasting peace through treaties that bound their nations through economic and political means. In 1950, this led the French Foreign Minister Robert Schuman to propose that the coal and steel industries of Western Europe join in a cooperative arraignment that furthered the national interests of cooperating countries. As a result, in 1951, the European Coal and Steel Community (ECSC) was established, with six members: Belgium, France, Italy, Luxembourg, The Netherlands, and West Germany.[10] This international cooperation led to the formation of an independent, supranational body called the *High Authority*, an authority to formulate polices and take actions on the coal and steel industries in the six member countries [ibid.].

The ECSC was such a success that the six member countries agreed to go further in economic and political coordination. In 1957, they signed the Treaty of Rome, creating the European Atomic Energy Community and the European Economic Community. The six member states effected removal of trade barriers between themselves and forming a "common market." In 1967, a single Commission, a single Council of Members, and the European Parliament were created. Since that year, the number of member countries has steadily increased in number.

The EU serves as an umbrella for its member countries to cooperate in areas that include a single trade market, unified foreign policy, mutual recognition of national credentials, and exchange of information bases.

The EU, which was created by the Treaty of Maastricht, came into existence in November 1993. In 1995, the European Community was renamed the EU when the organization grew from 12 to 15 member states [73]. Ten more member states joined the EU in 2004. As of 2004, the member states are: Austria, Belgium, Cyprus, Czech Republic, Denmark, Estonia, Finland, France, Germany, Great Britain, Greece, Hungary, Italy, Latvia, Lithuania, Luxembourg, Malta, Poland, Portugal, Republic of Ireland, Slovakia, Slovenia, Spain, Sweden, and The Netherlands. The EU has adopted a common currency, the Euro, which is used across member states.

The EU serves as an umbrella for its member countries to cooperate in areas that include a single trade market, unified foreign policy, mutual recognition of national credentials, and exchange of information bases. This cooperation is pursued through a complex organizational structure, consisting of five EU institutions and flanked by five other important bodies [74]. The five EU institutions are the following:

- European Parliament (elected by the peoples of the Member States;
- Council of the European Union (representing the governments of the Member States);

- European Commission (driving force and executive body);
- Court of Justice (ensuring compliance with the law);
- Court of Auditors (controlling sound and lawful management of the EU budget) [ibid.].

The other five bodies of importance to the EU's organization and its program of work are:

- European Economic and Social Committee (expresses the opinions of organised civil society on economic and social issues);
- Committee of the Regions (expresses the opinions of regional and local authorities);
- European Central Bank (responsible for monetary policy and managing the euro);
- European Investment Bank (helps achieve EU objectives by financing investment projects) [ibid.].

The EU, its predecessor European Community, and some member states have generated a substantial amount of environmental legislation. *Regulations* and *directives* are the two most common outcomes of EU legislation. Regulations become law throughout the EU on their effective date, generally enforceable in each member state. EU directives, in distinction, are not directly and generally enforceable in member states [75]. Rather, directives set goals for the EU's member states to achieve through national legislation.

As discussed in chapter 1, the EU has adopted the precautionary principle as policy. This is not surprising, given its origins in Sweden and Germany. Moreover, the slow adoption in Europe of quantitative risk assessment, in contrast to the United States, created something of a void in how to prevent adverse public health and environmental consequences of environmental hazards. In 2000, the European Commission, the administrative arm of the EU, published precautionary principle guidelines for the EU's member states [76]. The guidelines were discussed in chapter 1.

The EU's adoption of the precautionary principle as policy has already had public health and economic consequences. An example of the former is found in an EU proposed program of chemical testing. In October 2003 the European Commission adopted legislation for a new EU regulatory framework for chemicals [77]. Called the Registration, Evaluation and Authorisation of Chemicals (REACH) framework [78], the proposal would require chemical manufacturers to conduct extensive safety tests over a span of eleven years on approximately 30,000 of the most common chemicals in the EU market for which toxicity data are lacking [79]. The Precautionary Principle was the driving force behind REACH, which has been vigorously opposed by the U.S. government [80], asserting that the U.S. Toxic Substances Control Act is adequate for testing chemicals that reach the United States The global chemical industry also opposes REACH, primarily because of the high cost of conducting toxicity tests. In November 2005, the European Parliament approved a modified version of REACH. The amended REACH program reduced the overall number of chemicals that would be required for testing by chemical producers [81]. EU's member states must approve the Parliament's legislation before a final REACH program is adopted throughout the EU. As environmental health policy, better toxicological databases of substances already in commerce will help make better regulatory decisions and provide improved programs of public health.

The EU's body of environmental legislation covers a broad range of environmental concerns and issues, including: air and water pollution, solid and hazardous wastes, noise pollution, radioactive waste, conservation of wild fauna and flora, urban waste treatment, freedom of information on the environment, expanded waste regulations, conducting environmental impact statements of projects, and establishment of the European Environment Agency [75]. As of the end of 2004, the EU has adopted more than two hundred environmental protection directives that are applied in all member states. Most of the directives are designed to prevent air and water pollution and encourage waste disposal. Other major issues include nature conservation and the supervision of dangerous industrial processes. The EU "[w]ants transport, industry, agriculture, fisheries, energy and tourism to be organised in such a way that they can be developed without destroying our natural resources—in short, sustainable development. We already have cleaner air because of the EU decisions in the 1990s to put catalytic converters into all cars and to get rid of the lead added to petrol" [74]. Waste management is an example of one set of EU environmental directives.

In 1990, the EU created the *European Environment Agency* (EEA), based in Copenhagen. It has been operational since 1994. The agency does not have regulatory authority but is designated by the EU to collect and disseminate information about the environment. The EEA's information resources are used by the EU's member states, the European Commission, the European Parliament, and the public, among others, when developing, adopting, implementing, and evaluating environmental policies [82]. The EEA states its mission as, "[t]o support sustainable development and to help achieve significant and measurable improvement in Europe's environment through the provision of timely, targeted, relevant and reliable information to policy making agents and the public" [ibid.].

In 2004, the EEA announced the release of Europe's first industrial pollution register, a database similar in concept and purpose to the U.S. Toxics Release Inventory. The European Pollutant Registry (EPER) contains reports from the 15 European Union countries plus Norway (not an EU member). The first edition of the EPER provides data on emission releases of 50 pollutants to air, water, and to offsite wastewater treatment facilities by approximately 9,000 industrial plants in 36 industrial sectors [7]. The report is intended to report about 90 percent of pollution point sources released from Europe's largest and most polluting industrial sources. According to the EEA, the EPER report is being used by environmental organizations to focus attention on specific sources of pollution. In this regard, the policy of using emissions data, such as in the EPER or the TRI, to bring pubic pressure on industrial sources of pollution will continue to be effective public policy.

Among the EEA's significant accomplishments is the preparation and distribution of an overall assessment of Europe's environment. The documents are prepared using member states' environmental data and that maintained by the EEA itself. The assessments are candid and constitute a database for EU policy development and revision. For example, the Third Assessment contains the following summary findings, "This, the third assessment, shows that most progress on environmental improvement continues to come from 'end-of-pipe' measures, actions under well-established international conventions and legislation, or as a result of economic recession and restructuring. [m]oving towards more sustainable approaches seems to be more aspiration than reality in many parts of Europe. [T]here has been less progress on implementation and substantial barriers to real progress remain, both political and financial" [83]. This kind of candid, specific analysis would be of enormous help to policy makers serious about improving environmental quality, based on a foundation of sustainable development.

9.6   INTERNATIONAL RANKINGS OF ENVIRONMENTAL STATUS

Environmental pollution is no respecter of national boundaries. It is for this reason that international agencies and between-nation treaties have been created, as described in this chapter. As national governments' environmental programs and policies mature, there will be a need to assess their effectiveness in reducing pollution and protecting the health of humans and ecosystems. Nascent efforts are underway to objectively compare environmental performance between countries. Such comparisons can presumably identify environmental programs that are succeeding and those that are not. From such analysis can come changes in international policy making (e.g., directing funds to countries where environmental progress is deemed to be inadequate).

One source that ranks national environmental programs is the World Economic Forum, an independent international organization incorporated as a Swiss not-for-profit foundation and has Non-Government Organization (NGO) consultative status with the Economic and Social Council of the United Nations. A collaborative project between the Forum, Yale University, and Columbia University has developed two indexes to gauge a country's environmental status. One index, called the Environmental Sustainability Index (ESI), has been used to rank national environmental programs on the basis of five categories: "environmental systems, environmental stresses, human vulnerability to environmental risks, a society's institutional capacity to respond to environmental threats, and a nation's stewardship of the shared resources of the global commons" [84].

Twenty key indicators were calculated across the five ESI categories. Factors such as urban air quality, water, and the strength of environmental regulation are among the twenty indicators that constitute the ESI, which in turn is built upon sixty-eight underlying databases. "The ESI distills a country's capacity for sustained environmental strength into a single number ranging from 0 to 100" [84]. The top five countries, according to ESI ranking in descending order are Finland, Norway, Sweden, Canada, and Switzerland, with ESIs ranging from 73.9 to 66.5. Finland was top ranked because of low levels of air and water pollution, its high institutional capacity to handle environmental problems, and its comparatively low emission of greenhouse gases. The United States ranked 45th among 142 countries, achieving good marks on controlling water pollution, but lagging in controlling greenhouse gas emissions and underperforming in reducing waste [ibid.].

A second index, called the Environmental Performance Index (EPI), is designed to measure current environmental results at the national scale [85]. The EPI is a complement to the ESI, covering "[a] broader range of conditions aimed at measuring long-term environmental prospects" [ibid.]. Four indicators comprise the EPI: air quality, water quality, climate change, and land protection. Each indicator has two to four variables that can be calculated, e.g., dissolved oxygen, phosphorus concentrations, and biological oxygen demand comprise the variables for water quality calculations. Like the ESI, the EPI ranges between 0–100. Unlike the ESI, calculation of a country's EPI requires the existence of specific databases, such as concentration of sulfur dioxide in outdoor air. Few countries possess such databases, yielding a ranking of only twenty-three countries. The top five ranked countries were Sweden, Switzerland, Finland, Austria, and Denmark, with EPIs ranging from 74.9 to 60.6. The United States ranked 14th, with an EPI of 44.1, due in measure to low scores on climate change and air pollution control. In 2006, the EPI was modified to include more data on sustainability measurements. Using the modified EPI, the United States ranked 28th among 68 countries for which enough data were available to derive an EPI [86]. Most countries in Western Europe, Japan, Taiwan, Malaysia, Costa Rica, and Chile all ranked ahead of the United States According to the report, "The United States placed 28th in the rankings [...] indicates that the United States is under-performing on critical issues such as renewable energy, greenhouse gas emissions, and water resources" [ibid.].

In reflecting on the year 2002 ESI and year 2002 EPI rankings, the investigators concluded that environmental performance is strongly influenced by patterns of environmental governance, independent of levels of national wealth. Moreover, "[u]nderstanding the dynamics of environmental governance is enhanced by explicit consideration of the role of the private sector" [85]. It remains to be seen how these kinds of rankings might affect national environmental policies. Although rankings can reveal weaknesses in the performance of a country's environmental programs, it remains for each country to set its own environmental course, within the confines of international treaties.

9.7 SUMMARY

It is surely now a cliché to observe that environmental pollution is no respecter of national boundaries. Filth dumped into an ocean in one region becomes part of the pollution load in another region. Similarly, the greenhouse gas carbon dioxide produced in abundance by industrialized countries adds to that produced in developing countries, the sum contributing to global warming. In response to the globalization of environmental hazards, the United Nations, acting primarily through the United Nations Environment Programme, convened global summits on the environment and development in 1992 in Rio de Janeiro, Brazil, and in 2002 in Johannesburg, South Africa. These summits produced plans and policies to deal with environmental problems of global, regional, and national scales. The work done within the UN framework represents the best hope for controlling global environmental hazards.

One issue that has surfaced at UN environmental summits is the disparity in pollution generation between the industrialized countries and the developing world. As debated in international meetings, an issue of the disparity has been cast as environ-mental parity. In particular, developing countries have argued that pollution controls, as products of industrialization, should not be imposed on countries where industrialization is nascent. In effect, this is an issue of environmental equity, which will be discussed in the following chapter.

9.8 POLICY QUESTIONS

1. Discuss the significance to you of the two world summits on environment and development. Using material in this chapter and elsewhere, discuss what you think the U.S. government's role should be in these kinds of summits.

2. Consider the global health risk factors listed in Table 9.1: (A) Identify those of relevance to local health departments in the United States. Discuss any assump-tions and conditions inherent in your selection of specific health risk factors. (B) Using the material in chapter 1 on responsibilities of local health departments, as a local health department decision maker, discuss which of your selected risk factors would be amenable to disease and disability prevention programs. (C) Briefly discuss the elements of such prevention programs.

3. The International Labour Organization is the only United Nations body that operates on a tripartite policy, which requires government, industry, and labor participation in all matters of policy and practices. Discuss, using critical think-ing, whether the World Health Organization could also operate on a tripartite arrangement (e.g., government, commercial interests, at-risk populations).

4. Discuss and contrast the global environmental health impacts of the World Bank and the World Trade Organization. Does either organization's policies affect you as an individual? If so, how?

5. The International Agency for Research on Cancer (IARC) prepares monographs on individual chemicals of public health and environmental importance. The monographs are prepared by IARC appointed work groups. Should scientists with industrial affiliations become members of these committees? Why? Why not?

6. Using Internet resources, select a specific environmental topic and track it through the EU legislative process, including any actions taken by national legislative bodies.

7. It was asserted in this chapter that failure of the United States government to honor the Kyoto Protocol is inconsistent with the public's right-to-know policy. Select any two of the other public policy expectations (chapter 1) and discuss the implications of the U.S. government's inaction on each policy.

8. Using Internet and library resources, develop a database on global warming that would be purposeful for public health usage.

9. *Agenda 21* was the primary product of the 1992 United Nations Conference on Environment and Development, which became known as the Earth Summit. Using Internet and library resources, access a copy of *Agenda 21* and discuss its relevance to a local health department.

10. Discuss the Convention on Persistent Organic Pollutants and its public health significance.

NOTES

1. President George H.W. Bush led the U.S. delegation.
2. Carbon dioxide, methane, nitrous oxide, perflurocarbons (PFCs), hydroflurocarbons (HFCs), and sulfur hexafluoride (SF_6)
3. 1 tonne = 1,000 kilograms.
4. President George W. Bush did not attend.
5. Not shown are the Initiatives that accompany each set of commitments.
6. Aldrin, chlordane, DDT, dieldrin, endrin, heptachlor, hexachlorobenzene, mirex, toxaphene, PCBs, dioxins, furans [38].
7. The Rotterdam PIC Convention and the LRTAP POPs Protocol [42].
8. Riparian: relating to or living or located on the bank of a natural watercourse or of a lake or tidewater [65].
9. Gen. George C. Marshall served as Chief of Staff, U.S. Army, throughout WW II [67].
10. At the end of World War II, the Allies divided Germany into two parts: West Germany and East Germany, forming two independent countries. West Germany was generally allied with the U.S. and its European neighbors; East Germany was allied with the USSR. As the USSR weakened as a political structure in the early 1990s, the two German nations reunified in October 1990, constituting the current nation of Germany.

REFERENCES

1. Bruntland, G. ed., *Our Common Future*, Oxford University Press, Oxford, 1987.
2. WHO (World Health Organization), Draft Definition Developed at a WHO Consultation in Sofia, Bulgaria, World Health Organization, Geneva, 1993.
3. UN (United Nations, About the United Nations/history. Available at http://www.un.org/aboutun/history.htm, 2002.
4. UN (United Nations, The UN in brief: IV. Protecting our common environment. Available at http://www.un.org/overview/brief.html, 2002.
5. UNEP (United Nations Environment Programme), Mission Statement. Available at http://www.unep.org/about.asp, 2000.
6. UNEP (United Nations Environment Programme), United Nations Conference on Environment and Development: Agenda 21, United Nations, New York, 1992.
7. ENS (Environmental News Service), Europe publishes its first pollution register, London. February 25, 2004.
8. LaDou, J., Deadly migration, *Technology Reiew.* (July), 47, 1991.
9. UN (United Nations), United Nations Sustainable Development, Agenda 21–Chapter 6. Available at http://www.un.org/esa/sustdev/agenda21.htm, 2002.
10. CNN (Cable News Network), Air Force grounds fleet of B-2 stealth bombers. Available at http://www.cnn.com/US/9808/06/b2.grounded/, August 6, 1998.
11. PSR (Physicians for Social Responsibility), Global Climate Change and Human Health, Washington, D.C., 2001.
12. UNFCC (U.N. Framework Convention for Climate Change). Available at http://www.cop4.org/conv/beginner.html, 2002.
13. EPA (U.S. Environmental Protection Agency), Global warming—emissions. Available at http://yosemite.epa.gov/oar/globalwarming.nsf/content/Emissions.html, 2002.
14. Keppler, F. et al., Methane emission from terrestrial plants under aerobic conditions, *Nature*, 439, 187, 2006.
15. UNFCC (U.N. Framework Convention Climate Change). Text of the Convention. Available

at http://unfccc.int/resource/convkp.html, 2003.

16. UN (United Nations), Kyoto Protocol receives 100th ratification, press release, Framework Convention on Climate Change—Secretariat, Bonn, Germany, December 18, 2002.

17. Mydans, S. and Revkin, A.C., With Russia's nod, treaty on emissions clears last hurdle, *New York Times*, October 1, 2004.

18. EPA (U.S. Environmental Protection Agency), Global warming. Available at http://yosemite.epa.gov/oar/globalwarming.nsf/webprintview/ClimateUncertainties.html, 2005.

19. Edmonds, S., Analysis—global warming may take economic toll, PlanetArk. Available at http://www.planetark.com/avantgo/dailynewsstory.cfm?newsid=32025, 2005.

20. USDoS (U.S. Department of State), US Climate Action Report, Washington, D.C. May, 2002.

21. WHO (World Health Organization), Climate and Health Fact Sheet, July 2005. Geneva, 2005.

22. Royal Society,. Global Acidification due to Increasing Atmospheric Carbon Dioxide, London, 2005.

23. Corvalan, C.F. and Patz, J.A.,. Global warming kills trees, and people. *Bulletin of the World Health Organization*, 82(7), 401, 2004.

24. NSF (National Science Foundation), Climate change drives amphibian extinctions, scientists say, news release, Washington, D.C., January 12, 2006.

25. Revkin, A.C., Rise in gases unmatched by a history in ancient ice, *New York Times*, November 25, 2005.

26. Pickoff-White, L., Climate and storms break records in 2005, news release, The National Academies, Washington, D.C., December 30, 2005.

27. UN (United Nations), Press summary of the Secretary-General's Report on Implementing Agenda 21, United Nations Department of Public Information, New York, NY, September 4, 2002.

28. UN (United Nations), Johannesburg Summit 2002, Highlights of commitments and implementation, United Nations Department of Public Information, United Nations, New York, September 3, 2002.

29. UN (United Nations), Sustainable development summit concludes in Johannesburg: UN Secretary-General Kofi Annan says it's just the beginning, United Nations Department of Public Information, New York, September 4, 2002.

30. WWF (World Wildlife Federation), WSSD—looking back at events. Available at http://www.panda.org/about_wwf/what_we_do/policy_and events/wssd/, 2002.

31. ASIL (American Society of International Law), International legal materials. *American Society of International Law*, 28, 31, 1989.

32. ILM (International Legal Materials), United Nations document IG.80/3, final act and text of the Basel Convention, *American Society of International Law*, 28, 2, 1989.

33. UNEP (United Nations Environment Programme), Basel Convention on the Control of Transboundary Movements of Hazardous Wastes and Their Disposal, fact sheet, Geneva, 1996.

34. EPA (U.S. Environmental Protection Agency), Disposal of polychlorinated biphenyls; import for disposal, *Federal Register*, 61, 11095, 1996.

35. BNA (Bureau of National Affairs), Senate ratifies Basel Convention on transboundary shipments of waste, *Environmental Reporter*, August 21, 1992, 1255.

36. Johnson, S., The Basel Convention: The Shape of Things to Come for United States Waste Exports. Environmental Law 1991, Northwestern School of Law of Lewis & Clark College, Chicago, 1991.

37. Bowers, P., Developing nations used as dumps: Industrial waste proves harmful to inhabitants. *Washington Times*, December 9, 1996, A14.

38. EPA (U.S. Environmental Protection Agency), Persistent organic pollutants (POPs). Available at http://www.epa.gov/oppfead1/international/pops.htm, 2004.

39. POPs (Persistent Organic Pollutants), Stockholm Convention on Persistent Organic Pollutants. Available at http://www.pops.int/documents/context/convtext_en.pdf, 2001.

40. *New York Times*, Global treaty takes effect without U.S., May 17, 2004.

41. EPA (U.S. Environmental Protection Agency), Environmental news press release, Washington, D.C., April 11, 2002.

42. EPA Press Advisory, Legislation to reduce global persistent organic pollutants receives praise, Washington, D.C., December 19, 2005.

43. WHO (World Health Organization), Overview of WHO. Available at http://www.who.int/about/overview/en, 2002.

44. Yamey, G., WHO in 2002. Have the latest reforms reversed WHO's decline?, *British Medical Journal*, 325, 1107, 2002.

45. U.S. Senate. Appropriations. Available at http://appropriations.senate.gov/hearmarkups/07-14-05PRLaborHFull.htm-A663, 2005.

46. Stein, R., SARS prompts WHO to seek more power to fight disease, *Washington Post*, Washington, D.C., May 18, 2003, A10.

47. WHO (World Health Organization), The World Health Report 2002, Geneva, 2002.

48. World Bank, World Development Report: Investing in Health, Washington, D.C., 1993.

49. Homedes, N., The disability-adjusted life year (DALY) definition, measurement and potential use. Available at http://www.worldbank.org/html/extdr/hnp/hddflash/workp/wp_00068.html, 1995.

50. WHO (World Health Organization), Updated Status of the WHO Framework Convention on Tobacco Control, Geneva, 2004.

51. Kaufman, M., U.S. signs tobacco control treaty, *Washington Post*, Washington, D.C., May 12, 2004.

52. Gaudin, N., 2002. Personal communication.

53. IARC (International Agency for Research on Cancer). 12. Evaluation. Available at http://www.cie.iarc.fr/monoeval/eval/html, 2002.

54. Huff, J., IARC monographs, industry influence, and upgrading, downgrading, and undergrading chemicals, *International Journal Occupational & Environmental Health*, 8, 249, 2002.

55. IPCS (International Programme on Chemical Safety), About IPCS. Available at http://www.who.int/pcs/html, 2002.

56. WHO (World Health Organization), Assessing Human Health Risks of Chemicals: Derivation of Guidance Values for Health-Based Exposure Limits, Environmental Health Criteria No. 170, International Programme on Chemical Safety, Geneva, 1994.

57. IPCS (International Programme on Chemical Safety), Methodological Publications. Available at http://www.who.int/pcs/pubs, 2002.

58. IPCS (International Programme on Chemical Safety), INCHEM. Available at http://www.inchem.org, 2002.

59. IPCS (International Programme on Chemical Safety), INTOX. Available at http://www.intox.org, 2002.

60. WHO (World Health Organization), European Charter on Environment and Health, World Health Organization, Regional Office for Europe, Copenhagen, 1989.

61. ILO (International Labour Organization), About the ILO. Available at http://www.ilo.org/public/english/about/index.htm, 2002.

62. FAO (Food and Agriculture Organization), FAO at Work. Available at http://www.fao.org, 2004.

63. UNECE (United Nations Economic Commission for Europe), About the UNECE. Available at http://www.unece.org/about/about.htm, 2005.

64. UNECE (United Nations Economic Commission for Europe), Convention on the Protection and Use of Transboundary Wastecourses and International Lakes. Available at http://www.unece.org/env/water/pdf/watercon.pdf, 1992.

65. *Webster's Ninth New Collegiate Dictionary*, Merriam-Webster Publishers, Springfield, 1986.

66. WHO (World Health Organization), Treaty to prevent water-related diseases in Europe enters into force, press release EURO/12/5, WHO Regional Office for Europe, Copenhagen, August 3, 2005.

67. LOC (Library of Congress), Introduction to For European Recovery: The Fiftieth Anniversary of the Marshall Plan, Washington, D.C., 2002.

68. World Bank, Homepage. Available at http://web.worldbank.org, 2002.

69. AFL-CIO (American Federation of Labor-Congress of Industrial Organizations), *What is the WTO?* Available at http://www.aflcio.org/issuespolitics/globaleconomy/whatis.cfm?RenderFor Print-1, 2003.

70. WTO (World Trade Organization), The WTO. Available at http://www.wto.org, 2003.

71. FOE (Friends of the Earth), WTO scorecard. Available at http://www.foe.org, 2003.

72. EU (European Union), The History of the European Union, Brussels, 2004.

73. EU (European Union), Informational Profiles on European Union Environment Programmes, Brussels, 1997, 243.

74. EU (European Union). The European Union at a Glance. Available at http://europa.eu.int/abc/index_en.htm, 2004.

75. Smith, T.T. and Hunter, R.D., The European Community environmental legal system, in European Community Deskbook, Environmental Law Institute, Washington, D.C., 1992, 3.

76. EC (European Commission), Communication from the Commission on the Precautionary Principle, COM(2000) 1 final, Commission of the European Communities, Brussels, 2000.

77. EU (European Union), The new EU chemicals legislation—REACH. Available at http://europa.eu.int/comm/enterprise/reach/overview.htm, 2005.

78. EC (European Commission), REACH in Brief. Brussels, Belgium. September 9, 2004.

79. *Science*, E.U. starts a chemical reaction, *Science*, 300(5618), 405, April, 18, 2003.

80. WSJ (*Wall Street Journal*), U.S. opposes EU effort to test chemicals for health effects, September 9, 2004.

81. Reuters, Parliament backs new EU law on toxic chemicals, November 17, 2005.

82. EEA (European Environmental Agency), The European Environment Agency: Who We Are, What We Do, How We Do It, Copenhagen, 2004.

83. EEA (European Environmental Agency), Europe's Environment: The Third Assessment. Copenhagen, 2003.

84. CIESIN (Center for International Earth Sciences Information Network), Environmental sustainability index press release, Columbia University, New York, 2002.

85. Yale University,. Pilot Environmental Performance Index. An Initiative of the Global Leaders of Tomorrow Environment Task Force, World Economic Forum, Yale Center for Environmental Law and Policy, New Haven, 2002.

86. Yale University, New Zealand tops new environmental scorecard at World Economic Forum in Davos, Environment Task Force, World Economic Forum, Yale Center for Environmental Law and Policy. New Haven, 2006.

87. EPA (U.S. Environmental Protection Agency), Technology Transfer Network: National Air Toxics Assessment. Glossary of Key Terms, Office of Air and Radiation, Washington, D.C., 2006.

10 Policy Impacts of Environmental Equity/Justice

One man's justice is another's injustice.
Ralph Waldo Emerson, *Essays: First Series*

Injustice anywhere is a threat to justice everywhere.
Rev. Martin Luther King, Jr., *Letter from Birmingham City Jail*

10.1 INTRODUCTION

A major environmental health concern arose in the 1970s, as will be subsequently discussed in this chapter. The concern is called *environmental justice*, although other terms have been used. EPA's current definition of environmental justice is "[t]he fair treatment and meaningful involvement of all people regardless of race, color, national origin, or income with respect to the development, implementation, and enforcement of environmental laws, regulations, and policies" [1]. As the definition implies, fair treatment lies at the heart of environmental justice. But fair treatment in what and whose context? The answer to this essential question requires awareness of the history of environmental justice, which will be discussed later in this chapter. However, it can be stated at this point that fair treatment involves the prevention of inequitable distribution of environmental hazards across the segments that comprise a society.

As a prelude to comments on the environment's impact on people of color and on communities challenged by economic and social conditions, the generally inferior health status (compared with Caucasians) of people of color, particularly those of African-American descent, should be borne in mind. One way of depicting the disparity is to calculate the excess deaths experienced by African Americans beyond those for Caucasian Americans [2]. In 1990, 75,000 excess deaths occurred among African Americans. The primary causes for the excess are heart disease and stroke, cancer, cirrhosis, diabetes, homicides, unintentional injuries, infant mortality, and AIDS [ibid.]. This same source comments, "Although many associations with these excess deaths, such as income, education, access to care, and preventive practices, have been determined, the basic causal factors associated with ill health and race remain unanswered."

The consensus view is that much of the excess morbidity in people of color is still unexplained. This assertion includes the uncertain degree to which environmental hazards contribute to excess morbidity in these populations. What is known, as will be subsequently described, is that in the United States some environmental hazards are experienced more often by minority populations than by white populations.

This chapter will discuss the emergence of what is most commonly called environmental justice, although other terms have been used to describe the alleged imposition of environmental hazards on communities of color. As discussed in this chapter, the history of environmental justice is intertwined with issues of waste management, that is, allegations that hazardous waste, in particular, was deliberately targeted for storage or processing in communities of color. These allegations stimulated a series of demographic investigations of persons residing near hazardous waste sites. Findings from these studies are described, with a synthesis of them. The policy implications of environmental justice are presented and conclude the chapter.

10.2   A MATTER OF DEFINITION

Several terms have been used to characterize the social condition of unequal distribution of environmental hazards, especially when experienced by people of color or groups with low income. One term is called *environmental equity*. Other terms that have been used by community groups and some environmental organizations are *environmental racism* and *environmental inequity* [3]. The term *environmental justice* is the term now favored by many grassroots groups, some government agencies, and elected officials. Fairness is the core concern of both environmental equity and environmental justice. As a consequence of individuals' differences, a healthy democracy that treats people fairly may not always be able to treat them equally. Although equity and justice are both rooted in a concern for fair treatment, equity seems more directly synonymous with fairness than does justice. Justice has a more litigious image than does equity, and litigation does not always result in fair outcomes. Moreover, equity seems to connote a more prospective approach—actions that will guard against inequities in how environmental hazards are shared. Justice seems more retrospective—actions that give emphasis to redressing past actions that imposed disproportionate shares of environmental hazards.

Case example: Navajo Nation CERCLA sites

The Navajo Nation and the indigenous people of America hold a deep reverence for the earth and the environment [5]. The Navajo Nation consists of twenty-five thousand square miles of tribal land in Arizona, New Mexico, and Utah. Because of uranium mining, the Navajo Nation contains several hundred abandoned uranium mines. As a matter of environmental inequity, inadequate protection of tribal land occurred as mines were developed and later abandoned after mining operations ceased. The abandoned mines pose severe threats to human health and the environment [ibid.]. Uranium mining on Navajo lands occurred from the 1940s to the early 1990s. Open-pit surface mining was common, leading to abandoned pits. The Navajo Superfund Program has evaluated the hazards posed by these abandoned uranium mines. They include: exposure to ionizing radiation, inhalation of radon gas, ingestion of contaminated ground and surface water, direct ingestion of site contaminants, ingestion of contaminated food in contact with releases from sites, and physical hazards. Efforts are on-going to remediate abandoned mines and remove hazardous waste.

EPA gave a different argument in favor of the term environmental equity in their report on reducing risk for all communities [4]: "EPA chose the term environmental equity because it most readily lends itself to scientific risk analysis. The distribution of environmental risks is often measurable and can be quantified. The agency can act on inequities based on scientific data. Evaluating the existence of injustices and racism is more difficult because they take into account socioeconomic factors in addition to the distribution of environmental benefits that are beyond the scope of this report. Furthermore, environmental equity, in contrast to environmental racism, includes the disproportionate risk burden placed on any population group, as defined by gender, age, income, as well as race." In sum, EPA's preference in 1992 for the term environmental equity was based on their belief that measurability (i.e., of equity) was important.

The state of Washington, as did EPA in 1992, chose the term environmental equity "[b]ased upon the connotation that the word 'equity' better relates to something measured, as opposed to 'justice'" [6]. Both the state of Washington and the EPA define environmental equity as, "[t]he proportionate and equitable distribution of environmental benefits and risks among diverse economic and cultural communities. It ensures that policies, activities, and the responses of government entities do not differentially impact diverse social and economic groups. Environmental equity promotes a safe and healthy environment for all people."

However, the Protocol Committee that organized the National Environmental Justice Conference of 1994, held in Washington, D.C. asserted, "Environmental justice encompasses more than equal protection under environmental laws (environmental equity). It upholds those cultural norms and values, rules, regulations, and policies or decisions to support sustainable communities, where people can interact with confidence that their environment is safe, nurturing, and productive. Environmental justice is served when people can realize their highest potential, without experiencing sexism, racism and class bias. Environmental justice is supported by clean air, water and soil; sufficient, diverse and nutritious food; decent paying and safe jobs; quality schools and recreation; decent housing and adequate health care. Environmental justice is supported by democratic decision-making and personal empowerment; and communities free of violence, drugs, and poverty and where both cultural and biological diversity are respected" [7]. This definition of environmental justice is more expansive and idealistic than that used to define environmental equity. Moreover, the committee's concept of environmental justice does not seem as amenable to measurement as do the definitions of environmental equity.

The National Environmental Justice Conference had considerable impact on EPA, leading it to move away from the term environmental equity in favor of environmental justice. In 1995, as a consequence of a Presidential Executive Order on environmental justice that was signed by President Clinton during the conference, EPA developed working definitions for environmental justice and fair treatment [8]:

- Environmental Justice—means the fair treatment and meaningful involvement of all people, regardless of race, ethnicity, culture, income, or education level with respect to the development, implementation, and enforcement of environmental laws, regulations, and policies.
- Fair Treatment—means that no population, due to political or economic disempowerment, is forced to shoulder the negative human health and environmental impacts of pollution or other environmental hazards.

EPA's current definition of environmental justice, which can be found on their Web site, is "Environmental Justice is the fair treatment and meaningful involvement of all people regardless of race, color, national origin, or income with respect to the development, implementation, and enforcement of environmental laws, regulations, and policies." [1]. This definition is similar to the agency's 1995 definition. In addition to their definition, EPA elaborates what can be considered as their operational policy on environmental justice, "EPA has this goal [i.e., environmental justice] for all communities and persons across this Nation. It will be achieved when everyone enjoys the same degree of protection from environmental and health hazards and equal access to the decision-making process to have a healthy environment in which to live, learn, and work" [ibid.]. This policy statement implies a preference by the EPA for environmental equity, even though the policy statement follows the agency's definition of environmental justice.

The language of environmental equity and justice is apparently still evolving. For example, the Presidential/Congressional Commission on Risk Assessment and Risk Management (CRARM) defined environmental justice as, "Concern about the disproportionate occurrence of pollution and potential pollution-related health effects affecting low-income, cultural, and ethnic populations and lesser cleanup efforts in their communities" [9]. This definition seems directed to the Comprehensive Environmental Response, Compensation, and Liability Act's purposes, given the mention of cleanup efforts, and therefore would be more limited in application.

The preferred nomenclature seems dependent on the ethos and motives of the group using the term. Although this chapter is titled "Environmental Equity/Justice," much of what is summarized here and expressed elsewhere in reports and publications uses the term *environmental justice*, which will be the preferred term used throughout this chapter.

Environmental justice addresses the question of whether environmental risk factors are unfairly shared by all segments of a population. More specifically, are minority groups and persons of low income disproportionately exposed to environmental risks because of racist policies and actions by government and private sector agents (e.g., [10])? Some history is instructive.

10.3　HISTORY

Pinpointing any one event that triggered what became the environmental justice movement is difficult. As seen in much of the U.S. civil rights movement and social justice crusade, societal change occurs as the result of many events acting over time. A societal movement occurs if sufficient supportive public opinion develops and enough supporters are drawn to the cause. The movement then mobilizes its energy to prevail upon institutions, often governmental bodies, that can effectuate change. Federal voting rights legislation and ordinances that protect against sexual discrimination are examples of such changes.

The U.S. environmental justice movement is like a tree with many roots and branches. It is not yet a mature tree and its survival will depend on many factors. Certainly one factor will be public perception. Will the general public support environmental justice actions and at what cost? The tree has not grown enough to provide an answer. What are the seeds of our metaphoric tree called environmental justice?

10.3.1　Warren County, North Carolina, Protest

Several key events have shaped the environmental justice movement. If any one event can be termed "the" event, it occurred in Warren County, North Carolina, in 1982 [13]. Local opposition arose when the state announced its proposal to locate a hazardous waste facility in this county, which had a large African-American population. The landfill was targeted to receive PCB-contaminated waste [14] at a site near the community of Afton, which was 84 percent African American [10]. When protests very similar to those of the U.S. civil rights movement of the 1960s resulted, and more than five hundred persons were arrested, the protests and arrests caught the attention of national news media. Despite the protests and media attention, delivery began of hazardous waste to the landfill.

The final attempt to stop the landfill occurred in July 1982. The local chapter of the National Association for the Advancement of Colored People sought a preliminary injunction in federal court to prohibit placement of PCBs in the Warren County landfill. The court denied the request, stating that race was not an issue in siting the landfill because race was never mentioned as a motivating factor throughout all federal and state hearings and private party suits [15]. Although community opposition, national media attention, and legal proceedings did not halt construction and use of the landfill, the civil rights demonstrations and their aftermath mobilized attention on a new issue, which some called environmental racism [16].

The events in North Carolina led Walter Fauntroy (D-DC), District of Columbia Delegate to Congress, and Congressman James Florio (D-NJ) to ask the General Accounting Office (GAO)[1] in 1982 to assess the racial implications of facilities in the southern states that received hazardous waste. The GAO found African Americans were the predominate population living near three of the four largest facilities in the South [18]. That gave weight to the belief that landfills were deliberately being targeted for location in minority communities.

Issues that came from the Warren County, North Carolina, PCB landfill siting set into motion a series of events that have shaped the current environmental justice movement. Ten key events that shaped the movement are summarized in Table 10.1. Subsequent sections in this chapter describe the nature and importance of the events listed, but for historical perspective, one report and four events merit elaboration here.

Case example: Mojave Desert Hispanic community

Health studies of minority communities at potential health risk from exposure to hazardous waste must be conducted in ways that respect cultural sensibilities. An example is a study conducted by the California Department of Health Services of a small Hispanic community in the Mojave Desert [19]. The Department had found quite high levels of dioxin in the ash from a junkyard smelter where insulated wire was burned to reclaim the copper. This led to concerns that exposure to dioxin had occurred among the residents. A protocol was designed to collect blood samples for dioxin analysis. The state health department prepared all materials in both English and Spanish. A door-to-door survey was conducted bilingually. Small meetings were held with local residents to explain all blood sampling procedures. Special efforts were taken to overcome mistrust of government agencies. However, despite all these efforts by the study team, only 3 of 41 persons eligible for measurement of blood dioxin levels participated. Investigators believed that the large amount of blood, 1 pint, was a deterrent. Nonetheless, the culturally sensitive efforts of the research team were commendable.

10.3.2 BULLARD'S BOOK AND THESIS

The events in Warren County, North Carolina, gave impetus to research that Robert D. Bullard, a sociologist, published as *Dumping in Dixie: Race, Class, and Environmental Quality* [10,20]. The former book [10], which quickly became a cardinal, seminal work within the civil rights movement, focused on five African-American communities struggling with environmental problems: Houston, Texas; Dallas, Texas; Institute, West Virginia; Alsen, Louisiana; and Emelle, Alabama. Bullard described in detail the concerns of local residents and the potential health risk ascribed by them to the presence of hazardous waste landfills or operating chemical plants.

Bullard's book contains methods of dispute resolution and grassroots strategies that can be used to counter environmental inequities. This work places him at the center of scholars who developed the intellectual framework that constitutes environmental justice [10,14,16,20] and has contributed much to improving social and environmental policies intended to prevent environmentally discriminatory actions. Bullard's published work is essential reading for persons seeking basic information about environmental justice.

TABLE 10.1

Key Events That Shaped the Environmental Justice Movement

Event/Year	Impact of Event	Event/Year	Impact of Event
1. Warren County, NC, civil rights opposition to a proposed landfill/1982	Brought national attention via news media and made environmental concerns a matter of civil rights	6. ATSDR's National Conference on Minorities and Environmental Pollution/1990	First federal conference to focus on research findings and gaps in knowledge about health effects of environmental hazards on minorities
2. GAO study of 4 hazardous waste landfills/1983	Elevated to congressional attention the potential inequity of placing hazardous waste facilities in areas that have large minority populations	7. First National People of Color Environmental Leadership Summit I/1991	First national conference of environmental justice activists; developed first set of Principles of Environmental Justice
3. Release of Commission on Social Justice study of minorities and waste facilities/1987	Report gave credence to concerns that minorities were over represented in areas around waste sites; became seminal document within civil rights and social justice movements	8. National Conference on Environmental Justice/1994	Second federal conference specific to environmental justice; developed strategies for environmental justice pursuits; led to Presidential Executive Order
4. University of Michigan Natural Resources Conference and formation of the Michigan Coalition/1990	Michigan Coalition had major impact on EPA's recognition of environmental equity as a concern [11]	9. Issuance of Presidential Executive Order on environmental justice/1994	Provided resources, authority, and imprimatur of federal government
5. Publication of the book *Dumping in Dixie*/1990	First academically based report of patterns of environmental inequities according to race	10. Second National People of Color Environmental Leadership Summit II/[12]	Reaffirmed Principles of Environmental Justice from Summit I

10.3.3 KEY CONFERENCES/MEETINGS

Several conferences have had great impact on the evolution of environmental justice by helping identify disparities according to race, income, or culture. Each influenced subsequent actions that advanced environmental justice and justice policies, particularly those of federal government agencies.

10.3.3.1 University of Michigan Natural Resources Conference of 1990

The debate over environmental justice had advanced sufficiently by 1990 to warrant a conference on race and environmental hazards. In January of that year, the University of Michigan's School of Natural Resources convened scholar-activists in a national conference to address the distribution and management of environmental risk [11,21].

Nine of twelve scholars who presented papers were minorities, marking the first environmental justice conference where the majority of presenters of scholarly papers were people of color [21]. The Michigan conference resulted in a compilation of papers that advanced the debate about race and environmental justice.

However, the most important outcome of the Michigan conference was the creation of what became known as the Michigan Coalition, a subgroup of conferees who composed an agenda for environmental justice and conducted a series of meetings with senior federal government officials to present their agenda. According to EPA Administrator William Reilly, "It was the arguments of this group that prompted me to create the Environmental Equity Workgroup" [10]. In turn, the EPA Equity Workgroup evaluated environmental risk and race data and produced the report *Environmental Equity: Reducing Risk for All Communities*. This EPA report, which is described in this chapter, was the federal government's first official expression on environmental justice and became a primary resource for environmental justice advocates who lobbied for government support and action.

10.3.3.2   ATSDR Environmental Justice Conference of 1990

In 1990, ATSDR organized the first federal conference on minority health and environmental contamination [22]. The four hundred participants were primarily researchers and investigators from government agencies and universities. The conference concentrated on adverse health effects of hazardous substances in minority populations, educational needs of low-income communities, and improvements needed in risk assessment to account for potential disproportionate impact of hazardous substances on minorities. The meeting resulted in agreement that minorities were at increased health risk from various environmental hazards, that risk assessments should integrate concern for minorities and susceptible populations (e.g., children), and that additional research and data collection were warranted.

10.3.3.3   First National People of Color Environmental Leadership Summit of 1991

Two national meetings have been organized and conducted by a coalition of environmental justice advocates. The coalition included grassroots and community groups, labor organizations, civil rights organizations, tribal representatives, cultural representatives, and feminists, among others. The First National People of Color Environmental Leadership Summit occurred in 1991, held in Washington, D.C. [23]. In later years, the meeting became known as Summit I. Approximately 1,000 persons attended. Summit I was a historic meeting in several regards. It was the first national environmental justice meeting organized by, and focused on, activists concerned with the personal and social issues of environmental hazards imposed upon people of color. Summit I set into motion a national network of groups committed to environmental justice goals and practices.

Delegates to Summit I produced a set of Principles of Environmental Justice (Table 10.2), a sweeping declaration of socioeconomic and political statements. The 17

TABLE 10.2
Principles of Environmental Justice from the First National People of Color Environmental Leadership Summit [23]

We, the People of Color, are gathered together at this First National People of Color Environmental Leadership Summit, to begin to build a national movement of all peoples of color to fight the destruction of our lands and communities, do hereby reestablish our spiritual interdependence to the sacredness of our Mother Earth; we respect and celebrate each of our cultures, languages and beliefs about the natural world and our roles in healing ourselves; to insure environmental justice; to promote economic alternatives which would contribute to the development of environmentally safe livelihoods; and to secure our political, economic and cultural liberation that has been denied for over 500 years of colonization and oppression, resulting in the poisoning of our communities and land and the genocide of our peoples, do affirm and adopt these Principles of Environmental Justice.

1. Environmental justice affirms the sacredness of Mother Earth, ecological unity and the interdependence of all species, and the right to be free from ecological destruction.
2. Environmental justice demands that public policy be based on mutual respect and justice for all peoples, free from any form of discrimination or bias.
3. Environmental justice mandates the right to ethical, balanced and responsible uses of land and renewable resources in the interest of a sustainable planet for humans and other living things.
4. Environmental justice calls for universal protection from extraction, production and disposal of toxic/hazardous wastes and poisons that threaten the fundamental right to clean air, land, water and food.
5. Environmental justice affirms the fundamental right to political, economic, cultural and environmental self-determination to all peoples.
6. Environmental justice demands the cessation of the production of all toxins, hazardous wastes, and radioactive substances, and that all past and current producers be held strictly accountable to the people for detoxification and the containment at the point of production.
7. Environmental justice demands the right to participate as equal partners at every level of decision-making including needs assessment, planning, implementation, enforcement and evaluation.
8. Environmental justice affirms the right of all workers to a safe and healthy work environment, without being forced to choose between an unsafe livelihood and unemployment It also affirms the right of those who work at home to be free from environmental hazards.
9. Environmental justice protects the rights of victims of environmental justice to receive full compensation and reparations for damages as well as quality health care.
10. Environmental justice considers governmental acts of environmental injustice a violation of international law, the Universal Declaration on Human Rights, and the United Nations Convention on Genocide.
11. Environmental justice recognizes the special legal relationship of Native Americans to the U.S. government through treaties, agreements, compacts, and covenants affirming their sovereignty and self-determination.
12. Environmental justice affirms the need for an urban and rural ecology to clean up and rebuild our cities and rural areas in balance with nature, honoring the cultural integrity of all our communities, and providing fair access for all to the full range of resources.
13. Environmental justice calls for the strict enforcement of principles of informed consent, and a halt to the testing of experimental reproductive and medical procedures and vaccinations on people of color.
14. Environmental justice opposes the destructive operations of multi-national corporations.
15. Environmental justice opposes military occupations, repression and exploitation of lands, peoples and cultures.
16. Environmental justice calls for the education of present and future generations which emphasizes social and environmental issues, based on our experiences and an application of our diverse cultural perspectives.
17. Environmental justice requires that we, as individuals, make personal and consumer choices to consume as little of Mother Earth's resources and to produce as little waste as possible, and make the conscious decision to challenge and reprioritize our lifestyles to insure the health of the natural world for present and future generations.

principles contain elements of sustainable development, Native American rights, pollution prevention, workers' health and safety, victims' compensation, informed consent, cultural involvement in decision making, and antiwar sentiment, among others. The set of principles represent an idealistic statement of how communities and people of color expect to be respected and involved in environmental decisions that can affect them. To what extent these principles can be adopted as policy by environmental and public health policy makers is unclear in most instances.

10.3.3.4   National Environmental Justice Conference of 1994

The ATSDR meeting of 1990 led to a much larger federal conference in 1994 that brought together 1,100 environmental justice advocates, state and federal government representatives, university researchers, and others. The meeting was preceded by issuance of a set of 10 review papers that helped shape dialogue during the conference [23]. The conduct of the meeting ranged from confrontation to conciliation. From the meeting came agreement on a set of five recommendations designed to impact government actions on environmental justice [7]:

> "I. Conduct meaningful health research in support of people of color and low-income communities. Preventing disease in all communities and providing universal access to health care are major goals of health care reform. Effective preventive measures cannot be equitably implemented in the absence of a targeted process that addresses the environmental health research needs of high risk workers and communities, especially communities of color.
>
> II. Promote disease prevention and pollution prevention strategies. Although treating disease and cleaning up environmental hazards are essential, long-term solutions must rely upon truly preventive approaches.
>
> III. Promote interagency coordination to ensure environmental justice. Although at-risk communities and workers are most threatened by occupational and environmental hazards, government agencies (federal, regional, state, local and tribal) are also important stakeholders. Unfortunately, environmental problems are not organized along departmental lines. Solutions require many agencies to work together effectively and efficiently.
>
> IV. Provide effective outreach, education and communications. Findings of community-based research projects should be produced and shared with community members and workers in ways that are sensitive and respectful to race, ethnicity, gender, and language, culture, and in ways that promote public health action.
>
> V. Design legislative and legal remedies." [No narrative accompanied this recommendation.]

These five elements were accompanied by specific strategies and activities that should be pursued by government and private sector entities. The elements have had substantive impact on the federal government's environmental justice strategies, as described later in this chapter.

During the conference, President Clinton issued an executive order on environmental justice, which is described in a following section. The long-term effect of the executive order remains to be determined, but clearly the national environmental justice conference held in 1994 played a large role in the issuance of the executive order.

The events in Table 10.1 were seeds that planted environmental justice in the orchard of civil rights. A series of studies followed that tried to better define minority groups and persons of low income who are at risk because of exposure to environmental hazards. Most of these studies concentrated on the demographics of populations living near uncontrolled hazardous waste sites or those facilities permitted to treat, store, and dispose of hazardous waste (TSDFs).

10.3.3.5   Second National People of Color Environmental Leadership Summit of 2002

A decade after the First National People of Color Environmental Leadership Summit was held in 1991, the second summit occurred in October 2002 [12]. Summit II, held for four days in Washington, D.C., commencing on October 28, brought together more than 1,200 delegates who represented community and grassroots organizations, civil rights groups, organized labor, and academic institutions, among others. Delegates recommended that environmental justice must be a top priority in the twenty-first century. Delegates also reaffirmed the Principles of Environmental Justice (Table 10.2), which had been developed at Summit I. Summit II enlarged the networking of environmental justice organizations and gave special attention to the training of future environmental justice leaders, particularly young persons.

10.3.4   FIVE SEMINAL REPORTS

Concerning environmental equity, are hazardous waste treatment, storage, and disposal facilities (TSDFs) and uncontrolled hazardous waste sites (i.e., CERCLA sites) found more often in minority communities than elsewhere? Moreover, regarding environmental justice, do data support the assertion that minority populations and persons of low income have been targeted for placement of TSDFs in their communities? Five reports were important for shaping opinions and actions on environmental justice. They are described in the following sections.

10.3.4.1   General Accounting Office (GAO) Study of 1983

In December 1982, following the Warren County episode, District of Columbia Delegate to Congress Walter Fauntroy (D-DC) and Congressman James Florio (D-NJ) requested

GAO to "[d]etermine the correlation between the location of hazardous waste landfills and the racial and economic status of the surrounding communities" [15]. According to GAO, agreement with the study's requesters led to examining sites only in the eight southeastern states. The agreement also included examining only off-site landfills, those not contiguous to industrial facilities.

GAO identified four operating landfills in the Southeast that were receiving hazardous waste: Chemical Waste Management, Sumter County, Alabama; Industrial Chemical Company, Chester County, Alabama; SCA Services, Sumter County, South Carolina; and the Warren County PCB Landfill, North Carolina. For each site, Bureau of Census racial and economic data of 1980 were obtained for census areas in which the landfills were located and the census areas that had borders within about four miles of the landfill.

GAO found that African Americans were the majority population in census areas at three of the four sites: Chemical Waste Management (CWM), Industrial Chemical Company (ICC), and Warren County PCB Landfill. At all four sites, African-American populations in census areas containing landfills had mean incomes lower than the mean income for all races combined in the same census area. For example, data indicate that the Warren County PCB Landfill was located in a census area with a population of 804 persons, of which 66 percent were African American. The landfill was located in a county with a population of 16,232, of which 60 percent were African American. The mean family income for all races in the landfill's census area was $10,367, compared with $9,285 mean family income for African Americans in the same census area.

The data show that percentages of African Americans in census tracts containing landfills generally mirrored the minority population of the counties in which the landfills were located. For all four sites, the mean family income for African Americans living within the landfills' census area was lower than the mean income for all races in the landfills' census areas. These data suggest that African Americans living near the four landfills had lower incomes than did other persons in the areas. However, little in the GAO report sheds light on what factors led to each site's location. Nonetheless, GAO's findings added weight to the belief that areas with high percentage of minorities were being targeted for location of hazardous waste sites.

10.3.4.2   United Church of Christ Report of 1987

Drawing from the GAO report, the United Church of Christ's Commission for Social Justice (CSJ) conducted two studies to determine racial and socioeconomic characteristics of persons in the United States living (1) in residential areas surrounding commercial TSDFs and (2) near uncontrolled hazardous waste sites [25].

The first CSJ study sought to determine whether the variables of race and socioeconomic status played significant roles in the location of commercial TSDFs. The methodological approach compared geographic characteristics presumed to be relevant to the siting of commercial hazardous waste facilities. The study analyzed five sets of national data: (1) minority percentage of the population, (2) mean household income, (3) mean value of owner-occupied homes, (4) number of uncontrolled hazardous waste sites per 1,000 persons, and (5) pounds of hazardous waste generated per person. Racial classifications were taken from the 1980 U.S. Census, and data on TSDFs were obtained from EPA databases.

Minority percentage of the population was used to measure racial composition of communities. Mean household income and mean value of owner-occupied homes were included in the analysis to determine whether socioeconomic factors were more important than race in locating commercial hazardous waste facilities. Existence of uncontrolled waste sites was evaluated to see whether underlying historic or geographic factors were associated with siting of commercial hazardous waste facilities in ways that were not accounted for by other variables used in the analysis.

Results of discriminant function analysis showed that minority percentage of the population was statistically significant in relation to the presence of commercial hazardous waste facilities. The percentage of community residents that belonged to particular racial and ethnic groups was a stronger predictor of the level of commercial hazardous waste activity than was household income, the value of homes, the number of uncontrolled toxic waste sites, or the estimated amount of hazardous wastes generated by industry. A key finding was that in zip code areas[2] having one commercial TSDF operating in 1986, the percentage minority population, on average, was twice that of areas that did not contain TSDFs.

The second CSJ study was descriptive in nature. Its primary purposes were (1) measure the number of racial and ethnic persons who lived in residential areas where uncontrolled hazardous waste sites were located and (2) make comparisons between the extent to which uncontrolled waste sites were located among different racial populations. Investigators used U.S. Census data for 1980 and data in EPA's national list of uncontrolled hazardous waste sites, which is called the Comprehensive Environmental Response, Compensation, and Liability Information System (CERCLIS). At the time of the study, CERCLIS contained information on 18,164 uncontrolled toxic waste sites. Residential five-digit zip code areas were used to define "communities."

The CSJ's descriptive study found the presence of uncontrolled waste sites to be "highly pervasive." More than half the U.S. population lived in residential zip code areas with one or more uncontrolled hazardous waste sites. Moreover, three of every five African Americans and Hispanics lived in communities with uncontrolled hazardous waste sites, which amounted to more than fifteen million African Americans and eight million Hispanics. The investigators estimated that two million Asian/Pacific Islanders and 700,000 Native Americans also lived in such communities.

10.3.4.3   Mohai and Bryant Study of 1992

Mohai and Bryant examined the regional demographics of persons living near sixteen commercial TSDFs in three counties (Macomb, Oakland, Wayne) in the Detroit, Michigan, area [28]. Data on race and income were obtained from face-to-face interviews of persons in a sample of households selected with equal probability. An additional oversample was drawn of households within 1.5 miles of existing (n = 14) and proposed TSDFs (n = 2). Information about race and household income was obtained for 793 respondents; for analysis, all nonwhites were combined into one category "minority." Analyses were conducted of respondents living within 1 mile of a facility; from 1 to 1.5 miles of a facility; and persons living more than 1.5 miles from a facility. Results showed that percentage minority population and percentage below poverty level varied with distance from TSDF facilities. Within 1 mile, 48 percent were minority and 29 percent were below the poverty level; for 1 to 1.5 mile radial distance the corresponding numbers were 39 percent and 18 percent; and for more than 1.5 mile, 18 percent and 10 percent. Chi-square tests indicated all these percentage differences were statistically significant.

A second objective of the Mohai and Bryant study was to examine relationships between race and income on the distribution of commercial hazardous waste facilities. Multiple linear regression was used to test the strength of associations. Investigators tested whether race (coded as 1 = white and 0 = minority) and household income (measured in dollars) had independent relationships with the distance of residents from a TSDF. The investigators found that the relationship between race and location of TSDFs in the three-county area was independent of income. Moreover, race was the stronger predictor of proximity to a TSDF.

Mohai and Bryant concluded, "Review of 15 existing studies plus results of our Detroit area study provide clear and unequivocal evidence that income and racial biases in the distribution of environmental hazards exist. Our findings also appear to support the claims of those who have argued that race is more importantly related to the distribution of these hazards than income."

10.3.4.4 EPA Study of 1992

At the direction of Administrator William Reilly, an EPA Environmental Equity Workgroup was formed in July 1990 to review evidence that racial minority and low-income communities bore disproportionate burdens of environmental risks. The workgroup conducted a comprehensive evaluation of the scientific literature on environmental justice and related issues, examined environmental and human exposure databases, and reviewed data collected by federal health agencies on the health of minorities. They also reviewed socioeconomic data pertinent to environmental justice concerns.

Six findings came from these evaluations. Three were specific to risk communication and government policy issues, the other three were recommendations specific to human health and environmental hazards and are therefore more germane for purposes of this chapter. As quoted from the EPA Environmental Equity Workgroup's final report [4]:

1. There are clear differences between racial groups in terms of disease and death rates. There are limited data to explain the environmental contribution to these differences. In fact, there is a general lack of data on environmental health effects by race and income. For diseases that are known to have environmental causes, data are not typically disaggregated by race and socioeconomic group. The notable exception is lead poisoning. A significantly higher

percentage of African-American children compared with white children have unacceptably high blood lead levels.

2. Racial minority and low-income populations experience disproportionate exposures to selected air pollutants, hazardous waste facilities, contaminated fish and agricultural pesticides in the workplace. Exposure does not always result in an immediate or acute health effect. High exposures, and the possibility of chronic effects, are nevertheless a clear cause for health concerns.

3. Environmental and health data are not routinely collected and analyzed by income and race. Nor are data routinely collected on health risks posed by multiple industrial facilities, cumulative and synergistic effects, or multiple and different pathways of exposure. Risk assessment and risk management procedures are not in themselves biased against certain income or racial groups. However, risk assessment and risk management procedures can be improved to better take into account equity considerations.

EPA found major limitations in the environmental and health databases pertinent to environmental justice issues. However, for three hazards compelling data supported an EPA finding that environmental exposures were disproportionately borne by minorities. The three hazards, air pollution, children's exposure to lead, and hazardous waste sites, and others in the EPA report are discussed in the following sections.

10.3.4.4.1 Air Pollution

As EPA noted, air pollution is primarily a problem of urban areas, where pollution emission densities are greatest [4]. EPA noted that the percentages of various populations living in polluted urban areas differed by ethnic category: White (70.3%), Black (86.1%), Hispanic (91.2%), and Other (86.5%). Citing the work of Wernette and Nieves [29], who analyzed the demographics of areas that EPA had designated as being out of compliance with the Clean Air Act, EPA concluded that minorities were disproportionately exposed to air pollutants. Table 10.3 contains the data that undergirded EPA's conclusion. These data show the importance to minorities of attaining urban air quality standards under the Clean Air Act.

TABLE 10.3
Percentages of Populations Living in Air Quality Nonattainment Areas [4]

Air Pollutants	Whites (70.3% urban)	Blacks (86.1% urban)	Hispanics (91.2% urban)
Particulate matter	14.7	16.5	34.0
Carbon monoxide	33.6	46.0	57.1
Ozone	52.5	62.2	71.2
Sulfur dioxide	7.0	12.1	5.7
Lead	6.0	9.2	18.5

The impact of air pollution on Hispanics is also a matter of concern. The National Coalition of Hispanic Health and Human Services Organizations has pointed out that reducing exposure to air pollution is a priority issue for Hispanic communities because, in an update of the data in Table 10.3, about 80 percent of Hispanics live in areas that did not attain EPA air quality standards [30]. By updated comparison, about 65 percent of non-Hispanic African Americans and 57 percent of non-Hispanic whites still live in nonattainment areas. The implications for Hispanics is a greater rate of respiratory morbidity and mortality and other adverse health effects than for other groups.

10.3.4.4.2 Children's Lead Exposure

EPA concluded that children's exposure to lead was the environmental hazard for which the strongest evidence supported a disproportionate effect on minority populations [4]. Drawing upon data assembled by ATSDR [31], EPA concluded that the evidence was unambiguous: children of color had higher blood lead levels than did white children. Moreover, all socioeconomic and racial groups had children with lead in their blood high enough to cause adverse health effects. This was found to be particularly true for African-American children. Lower family income was associated with higher prevalence of elevated blood lead levels in children.

Subsequent to the EPA report [4], the Centers for Disease Control and Prevention (CDC) updated their data on blood lead levels in young children [32]. CDC's National Health and Nutrition Examination Survey (NHANES) is a population-based, periodic series of national examinations of the health and nutritional status of the civilian, noninstitutionalized U.S. population. Geometric mean blood lead levels in the U.S. population, age one year and older, declined from 12.8 μg/dL in 1976–1980 to 2.3 μg/dL in 1991–1994, and further declined for the period 1999–2002 to 1.6 μg/dL [33]. This remarkable outcome is attributed largely to: a) removing lead from gasoline, which in turn reduced ambient air levels of lead, and b) removal (or containment) of lead-containing paint in older housing. The reduction of lead levels in the U.S. population represents the single most successful environmental health outcome.

However impressive the decrease in the national mean blood lead level, disparities continue to exist across racial/cultural and income lines in the U.S. population. As shown in Table 10.4, the percentages of one- to five-year-old children (the age span of children's greatest blood lead levels), who have blood lead levels (BLLs) ≥10 μg/dL (i.e., CDC's action level)[3] were highest in urban, low income, African-American, non-Hispanic children [32]. Older housing containing lead-based paint accounts for much of this elevation in blood lead levels. In a subsequent update, CDC reported a continued decrease in the prevalence of BLLs ≥10 μg/dL in one- to five-year-old children living in the United States [33] (Table 10.4). CDC's analysis was based on BBL surveillance data for 1997–2001 [35] and 1999–2002 [33], using two data sources: NHANES data and state child blood lead surveillance data. According to the CDC study, the number of one- to five-year-old children in the United States reported with confirmed elevated BLLs ≥10 μg/dL decreased from an estimated 930,00 in 1991–1994 [32] to an estimate of about 310,000 in 1999–2002 [33]. As a matter of environmental health policy, having in place surveillance systems like NHANES and state-based reporting systems for

TABLE 10.4
Percentages of U.S. Children (One to Five Years Old)
with BLLs ≥ 10 µg/dL [32]

Race/Ethnicity Percentage	Years	
	1997–1994	2001–2002
African Americans, non-Hispanic	11.2	3.1
Mexican-American	4.0	2.0
White non-Hispanic	2.3	1.3
Income		
Low	8.0	
Middle	1.9	
High	1.0	

collection of health data is an enormous resource for public health officials and decision makers. Goals for achieving disease and disability prevention goals can be established and monitored through surveillance systems. Unfortunately, health surveillance systems are costly and some decision makers (e.g., legislators) must be convinced of the systems' efficacy.

10.3.4.4.3 Waste Sites

EPA's [4] analysis of the impact of environmental hazards on minorities and low-income groups identified residence near CERCLA sites and operating hazardous waste facilities as a matter of environmental inequity. Their conclusion was based on studies conducted by the United Church of Christ [29] and GAO [15]. These two reports are described elsewhere in this chapter.

10.3.4.4.4 Water Contamination Problems

Scientists from EPA and public health agencies formed a panel to review the impact of contaminants in water on minorities and low income populations [36]. The panel used the Safe Drinking Water Act and the Clean Water Act as background information against which relevant studies and reports were evaluated. The panel reviewed information about microbial content of water on tribal lands, drinking water quality in migrant worker camps, groundwater contamination from hazardous wastes in poor rural counties, drinking water quality and sanitation along the U.S./Mexico border, lead in drinking water, case studies on water quality problems on Navajo lands, and the consumption of fish from contaminated bodies of water. The panel found that most information was anecdotal or case studies and did not therefore lend itself to quantitative comparisons or analyses.

However, the panel concluded, "Despite the sparseness and limitations of the data, the existing data suggest that environmental inequities exist. While the existing data do not support any broad nationwide pattern of inequity, there are, however, clear situations where certain populations are exposed to higher levels of contaminants in water." The panel did not separate factors of low income and race/ethnicity in arriving at their conclusion. The panel advocated collection of additional data on water contamination and populations at health risk. Amendments to the Safe Drinking Water Act in 1996 contain the statutory directive to collect this kind of data.

10.3.4.4.5 Other Environmental Hazards

In addition to EPA's [4] analysis of air pollution, children's lead exposure, waste sites, and water contamination problems, consideration was also given to minorities' exposure to pesticides and the consumption of fish caught in bodies of water contaminated with toxicants. However, data were generally lacking that might relate those hazards to any inequities experienced by minorities.

EPA's Environmental Equity Workgroup developed recommendations to the EPA Administrator on environmental justice issues [ibid.]. They published their findings in a two-volume report entitled *Environmental Equity: Reducing Risk For All Communities*. The findings led to the establishment of environmental justice policies and activities at EPA.

10.3.4.5 Environmental Hazards and Hispanics' Health

Metzger et al. [37] used EPA's rankings [4] to extrapolate the effect of environmental hazards on Hispanics' health. Investigators noted that 22.4 million Hispanics are in the U.S. population, and the number will grow to 31 million by the year 2010. Metzger et al. state, "There are numerous indicators that Hispanics face a disproportionate risk of exposure to environmental hazards. Ambient air pollution, worker exposure to chemicals, indoor air pollution, and drinking water are among the top four threats to human health and all are areas in which indicators point to elevate risk for Hispanic population."

Metzger and colleagues cite EPA data indicating that Hispanic populations experienced higher risk. They note that 80 percent of Hispanics live in areas that fail to meet at least one EPA air quality standard compared with 65 percent of African Americans and 57 percent of whites. According to EPA data on air quality nonattainment areas, Hispanics are also more than twice as likely as either African Americans or whites to live in areas that have elevated levels of particulate matter. Concerning workers' exposure to chemicals, Metzger et al. observe that 71 percent of all seasonal agricultural workers are Hispanic, compared with 23 percent who are white and 3 percent African American. The use of pesticides in agricultural applications places Hispanics at elevated health risk. Concerning indoor air pollution, Metzger and colleagues cite the failure to communicate to Hispanics the risk of radon in indoor air. They note that 61 percent of Hispanics have never heard of radon compared with 21 percent of whites. Metzger et al. include the presence of lead and biologic contaminants in drinking water supplies as two examples of problems that some Hispanic populations face.

10.3.4.6   *National Law Journal* Study of 1992

Staff of the *National Law Journal* examined 1,177 of the 1,206 National Priorities List (NPL) sites as of March 1992 [38]. They found that it took on average 5.6 years from time of waste site discovery until the site was placed on the NPL, but placing uncontrolled hazardous waste sites on the NPL took 20 percent longer in minority communities than in white communities. Lavelle and Coyle also analyzed outcomes of all environmental lawsuits filed in federal courts over a seven-year period. The average fine imposed for violating federal toxic waste laws in white residential areas, $335,566, was more than six times the average fine imposed in minority residential areas. The disparity occurred by race alone, not income; the average penalty in areas with the lowest median incomes was only 3 percent greater than the average penalty in areas with the highest median incomes.

10.3.4.7   Municipal Zoning Laws and Environmental Justice Study of 2002

How land is zoned for use can have serious environmental health consequences. For example, when polluting industries or waste management facilities are permitted by zoning laws to locate within residential areas, releases of hazardous substances can occur, placing residents at increased risk of adverse health effects. Maantay [39] reviewed New York City's zoning decisions between the years 1958–1990 in regard to potential environmental inequities. In particular, changes in land designated as "M" zones (manufacturing zones), were reviewed for location within the city's boroughs. Maantay's investigation concluded, "[T]hese zoning changes have had the effect of concentrating the noxious uses in the poorest and more minority neighborhoods." For example, the Bronx, the city's least affluent borough, had the most major increases in M zones. This study suggests, but does not prove, that zoning decisions discriminated against residents of poor and minority communities.

10.3.4.8   Demographics Investigations

The preceding section summarized several key reports that gave impetus to assessing environmental inequity concerns. However, these reports often had important methodological limitations such as the use of zip code areas for geographic analyses. Because the U.S. Postal Service developed zip codes to facilitate mail delivery, they are subject to change as the Postal Service refines mail delivery patterns. Zip codes therefore represent variable geographic areas that can lead to uncertainties in demographic analyses. To avoid such methodological shortcomings, several researchers have conducted more in-depth demographic studies on associations among race, ethnicity, and socioeconomic variables as they relate to siting of hazardous waste facilities and location of CERCLA sites. The key studies are summarized in this section.

10.3.4.8.1 Hird Study of 1993

Hird [40] looked at three broad equity implications of the EPA CERCLA program for environmental policy analysis. He examined three elements of environmental justice: geographic distribution of NPL sites (chapter 8), who pays for site cleanups, and the pace of cleanups. Only the first and third elements of his study will be summarized.[4]

To examine the distributional equity of NPL sites, data were collected on the socioeconomic characteristics of each county in the United States (n = 3,139) and the number of current or proposed NPL sites (n = 788) in each county as of January 1, 1989. This permitted a determination of whether the number of NPL sites in each county was correlated with the socioeconomic characteristics of the surrounding area. Hird argues: "The county is both large enough to include the effects of hazardous waste sites, and small enough to record significant socioeconomic variation." His county-level socioeconomic data were obtained from the U.S. Census Bureau. The number of NPL sites per county was the dependent variable in a multivariate Tobit statistical analysis. Independent variables in the Tobit analysis included quantity of hazardous waste generated in each state; percentage of each county's economy attributable to manufacturing; percentage of college educated residents; percentage of housing units occupied by owners; the median housing value; and percentages of county residents who were unemployed, nonwhite, and below the mean poverty level.

Results showed the mean number of NPL sites per county was 0.37 (sd = 1.28, n = 3,139). Manufacturing presence was strongly associated with more county NPL sites. Hird noted: "[t]he results indicate that more economically advantaged counties (in terms of both wealth and the absence of poverty) are likely to have more Superfund sites." For all 3,139 counties, Hird found no statistically significant association nationally between poor counties and the number of NPL sites they contain. However, counties with high concentrations of nonwhites had *more* NPL sites than did others (holding other socioeconomic factors constant), an outcome that Hird characterized as "[c]orroborating the United Church of Christ findings for all hazardous waste sites."

Because multivariate Tobit analysis may have obscured simple relationships between the presence of CERCLA sites and measures of ethnicity and socioeconomic factors, simple bivariate Tobit estimates were calculated. The dependent variable was the number of county NPL sites and the independent variable was either poverty or unemployment rates, median housing values, or percentage of nonwhites in the county. The results of the bivariate analysis showed high numbers of NPL sites strongly related to higher housing values, lower percentages of nonwhites, and the lack of poverty and unemployment.

A different picture emerged about distribution of NPL sites when subsets of all counties were evaluated on the basis of their exceeding high rates of poverty (n = 1,292 counties), unemployment (n = 1,274), ethnicity (n = 1,195), or median housing values (n = 1,254). The average number of NPL sites per county was 0.11 in counties highly represented by persons of low income. For counties with high percentages of unemployed, the average number of NPL sites per county was 0.23; for counties with high percentages of nonwhites, the figure was 0.33 NPL sites per county. All three averages are therefore *below* the national average of 0.37 NPL sites per county. For the subset of counties with high median housing value, NPL sites per county was 0.74, which was higher than the national county average. Hird concluded, "Therefore, these results indicate that NPL sites are located predominately in affluent areas, and generally irrespective of race."

Regarding equity in cleanup of NPL sites, three measures of site remediation speed were used. Hird examined data from Remedial Investigation and Feasibility studies, Records of Decision for NPL sites, and actual remedial actions. Because these three events occur in temporal sequence, some indication of the remediation speed can be evaluated. The most important indicator of a site's cleanup stage was found to be the Hazard Ranking Score, that is, the higher the hazard score the faster the cleanup. No association was found between pace of site cleanup and the county's socioeconomic characteristics (which included percentage of nonwhite population).

10.3.4.8.2   Anderton et al. Study of 1994

Investigators at the Social and Demographic Research Institute, University of Massachusetts, conducted a comprehensive study of the racial and cultural demographics and income levels of persons living near commercial, controlled hazardous waste facilities [27,41]. These are sites permitted to operate under the Resource Conservation and Recovery Act. The investigators focused on commercial facilities that treated, stored, and disposed of hazardous wastes (TSDFs). Note the important difference between TSDFs, which are controlled hazardous waste facilities, and CERCLA NPL sites, which are uncontrolled hazardous waste sites.

The Anderton and colleagues study is noteworthy because investigators examined the effect of using different geographic units on the outcome of demographics analyses. The investigators chose the census tract[5] as their primary geographic unit to avoid aggregation errors inherent in larger geographic units, such as zip code areas.

Commercial TSDFs were identified within census tracts for facilities that had opened for business before 1990 and were still operating in 1992. The investigators defined a TSDF as being privately owned and operated and receiving waste from firms of different ownership; TSDFs were excluded if they were the primary producers of waste. Before the 1990 census, tracts were defined only for Standard Metropolitan Statistical Areas (SMSAs).[6] About 15 percent of TSDFs are located outside SMSAs and hence were not included in the analysis. Using these criteria, 454 facilities were identified for demographic analysis.

The investigators' first analysis examined how census tracts with TSDFs differ from those without TSDFs. Comparisons were made of census tracts containing TSDFs with tracts that had TSDFs but within SMSAs that contained at least one facility inside their borders. This resulted in analysis of 408 tracts with TSDFs and 31,595 without. The mean percentages of African Americans were 14.5 percent in census tracts with TSDFs and 15.2 percent in tracts without TSDFs; the difference was not statistically significant. The mean percentages of Hispanics were 9.4 percent in tracts with TSDFs and 7.7 percent for tracts without facilities, which was not statistically significant. Similarly, no statistically significant difference was found in the median percentage of African Americans residing in tracts with and without TSDFs, which led Anderton and colleagues [41] to observe, "This single finding is sufficient to raise substantial questions about the previously cited research conducted at a zip code level, and about its substantial influence on national policy."

The investigators found a substantially higher mean percentage of persons employed in precision manufacturing located in TSDF tracts (38.6%) than in surrounding areas (30.6%). This suggested that TSDF facilities are located in industrial areas for reasons unrelated to issues of race and ethnicity.

To determine whether environmental inequities differed for large cities, the census tract comparison was repeated for the twenty-five largest SMSAs. For these areas, TSDF tracts were found to have significantly lower percentages of African Americans, but larger percentages of Hispanics, compared with census tracts without TSDFs. For the twenty-five SMSAs, TSDF tracts had significantly higher levels of industrial employment, with less expensive and older houses.

Anderton and colleagues next constructed larger aerial units of analysis, consisting of all tracts with at least 50 percent of their areas falling within 2.5-mile radii of the center of tracts in which TSDFs were located. The percentage of African Americans in these larger areas (25.7%) was significantly higher than in other tracts (14.5%). For Hispanics, the comparable numbers were 11 percent versus 7 percent for other tracts. Furthermore, in these larger areas, industrial development remained significantly higher than in other tracts. A multivariate analysis using "Being a TSDF Tract" as the dependent measure showed, "[t]he most significant effects in each case are not those of percentage black or percentage Hispanic, but of unemployment and industrial employment within the area. For census tracts, the effects of percentage black and percentage Hispanic are not significant. However, in much larger areas [i.e., 2.5-mile radius areas], both variables appear to be associated with the presence of TSDFs."

In summary, Anderton and colleagues [41], using census tract-level data, found no nationally consistent and statistically significant differences between the racial or ethnic composition of tracts that contain commercial TSDFs and those that do not. The investigators noted that TSDFs were more likely to be found in tracts with Hispanic groups. In a companion paper [27], they concluded: "We believe our findings show that TSDFs are more likely to be attracted to industrial tracts and those tracts do not generally have a greater number of minority residents."

10.3.4.8.3 Zimmerman Study of 1993

Zimmerman [42] used a unit of geographic analysis different from that used by Anderton et al. [41] to examine equity issues of relevance to NPL sites. She focused on social and economic characteristics at the geographic level of communities, which she defined as U.S. Census "Places," or, where places do not exist, as "Minor Civil Divisions" (MCDs). She observes that these communities represent political subdivisions and are the smallest formal level of political decision making.

In addition to assessing the demographics of populations living near NPL sites, Zimmerman evaluated whether the time taken to develop a Record of Decision (ROD)[7] for a site was associated, as a matter of environmental inequity, with minority communities. Zimmerman initially obtained demographics data for the 1,090 sites on the NPL at the time of her analysis. Her list excluded sites in extremely rural areas whose community populations in 1980 were fewer than 2,500. This resulted in excluding 260 sites, which she asserted, had minimal effect on her overall analysis. Characteristics of NPL communities were compared with the Nation and the four Census regions (Northeast, Midwest, South, West), based on 1990 Census data. Two methods of portraying average percentages for race, ethnicity, and poverty were used. One method was an unweighted averaging of means, counting each community equally regardless of its population. The second method weighted communities according to each community's population (total population as well as minority population). Zimmerman's findings differed according to which method she used.

Using the unweighted averaging method, African Americans represented an average of 9.1 percent of the population in 1990 for the approximately 800 NPL communities evaluated. This was lower than the national average of 12 percent. The percentages of African Americans in NPL communities were lower in three of the four U.S. Census regions. In the South census region, percentage of African Americans in NPL communities was 23.7 percent compared with 18.5 percent of African Americans living in the census region. The percentage of Hispanics in NPL communities was 6.6 percent compared with 9.0 percent nationally; no notable regional differences were found. The mean percentage of persons below the poverty level was 10.6 percent in NPL communities compared with 13.5 percent for the nation.

When Zimmerman used the alternate approach of weighting minority populations by total population, the mean percentage of African Americans in NPL communities was 18.7 percent and 13.7 percent for Hispanics. This was based on 622 Census Places and MCDs that contained 825 NPL sites. These percentages are greater than national percentages for African Americans (12.1%) and Hispanics (9.0%). She noted that differences in race and ethnicity between the two kinds of population analysis (i.e., non-weighted vs. weighted) reflects the effect of a relatively few large communities with NPL sites that have large African-American populations. She concluded, "Thus, racial and ethnic disproportionalities with respect to inactive hazardous waste site location seem to be concentrated in a relatively few areas."

To explore the relationship between Record of Decision (ROD) status for NPL sites and socioeconomic characteristics, a Probit statistical analysis was conducted of all 1,090 NPL sites. In general, the set of independent variables (including race and ethnicity) taken as a group did not contribute much to the variance in the independent variable—existence of a ROD. However, when the analysis was confined to a subset of NPL sites in poor communities with relatively high African-American populations, about 20 percent of sites were found to be without RODs. On this point, Zimmerman concluded, "Disproportionalities with respect to cleanups do exist, but appear to be more a function of the nature of the process of designation of NPL sites in the early 1980s rather than a result of actions connected with cleanup plans *per se*."

10.3.4.8.4   GAO Study of 1995

The General Accounting Office (GAO) conducted a multipurpose study in 1995 of persons living near nonhazardous municipal landfills [44]. The study was requested by Senator John Glenn (D-OH) and Congressman John Lewis (D-GA). The primary objective of the study was to evaluate demographics and income levels of persons living near the examined facilities. Another objective was to evaluate ten published demographics studies of persons living near hazardous waste facilities. Other objectives were to examine EPA's efforts to address environmental justice in their regulations, and to provide information on the extent of data that measure human health effects of waste facilities on minorities and persons of low income. Only findings from the study's first two objectives are summarized here.

To address demographics and income, GAO identified a potential universe of 4,330 landfills in the United State. This universe was subdivided, using zip codes of landfills, into categories of 1,498 metropolitan and 2,832 nonmetropolitan landfills. GAO used a questionnaire to survey landfill operators with equal probability in each landfill category. The survey elicited information on the geographic location and other characteristics of respondents' landfills. The final sample consisted of 190 metropolitan and 105 nonmetropolitan landfills. The demographics and income levels of persons living within one and three miles of these 295 landfills were evaluated.

Using a Geographic Information Systems (GIS) technique, the latitude and longitude for each site permitted defining two areas that separated landfills from the rest of the county. These areas were within one and three miles from the boundary of the landfill. To determine the demographics of persons within these two boundaries, GAO used the smallest level of aggregation possible, census block groups,[8] as their units of geographic analysis. GAO did not use census blocks[9] as their index because Bureau of Census data do not include information on residential income at the census block level.

The number of minorities and nonminorities living in complete and partial block groups was summed, using 1990 U.S. Census data, to determine the total number of persons living in the one- and three-mile areas. These numbers were subtracted from the number of persons living in each county that contained a landfill. GAO excluded block groups within the one- and three-mile areas that fell outside the county in which the landfill was located. For each one- and three-mile area and the corresponding rest of the county, GAO developed demographic information in five categories: race/ethnicity, poverty status, median household income, poverty status by race/ethnicity, and median household income by race/ethnicity.

GAO found that minorities and persons of low income were not generally over represented near nonhazardous municipal landfills. For 73 percent of metropolitan landfills and 63% of nonmetropolitan landfills, percentages of minorities living within one mile of landfills were lower than percentages of minorities living in the rest of the counties. GAO estimated that people living within one mile of about half the landfills analyzed had median household incomes higher than the incomes of residents in the rest of the county. The same result occurred when the three-mile areas were used for analysis.

GAO's second objective was to evaluate ten published reports on demographics of communities located near waste sites [44]. The ten studies are included in this chapter. GAO noted that three of the ten studies found minorities more likely than nonminorities to live near hazardous waste sites, four showed either no significant association between the location of a waste site and minority populations or that minorities were *less* likely to live nearby, and three yielded multiple conclusions on whether a disproportionate percentage of minorities lived near the waste sites and facilities.

Seven of the ten studies also assessed economic factors. GAO concluded that three studies had found incomes of persons living near hazardous waste facilities were lower than incomes of persons living distant; two reported no significant differences in incomes between those near and those distant from hazardous waste facilities; and two reported multiple results, depending on the analytic method used.

GAO noted that actual data on exposure to hazardous substances of minorities and persons of low income were generally lacking. Furthermore, they cautioned that comparing demographics and income data across research studies was difficult, because investigators used U.S. Census databases and geographic units of analysis that differed across studies. Overall, GAO found only marginal support for the argument that minorities and persons of low income are disproportionately located near nonhazardous waste facilities, that is, TSDFs.

10.3.4.8.5 Been Study of 1995

A set of particularly noteworthy papers on environmental justice was published by Vicki Been, New York University School of Law [45–47]. The papers are noteworthy because of the clarity of writing, presentation of clear and logical arguments, and close attention to data analysis. In particular, Been's work on environmental justice issues should be compared with the work of Anderton et al. [27,41] because both investigators used similar databases and methods, but with somewhat different outcomes.

Been compared various characteristics of census tracts with TSDFs versus non-TSDF census tracts. In distinction to Anderton and colleagues [27,41], Been identified census tracts hosting TSDFs by examining TSDF listings in the 1994 edition of *Environmental Services Directory* (ESD), which she supplemented with an EPA database, the *Resource Conservation and Recovery Information System*. She used telephone contacts to verify the location and kind of operation of individual TSDFs. From these efforts, Been identified 608 TSDFs for analysis, which she asserts is more accurate than that used by other investigators. Been comments that Anderton's group [ibid.] limited their comparison to non-TSDF census tracts in Metropolitan Statistical Areas or rural counties that had at least one facility. Been made no similar restrictions and compared all TSDF tracts with all approximately 42,000 populated non-TSDF tracts within the continental United States, using 1990 census data in her demographics analyses.

Been conducted both univariate and multivariate analyses of characteristics between TSDF and non-TSDF census tracts. Landfills, incinerators, and kilns were separated as a group from other kinds of TSDFs to address whether different kinds of TSDFs were associated with racial, ethnicity, or socioeconomic characteristics. The breakout revealed no statistically significant differences in the mean percentages of African Americans and lower-income persons living near this particular grouping of TSDF facilities.

To determine whether a smaller geographic comparison would change the nature of differences between TSDF and non-TSDF tracts, Been calculated ratios of the demographics of TSDF sites to demographics of all non-TSDF sites within a state and within a Metropolitan Statistical Area (MSA). The mean of the ratios was tested for significance from unity, the ratio that would occur if the TSDF tracts' characteristics were identical to the non-TSDF tracts' characteristics. Using this approach, and using national demographics data, she found that percentages of African Americans in TSDF tracts did not differ significantly from percentages in non-TSDF tracts. The percentage of Hispanics, however, was significantly greater for TSDF tracts than for non-TSDF tracts. These results for African Americans and Hispanics maintained when comparisons were made within states. However, when comparisons were made within MSAs, African-American differences remained statistically insignificant and differences in percentages of Hispanics narrowed. Been noted that differences between median housing values in TSDF and non-TSDF tracts narrows considerably when only the host MSA is studied.

Been extended her own work and that of others by examining not only the means of demographics variables (e.g., racial percentages), but also whether the distribution of TSDF facilities matched the distribution of populations around the mean. Been [47] assumed that a "fair" distribution of TSDF facilities would be proportionate to the distribution of the population. Using this assumption, she calculated the number of facilities that would be located in particular kinds of neighborhoods if distribution of TSDF facilities were proportionate. The results are fascinating. According to Been [ibid.], "In terms of raw numbers, if the distribution of facilities followed the distribution of the population, there would be twenty-four more facilities sited in the neighborhoods with no or very few African Americans. In neighborhoods where African Americans made up more than 10% but less than 70% of the population, there would be thirty-four fewer facilities. Neighborhoods with African-American populations of more than 70% would have ten more facilities. Similarly, neighborhoods with Hispanic populations of more than 20% are bearing more facilities than they should if facilities were distributed in the same way in the population." Been also found that neighborhoods with median family incomes of $10,001 to $40,000 bear sixty-two more facilities than would be proportionate.

Been [47] concluded, "[a] more sophisticated comparison of the distribution of facilities to the distribution of neighborhoods with particular demographic characteristics reveals that certain kinds of neighborhoods—those with median family incomes between $10,001 and $40,000, those with African-American populations between 10% and 70%, those with Hispanic populations of more than 20%, and those with lower education attainment—are being asked to bear a disproportionate share of the Nation's facilities. Analysis of the joint distribution of income and percentage of African-Americans in the population suggests that income explains most of the disparity. Multivariate analysis, however, suggests that race is a better predictor of facilities than income. In total, the analysis reveals that environmental injustice is not a simplistic PIBBY— "put it in black's backyards." It suggests instead, a much more ambiguous and complicated entanglement of class, race, educational attainment, occupational patterns, relationships between the metropolitan areas and rural or non-metropolitan cities, and possibly market dynamics."

10.3.4.8.6   Heitgerd et al. Study of 1995

Researchers at ATSDR used a GIS approach to assess the demographics of populations living near NPL sites [48]. GIS is a powerful tool that permits the analysis of datasets that have been overlaid on a geographic database. For example, census data can be overlaid on a geographic base to determine housing patterns according to race, income, and geographic location.

Racial and Hispanic origin subpopulations living within one mile of NPL sites were compared with subpopulations in the same county but living outside the one mile border. The investigators extracted census block boundaries from the Bureau of Census' 1990 Topologically Integrated Geographic Encoding and Referencing/Line files and linked them with information on total population, race, and Hispanic-origin data in 1990 census block data. EPA-defined site boundaries were used to specify the boundaries of 1,200 NPL sites.

Heitgerd and colleagues found that 670 counties had parts of their area located within one mile of the 1,200 NPL sites they examined. This represented 22 percent of all counties in the contiguous states of the United States. The areas within one-mile borders of the NPL sites comprised 184,191 census blocks, of which 10.6 percent were within one mile of two or more NPL sites. Each site's demographics were derived for each county by summing over all census blocks within the one-mile range. The investigators assert this shifts the focus of the demographics analysis from NPL sites *per se* to the counties within one mile of NPL sites while retaining block-level data. The comparison population was spatially defined as persons living in the 670 impacted counties but at distances greater than 1 mile from NPL sites. Population data for the comparison area were obtained by subtracting the site area data from county totals.

A three-factor analysis of variance (ANOVA) model served as the investigators' statistical method. The factors used in the ANOVA analysis were NPL (two levels-within or outside one-mile buffers of NPL sites), state (48 contiguous states), and county (670 counties).

The investigators found approximately 11 million persons resided within one-mile boundaries of the NPL sites they assessed. As was expected, considerable variation was found in the population figures according to state and regional factors. For example, California ranked highest in the average population per NPL site and average population density, while ranking eighth in average site area per square mile. Furthermore, fewer NPL sites were in the Great Plains states, and those sites accounted for relatively fewer persons compared with other regions [48].

A separate ATSDR analysis of 972 NPL sites found that 949,000 children six years or younger resided within one mile of sites' borders, which represented 11 percent of the population. This is an average of 980 children six years or younger per NPL site. Given a current total of about 1,300 NPL sites, one can calculate that approximately 1.3 million young children resided within one mile of NPL sites [49].

Concerning race and ethnicity, ANOVA analysis revealed a statistically significant difference ($p \leq 0.001$) in the mean percentage for each racial group and persons of Hispanic origin in populations living within 1-mile boundaries of sites compared with the remainder of the counties. The analysis controlled for state and county of NPL site. The largest percentage difference was found for African Americans. They represented 8.3 percent of the comparison area but 9.4 percent of the population living within one mile of the sites [48].

To put these percentages in perspective, ATSDR considered the 33 counties identified as the top 5 percent of counties adjacent to NPL sites when ranked by the percentages of African Americans living within one mile of sites [49]. There were approximately 221,000 persons of whom 145,000 (66%) were African Americans. When the racial composition of the remaining areas (greater than one mile from site) of the 33 counties was examined, there were about 7,365,000 persons of whom 2,212,000 (30%) were African Americans. If the same county rate outside the one-mile limit (i.e., 30% African Americans) were applied to areas within the one-mile borders of NPL sites, one would expect to find approximately 66,000 African Americans rather than the 145,000 actually identified. This reveals an additional approximately 79,000 African Americans who live near NPL sites in the 33 counties considered by Burg [ibid.].

Heitgerd and colleagues [48] concluded: "If it is assumed that the NPL sites are representative of all uncontrolled hazardous waste facilities, then the results support existing environmental inequity research that suggests the location of hazardous waste facilities is more burdensome for minority communities."

Because the study relies on a GIS approach and uses county-based comparison data, it is an important contribution to the literature on environmental justice. Its limitations are the lack of control for sociodemographic variables (e.g., are the observed disparities the result of economic conditions?) and uncertainty about whether the demographic results might be a consequence of the one-mile buffers chosen (e.g., would the results change if some other measure, perhaps, 0.5 mile, had been used?).

10.3.4.8.7 Oakes et al. Study of 1996

Building upon their previous cross-sectional work [27,41], researchers at the University of Massachusetts conducted the first national longitudinal study of residential characteristics in census tracts that contain TSDFs [50]. This study addresses the central issue in environmental justice as it relates to TSDFs, the issue of alleged racism in the deliberate siting of waste facilities in minority neighborhoods. Oakes and colleagues used rigorous statistical analysis and current demographics databases thereby making their study an important contribution to the scientific literature on environmental justice.

Oakes and colleagues evaluated community characteristics over a 20-year period before and after new TSDFs were sited. In a follow-up to previous findings [27,41] that indicated a relationship between TSDFs and the level of industrialization in a community, they compared trends within TSDF communities to other similar industrial communities and to less industrialized communities. Data on 476 commercial TSDFs were compiled from the 1992 edition of *Environmental Services Directory*, using a telephone survey of each facility. Analysis was restricted to census tracts within metropolitan statistical areas and rural counties that each contain at least one TSDF. There were 35,208 census tracts without TSDFs. Data on residential communities came from the 1970, 1980, and 1990 tract-level U.S. Census files. Census data files contained more than 130 census variables that summarized tract composition. Special efforts were made to reconcile any changes in tract locations or TSDF locations due to changes in census track identification.

Oakes and colleagues first conducted a cross-sectional analysis using 1990 census tract data. They compared racial and economic indicators between tracts that had TSDFs and tracts that did not. Findings showed the average percentages of persons living in TSDF tracts who identified themselves as African American or Hispanic were 17.09 and

10.75, respectively, and for non-TSDF tracts, 16.26 percent African American and 9.74 percent Hispanic. These percentages were not statistically significant between TSDF tracts and non-TSDF tracks. This result agrees with the researchers' prior report that used 1980 census data [27,41]. The Oakes group concluded, "The largest significant differences that were found between tracts with and without commercial TSDFs were in the average percentage (33.32% in TSDF tracts and 25.28% in non-TSDF tracts) and the median percentage of persons employed in industrial and manufacturing occupations." This again supports the conclusion of Anderton and colleagues [ibid.] that TSDF tracts are somewhat more likely to be found in industrial working-class neighborhoods.

Oakes and colleagues then analyzed communities' characteristics across two decades, using 1970, 1980, and 1990 census data. They first assessed demographic and socioeconomic characterizations of communities before TSDFs were sited. They found that TSDFs located in the 1970s and 1980s were, on the average, not systematically sited in areas with unusually high percentages of African-American or Hispanic populations, when compared with other areas with significant industrial development. Moreover, results showed the characteristics of communities, after siting of TSDFs, have trends that parallel those in the population at large.

The researchers performed multivariate methods of analysis to examine whether siting of TSDFs was associated with racial or ethic disparities and other indicators of environmental inequity. No evidence was found to support environmental inequity claims that TSDFs were sited in areas because of racial or ethnic bias. They concluded, "We believe this research, in concert with our earlier findings, suggests that commercial TSDF census-tract communities are best characterized as areas with largely white and disproportionately industrial working-class residential areas, a characterization consistent with what one might historically expect near industrial facilities."

10.3.4.8.8 Anderton et al. Study of 1997

Investigators at the University of Massachusetts conducted a comprehensive three-part examination of CERCLIS and NPL sites for evidence of environmental injustices. CERCLIS hazardous waste sites are those reported to EPA, but not placed on the National Priorities List (NPL). First, the investigation compared demographic characteristics of CERCLIS sites with those without such sites. Second, the demographics of neighborhoods with NPL sites were compared to those with CERCLIS sites. Third, possible bias in how CERCLIS sites were prioritized as NPL sites was evaluated. For the purposes of this chapter, only the first study is described [51].

In the first study, 1990 U.S. Census data and EPA CERCLIS records were used to compare neighborhoods with CERCLIS sites against neighborhoods without any such sites. Data analysis showed the percentage of African Americans residing in census tracts with CERCLIS sites was 11.6 percent, whereas for non-CERCLIS sites the percentage was 13.7 percent. Similarly, fewer Hispanics resided in CERCLIS tracts than in tracts without CERCLIS sites. However, the CERCLIS sites in nonmetropolitan neighborhoods did contain larger percentages of Native Americans than in comparison tracts. Anderton and colleagues concluded, "[t]he early discovery of CERCLIS sites in minority neighborhoods does not support the hypothesis of bias in discovery processes."

10.3.4.8.9   Baden and Coursey Study of 1997

The environmental justice implications of locating waste sites within the city of Chicago were investigated through demographic, social, and economic analysis [52]. The investigators examined locations of three kinds of waste sites: all CERCLIS sites and TSDFs within the city limits, RCRA hazardous waste generators, and historical hazardous waste sites. Sites were examined with regard to racial, ethnic, and income variables; access to transportation; and waste disposal. Sites were linked with corresponding census tract information. Regression analysis was the primary statistical method used to associate geographic and demographic variables. Different kinds of regression analyses were used to investigate which demographic and physical features predicted the location of sites within limited communities, the location of sites within larger neighborhoods, and the geographic concentration of sites. Census and waste site data were specific to the years 1960 and 1990 for comparison.

Baden and Coursey found that waste sites in 1990 tended to be located in Chicago areas of low-population density near commercial waterways and commercial highways. They found no evidence of environmental racism against African Americans for either CERCLA sites or TSDFs. There was no indication that African Americans lived in areas with higher concentrations of hazardous waste than did whites or Hispanics. Evidence showed that the percentage of Hispanics in an area was significant with regard to the location of CERCLA sites and solid waste disposal facilities, perhaps because of recent migration of Hispanics into white ethnic neighborhoods. Baden and Coursey observed: "Surprisingly, areas where RCRA and solid waste disposal sites are located tend to have higher incomes; this is likely the result of the recent trend in construction of high price river-front residences in previously industrial areas." In summary, Baden and Coursey found little to no indication of environmental injustice in the location of hazardous waste sites and facilities within Chicago census tracts.

10.3.4.8.10   Carlin and Xia Study of 1999

Carlin and Xia [53] used Bayesian hierarchical models to investigate geographic and racial associations between ambient ozone levels and rates of pediatric asthma emergency room (ER) visits in Atlanta, Georgia, for years 1993 through 1995. Local ozone air quality data were used to calculate one-hour and eight-hour (maximum) averages. Results showed "[t]he association between pediatric asthma ER visits and the percent blacks in a zip (i.e., zip code) is significant and positive; the fitted relative risk shows that a theoretical all-black zip would have relative risk nearly three times that of a comparative all-non-black zip." Further, an increase of about 2.6 percent for every 20 ppb increase in ozone level (eight-hour max value) was found. This study's design did not permit the identification of factors that might explain the racial disparity in asthma ER visits.

10.3.4.8.11   Davidson and Anderton Study of 2000

Davidson and Anderton [54] investigated the association between RCRA-governed facilities and indicators of environmental inequity. This work extended previous work by Anderton and associates by considering additional TSDFs beyond their prior investigations. The census tract locations of 6,550 RCRA facilities were examined according to minority community composition and other demographic data. Tracts with at least one RCRA facility were compared with tracks without a RCRA facility. The authors found, "[t]hese findings suggest that tracts with RCRA facilities may be described as working-class neighborhoods with a lower percentage of Hispanic and black residents (except in nonmetropolitan regions), higher levels of industrial employment, lower average levels of education, and more modest housing." This study's findings are similar to previous investigations of TSDFs, suggesting no environmental inequities according to racial and ethnic criteria.

10.3.4.8.12   Morello-Frosch et al. Study of 2002

Morello-Frosch and colleagues [55] summarized the findings of environmental justice investigations undertaken by an academic and community-based collaborative. In one study, the locations of TSDFs in Los Angeles, California, were examined for evidence of environmental inequalities. Results showed those census tracks containing a TSDF or located within a one-mile radius of a TSDF had significantly higher percentages of residents of color, lower per capita and household incomes, and a lower proportion of registered voters. In a second study of similar design and purpose, an analysis was conducted of air emissions reported to EPA's Toxics Release Inventory (TRI)[10] by sources in southern California. The study distinguished between all TRI facilities and those facilities releasing toxicants classified by EPA as being priority hazardous substances. Logistic regression analysis that controlled for income, industrial land use, and population density found that proportion of minority residents was significantly associated with proximity to a TRI facility. Potential bias in the study due to the nature of the collaborative project were not discussed.

10.3.4.9   Tabulation of Studies

Summarized in Table 10.5 are the environmental justice findings from the demographic and other studies described in this chapter. The table is broadly structured into investigations that focused on uncontrolled hazardous waste sites (HWS) and those focused on treatment, storage, and disposal facilities (TSDFs). As shown in the table, the Anderton group [51] study, which examined CERCLIS sites, was unique and is highlighted by dash lines. Above it are studies of HWS; below are investigations of TSDFs. If one disregards those studies that were not independently peer reviewed, an expectation of contemporary science, the remaining studies cited in Table 10.5 point to a difference in environmental justice findings between HSW and TSDFs. There is an indication of excess numbers of minorities who reside near HWS. No similar pattern exists in regard to TSDFs. Of the five studies that attempted to assess evidence of environmental injustice, i.e., determine if waste sites had been deliberately located in minority or low-income areas, no evidence of such injustice was reported.

In summary, the tabulated studies cited in Table 10.5 point to environmental in-

TABLE 10.5
Key Waste Site Studies Bearing on Environmental Equity

Author (year)	Scope/Sites Studied	Geographic Unit	Excess in Minorities?	Environmental Injustice?[a]	External Peer Review?	Published?
CSJ (1987)	National/ HWS, TSDFs	ZIP code	Yes	NI	No	No
Lavelle and Coyle (1992)	National/ HWS (NPL)	None	NI	Yes?	?	Journal
Hird (1993)	National/ HWS (NPL)	counties	Yes	NI	Yes	Journal
Zimmerman (1993)	National/ HWS (NPL)	census places	Yes	NI	Yes	Journal
GAO (1995)	National/ HWS	census block group	No	NI	No	No
Heitgerd et al. (1995)	National/ HWS (NPL)	census block	Yes	NI	Yes	Journal
Anderton et al. (1997)	CERCLIS sites	census tract	No	No	Yes	Journal
GAO (1983)	Regional/ TSDFs	census area	Yes	NI	No	No
Mohai & Bryant (1992)	Local/ TSDFs (Detroit)	individual TSDFs	Yes	NI	No?	Journal
Anderton et al. (1994)	National/ TSDFs	census tract	No	No	Yes	Journal
Been (1995)	National/ TSDFs	census tract	No	NI	Yes	Journal
Oakes et al. (1996)	National/ TSDFs	census tract	No	No	Yes	Journal
Baden & Coursey (1997)	National/ HWS, TSDFs	census tract	Yes (Hispanics)	No	Yes	Journal
Davidson & Anderton (2000)	RCRA facilities	census tract	No	No	Yes	Yes
Morello-Frosch et al. (2002)	TSDFs (Los Angeles)	census tract	Yes	NI	Yes	Journal

[a] Environmental justice refers to whether waste sites were deliberately sited in minority communities; NI = Not investigated; HWS = uncontrolled hazardous waste sites; NPL = National Priorities List; TSDFs = treatment, storage, and disposal facilities.

equity in regard to minorities who reside near HWS, but not for TSDFs. The matter of environmental injustice, at least on a national scale, has been inadequately researched, but studies to date have shown little, if any, such evidence.

10.4   PRESIDENTIAL EXECUTIVE ORDER ON ENVIRONMENTAL JUSTICE

Any injustices that occur across cultural and racial groups related to exposure to environmental hazards must be prevented as a matter of fairness and social justice. On February 11, 1994, at the urging of environmental justice advocates, President Clinton signed Executive Order 12898 entitled "Federal Actions to Address Environmental Justice in Minority Populations and Low-Income Populations." The order is a prime example of the PACM policy-making model, as discussed in chapter 2. As with all executive orders, this order applies only to federal agencies. Although state and local governments and private sector entities are not directly subject to executive orders, actions that federal agencies take under executive order can have substantial ripple effects on other levels of government and the private sector.

The Clinton Executive Order directs each federal agency to "[m]ake achieving environmental justice part of its mission by identifying and addressing, as appropriate, disproportionately high and adverse human health or environmental effects of its programs, policies, and activities on minority populations and low-income populations" [56]. The several responsibilities prescribed in the executive order for federal agencies are outlined in the following section.

Creation of an interagency working group—The administrator of EPA was directed to convene and chair an interagency federal working group on environmental justice. Members of the group include EPA, the Departments of Defense, Energy, Health and Human Services, Commerce, Housing and Urban Development, Agriculture, Transportation, Labor, Justice, Interior, and various White House offices.

The working group was to serve as the federal government's primary environmental justice body with responsibility for: (1) guiding federal agencies on criteria for identifying human health or environmental effects that are disproportionately high and adverse on minority and low-income populations; (2) being a clearinghouse for federal agencies as they develop their environmental justice strategies; (3) assisting in coordinating research, (4) assisting in the collection of data required by the executive order; (5) examining existing data and studies on environmental justice; (6) holding public meetings; and (7) developing interagency model projects on environmental justice. Elaboration on some of these responsibilities follows.

Development of agency strategies—The executive order directs each federal agency to develop an agency-wide environmental justice strategy. Each agency's strategy must identify and address disproportionately high and adverse human health or environmental effects of its programs, policies, and activities on minority populations and low-income populations. Each strategy must also list programs, policies, planning and public participation processes, enforcement, or rulemakings related to human health or the environment that should be revised in light of environmental justice concerns.

Federal agency responsibilities for federal programs—Each federal agency is directed by the executive order to conduct its programs, policies, and activities that substantially affect human health or the environment in a manner that ensures against the effect of excluding persons from participation on the basis of race, color, or national origin.

Research, data collection, and analysis—The executive order mandates each federal agency, whenever practicable and appropriate, to collect, maintain, and analyze information assessing and comparing environmental and human health risks borne by populations identified by race, national origin, or income. Furthermore, agencies are directed to collect such population data for areas around federal facilities and similar areas expected to have a substantial impact on the environment, human health, or economy. Federal agencies must use this population-based data to determine whether their programs, policies, and activities have disproportionately high and adverse human health or environmental effects on minority populations and low-income populations.

Subsistence consumption of fish and wildlife—The executive order requires federal agencies, whenever practicable and appropriate, to collect, maintain, and analyze information on the consumption patters of populations who principally rely on fish or wildlife for subsistence. Agencies are directed to communicate to the public the risks of these consumption patterns. This directive is intended to address the human health issues of eating fish and wildlife that contain hazardous substances in their tissues.

10.4.1 Observations on the Executive Order

Although the federal government's environmental justice strategies are still evolving, some preliminary observations can be made. First, the executive order seems to give priority to collecting data that can be used to assess disproportionate impacts of environmental hazards on minority and low-income populations. How federal agencies will use these data remains to be determined, but they could have considerable effect on future environmental policies of the federal government. Federal policies, in turn, are often emulated by state and local government policies.

There has been to date no overall evaluation of federal agencies' compliance with Executive Order (EO) 12898. However, in 2004 the EPA's Inspector General (IG) completed a review of EPA's compliance with the EO [57]. This report, dated March 1, 2004, is significant because of EPA's primacy in the EO for providing environmental justice leadership within the federal government. The IG's review found, "EPA has not fully implemented Executive Order 12898 nor consistently integrated environmental justice into its day-to-day operations" and "[A]lthough the Agency has been actively involved in implementing Executive Order 12898 for 10 years, it has not developed a clear vision or a comprehensive strategic plan, and has not established values, goals, expectations, and performance measurements" [ibid.]. These are quite serious assertions, given that a decade had passed since the issuance of EO 12898.

The response to the IG's report was given on March 3, 2004, by EPA's Office of Environmental Justice [58]. While agreeing with some of the IG's criticisms, there is considerable disagreement over key issues. Of key policy significance is the statement "[T]he Agency believes the Inspector General's report reflects a mistaken interpretation of E.O. 12898 by placing an emphasis on identifying communities based on race and/or income" [ibid.], asserting that the rights to a clean environment are rights "[t]hat belong to all people, regardless of race or income." This statement implies a dismissal by EPA management of the historical development of the U.S. environmental justice movement and its aspirations.

10.5   INTERNATIONAL PERSPECTIVE

As described in this chapter, environmental justice has developed as an agenda within the U.S. civil rights movement. On reflection, linkage between social justice and environmental justice seems a natural relationship within the U.S. political arena because both movements are rooted in issues of fairness and social change. But is environmental justice confined to the U.S. landscape? Do similar concerns exist in other countries about the impact of environmental hazards on the health and well-being of minority groups and persons of low income? To seek answers to these questions, an analysis was conducted of the single, most comprehensive international statement on environmental protection—*Agenda 21* (chapter 9).

Agenda 21 was developed at the United Nations Conference on Environment and Development, held in Rio de Janeiro in 1992. *Agenda 21* is a document of 40 chapters, consisting of about 600 pages. A review of relevant chapters that might address environmental justice as a matter of national or international policy reveals no statements specific to this issue. In particular, *Agenda 21*'s chapter 20, entitled "Environmentally Sound Management of Hazardous Wastes Including Prevention of Illegal International Traffic in Hazardous Wastes," makes no mention of the importance of ensuring that hazardous waste generation, transportation, storage, and disposal do not cause environmental injustices. A review of *Agenda 21*'s other chapters reveals a concern for the rights of indigenous peoples and acknowledges their historical commitment to protecting the environment, although this support for respecting the rights of indigenous peoples was not couched in the language of environmental justice.

It is curious why environmental justice did not elicit concern in the major international statement on the environment. However, assuming that environmental inequities can be prevented through the actions of informed persons who have access to political processes, *Agenda 21* provides some tangential support for environmental justice concerns. Specifically, the Preamble to Section III of *Agenda 21* states [59]:

> "One of the fundamental prerequisites for the achievement of sustainable development is broad public participation in decision-making. Furthermore, in the more specific context of environment and development, the need for new forms of participation has emerged. *This includes the need of individuals, groups and organizations to participate in environmental impact assessment procedures and to know about and participate in decisions, particularly those which potentially affect the communities in which they live and work* (emphasis added). Individuals, groups, and organizations should have access to information relevant to environment and development held by national authorities, including information on products and activities that have or are likely to have a significant impact on the environment, and information on environmental protection measures.

Adherence to the philosophy highlighted in this statement from the United Nations Environment Program would help prevent environmental injustices.

Environmental justice has become a subject of concern in Australia, according to Lloyd-Smith and Bell [60]. To them, "[T]he term "environmental justice" refers to the distribution and impacts of environmental problems as well as the policy responses to address them." Further, they assert "[E]nvironmental injustice focuses on the inequitable distribution of those who bear the risks." Both definitions, unlike those in the United States, are not restricted to environmental hazards that are specific to racial or ethnic populations. Lloyd-Smith and Bell present two case studies of "toxic disputes" in Australia. One dispute pertains to residents' concerns about the effects of living near a hazardous waste dump. The other dispute pertained to the destruction of a huge stockpile of hazardous hexachlorobenzene waste. The authors' analysis of the two disputes asserts that both represented cases of environmental injustice, primarily because residents at each site did not have information equal to that of the government and private industry. Further, inequalities were manifest because residents living distant from the two hazardous waste sites derived benefits from industrial operations formerly on the sites, whereas those living near the sites had not.

The Australian authors' [ibid.] notion of environmental justice and inequality illustrate how difficult it is to transfer the environmental concerns, policies, and practices of one country (e.g., the U.S.) to another. Each country, certainly those with democratic governments, will likely need to develop its own legal and ethical structure to address environmental justice and injustice issues. One country's issues may be quite different from those of another country. What constitutes "environmental justice" will be a key policy issue. Consider, for example, the "environmental justice" issues in Central and Eastern Europe, where one source asserts that "[N]ational minorities are often subjects of environmental injustice" [61]. They claim that residents frequently became second class citizens when their traditional national borders changed and they became minority populations with a new country, citing such areas in the Balkans. These demographic changes occurred because of wars within the region.

Varga and colleagues [ibid.] also cite "manipulated industrialization" as a characteristic feature of the fallen national communist regimes in Central and Eastern Europe, leading to nonremediated environmental hot spots that impact local residents and, sometimes, adjacent countries. The national governments in this region of Europe currently lack the political imperatives and socioeconomic resources to develop and implement environmental justice infrastructure. Such an infrastructure will appear only when environmental activists and community groups bring sufficient pressure on their governments.

10.6   OBSERVATION

The conferences, studies, and reports cited in this chapter resulted from serious questions about inequities along racial and low-income lines in placing TSDFs and in remediating CERCLA sites. Bullard [10,20], who conducted comprehensive sociologic evaluations of several communities near waste sites and waste disposal facilities, concluded: "[t]here is mounting empirical evidence that people-of-color and low-income communities suffer disproportionately from facility siting decisions involving municipal landfills, incinerators, and hazardous-waste disposal facilities." Studies by the GAO [18] and the United Church of Christ [25] raised similar concerns about racially or culturally discriminatory actions that allegedly led to locating waste facilities in minority and low-income communities.

More than 20 years have passed since the GAO study of 1983. What do the data now reveal about environmental inequities and hazardous waste sites? First, data are sufficient to show that some environmental hazards are not shared equally across racial and ethnic groups in the United States For example, data are compelling that African-American and Hispanic children are exposed to environmental lead sources in percentages that exceed their white counterparts. The presence of lead in the paint of older housing is a primary reason, and older housing is associated with low household income. As the problem of lead exposure for young children illustrates, race and ethnicity are intertwined with socioeconomic factors for some environmental hazards. From a public health perspective, these intertwined relationships should be understood and factored into public health interventions that eliminate or reduce health hazards.

Second, environmental justice issues about the location of hazardous waste sites have led to several demographic studies conducted since the late 1980s. The siting of hazardous waste treatment, storage, and disposal facilities has stimulated a debate about the inequity of site locations. Research on the demographics and socioeconomic characteristics of populations near hazardous waste sites has led to a better picture of what kinds of hazardous waste facilities and sites disproportionately impact minority groups and persons of low income.

The work of several investigators indicates that the method of selecting the geographic unit of analysis and the comparison data will determine how environmental justice questions are answered. Investigations that used large geographic units (e.g., zip code areas) generally show that African-American and Hispanic populations live near hazardous waste facilities (TSDFs and uncontrolled hazardous waste sites) in percentages that exceed their percentages in the U.S. population.

As the geographic unit of analysis is reduced in size from zip code areas to census tracts and census blocks, the question of minorities and low-income populations living near uncontrolled hazardous waste sites and TSDFs becomes more clear. Minorities appear to live disproportionately near uncontrolled hazardous waste sites [40,42,48]. The finding of inequity in the percentages of African Americans, and to a lesser extent, Hispanics, living near NPL sites may, however, be primarily because of a subset of NPL sites with large minority populations [42].

However, in distinction to uncontrolled hazardous waste sites, using geographic units smaller than zip code areas—primarily census tracts—investigators found no disproportionate association of minorities with living near TSDFs versus living elsewhere [10,20,44].

Moreover, how investigators of environmental justice communities develop toxicity indicators of putative exposure to hazardous substances can lead to different study outcomes. For example, Cutter and colleagues [62] evaluated six different toxicity indicators, as applied to substances released from hazardous waste sites. Indices included

Threshold Limit Values®, EPA Priority Chemical List, and the Environmental Defense Fund's Toxicity Equivalent Potential. Using the six toxicity indices, relative risk scores were calculated for the 426 facilities in South Carolina that comprised the state's Toxics Release Inventory. Findings from the study showed that the choice of toxicity indicator can result in different statistical and spatial variations in the results of investigations such as those investigating environmental justice claims.

The association between household income and persons' residential proximity to hazardous waste sites does not appear to be particularly significant. That is, household income is not a good predictor of a population's proximity to uncontrolled hazardous waste sites and TSDFs. The most consistent socioeconomic characteristic of communities near hazardous waste facilities and waste sites is the presence of manufacturing facilities, that is, hazardous waste facilities are located in industrial areas. What this finding portends for questions of environmental justice is unclear.

Only one study summarized in this chapter directly addressed the matter of racism and discriminatory actions putatively related to how TSDF facilities were sited. Oakes and colleagues analyzed communities' characteristics across two decades, using 1970, 1980, and 1990 census data. When they assessed the composition of communities before TSDFs were sited, they found, on average, that TSDFs sited in the 1970s and 1980s were not systematically in areas with unusually high percentages of African-American or Hispanic populations, compared with other areas that had significant industrial development. Moreover, results showed that characteristics of communities, after siting of TSDFs, have trends that mirror those in the population at large.

Other reviewers have assessed the adequacy of the environmental health literature and presented their conclusions. One reviewer, who examined the literature on race, class, and environmental health including much of the material described in this chapter [63], overlaid his review with editorial comments and recommendations on study designs and future research directions pertaining to environmental justice concerns. Brown concluded: "The overwhelming bulk of evidence supports the 'environmental justice' belief that environmental hazards are inequitably distributed by class, and especially race." He recommends that investigations into class and race issues move away from traditional epidemiological designs and toward in-depth ethnographic analysis of communities and neighborhoods. Brown's recommendation is predicated on the belief that traditional epidemiological designs exclude community input as a means of minimizing bias in the conduct of the epidemiological investigation. He offered no data to support his belief.

Bowen critiqued forty-two environmental justice publications that spanned three decades [64]. He concluded, "[t]he empirical foundations of environmental justice are so underdeveloped that little can be said with scientific authority regarding the existence of geographical patterns of disproportionate distributions and their health effects on minority, low-income, and other disadvantaged communities." The conclusion was based on such alleged factors as: inadequate study designs, inconsistent geographic units of analysis, and lack of health data for investigated minority communities.

While Brown [63] and Bowen [64] expressed different opinions on the adequacy of research underpinnings environmental justice claims, both agree that the scientific database should be improved and expanded.

Some persons have expressed skepticism about the future of environmental justice. For example, Foreman [13] described what he considers to be a crucial problem facing the environmental justice movement. He characterized the movement as being a "big tent" that embraces African American, Latino, Asian, and Native American grassroots organizations and their allies. He contends that maintaining harmony within the environmental justice movement has led to avoiding difficult but necessary decisions. He states, "The movement presumes that any person of color voicing any environmental-related anxiety or aspiration represents a genuine environmental justice problem. Indeed, a broader redistributative and cultural agenda, as well as a profound discomfort with industrial capitalism generally, lurks just behind the concerns over unequal pollution impacts." This strong assertion is given without supporting factual data. Foreman casts environmental justice as a movement with fragile political support and an uncertain future.

Although there is room for scholarly debate over the nature and extent of environmental injustices, there is no room for debate that such injustices are morally wrong and societally unethical when and where they exist. This leads to a simple policy: environmental injustice is wrong and must be prevented.

10.7　POLICY IMPLICATIONS OF ENVIRONMENTAL JUSTICE

Environmental justice has become an integral part of social justice policy and practice in the United States and elsewhere. The deliberate placement of environment hazards in communities of minority populations or low income is a breech of any framework of ethics. Although there is no comprehensive federal statute to prevent environmental injustices, litigation can be brought under civil rights statutes, both criminal and civil, and also litigated under common law. For example, some communities have successfully litigated against chemical companies that desired to expand their production facilities, thereby potentially increasing pollution in adjacent residential communities.

In addition to federal environmental justice policies, as expressed in Executive Order 12898, states have adopted various environmental justice policies. These policies vary according to each state's needs and conditions, although it is unclear if all states have developed environmental justice policies and resources. The following examples from the states of California, Illinois, Maryland, New York, and Washington will illustrate state-developed environmental justice policies.

California—"The Department of Toxic Substances Control is committed to ensuring that all of the state's population, without regard to color, national origin or income, are equally protected from adverse human or environmental effects as a result of the Department's policies, programs or activities. The Department will: 1. Ensure that, to the extent feasible, its decisions, actions and rulemaking avoid adding to disproportionate environmental and/or health impacts on affected communities and reduce disproportionate environmental and related health impacts on such communities. [...] 4. Allocate its permitting, enforcement and clean-up resources, to the extent feasible, so as to reduce disproportionate environmental and related health impacts on ethnic minority and low-income communities. [...] [65].

Illinois—"The Illinois Environmental Protection Agency (Illinois EPA or Agency) is committed to protecting the health of the citizens of Illinois and its environment, and to promoting environmental equity in the administration of its programs to the extent it may do so legally and practicably. The Illinois EPA supports the objectives of achieving environmental justice for all of the citizens of Illinois. Key goals of this policy are as follows: 1) to ensure that communities are not disproportionately impacted by degradation of the environment or receive a less than equitable share of environmental protection and benefits; 2) to strengthen the public's involvement in environmental decision-making, including permitting and regulation, and where practicable, enforcement matters; 3) to ensure that Agency personnel use a common approach to addressing EJ issues; and 4) to ensure that the Illinois EPA continues to refine its environmental justice strategy to ensure that it continues to protect the health of the citizens of Illinois and its environment, promotes environmental equity in the administration of its programs, and is responsive to the communities it serves." [...] [66].

Maryland—"In March 2001, the Governor created the Commission on Environmental Justice and Sustainable Communities [...]. The Commission advises State agencies on issues related to environmental justice and sustainable communities. With the Children's Environmental Health and Protection Advisory Council, it coordinates recommendations on such issues. The Commission analyzes and reviews what impact current State laws, regulations, and policy have on the equitable treatment and protection of communities threatened by development or environmental pollution, and determines what areas in the State need immediate attention. The Commission assesses the adequacy of current statutes to ensure environmental justice, and develops criteria to pinpoint which communities need sustaining." [...] [67].

New York—"This policy provides guidance for incorporating environmental justice concerns into the New York State Department of Environmental Conservation (DEC) environmental permit review process and the DEC application of the State Environmental Quality Review Act. The policy also incorporates environmental justice concerns into some aspects of the DEC's enforcement program, grants program and public participation provisions. The policy is written to assist DEC staff, the regulated community and the public in understanding the requirements and review process. [...] It is the general policy of DEC to promote environmental justice and incorporate measures for achieving environmental justice into its programs, policies, regulations, legislative proposals and activities. This policy is specifically intended to ensure that DEC's environmental permit process promotes environmental justice." [...] [68].

Washington—"The Washington State Board of Health's Environmental Justice Committee recognizes the progress that many community organizations, businesses, industries, and agencies have made in reducing pollution sources and minimizing the impact these sources have on human health. These guidelines are intended to build on these successes and promote an even more healthy and equitable environment for everyone. [...] The Board's Environmental Justice guidelines ask that: 1. All environmental and public health laws and regulations are equitably enforced in all communities. 2. Policy-makers use a combination of scientific evidence, traditional knowledge, and public testimony in their decision-making process." [...] [69].

Although these five states are but 10 percent of all states, some interesting policy observations can be gleaned from their environmental justice policies. First, environmental justice responsibilities are housed in different administrative offices across states, reflecting each state's sense of its environmental justice responsibilities. It is also noteworthy that three of the five states use their permit authorities as a means to identify environmental justice issues. This is an example of how a regulatory mechanism, permits, can be used for larger social purposes.

10.8   SUMMARY

If one accepts the premise that the search for environmental justice, framed as a political and societal movement, leads inevitably to confrontation raises the question whether a better approach might exist. A lesson from the Comprehensive Environmental Response, Compensation, and Liability Act is instructive. The statute contains the "polluter-pays" principle, which necessarily leads to assessing blame; that is, identifying which parties are responsible for paying the costs of remediating hazardous waste sites (chapters 1 and 8). This has resulted in assessing blame, which has caused confrontation, conflict, and litigation. Will environmental justice, with its tendency to assign blame for allegedly targeting minority communities as sites for hazardous waste sites, suffer the same fate as the act? The question cannot be answered yet, but some would argue that a philosophical focus on equity rather than justice might achieve better results.

Given the level of concern that environmental injustices be prevented, the policies and actions already set in motion will likely result in fewer minorities and persons of low income being exposed to environmental hazards. This is a very desirable outcome in terms of both public health and social justice. Government efforts to collect data on the impacts of environmental hazards on minorities and to ensure that federal policies do not create environmental inequities will be important. Furthermore, strengthening the training and education programs that inform communities about their environmental status will be required. A commitment to enhance the number and diversity of minority health and environmental professionals must be allied with such efforts.

The importance of fairness will be the centerpiece of whatever gets done. The U.S. public gives strong support to actions predicated on what is fair. Changes in voting rights legislation and fairness in job opportunities are examples of fairness translated into statutory action. Having minorities and persons of low income experience a disproportionate burden of environmental health hazards is unfair. Environmental justice can redress many of these inequities through improved data collection, culturally sensitive regulations, and heightened consciousness.

10.9   POLICY QUESTIONS

1.   Using the material in chapter 10, together with any auxiliary material, discuss which of these terms is preferable: *environmental racism, environmental justice,* or *environmental equity* for each of the following situations. (A) Assume you are a member of a community group that is concerned about a local landfill. (B)

Assume you are a senior policy maker at the U.S. EPA. (C) Assume you are a local public health official. Use critical thinking (as described in chapter 1) for each of these three cases.

2. Examine the seventeen elements of Principles of Environmental Justice (Table 10.2), as developed in 1991 at the First People of Color Environmental Leadership Summit. Select any three of the seventeen elements and critically discuss how they could be made into policy by local governments (e.g., county commissioners). Discuss how the selected elements could be made operational as county policy.

3. Draw a random sample of seven U.S. states, then use Internet resources to determine each state's environmental justice/justice (EJ) resources and policies. Rank the seven states in terms of their commitment to achieving EJ goals.

4. Select three federal government agencies and query them for copies of their actions in compliance with Presidential Executive Order 12898. Compare and rank each agency's responsiveness to the Order.

5. Using Internet resources, assess the policies and products of EPA's National Environmental Justice Advisory Committee (NEJAC). What policies seem particularly important to you in regard to preventing environmental injustices to minority or low-income populations?

6. Assume that you are a senior environmental health advisor to the director of the local health department. A community EJ group has asked to meet with the director for purpose of presenting health concerns about odors emanating from a local landfill. The director has asked for your advice as how to respond to the group's request. Detail your advice. Be specific, stating reasons for each recommendation that you provide to the director.

7. How do the demographics investigations of persons residing near hazardous waste sites differentiate between residential proximity to NPL. sites and RCRA facilities?

8. Several reports on EJ were not peer reviewed nor published in the peer-reviewed scientific literature. Yet these reports had, in their time, a great impact on EJ concerns and directions. Given their lack of peer review, should the reports have been ignored?

9. Choose any of the demographics investigations described in chapter 10 and evaluate its strengths and limitations.

10. How would you strengthen *Agenda 21* for relevance to EJ?

NOTES

1. The GAO Human Capital Reform Act of 2004 changed the agency's name to Government Accountability Office, effective July 7, 2004 [17].

2. Zip code areas are administrative units established by the U.S. Postal Service for the distribution of mail and do not generally respect political or census statistical area boundaries [26]. Zip code areas for 1980 contained, on the average, about twice as many persons (6,500) as did census tracts (3,900) [27].

3. CDC considers children to have an elevated level of lead if the BLL is at least 10 μg/dL.

Further, the agency recommends medical evaluation and environmental investigation and remediation should be done for all children with BLLs greater than 20 µg/dL [34].

4. This does not imply that who pays for site remediation is unimportant. Indeed, much of the controversy attending CERCLA is about the "polluter pays" principle that undergirds the statute (See [40]). However, who pays for site cleanups has not been part of the debate on environmental justice.

5. Generally a census tract is a small statistical subdivision of a county. Census tracts have identifiable boundaries and average about four thousand persons [27].

6. SMSAs consist of cities with populations of fifty thousand or more persons including surrounding counties or urbanized areas but omitting many rural areas and small cities and towns [27].

7. An EPA Record of Decision discusses the various cleanup techniques that were considered for a site and explains why a particular course of action was selected [43].

8. Census block groups can be geographic block groups or tabulation block groups. The former are clusters of blocks having the same first digit of their three-digit identifying numbers within census tracts or block numbering areas. Tabulation block groups and geographic block groups may be split to present data for every unique combination of county subdivision, place, American Indian and Alaska Native area, urbanized area, voting district, urban/rural and congressional district shown in the data product [26].

9. Census blocks are small areas bounded on all sides by visible features such as streets, roads, streams, and railroad tracks, and by invisible boundaries such as city, town, township, and county limits, property lines, and short, imaginary extensions of streets and roads [26].

10. The TRI database was established under provisions of Title III of the Superfund Amendments and Reauthorization Act of 1986, as discussed in chapter 8.

REFERENCES

1. EPA (U.S. Environmental Protection Agency), Environmental Justice. Available at http://www.epa.gov/ebtpages/environmentaljustice.html, 2005

2. Warren, R.C., The morbidity/mortality gaps: What is the problem?, *Annals of Epidemiology* 3, 127, 1993.

3. Sexton, K., Olden, K., and Johnson, B.L., Environmental justice: The central role of research in establishing a credible scientific foundation for informed decision making, *Toxicology and Industrial Health*, 9, 685, 1993.

4. EPA (U.S. Environmental Protection Agency), Environmental Equity: Reducing Risk for All Communities, Volumes 1 and 2, Office of Solid Waste and Emergency Response, Washington, D.C., 1992, 2, 8, 9.

5. Antonio, P. et al., Assessments of hazards posed by abandoned uranium mines on the Navajo Nation, in *National Minority Health Conference: Focus on Environmental Contamination*, Johnson, B., Williams, R., and Harris, C., eds,, Princeton Scientific Publishing, Princeton, NJ, 37, 1992.

6. WDOE (Washington Department of Ecology), A Study on Environmental Equity in Washington State, Report No 95-413, Olympia, 1995.

7. Protocol Committee, Symposium on Health Research and Needs to Ensure Environmental Justice, National Institute for Environmental Health Sciences, Research Triangle Park, NC, 1994.

8. ATSDR (Agency for Toxic Substances and Disease Registry), *Mississippi Delta Project: Health and Environmental Prospectus*, Atlanta, GA, 1995, 6.

9. CRARM (Commission on Risk Assessment and Risk Management), Framework for Environmental Health Risk Management, U.S. Environmental Protection Agency, Washington, D.C., 1997.

10. Bullard, R.D., *Dumping in Dixie: Race, Class, and Environmental Quality*, Westview Press, Boulder, CO, 1990, 139.

11. Reilly, W., Environmental equity: EPA's position, *EPA Journal*, March/April, 18, 1992.

12. EJRC (Environmental Justice Research Center), Environmental justice summit draws over 1,200 delegates, Environmental Justice Center, Clark-Atlanta University, Atlanta, GA, 2002.

13. Foreman, C.H. Jr., A winning hand? The uncertain future of environmental justice, *The Brookings Review*, Spring, 22, 1996.

14. Bullard, R.D., Environmental justice: Strategies for achieving health and sustainable communities, Paper presented at the International Congress on Hazardous Waste and Public Health and Ecology, Atlanta, GA, June, 1995.

15. GAO (General Accounting Office), Siting of Hazardous Waste Landfills and Their Correspondence with Racial and Economic Status of Surrounding Communities, Washington, D.C., 1983.

16. Bullard, R.D., 1992. Use of demographic data to evaluate minority environmental health issues: The role of case studies, in *National Minority Health Conference: Focus on Environmental Contamination*, Johnson, B., Williams, R., and Harris, C., eds., Princeton Scientific Publishing, Princeton, NJ, 1992, 161.

17. GAO (Government Accountability Office), GAO Human Capital Reform Act of 2004. Available at http://www.gao.gov, 2005.

18. GAO (General Accounting Office), Siting of Hazardous Waste Landfills and Their Correspondence with Racial and Economic Status of Surrounding Communities, Washington, D.C., 1983.

19. Teran, S.P. et al., Is biologic monitoring for toxic chemicals always what exposed communities want? A practical ethical approach to informed consent, in *Hazardous Waste and Public Health*, Andrews, J., et al., eds., Princeton Scientific Publishing, Princeton, NJ, 1994, 90..

20. Bullard, R.D., ed., Unequal Protection: Environmental Justice and Communities of Color. Sierra Club, San Francisco, 1994.

21. Bryant, B. and Mohai, P., 1992. The Michigan conference: A turning point. *EPA Journal*, March/April, 9, 1992.

22. Johnson, B.L., Williams, R., and Harris, C., eds., *National Minority Health Conference: Focus on Environmental Contamination*, Princeton Scientific Publishing, Princeton, NJ, 1992.

23. First National People of Color Environmental Leadership Summit. Available at http://www. apcd.org/permit/t5tutorial/t5ej/tsld005.htm, 1991.

24. Sexton, K. and Anderson, Y., eds., Equity in environmental health: Research issues and needs, *Toxicology and Industrial Health*, 9, no. 5, 1993.

25. CSJ (Commission on Social Justice), Toxic Waste and Race in the United States, United Church of Christ, New York, 1987.

26. BOC (Bureau of Census), Census of Population and Housing, 1990: Summary Tape File 3 on CD-ROM Technical Documentation, Department of Commerce, Washington, D.C., 1992.

27. Anderton, D. et al., Environmental equity issues in metropolitan areas, *Evaluation Review*, 18, 123, 1994.

28. Mohai, P. and Bryant, B., Environmental racism: Reviewing the evidence, in *Race and the Incidence of Environmental Hazards: A Time for Discourse,* Bryant, B. and Mohai, P., eds., Westview Press, Boulder, CO, 1992, 163.

29. Wernette, D. and Nieves, L., Minorities and air pollution: A preliminary geo-demographic analysis, Paper presented at the Socioeconomic Research Analysis Conference, Baltimore, MD, June 27–28., 1991.

30. COSSMHO (Coalition of Hispanic Health and Human Services Organizations), Hispanic environmental health: Ambient and indoor air pollution, *Otolaryngology Head Neck Surgery,* 114, 256, 1996.

31. ATSDR (Agency for Toxic Substances and Disease Registry), The Nature and Extent of Lead Poisoning in Children in the United States: A Report to Congress, Atlanta, GA, 1988.

32. CDC (Centers for Disease Control and Prevention), Update: Blood lead levels-United States, 1991–1994, *Morbidity & Mortality Weekly Report,* 46, 141, 1997.

33. CDC (Centers for Disease Control and Prevention), Blood Lead Levels—United States, 1999–2002, *Morbidity & Mortality Weekly Report,* 54, 513, 2005.

35. CDC (Centers for Disease Control and Prevention), Surveillance for elevated blood lead levels among children—United States, 1997–2001, *Morbidity & Mortality Weekly Report,* 52, 1, 2003.

36. Calderon, R.L. et al., Health risk from contaminated water: Do class and race matter?, *Toxicology and Industrial Health,* 9, 879, 1993.

37. Metzger, R, Delgado, J.L., and Herrell, R., Environmental health and Hispanic children, *Environmental Health Perspectives,* 103(Suppl 6), 25, 1995.

38. Lavelle, M. and Coyle, M., Unequal protection. The racial divide in environmental law, *National Law Journal,* September 21, 1992.

39. Maantay, J., Zoning law, health, and environmental justice" What's the connection?, *Journal of Law, Medicine & Ethics,* 30, 572, 2002.

40. Hird, J., Environmental policy and equity: The case of Superfund, *Journal of Policy Analysis and Management,* 12, 323, 1993.

41. Anderton, D. et al., Environmental equity: The demographics of dumping, *Demography,* 31, 229, 1994.

42. Zimmerman, R., Social equity and environmental risk, *Risk Analysis,* 13, 649, 1993.

43. EPA (U.S. Environmental Protection Agency), Superfund Progress—Aficionado's Version, Report PB92-963267, Office of Solid Waste and Emergency Response, Washington, D.C., 1992, 11.

44. GAO (General Accounting Office), Demographics of People Living Near Waste Facilities, Washington, D.C., 1995.

45. Been, V., What's fairness got to do with it? Environmental justice and the siting of locally undesirable land uses, *Cornell Law Review,* 78, 1001, 1993.

46. Been, V., Locally undesirable land uses in minority neighborhoods: Disproportionate siting or market dynamics?, *Yale Law Journal,* 103, 1383, 1994.

47. Been, V., Analyzing evidence of environmental justice, *Journal of Land Use and Environmental Law,* 11, 1, 1995.

48. Heitgerd, J.L., Burg, J.R., and Strickland, H.G., A geographic information systems approach to estimating and assessing National Priorities List site demographics: Racial and Hispanic origin composition, *International Journal of Occupational Medical Toxicology.,* 4, 343, 1995.

49. Burg, J.R., 1996. Personal communication.

50. Oakes, J., Anderton, D., and Anderson, A., A longitudinal analysis of environmental equity in communities with hazardous waste facilities, *Social Science Research*, 25, 125, 1996.

51. Anderton, D.L., Oakes, J.M., and Egan, K.L., Environmental equity in Superfund: Demographics of the discovery and prioritization of abandoned toxic sites, *Evaluation Review*, 21(1), 3, 1997.

52. Baden, B. and Coursey, D., The locality of waste sites within the city of Chicago: A demographic, social, and economic analysis, Harris School of Public Policy Studies, University of Chicago, Chicago, 1997.

53. Carlin, B.P. and Xia, A.H., Assessing environmental justice using Bayesian hierarchical models: Two case studies, *Journal of Exposure Analysis Environmental. Epidemiology*, 9, 66, 1999.

54. Davidson, P. and Anderton, D.L., Demographics of dumping II: A national environmental equity survey and the distribution of hazardous materials handlers, *Demography*, 37(4), 461, 2000.

55. Morello-Frosch, R. et al., Environmental justice and regional inequity in southern California: Implications for future research, *Environmental Health Perspectives*, 110(Suppl 2), 149, 2002.

56. Clinton, W.J., Federal actions to address environmental justice in minority populations and low-income populations, *Federal Register*, 59, 7629, 1994.

57. EPA (U.S. Environmental Protection Agency), Evaluation Report. EPA Needs to Consistently Implement the Intent of the Executive Order on Environmental Justice, Office of Inspector General, Washington, D.C., March 1, 2004.

58. EPA (U.S. Environmental Protection Agency), Memorandum. Agency Statement on the Inspector General's Report on EPA's Environmental Justice Implementation, Office of Enforcement and Compliance Assurance, Office of Environmental Justice, Washington, D.C., March 3, 2004.

59. UNEP (United Nations Environment Programme), Report of the United Nations Conference on Environment and Development, United Nations, New York, 1992.

60. Lloyd-Smith, M.E. and Bell, L., Toxic disputes and the rise of environmental justice in Australia. *International Journal of Occupational Environmental Health*, 9, 14, 2003.

61. Varga, C., Kiss, I., and Ember, I., The lack of environmental justice in Central and Eastern Europe, *Environmental Health Perspectives*, 110, 2000, A662.

62. Cutter, S.L., Scott, M.S., and Hill, A.A., Spatial variability in toxicity indicators used to rank chemical risks, *American Journal of Public Health*, 92(3), 420, 2002.

63. Brown, P., Race, class, and environmental health: A review of systematization of the literature, *Environmental Research*, 69, 15, 1995.

64. Bowen, W., An analytical review of environmental justice research: What do we really know? *Environmental Management*, 29(1), 3, 2002.

65. State of California, Draft environmental justice policy, Available at http://www.dtsc.ca.gov/PolicyAndProcedures/env_justice/OEA_POL_DRAFTEJ.pdf, 2003.

66. State of Illinois, Interim environmental justice (EJ) policy, Available at http://www.epa.state.il.us/environmental-justice/policy.html, 2005.

67. State of Maryland, Commission on Environmental Justice & Sustainable Communities, Available at http:www.mdarchives.state.md.us/msa/mdmanual/26excom/html/13envju.htm/, 2003.

68. State of New York, Commission Policy—29, Environmental Justice and Permitting, Available at http://www.dec.state.ny.us/website/ej/ejpolicy.html, 2005.

69. State of Washington, Proposed guidelines on environmental justice, Washington State Board of Health, Available at http://www.doh.wa.gov/sboh/Priorities/EJustice/EJGuidelinesRev.pdf, 2005.

11 Policy Impacts of Risk Assessment

11.1 INTRODUCTION

The development of a comprehensive set of federal laws to protect the environment and human health was discussed in chapters 4–8. The key laws are built upon a regulatory structure, using a command and control approach. In particular, the Clean Air Act, Clean Water Act, Safe Drinking Water Act, Superfund Act, and their amendments, direct EPA to establish standards for pollutants in air, water, and soil, respectively. Over time, EPA and other regulatory agencies (e.g., OSHA) have turned to risk assessment as the method of choice for developing their standards. The evolution of risk assessment from a regulatory method of convenience to a driver of national environmental policies is best appreciated by reviewing how risk assessment evolved in U.S. regulatory agencies, followed by how risk assessment is applied to reducing the consequences of environmental hazards.

"Risk assessment means the characterization of the potential adverse health effects of human exposures to environmental hazards" [1].

As defined by the National Research Council (NRC), "[r]isk assessment means the characterization of the potential adverse health effects of human exposures to environmental hazards" [1]. This definition was developed for application to situations of human exposure. As the concept and methods of risk assessment have expanded to include ecological situations and occupational injuries, other definitions have emerged that are specific to their situations. These condition-specific situations of risk assessment will be discussed later in this chapter.

This chapter describes the evolution of risk assessment as a means to make environmental health policy, the different kinds of risk assessment currently in practice, uses of risk assessment for management of environmental hazards, and the debate between risk assessors and public health practitioners on how to assess and manage environmental hazards. Application of risk assessment as a means to compare the importance of individual environmental hazards is discussed, along with differences in perception between public health specialists and risk assessors as to the applicability of risk assessment for public health purposes.

Before proceeding, it is important to define two important words: hazard and risk. *Hazard* is a factor or exposure that may adversely affect health. *Risk* is the probability that an event will occur [2]. Examples of environmental hazards are air pollutants, adulterated food, and radiation. We all experience some environmental hazards in our everyday life (e.g., minute or trace amounts of air contaminants). Whether a hazard causes harm to human or ecological health depends on such factors as extent of exposure, potency of a toxicant, and susceptibility factors such as age and health condition. The extent to which exposure to a hazard has the potential to cause harm can be expressed as the risk (i.e., the potential that the exposure will be harmful).

HAZARD: A factor or exposure that may adversely affect health.
RISK: The probability that an event will occur. [2]

11.2   EVOLUTION OF RISK ASSESSMENT

While the term *risk assessment* is a relatively new term with respect to environmental regulations and public health practice, the concept of trying to determine the consequences of exposure to environmental hazards and their potential impact on human health is not new. Indeed, as humankind evolved, our distant ancestors had to constantly assess the consequences of hazards around them. Will wild animals harm us? Will the tribes migrating into our area be warlike? Will forest fire destroy us? These kinds of questions relate to survival of individuals and groups of people and assessment of hazards remains a human endeavor still today.

Hazards to public health have long been of great interest to governments and health specialists. Problems of contaminated water, in particular, and waste management were issues of concern in the United States in the eighteenth century and continued thereafter. More recently, in the twentieth and twenty-first centuries, health hazards from industrialization, pests (e.g., West Nile fever from migratory mosquitoes) and biological weapons of terror have been added to the public health agenda. These kinds of threats to the public's health have historically been approached using the traditional public approach of *hazard evaluation* (i.e., assess the nature and severity of a hazard but not calculate a risk estimate). Public health hazard evaluations conducted by federal authorities consist of: assess the relevant scientific findings, develop consensus about the threat to public health, provide services to potentially affected populations, particularly through assistance to state and local health departments. In distinction, federal regulatory risk assessment consists of: risk assessment, risk management via regulatory actions, and services to affected populations, particularly state environmental departments.

The professional disciplines of risk assessors have expanded over time. Probably the earliest risk assessors (but without being called such) were radiation biologists, persons interested in the biological hazard of radiation. This kind of risk assessment began early in the twentieth century and continues today. As U.S. industry expanded during and following World War II, concerns about workplace hazards grew, but at a rate slower than industrial expansion. Industrial hygienists, persons with specialty training in chemistry, toxicology, and industrial processes, conducted *safety assessment* of workplace substances. Such assessments considered the toxicology of a substance or agent of concern, sometimes using the results from their company's own toxicology studies, and workplace exposure data.

The need for assessment of occupational hazards contributed to the creation in 1938 of a key organization, the National Conference of Governmental Industrial Hygienists. The organization originally limited its membership to two representatives from each federal government agency with industrial hygiene programs. The Threshold Limit Values Chemical Substances Committee was established in 1941 [3]. This group was charged with recommending exposure limits for chemical substances in the workplace. This remains this committee's primary work. In 1946, the parent organization changed its name to the American Conference of Governmental Industrial Hygienists (ACGIH®) and has broadened its membership from only industrial hygienists to other relevant disciplines (e.g., toxicologists). Moreover, government employees who served on ACGIH's committees represented only themselves, not the government agencies with which they were affiliated.

ACGIH's Threshold Limit Values (TLVs®)[1] are developed by scientists from government, private industry, and academia. They are advisory guidelines (i.e., do not carry the weight of law). The TLVs have had a significant impact on workplace standards. TLVs were adopted by the newly-created OSHA in the early 1970s as Permissible Exposure Limits (PELs), owing to OSHA's quick need for legally enforceable workplace standards for workplace hazardous substances. PELs are legally enforceable by OSHA under authorities in the Occupational Safety and Health Act. Some states and other countries have adopted specific TLVs for their own uses, although ACGIH cautions that TLVs should not be adopted as standards without an analysis of other factors necessary to make risk management decisions [3]. Some persons have questioned the substantial role of industry representatives in the work of TLV committees [4,5]. They assert that corporate scientists were given primary responsibility for developing TLVs on proprietary chemicals produced by their employers, suggesting a possible conflict of interest.

As a matter of policy, should government agencies, such as OSHA, adopt exposure limits (e.g., TLVs) that are developed by nongovernment organizations (e.g., ACGIH) that do not provide opportunities for public participation (e.g., the opportunity to review and comment on proposed exposure limits)? This is a difficult question, because a case can be made for both adoption or rejection of non-government derived exposure limits. Adoption en masse of the TLVs gave a nascent OSHA a set of PELs circa 1970 that could be enforced by the agency. Rejection of the TLVs would have forced OSHA to establish, working through a process that would allow the public's involvement, PELs for individual chemicals on a one-by-one basis This mechanism would have been met with industry opposition and fallen victim to protracted litigation, chemical-by-chemical.

11.3   FEDERAL GOVERNMENT'S INVOLVEMENT

Though the early 1950s, federal public health authorities conducted hazard evaluations which were advisory to states and local health departments. The advisories did not bear the weight of regulatory imperative, but were effective to some degree because they came from a trusted source, often the Office of Surgeon General. This lack of legal authority began to change with the enactment of the Air Pollution Control Act Amendments of 1963. The federal Department of Health, Education, and Welfare[2] was directed to develop air quality criteria that were to be "an expression of the scientific knowledge of the effect of various concentrations of pollutants depending on the intended use of a particular [m]ass of air" [6]. However, these criteria were merely advisory and states' air pollution control agencies were not mandated to adopt the air quality criteria. Nevertheless, the die had been cast, with later congressional environmental legislation that mandated EPA to develop environmental standards, using risk assessment methods.

So the concepts of hazard and risk are nothing new. However, what *is* new are the development and formalization of risk assessment as a body of both scholarship and practice used to aid decisions on the consequences of exposure to individual hazards in the environment. In particular, federal government regulatory agencies such as EPA and OSHA use risk assessment to evaluate the importance of various levels of toxicants in the atmosphere and the workplace, respectively.

Current risk assessment policies and practices at EPA have evolved over two decades. With the establishment of EPA in 1970 came the statutory responsibility to set air and water quality standards for toxicants in the environment. At the same time, the OSH Act of 1970 required OSHA to regulate workplace conditions, including control of workplace hazards through Permitted Exposure Levels (PELs). PELs are enforceable under federal and state laws (i.e., employers can be fined if PELs are exceeded). The same act directed the National Institute for Safety and Health (NIOSH)[3] to develop and disseminate criteria documents that would contain Recommended Exposure Levels (RELs) for specific toxicants and physical agents in U.S. workplaces. Therefore, EPA, NIOSH, and OSHA were all faced with how to evaluate environmental hazards and develop exposure recommendations (NIOSH) or standards (EPA, OSHA). An early challenge to EPA and OSHA was how to establish standards in ways that would withstand judicial review. Other federal regulatory agencies had the same need, leading to the formation of the Interagency Liaison Regulatory Group (ILRG) in 1977.

The ILRG consisted of representatives from the four primary regulatory agencies involved with particular environmental hazards: EPA, OSHA, FDA, and the CPSC. The ILRG was an attempt by the Carter administration to coordinate federal government programs and to develop and implement consensus policies and procedures on matters of interagency common interests. The ILRG's mission "was to coordinate the activities of its members and help them reach their common goals more effectively" [7]. The ILRG formed work groups around particular issues. Each group's effort was supervised, through surrogates, by one of the heads of the four member agencies comprising the ILRG.

One of the ILRG work groups was the Risk Assessment Work Group (RAWG). Composed of scientists from the four member agencies, this group undertook the task of developing a document that would be a state-of-the-science discussion on cancer risk assessment. Despite political challenges inherent in the work group's efforts, primarily differences between OSHA and EPA on matters of risk assessment practices, the RAWG produced a document entitled "Scientific Bases for Identification of Potential Carcinogens and Estimation of Risks," which was published in the *Federal Register* in 1979 [7]. The ILRG guidelines on cancer risk assessment later appeared in the *Journal of the National Cancer Institute* in 1979. The guidelines, developed over a two-year period, were a product of compromise between ILRG member agencies but, in the end, had limited impact on the cancer risk assessment policies of EPA, OSHA, and other agencies responsibilities for environmental health policies. Each agency simply adopted its own cancer risk assessment policies, as they interpreted their particular statutory responsibilities.

With the passing of the Carter administration, the incoming Reagan administration tried to revise federal policies and practices on environmental regulations. Reagan's presidential campaign of 1980 included a goal of eliminating several federal departments and agencies, including the Department of Education, EPA, and OSHA, as matters of "regulatory reform." With Democrats in control of both houses of Congress, elimination of these federal agencies did not occur. While some of the Reagan agenda can be ascribed to campaign rhetoric, in fact, political appointees at EPA and OSHA were able to make significant changes in how workplace and environmental standards were determined [8]. Policy and procedural changes made standards setting more problematic through protracted reviews of documents and reduction in staff. The second Reagan term in office saw the appointment of officials more willing to harmonize their political agenda with the statutory requirements of standards development and regulatory implementation.

11.4 HUMAN HEALTH RISK ASSESSMENT

Beginning in the mid-1970s, U.S. federal regulatory agencies have struggled to develop policies to control workplace and environmental hazards. As discussed in chapters 4–8, federal legislation such as the Clean Air Act, the Clean Water Act, the Occupational Safety and Health Act, and the Food, Drug, and Cosmetics Act directed EPA, OSHA, and FDA, respectively, to promulgate regulations and standards to control various environmental hazards. Not the least of the challenges has been how to regulate chemical carcinogens. The task undertaken by federal regulators was—and remains—daunting: how to extrapolate findings from chemical-specific laboratory toxicology studies (and occasional epidemiological investigations, e.g., workers' health studies) to apply to humans in occupational or community conditions. In the late 1970s, OSHA's cancer policy embodied both hazard evaluation and control strategies. Regulatory action was predicated on hazard identification—usually a positive laboratory animal carcinogenesis study—coupled with supporting evidence such as a positive mutagenicity assay [9]. Hazard control was then triggered, targeted at determining and regulating the lowest feasible workplace exposure to the carcinogen under regulation.

In 1980, a Supreme Court decision changed how U.S. regulatory agencies developed risk assessment in support of environmental and workplace standards. As background to the Court's decision, industry had sued OSHA over the agency's reduction of its workplace limit on workers' exposure to benzene from 10 ppm to 1 ppm. The affected industries had argued successfully in a lower court that the new standard should be set aside because OSHA had not shown that its health benefits outweighed its substantial costs. The Supreme Court, however, found it unnecessary to resolve the dispute over whether the law required the agency to balance risks and benefits, ruling instead that the law demanded that OSHA must first establish that current allowed exposures posed a significant risk to workers' health [10]. OSHA, EPA, and other federal regulatory agencies interpreted the Court's decision as a mandate for them to undergird their regulatory actions through use of risk assessment. But how should this occur was the key question facing the agencies circa 1980. Congress was aware of this challenge and intervened by asking advice from the National Research Council.

In 1982, the National Research Council (NRC) was directed by Congress to conduct a study of how federal regulatory agencies should manage their risk assessment processes. The ensuing NRC report *Risk Assessment in the Federal Government: Managing the Process* [1] provided the framework being sought by the federal regulatory agencies. According to the NRC, risk assessment consists of four elements:

- Hazard identification—The qualitative evaluation of the adverse health effects of a substance(s) in animals or in humans.
- Dose-response assessment—The process of estimating the relation between the dose (i.e., the amount of the substance received by the target organism) of a substance(s) and the incidence of an adverse health effect.
- Exposure assessment—The evaluation of the types (routes and media), magnitudes, time, and duration of actual or anticipated exposures and of doses, when known; and, when appropriate, the number of persons who are likely to be exposed.
- Risk characterization—The process of estimating the probable incidence of an adverse health effect to humans under various conditions of exposure, including a description of the uncertainties involved.

The four-step risk assessment paradigm and other recommendations in the 1983 NRC report set into motion a policy change in how federal agencies regulate environmental hazards. In some sense, this was due to serendipity. In 1983, William Ruckelshaus returned to become EPA administrator, following the resignation of an administrator who had become a political liability to the Reagan administration. Ruckelshaus inherited a demoralized EPA staff and a legacy of adverse public opinion about the attempted elimination of the agency by the Reagan administration. According to one source, he realized the need to reorient EPA around a process that would reinvigorate the science base of the agency and reposition EPA to resume its statutory mandates of regulating environmental hazards, as required by existing environmental statutes [11]. The process that Ruckelshaus chose was the risk assessment and risk management paradigm outlined in the NRC report. It became EPA policy to conduct risk assessments of environmental hazards in accord with the NRC paradigm and to keep risk management separate from the risk assessment process in order to let the science of risk assessment not be unduly influenced by the political and societal aspects of risk management. Ruckelshaus's policy decision served EPA well and righted the policy structure of the agency.

11.4.1   CASE EXAMPLE: ARSENIC RISK BROUHAHA

Reassessing the U.S. arsenic standard is an example of a risk assessment that got caught in the political thickets of environmental politics and policies. Arsenic is a well-known poison when ingested in a sufficient dose. Avoidance of exposure to arsenic is good advice, but in many areas of the world, including the United States, arsenic naturally occurs in soil and contaminates groundwater supplies. The concentration of arsenic in soil varies widely, generally ranging from 1 to 40 ppm,[4] with an average of 5 ppm [12]. The concentration of arsenic in natural surface and groundwater is generally about 1 ppb,[5] but may exceed 1,000 ppb in mining areas or where arsenic levels in soil are high [ibid.]. Groundwater supplies are primary sources of drinking water, whether ancient wells in Mongolia or municipal water supplies in some areas of the western United States As a hazard to human health, the question arises, how much arsenic in drinking water presents an unacceptable risk to the public's health?

In the United States, the arsenic in drinking water standard was set at 50 ppb by the Public Health Service in 1947, which precedes the establishment of EPA in 1970. This level was maintained for more than fifty years, even though scientific evidence had mounted that lower levels of arsenic in drinking water were necessary to protect the public's health (Table 11.1). Decreasing the standard from 50 ppb to a lower level

TABLE 11.1
Key Events in the Revision of EPA's Arsenic in Drinking Water Standard

Year	Event
1993	WHO decreased from 50 ppb to 10 ppb
1995	EPA chose not to adopt WHO's recommendation
1996	Safe Drinking Water Act Amendments directed EPA to update their arsenic standard by January 1, 2001
1999	NAS recommended that the arsenic standard be lowered from 50 ppb
2001 (January)	Clinton administration published rules to lower the arsenic standard to 10 ppb
2001 (March)	Bush administration withdrew the proposed rule
2001 (July)	U.S. House of Representatives voted to support a standard of 10 ppb
2001 (September)	NAS issues a report, declaring arsenic in drinking water to be a carcinogen at levels of 10 ppb
2001 (November)	Bush administration announced its reinstatement of the Clinton administration's proposed arsenic standard of 10 ppb
2002 (November)	EPA announced industry's voluntary action to phase out pressure-treated wood containing arsenic as a preservative
2003 (June)	Federal Circuit Court upholds EPA's revised arsenic standard of 10 ppb

led to a brouhaha that involved EPA, Congress, two presidents' administrations, environmental groups, and business associations (Table 11.1).

In 1993, the World Health Organization (WHO), recommended that arsenic levels in drinking water not exceed 10 ppb. EPA chose not to follow WHO's recommendation, an outcome not favored by environmental groups. EPA chose not to act upon WHO's recommendation, believing they had higher priority water contaminants for which to develop standards, together with pressure being brought upon the agency by mining companies opposed to any revision of the arsenic standard in the United States In response to EPA's inaction, environmental groups, joined by public health advocates, brought pressure upon Congress to force EPA to revise its arsenic standard. Opposed to any change in the arsenic standard were mining companies and some municipal water suppliers in western states. Their opposition was based on the economic impact of a lower arsenic standard. Arsenic in mining waste is a source of groundwater contamination; remediation of the waste would be costly. Water suppliers objected to a lower arsenic standard because new, costly equipment would need to be purchased to remove arsenic from drinking water.

Congress sided with those groups advocating a lower arsenic standard. The Safe Drinking Water Amendments of 1996 directed EPA to update the arsenic standard by January 1, 2001. As a matter of science, EPA asked the U.S. National Academy of Sciences (NAS) to review the toxicology and human health literature on arsenic and recommend courses of action. In 1999, the NAS recommended that the EPA standard be lowered from its then current level of 50 ppb. Two years later, January 2001, as the Clinton administration was preparing to leave office, EPA published rules that would lower the arsenic standard to 10 ppb. On January 20, 2001, Democrat Bill Clinton was succeeded in office by Republican George W. Bush. In March, as its first environmental policy action, the new administration withdrew the proposed arsenic rule. What followed was considerable criticism of the Bush administration's action, providing environmental and public health groups with an easy platform to criticize the new administration. Remarkably, the ensuring clamor led the U.S. House of Representatives, under Republication control, to support the Clinton-era 10 ppb proposed rule on arsenic in drinking water. In July, 2001, the NAS issued a second report on arsenic, finding it to be carcinogenic and recommending a level of less than 10 ppb as a standard. Subsequent to the NAS report, the Bush administration reversed itself and in November 2001 reinstated the arsenic rule proposed by the Clinton administration.

The revised EPA water quality standard of 10 ppb for arsenic in drinking water was litigated by the state of Nebraska and the city of Alliance, Nebraska. They argued that regulating drinking water quality was a state, not federal, responsibility. In June 2003, a three-judge panel of the U.S. Court of Appeals, District of Columbia Circuit, held in favor of EPA, concluding that plaintiffs had failed to show that EPA's actions were in violation of the U.S. Constitution [13]. The EPA arsenic standard was defended before the court by the Justice Department and the Natural Resources Defense Council, a national environmental advocacy organization. The court's decision removed barriers to enforcing the new, lower standard adopted by EPA, ending the political brouhaha between the Clinton and Bush administrations.

As a matter of environmental health policy, how should a new administration, with political ideals and values different from the replaced administration, approach proposed regulations held over from the previous administration? The arsenic case illustrates how *not* to proceed. The incoming Bush administration's quick rejection of the Clinton administration's arsenic standard of 10 ppm gave critics an easy platform. The quick, political rejection of the standard occurred without substantive public debate and inadequate political vetting, thereby providing critics of the Bush administration with an easily-characterized issue of public health (i.e., the public's health is put in jeopardy by allowing too much of a carcinogen in drinking water). Incoming administrations would be wise to understand the arsenic brouhaha in the context of how not to manage environmental health policy issues.

In addition to the public health concern over arsenic levels in drinking water supplies, environmental groups have expressed concern about arsenic contamination of soil attributable to pressure-treated wood. Pressure treating wood is a way to extend its life through introduction of preservatives that protect against termites, molds, fungi, and dry rot. Chromated copper arsenate (CCA) has been used as a preservative in wood, but as the wood ages, CCA migrates into soil, causing focal areas of arsenic contamination. EPA has determined that CCA-containing products cannot be produced after January 2004, announcing that industry has voluntarily agreed to phase-out such products [14].

As a matter of environmental health policy, risk assessment was used by EPA, and earlier by WHO, to develop the standard for arsenic in drinking water supplies. However, changes in political direction at EPA led to conflicts in whether or not to accept a lower standard. This raises the question of whether risk-based standards should be spared the political gauntlet of partisan politics. Is it possible to isolate the standards process from pressure from special interest groups? The arsenic in drinking water issue illustrates the difficulty in revising existing environmental standards and the intrusion of changed political bases. Risk assessment of a hazard—arsenic, in this case—that results in a significantly lower standard will be a challenge if economic impacts on the regulated parties are sufficiently great.

11.4.2 What Is the Value of a Human's Life?

In the mid-1990s, Congress began adding cost-benefit analysis as a requirement for federal regulatory agencies when developing proposed environmental regulations and standards. In theory, benefits of a proposed regulation (e.g., a lowered standard for arsenic in drinking water) should outweigh its costs. Regulated entities were the principal advocates of cost-benefit analysis. While cost-benefit analysis seems, on the surface, a reasonable and prudent public policy, in fact, it has become a matter of some controversy when put into practice. Much of the controversy concerns how costs and benefits of a proposed regulation are calculated. An example is the calculation of the value of a human life.

What is the value of your life? Or that of a child? Or an elderly person? Is the value of an elected official's life different from, say, a young medical doctor's? Or should all human life be valued as equal? These are questions, without easy answers, inherent in the calculus of cost/benefit analyses. The controversy arising from these kinds of questions divides into two camps of supporters. One camp considers, for regulatory purposes, all human life to be equal in value. One could call this the *egalitarian* camp.

An egalitarian approach was developed and subsequently practiced by the George H.W. Bush administration and continued as policy by the Clinton administration. Using the egalitarian approach, EPA determined that each life saved by a change in a regulation was worth $6.1million per life [15].

The *utilitarian* approach attempts to value human life on the bases of an individual's age and health status. This approach, advocated by the Office of Management and Budget (OMB) of the George W. Bush administration, produces life value estimates that differ across age and statistical lines. In one permutation of the utilitarian approach, OMB advocated that $3.7 million should be allocated for the life of a person younger than seventy years old and $2.3 million for persons older than 70 [16]. Critics of this approach quickly labeled it as the "senior death discount." In 2003, EPA used both the egalitarian and the utilitarian approaches to estimate the health benefits of proposed changes in the Clean Air Act. EPA estimated that approximately 12,000 lives would be saved by the proposed changes. But what would be the economic impact? Using the egalitarian approach ($6.1 million per life saved) resulted in an overall health benefit of $93 billion by year 2020. Using OMB's utilitarian approach yielded an estimated health benefit of $14.1 billion, a figure reasonably close to EPA's cost estimate of $6.5 billion [15]. Of note, when costs of projected regulations are close to the value of benefits, policy makers are loath to support regulatory actions, given that policy makers prefer situations where the outcome is more clear cut, thereby lessening the possibility that an unpopular decision will be made.

It will be a matter of policy, not science, that decides which approach, egalitarian or utilitarian, will be adopted by federal regulatory agencies. It can be predicted that regulated communities will bring pressure to bear on legislative bodies to adopt the utilitarian approach, predicating their argument on the cost savings of such an approach.

11.5   ECOLOGICAL RISK ASSESSMENT

Risk assessment (or what now would be called *hazard eval*uation) was first directed to hazards to human health. This was due to the need for EPA to set air and water quality standards under the Clean Air Act, Clean Water Act, and Safe Drinking Water Act, and their amendments. Similarly, the OSHA act directed OSHA to promulgate standards to protect workers' health and control of hazards to safe work conditions. Also, FDA's regulatory responsibilities under the Food, Drug and Cosmetic Act were specific to controlling hazards to human health. However, the passage of the Comprehensive Environmental Response, Compensation, and Liability Act, as amended, expanded federal environmental policy to include assessment of hazards to natural resources and ecosystems. This created the need for ecological risk assessment.

"Ecological risk assessment is a process that evaluates the likelihood that adverse ecological effects may occur as a result of exposure to a stressor" [17].

Distinct from human risk assessment, ecological risk assessment has evolved into a discipline of its own, not merely as an adjunct to human risk assessment—even though some methodological approaches are in common. In the most common situation, ecological risk assessment is used to estimate the impact of a hazard (e.g., waste site toxicants that leached into a lake or river) on a natural resource. The Superfund Law, as amended, requires polluters to pay for the consequences of damage to ecosystems. The damages can be considerable and quite costly. For example, waste released into a river can reduce or eliminate fish populations, causing long-lasting loss of commercial and recreational fishing. Remediation of the river's water and sediment can bring huge expenses to those responsible for the pollution.

EPA defines ecological risk assessment as "[a]n emerging science that identifies stressors that may alter ecosystems and quantifies the probable severity of adverse effects on those ecosystems" [18]. The agency considers ecological risk assessment as comprising three primary phases: problem identification, analysis, and risk characterization. Other ecological risk assessors generally follow the four steps used in human risk assessment: hazard evaluation, dose response analysis, exposure assessment, and risk characterization. An example of an ecological risk assessment will illustrate their nature and conduct.

11.5.1 Ecological Risk of Chlorpyrifos

Solomon and colleagues [19] conducted an ecological risk assessment of the use of chlorpyrifos, an insecticide, to determine the probability and significance of effects to wildlife from chlorpyrifos use in terrestrial ecosystems, particularly corn agrosystems. The following were features of the risk assessment:

- Hazard Identification: Chlorpyrifos is an organophorphorus insecticide. Under the environmental conditions of most often agricultural use, it is not considered as persistent in the environment. In the context of the risk assessment, chlorpyrifos was used for insect control in corn agrosystems. Under certain conditions of exposure, chlorpyrifos can be toxic to birds and mammals. Young birds and animals are more susceptible to chlorpyrifos toxicity than adults of their species.
- Exposure Assessment: Birds and mammals are exposed to chlorpyrifos through food contamination (insects, worms) and through contact with soil and water contaminated with the insecticide.
- Risk Characterization: Using toxicity data and assessment of various acute and chronic exposure scenarios, risk characterization of both granular and liquid administration of chlorpyrifos showed overall negligible consequences to exposed birds and mammals. Granular chlorpyrifos was less hazardous than liquid administration. Liquid application was found to present greater risks to young birds than adult birds feeding on insects from fields treated with chlorpyrifos.

The findings from this assessment have the potential to improve how farmers apply chlorpyrifos to their fields. If lesser amounts of this hazardous substance are applied to crop fields, the public's health will benefit because less of the substance will enter environmental pathways, including those portending human exposure, such as groundwater used as drinking water sources.

11.6 OCCUPATIONAL INJURY RISK ASSESSMENT

The methods and practices of risk assessment were originally applied to assess the risk of radiation, later applied to human exposure to chemicals. A more recent development has been the use of risk assessment in programs of occupational injury control. The U.S. National Institute for Safety and Health states that "occupational injury risk assessment concerns the estimation of risk in hazardous occupational environments that lead to traumatic injury," noting, "The risk assessment paradigm provides a useful framework to address problems resulting from workplace exposures that cause traumatic disabling and fatal injuries [20]." For instance, occupational injury risk assessment has been applied to workplace conditions that produce musculoskeletal injuries [21], cause occupational fatalities [22], affect machine design [23], and produce injuries from farming [24]. At the heart of occupational injury risk assessment is the prevention of traumatic injury and fatalities.

According to one source, the NRC's four-step risk assessment paradigm must be modified as follows when used in assessing the risk of occupational injuries [20]:

- Hazard identification—requires an evaluation of data on injuries or fatalities associated with specific workplace factors such as tools, workplace practices, or environmental conditions.
- Exposure assessment—the frequency, variability, and duration of workplace conditions associated with occupational injuries are represented through the use of statistical distributions.
- Dose-response assessment—exposure for injuries can be based on duration and frequency of such factors as biomechanical stress, workers' fatigue, and workplace practices.
- Risk characterization—probability models "[p]rovide the basis to describe the random nature of injuries and define a stochastic mechanism for injury incidence that can be useful for measuring and characterizing risk" [ibid.].

These four steps provide a framework for systematic assessment of occupational injuries. An example of an occupational injury risk assessment follows.

"Occupational injury risk assessment concerns the estimation of risk in hazardous occupational environments that lead to traumatic injury" [20].

Kines [25] assessed the risk of injuries in the Danish construction industry for the period 1993–1999. Hazard identification was conducted by examining lost-time injury incidents reported by employers to the National Working Environment Authority in Denmark for the period of interest. From this database, injury incidents resulting in amputations, bone fractures, and multi-trauma injuries in construction work were extracted for analysis. Risk analyses included calculations of proportions, relative rates, fatal injury incidence rates, nonfatal injury incidence odds ratios, and injury severity odds ratios. These statistics were used for both exposure assessment and risk characterization. Findings from the study showed: (1) carpenters had excessively high proportions, rates, and hazards for falls from heights, compared to the entire Danish construction industry, and (2) rates of serious injury falls from heights increased with increasing age of workers. From such findings, programs of injury prevention can be designed and implemented.

11.7 OTHER APPLICATIONS OF RISK ASSESSMENT

Risk assessment, as previously discussed in this chapter, has become the most frequently applied method for estimating the potential harm from exposure to environmental hazards. It has been utilized to assess both human health risk and risk to ecological systems. But other applications of risk assessment have arisen. The following section discusses: (1) the use of risk estimates for comparing and ranking individual environmental hazards, (2) the application of risk assessment to make risk management actions, and (3) the process of deriving data-driven safety factors. Each of these three applications of risk assessment has implications for policy makers, because each application has implications for risk management. Risk management, in turn, has such societal impacts as technology costs (e.g., pollution controls), levels of regulated pollutants, and how risk is communicated to the public.

11.7.1 COMPARATIVE RISK ASSESSMENT[6]

How individuals and societal structures, such as legislatures, compare environmental hazards has become a subject of great interest. As individuals, we make personal decisions that, knowingly or not, constitute a comparison of health risks. Some persons unwisely choose to smoke tobacco, perhaps unaware that nicotine in tobacco smoke is addictive. How persons select their living arrangements can result from comparing health risks. Some locales, such as in urban areas, present higher health risks because of higher levels of air pollutants there than in rural areas. However, health risks due to urban environmental pollution may be outweighed by risks that come with commuting to work from less environmentally polluted suburban or rural areas. Sometimes these personal choices are deliberate and based on factual information; other times, personal choices stem from fears or perceptions not based on scientific data or fact.

What priorities exist for evaluating the nation's environmental hazards? Is there a consensus method for ranking them? These two questions have become increasingly important to legislators, government officials, and public service groups. Legislators, in particular, assert their need to match legislative actions with environmental priorities. One approach, called *comparative risk assessment*, compares risks across various environmental hazards. Some persons consider this the best approach to establishing priorities for environmental and public health programs. How did comparative risk assessment become a prominent method of setting priorities for environmental hazards? In 1987, EPA released its *Unfinished Business* report, which compared and prioritized thirty-one national environmental hazards for which EPA had regulatory jurisdiction [29]. Since then, several states, some counties and cities, and at least one tribal organization, have conducted similar comparative risk assessments of environmental hazards.

"Comparative risk assessment is an environmental policy and planning process that attempts to bridge the gap between technical environmental risk analysis and public values and perceptions about environmental risk" [28].

Like individuals, societal structures such as legislatures must make decisions about environmental hazards. This occurs when legislators craft environmental and public health legislation and appropriate budgets to government programs. Private industry performs similar comparisons of environmental hazards. Companies must budget according to their own environmental priorities and those required to meet government regulations and community concerns. Because environmental protection and remediation programs are costly, legislators must seek better methods on which to predicate legislative actions. Comparative risk assessment has most often been suggested as the lamp to light the way to improved legislative decisions. This stems from the belief that because risk assessment is a quantitative and systematic approach to characterizing risks, using it to compare individual environmental hazards would lead to more precise ways of prioritizing hazards.

Two examples illustrate how some government officials view comparative risk assessment of environmental hazards. Former Governor John Engler of Michigan said, "Too often in the past, Michigan's environmental priorities have been set by the crisis of the moment, budget uncertainties, media attention, or conflicting data. I am convinced that it is time to carefully review and evaluate our priorities and base those priorities on careful thought and scientific information. We must do this in order to efficiently apply our limited resources to addressing the most serious environmental risks that our state faces" [30]. In a similar vein, Jan Eastman, Secretary, Agency of Natural Resources, State of Vermont, stated, "The Agency will seek to reduce risks to Vermont and Vermonters by exploring approaches to environmental management, including pollution prevention, toxics reduction, market incentives, and the continued use of public information and education. These approaches may help the Agency make the best possible use of its increasingly scarce resources. The Advisory Committee's ranking of the relative severity of Vermont's environmental problems helps to provide a useful foundation for action" [31]. In both examples, linkage between government resources and ranking of environmental hazards is evident.

"[t]he comparative risk process attempts to identify those aspects of the environment which both technical and public groups feel are of top priority" [28].

A perception that "low priority" risks have received too much attention and too many resources underlies the desire for better legislative decisions on environmental hazards. Proponents of this thesis often cite the cost of remediating uncontrolled hazardous waste sites as an example of the imbalance between environmental benefits and costs. They assert that the multibillion dollar costs of remediating uncontrolled hazardous waste sites outweigh the beneficial effects to human health, ecologic systems, and environmental quality. To what extent is this assertion legitimate and where do the risks of hazardous waste sites rank in comparison to those of other environmental hazards? For example, do hazardous waste sites present greater risks to human health than do indoor air pollutants?

Lee Thomas, when serving as EPA administrator, summarized his views on risk comparison by observing, "Although EPA's mission enjoys broad public support, our agency nonetheless must operate on finite resources. Therefore, we must choose our priorities carefully so that we apply those resources as effectively as possible. While we have made much progress to date, the cost of further environmental improvements in many areas will be high. For example, removing additional increments of toxics from industrial effluents or cleaning up contaminated ground water to background levels can be enormously expensive. The unit cost of moving ever closer to the point of zero discharge, zero contamination, and zero risk increases exponentially. Yet this agency must proceed to carry out its mandates and to set its priorities" [32].

What then is comparative risk assessment? One source states, "Comparative risk [assessment] is an environmental policy and planning process which attempts to bridge the gap between technical environmental risk analysis and public values and perceptions about environmental risk. Simply stated, the comparative risk process attempts to identify those aspects of the environment which both technical and public groups feel are of top priority. These issues, in turn, constitute the basis for public environmental expenditure allocations and environmental planning and risk reduction strategies" [28]. This broad statement holds out the promise of an orderly, systematic approach to ranking and controlling environmental hazards. As subsequent sections describe, this promise is only being partially realized by states and others. An example of an EPA-conducted comparative risk assessment follows.

◊ ◊ ◊

In 1987, EPA released its *Unfinished Business* report [29]. The report was prepared by seventy-five career EPA managers who ranked thirty-one environmental hazards for which EPA had regulatory jurisdiction. Because of lack of jurisdiction, EPA did not rank some key environmental hazards that state and local governments later considered important in their comparative risk assessments, including food safety, lead contamination, and natural hazards.

For the 31 environmental hazards, EPA assessed four different kinds of risk: cancer, non-cancer health effects, ecologic effects, and welfare effects (e.g., materials damage, aesthetic degradation).[7] Quantitative cancer risk estimates were developed where possible; other risks were qualitatively estimated through a process involving professional judgment and consensus. An overall risk assessment priority for each environmental hazard resulted when rankings from the four risk criteria were combined. The ten top-ranked (not in ranked order, according to the EPA report) environmental hazards in terms of overall risk are shown in Table 11.2.

EPA's Science Advisory Board (SAB) endorsed the process of how the agency had prepared the *Unfinished Business* report and the report's findings. Moreover, the SAB recommended that EPA should adjust its program priorities in accord with the report's comparative risk findings. As a matter of policy, EPA has never implemented the SAB's proposal. The primary hurdle for EPA has been the reality of congressional budgeting. EPA's priorities, whether the product of comparative risk assessment or not, are not necessarily the same as those of congressional appropriations committees. For example, efforts at EPA and in other federal agencies to give greater emphasis to research on climate change have been only partially successful. Groups opposed to any change in United States policy on greenhouse gases bring pressure to bear on Members of Congress, encouraging them to proceed slowly on funding climate change research, concerned that research findings could lead to greater regulatory control over emissions of greenhouse gases.

◊ ◊ ◊

In summary, comparative risk analysis is an environmental decision-making tool used to systematically measure, compare, rank, and act upon environmental hazards or issues. The process typically focuses on the risks an environmental problem poses to human health, the natural environment, and quality of life. The outcome is a list of environmental hazards or issues that are ranked in terms of relative risks. Comparative risk analysis typically investigates what are called "residual risks," the risks remaining

---

TABLE 11.2

Top-Ranked Environmental Hazards In EPA's Comparative Risk Assessment Study (not listed in rank order) [29]

Criteria air pollutants

Hazardous/toxic air pollutants

Other air pollutants

Radon—indoor

Indoor air pollutants other than radon

Radiation from sources other than indoor radon

Substances suspected of depleting the stratospheric ozone layer

CO_2 and global warming

Direct, point-source discharges to surface waters

Indirect, point-source discharges to surface waters

after an environmental problem is addressed by current regulatory controls or other administrative means. For example, a state environmental department may determine that its current food safety actions may leave little residual risk to the public's health, as compared to higher level of residual health risk presented by uncontrolled hazardous air pollutants. As a matter for environmental health policy makers, it is important that effective programs (and that have a low residual risk) not be shorted in resources and authorities in deference to environmental hazards that have a higher residual risk. What is working effectively to protect the public's health should be changed or revised only with great caution.

11.7.2   RISK-BASED CORRECTIVE ACTION

Another practical use of risk assessment appeared in the mid-1990s as the result of congressional concern about underground storage tanks (USTs). EPA estimates that more than one million USTs have been in service in the United States, primarily used for fuel storage at gasoline stations [33]. Of these, more than 400,000 have been confirmed as leaking USTs. The median cost to investigate and remediate a leaking UST is more than $100,000 [ibid.]. In addition to gasoline, USTs hold such liquid hazardous substances as pesticides, fertilizers, and industrial chemicals. The public health relevance is that leaking USTs contaminate the environment, including groundwater and surface waters used as sources of drinking water.

Under Subtitle I of the RCRAct, Congress directed EPA to establish regulatory programs that would prevent, detect, and remediate releases from USTs. In 1988, EPA released the required regulations and directed their implementation by state and local agencies. The EPA regulations do not specify cleanup levels or administrative procedures that the states must follow, requiring only that state or local remediation programs must be protective of human health and the environment. EPA's regulations allow states to make choices about how they will design and conduct their corrective action programs. As applied to corrective action at UST release sites, risk-based corrective action (RBCA) is a process that utilizes risk and exposure assessment methodology to help UST implementing agencies to make determinations about the extent and urgency of corrective action and about the scope and intensity of their oversight of corrective action by UST owners and operators [34].

In 1993, in large measure because individual states were struggling to develop their UST programs, the American Society of Testing & Materials (ASTM) began work on development of a streamlined process for assessment and response to subsurface contamination associated with petroleum hydrocarbon releases. The standard was reissued in final form in December 1995 as ASTM E 1739-95 *Standard Guide for Risk-Based Corrective Action Applied at Petroleum Release Sites* [35] and was later expanded and reissued as ASTM PS 104-98 *Provisional Standard Guide for Risk-Based Corrective Action* [36], addressing all types of chemical releases to the environment. The RBCA process, as defined in the ASTM standard, is a flexible, science-based, decision management framework that may be customized by individual regulatory agencies to design or revise their corrective action programs [37]. In simple terms, the RBCA process entails: (1) identification of applicable risk factors on a site-specific basis and (2) implementation of appropriate corrective measures in a time frame necessary to prevent unsafe conditions.

The goal of RBCA programs is to identify those leaking USTs of greatest hazard to human health and the environment, relegating those of lesser risk to categories of environmental monitoring or inactivity. To examine the performance of states' RBCA programs that utilize the ASTM standard (or similar standard), Cooper and McHugh [37] conducted detailed evaluations of five state environmental agencies. Comparison of pre-RBCA to post-RBCA program management statistics found an increase in the number of case closures (ranging from 46% to 134%), reduced environmental cleanup costs (e.g., in Texas the median cost was reduced from $250,000 to $107,000 for low-risk groundwater sites), and more effective targeting of resources toward responding to higher-risk sites.

As a matter of environmental health policy, risk-based correction action has been developed and applied to the problem of leaking underground storage tanks. In principle, this approach emulates the process of risk assessment and management of Superfund sites. In theory, those sites (USTs or Superfund sites) that are of greatest hazard to human and ecological health are remediated before sites of lesser risk are remediated. As policy, this makes sense in a public health context. The sites of greatest urgency are responded to first, lessening the likelihood that humans will be exposed to hazardous substances released from the sites. However, what is lacking is any retrospective analysis that proves that the worst sites were accurately characterized in the first place. Public polices that are constructed on theoretical constructs or "common sense" bases require an evaluation of their effectiveness. This is infrequently done by public agencies due to lack of interest and sometimes, limited resources needed to conduct policy effectiveness reviews. One possible fix for this lack of assessment would be for periodic policy reviews conducted by committees of Congress or state legislatures, depending on whether the policies are federal or state based.

11.7.3   DATA-DERIVED SAFETY FACTORS

Another example of risk assessment being used for practical purposes is the evaluation of safety factors used to calculate toxicity thresholds. Thresholds occur for non-carcinogenic substances. For science policy purposes, carcinogens are considered to have no threshold exposure. Two kinds of threshold studies have been developed by toxicologists and the study results used in risk assessment calculations. A no observed adverse effect level (NOAEL) is the point on a dose-effect curve at which a threshold is reached. A lowest observed adverse effect level (LOAEL) is the lowest level for which a toxic response is observed. NOAELs and LOAELs are usually determined from laboratory animal toxicology studies that use a range of exposure levels. As background, the risk assessment process involves the extrapolation to humans of toxicological animal data and associated science. Because there may be a range of sensitivities in humans to a specific toxic exposure, uncertainty factors (UFs) are used to account for scientific uncertainties in underlying databases. How the UFs are determined has changed over time. The overall goal has been to reduce the range of UFs, relying less on default values,[8] leading to more precise risk estimates. Some background on how UFs have evolved is instructive.

The policy of using data-derived uncertainty factors (UFs), rather than default values of 10 or other values, leads to science rather than speculation.

Government agencies have historically been responsible for developing recommendations or regulations to protect against human consumption or other exposure to hazardous substances. An early concept used for substances with toxicity thresholds is called an *acceptable daily intake* (ADI). According to one source [28], the concept of using UFs in risk assessment was first proposed by Lehman and Fitzhugh [39]. They advocated a "100-fold margin of safety" to provide safety factors when extrapolating animal toxicological data to sensitive human populations. In 1977, the National Research Council's Safe Drinking Water Committee recommended that the safety factor be increased from 100-fold to 1,000-fold when toxicity data are found inadequate (as cited in [40]). The committee's recommendation was an expression of concern that in the absence of satisfactory data, water quality standards should be conservatively based. Later, the use of Reference Dose (RfD) and UF replaced Acceptable Daily Intake and Safety Factor, respectively [40]. The RfD is equal to the No Observed Adverse Effect Level (NOAEL) divided by the product of UFs (i.e., factors of 10) and an additional modifying factor (MF). This can be expressed as

$$RfD = [NOAEL (or LOAEL)]/[(UF)(MF)].$$

The UF components have been categorized as follows: subchronic to chronic exposure extrapolations, interspecies differences between animals and humans, and variability in sensitivity among humans, and incomplete database [38]. Based on their evaluation of relevant databases (e.g., data from animal carcinogenicity studies) risk assessors will assign the ten-fold default factor, unless data indicate a lesser value. The Agency for Toxic Substances and Disease Registry has developed Minimal Risk Levels (MRLs) that are similar in concept to EPA's RfDs.

Example Calculation of MRL for Chronic Exposure to Ethylene Glycol (EG) [41].

- ATSDR used the study of DePass et al. [42] of rats fed diets with EG at 0, 40, 200, or 1000 mg EG/kg/day
- Rats exhibited chronic nephritis at 1,000 mg EG/kg/day; no effects at lower doses
- NOAEL = 200 mg EG/kg/day
- Uncertainty factors applied by ATSDR: 10 for extrapolation from rats to humans and 10 for human variability
- MRL - 200/(10)(10) = 2 mg EG/kg/day for chronic oral dose of EG

One uncertainty factor, the extrapolation of toxicological data to account for differences between and within species, has become the subject of data-derived methodology. To be more specific, the relative magnitude of toxicokinetic and toxicodynamic variations between or within species have been examined (e.g., [43–45]). In one study, composite factors were all lower than 10 for 6 pharmaceuticals and 8 of the 12 composite factors were less than 5.5. When UFs are less than the default value of 10, the RfD is increased, which, in effect, is a lessening of the risk estimate. This observation means that, in a public health context, it is vital to ensure that accurate and unbiased estimates of UFs are derived.

The policy implications of using data-derived UFs, rather than default values of 10 or other value, are considerable. It can be argued that data-derived factors, which are based on findings from experimental research, reduce the uncertainty that attends default values. In other words, science replaces speculation. On the other hand, what are the guarantees that data-derived UFs are unbiased and accurate? How regulatory agencies incorporate data-derived UFs into their regulatory processes remains to be determined, but will surely be influenced by the requirements of the Information Quality Act of 2001 (discussed in chapter 4).

11.8 PUBLIC HEALTH CONCERNS REGARDING RISK ASSESSMENT

Risk assessment in the United States has evolved since the 1970s. It has become the engine that drives much of environmental policy making. But this has not occurred without criticism. In particular, public health specialists and some environmentalists have questioned whether risk assessment has failed to advance public health goals. It is therefore useful to reflect on the relationship between risk assessment, as practiced by regulatory agencies, and its impact on the public's health.

The public health community (i.e., public health officials and practitioners) has been slow to embrace risk assessment [46]. The reasons are complex, but can be distilled into three broad categories: the public health tradition, prevention ethos, and public health resources.

The public health tradition can be stated to comprise science, consensus, and services. In this context, science includes laboratory and field research, epidemiologic investigations, and studies of causal mechanisms of disease. Findings from contemporary research, when added to an existing body of knowledge, can create a science foundation that can be applied to a public health problem at hand. For example, findings from studies of children exposed prenatally to lead released from maternal tissues revealed a public health problem. The children evidenced developmental disorders (e.g., delays in cognitive processes) due to their fetal exposure to lead. In the public health tradition, science of import, such as the lead findings in children, must be vetted through peer reviewed publications of research findings and disclosure to the scientific community. Consensus formation on the gravity of a body of science is a key step in the public health tradition. This is necessary because public health resources are limited and must be directed to prevention of significant, not trivial, public health problems. Consensus is often pursued by reliance on advisory organizations (e.g., the National Academy of

Sciences) or issue-specific advisory committees (e.g., the CDC's Advisory Committee on Childhood Lead Poisoning Prevention). When a public health problem has been identified, it becomes essential to obtain cooperation among health agencies on how to prevent or contain the problem. A significant part of public health agencies' cooperation is the sharing of resources and services. For example, federal health agencies can provide states with grants (e.g., childhood lead exposure surveillance) for purpose of disease prevention, assuming that the federal agencies have granting authority under their authorizing statutes.

In distinction to the public health approach, the regulatory approach comprises science, regulations, and enforcement. Much like in the public health tradition, regulatory agencies require a body of science in order to develop regulations and standards. As with public health organizations, regulatory agencies cannot act on a whim (i.e., in the absence of scientific data). Regulatory agencies utilize scientific data (e.g., findings from toxicology studies) to develop proposed regulations and standards. Enforcement of regulations, when authorized by law, completes the regulatory approach to controlling environmental hazards.

Unlike public health authorities, regulatory agencies have the weight of law buttressing their actions, such as environmental standards. Strictly speaking, regulatory agencies do not require consensus on their proposed regulatory actions. Rather, federal regulatory agencies hold public meetings and place public notices in the *Federal Register* to solicit comments from the public and the targets of proposed regulations. A consequence of these meetings and public notices is often the revision of proposed regulations. Therefore, to some extent, such actions constitute a kind of consensus formation.

In summary, the public health tradition differs from the regulatory approach in response to control of environmental hazards. This duality in approach has been uncomfortable for some public health practitioners, contributing to a slow acceptance or outright rejection of quantitative risk assessment. The second area that contributes to the public health community's slow acceptance of risk assessment concerns the ethos of prevention of disease and disability, the centerpiece of public health theory and practice. Some persons with experience in both public health practice and regulatory agency responsibilities [46,47] have opined that prevention (in a public health context) is not prominent in the regulatory approach. Goldman [47] has observed that prevention as a possible anchor for risk assessment was not addressed in the National Research Council's (NRC's) seminal report *Risk Assessment in the Federal Government: Managing the Process* [1]. Because federal regulatory agencies adopted the NRC's recommendations on how to conduct risk assessments, any lack of prevention perspective was passed along to regulatory agencies.

Whether or not risk assessment lacks a prevention thrust is open to interpretation. It must be acknowledged that risk assessment of a hazard generally occurs with the hazard already present in the environment. Risk assessment is employed to address whether existing levels of exposure to a hazard should be lowered or eliminated because the risk of adverse health effects is unacceptable. This is a kind of post hoc disease prevention. However, it is possible to use risk assessment in a prospective manner. For example, some business enterprises, as they develop new products, conduct risk assessments during the product development process. Company risk assessors attempt to determine if their product could harm humans or ecosystems. Keeping harmful products out of commerce is a matter of primary prevention. However, it is unknown to what extent this kind of self-censoring of products under development occurs, since such actions are usually considered "business confidential." The belief by some public health practitioners that risk assessment does not readily comport with the principle of disease prevention has led them to recommend that it be replaced by application of the precautionary principle (discussed in chapter 2).

"[o]n the state level the traditional roles of health agencies, including epidemiology and health surveillance, became more distant from the evaluation of population health to the quantification of population risk" [46].

The third area that contributes to public health authorities' skepticism about risk assessment can loosely be termed "lack of public health resources" found in the risk assessment process. In the context used here, public health resources refers to both people and databases. Goldman [47] observed, referring to the establishment of regulatory agencies, "[p]ublic health decision-making was moved into realms where there were very few people with public health training and experience." This situation has changed little over the years following EPA's and OSHA's establishment. Although both agencies employ epidemiologists and public health specialists, their numbers are few in comparison to the numbers in public health agencies. As a consequence, some have asserted that regulatory risk assessments have suffered from lack of depth in epidemiologic and health surveillance perspectives [46].

In regard to public health databases, some public health spokespersons have lamented the absence of epidemiologic databases and health surveillance systems both within regulatory agencies and in individual risk assessments. Burke [46], commenting on the adoption of risk assessment by federal and state agencies, stated, "On the Federal level EPA support for epidemiology declined, and on the state level the traditional roles of health agencies, including epidemiology and health surveillance, became more distant from the evaluation of population health to the quantification of population risk." As a consequence, fewer data from epidemiological investigations and health surveillance systems are available for risk assessments of environmental hazards.

11.9 OTHER CRITICS OF RISK ASSESSMENT

In addition to the community of public health specialists, others have also expressed concerns about the use of risk assessment in aspects of environmental health policy making. For example, the United Kingdom's Royal Commission on Environmental Pollution advocated that risk assessment, as currently practiced, be replaced [48,49]. The commission concluded that current risk assessments were inadequate, cumbersome, and slow. Moreover, they expressed criticism of current risk assessment methodologies, as practiced within the European Union, that involve "[a] range of criteria, including toxicity, persistence, bioaccumulation and, importantly, exposure as a basis for management" [49]. In lieu of the current methodology, the commission recommended a new paradigm that focuses on environmental persistence of a toxicant and its bioaccumulation, leading to decision making that is hazard-based, not risk-based. Further, the commission recommended that Qualitative Structure Activity Relationships (QSARs) be used to predict toxicity results, based on chemical structures, rather than conducting toxicity tests directly.

The commission's recommendations, if adopted in lieu of risk assessment, would, they assert, reduce delays in policy making that is currently based on risk assessment. As noted by Calow and Forbes [49], the commission's new paradigm seems willing to accept less scientific information in order to reduce delays in decision making. This approach is in the spirit of a precautionary approach for dealing with environmental hazards. Whether or not the commission's recommendations become policy in Europe remains to be seen.

11.10   SUMMARY

Quantitative risk assessment did not just drop from the sky into the awaiting nets of U.S. regulatory agencies. Rather, risk assessment of toxicants in community and workplace environments evolved in reaction to court decisions in the early 1980s that required regulatory agencies to quantify levels of risk posed by specific federal regulations. The court decisions set into motion on-going efforts at EPA and OSHA to systematize how risk assessment and risk management polices and practices are developed and implemented. In this regard, a study conducted by the U.S. National Research Council [1] was particularly influential on how risk assessment was systematized by EPA and other regulatory agencies, federal and state, as discussed in this chapter. The NRC risk assessment structure of hazard identification, dose-response assessment, exposure assessment, and risk characterization was widely accepted in the United States as a sensible roadmap for use by regulatory agencies and remains in current use.

Systematization of regulatory risk assessment policies (e.g., a zero threshold for carcinogens) and practices (e.g., public hearings on proposed regulations) was important. It informs both a regulatory agency's staff and the general public about what to expect. This is laudable in the context of the public's right-to-know policy, but has also contributed to almost automatic litigation over environmental regulations and standards. All risk assessments require that risk assessors deal with uncertainties in scientific data, leading to choices that can be litigated.

Risk assessment, as the core of regulatory agencies' programs to control environmental hazards has given rise to criticism by some public health practitioners and environmentalists. Critics have voiced concern about the inordinately long time to establish a regulation (e.g., the OSHA ergonomics rule that was in development for ten years, the litigious nature of current risk assessments, and the politicalness of proposed regulations). These concerns have led public health advocates and environmentalists to propose using the precautionary approach (discussed in chapter 2) in lieu of risk assessment to control environmental hazards, believing that less time would be required to take action to control environmental hazards. However, given the U.S. investment in risk assessment-based regulations and standards, it is unlikely that the precautionary approach will soon be adopted as a replacement.

11.11   POLICY QUESTIONS

1.   Compare the merits and drawbacks of the use of risk assessment-based regulatory strategies versus a precautionary-approach strategy, as discussed in chapter 2. Which of these two strategies do you prefer, and why?
2.   Does ecological risk assessment have any relevance for public health? If so, discuss why. If not, discuss why.
3.   What is the dollar value of your life? How did you determine the value? Do you favor the utilitarian or the egalitarian approach to determining the value of human life? Defend your selection.
4.   Discuss each of the four elements of risk assessment in terms of importance to public health practice.
5.   Assume that you are a member of a local health department. The director of the department asks for your advice on the following two topics: (A) How could the county's hazards be compared and ranked? (B) Should a survey of the public's risk perceptions be undertaken and used as a factor on which to allocate scarce budgetary funds?
6.   Discuss the relevance of the U.S. Supreme Court's benzene decision on public health.
7.   How does hazard differ from risk?
8.   Discuss how the American Conference of Governmental Hygienists's (ACGIH) Threshold Limit Values have had a substantive impact on occupation health.
9.   How could the arsenic brouhaha (Table 11.1) have been avoided?

NOTES

1.   TLV is a registered trademark of the American Conference of Governmental Industrial Hygienists, Cincinnati, Ohio.
2.   This department was split into the Department of Health and Human Services and the Department of Education in 1979.
3.   See chapter 3 for a discussion of the administrative relationship between NIOSH and OSHA.
4.   One part arsenic in one million parts of soil
5.   One part arsenic in one billion parts of water
6.   Material in this section is adapted from Johnson [26] and Johnson [27].
7.   Not all thirty-one hazards were ranked on all four risk dimensions.
8.   Default values are those used in lieu of values otherwise available.

REFERENCES

1. NRC (National Research Council), *Risk Assessment in the Federal Government: Managing the Process*, National Academy Press, Washington, D.C., 1983.
2. Yassi, A. et al., *Basic Environmental Health*, Oxford University Press, Oxford, 2001.
3. ACGIH (American Conference of Governmental Industrial Hygienists), *History*. Available at http://www.acgih.com/About/History.htm, 2002.
4. Castleman, B.I. and Ziem, G.E., Corporate influence on threshold limit values, *American Journal of Industrial Medicine*, 13(5), 531, 1988.
5. Ziem, G.E. and Castleman, B.I., Threshold limit values: Historical perspectives and current practice, *Journal of Occupational Medicine*, 31, 910, 1989.
6. Fromson, J., A history of federal air pollution control, in *Environmental Law Review–1970*, Sage Hill Publishers, Albany, NY, 1970, 214.
7. Landy, M.K., Roberts, M.J., and Thomas, S.R., *The Environmental Protection Agency: Asking the Wrong Questions from Nixon to Clinton*, Oxford University Press, New York, 1994.
8. Rosenbaum, W.A., *Environmental Politics and Policy*, 4th ed., Congressional Quarterly, Inc., Washington, D.C., 1998, 10, 142.
9. Mirer, F., Distortions of the "mis-read" book: Adding procedural botox to paralysis by analysis, *Human and Ecological Risk Assessment*, 9, 1129, 2004.
10. Merrill, R.A., The red book in historical context, *Human and Ecological Risk Assessment*, 9, 1119, 2003.
11. Omenn, G., On the significance of "the red book" in the evolution of risk assessment and risk management, *Human and Ecological Risk Assessment*, 9, 1155, 2004.
12. ATSDR (Agency for Toxic Substances and Disease Registry), *Toxicological Profile for Asbestos*, Department of Health and Human Services, Public Health Service, Atlanta, GA, 2000, 3.
13. *New York Times*, Court upholds tougher rule on arsenic limits in water, June 21, 2003.
14. EPA (U.S. Environmental Protection Agency), Whitman announces transition from consumer use of treated wood containing arsenic, Communications, Education, and Media Relations (1703A), Washington, D.C., February 12, 2002.
15. Barnett, J., Method of valuing life sparks debate, Newhouse News Service. Available at http://www.newhousenews.com/archive/barnett031703.html, 2003.
16. Seelye, K.Q. and Tierney, J., E.P.A. drops age-based cost studies, *The New York Times*, May 8, 2003.
17. EPA (U.S. Environmental Protection Agency), Guidelines for Ecological Risk Assessment, EPA/630/R-95/002, Risk Assessment Forum, Washington, D.C., 1998.
18. EPA (U.S. Environmental Protection Agency), What is Ecological Risk Assessment? Available at http://www.merac.umn.edu/whatisera/default.htm, 2002.
19. Solomon, K.R. et al., Chlorpyrifos: Ecotoxicological risk assessment for birds and mammals in corn agrosystems, *Human and Ecological Risk Assessment*, 7, 497, 2001.
20. Wassell, J.T., Occupational injury risk assessment: An unintended and unanticipated consequence of the red book, *Human and Ecological Risk Assessment*, 9, 1383, 2003.
21. Melhorn, J.M., Wilkinson, L.K., and O'Malley, M.D., Successful management of musculoskeletal disorders, *Human and Ecological Risk Assessment*, 7, 1801, 2001.
22. Chen, G.-Y. et al., Work-related and non-work-related injury deaths in the U.S.: A comparative study. *Human and Ecological Risk Assessment*, 7(7), 1859, 2001.
23. Etherton, J. et al., Machinery risk assessment for risk reduction, *Human and Ecological Risk Assessment*, 7, 1787, 2001.

24. Josefsson, K.G. et al.,. A hazard analysis of three silage storage methods for dairy cattle. *Human and Ecological Risk Assessment*, 7, 1895, 2001.

25. Kines, P., Occupational injury risk assessment using injury severity odds ratios: Male falls from heights in the Danish construction industry, 1993–1999, *Human and Ecological Risk Assessment*, 7, 1929, 2001.

26. Johnson, B.L., *Impact of Hazardous Waste on Human Health*, CRC Press, Lewis Publishers, Boca Raton, FL, 1999, 311.

27. Johnson, B.L., A review of health-based comparative risk assessments in the United States. *Reviews on Environmental Health*, 15(3), 273, 2000.

28. Ohio University, Survey Results for the Athens County Comparative Risk Survey, Institute for Local Government Administration, Ohio University, Athens, 1995.

29. EPA (U.S. Environmental Protection Agency), Unfinished Business: A Comparative Assessment of Environmental Problems, Washington, D.C., 1987, xix.

30. MDNR (Michigan Department of Natural Resources), Michigan's Environment and Relative Risk, Lansing, 1992.

31. VANR (Vermont Agency of Natural Resources), Environment 1991: Risks to Vermont and Vermonters, Waterbury, 1991.

32. Thomas, L.M. et al., Preface, Unfinished Business: A Comparative Assessment of Environmental Problems, Report PB88-127048, Department of Congress, National Technical Information Service, Springfield, VA, 1987.

33. Connor, J.A. and McHugh, T.E., Impact of risk-based corrective action (RBCCA) on state corrective action programs, *Human and Ecological Risk Assessment*, 8, 573, 2002.

34. EPA (U.S. Environmental Protection Agency), RBCA leaflet, EPA 510-F-95-001, Office of Underground Storage Tanks, Washington, D.C., 1995.

35. ASTM (American Society for Testing and Materials), *Standard Guide for Risk-Based Corrective Action Applied at Petroleum Sites*, ASTM E-1739-1795, West Conshohocken, PA, 1995.

36. ASTM (American Society for Testing and Materials), *Provisional Standard Guide for Risk-Based Corrective Action*, ASTM PS-104-98, West Conshohocken, PA, 1998.

37. Cooper, J.A. and McHugh, T.E., 2002. Impact of risk-based corrective action (RBCA) on state corrective action programs, *Human and Ecological Risk Assessment*, 8, 573, 2002.

38. Riyad, C.E. et al., A study of safety factors for risk assessment of drugs used for treatment of attention deficiency hyperactivity disorder in sensitive populations, *Human and Ecological Risk Assessment*, 8,:823, 2002.

39. Lehman, A.J. and Fitzhugh, O.G., 100-Fold margin of safety, *Association of Food and Drug Officials' USQ Bulletin*, 18, 33, 1954.

40. Barnes, D.G. and Dourson, M., Reference dose (RfD): Description and use in health risk assessments, *Regulatory Toxicology and Pharmacology*, 8, 471, 1988.

41. ATSDR (Agency for Toxic Substances and Disease Registry), *Toxicological Profile for Ethylene Glycol and Propylene Glycol*, U.S. Public Health Service, Atlanta, GA, 1995, Appendix A.

42. DePass, L.R. et al., Chronic toxicity and oncogenicity studies of ethylene glycol in rats and mice, *Fundamental Applied Toxicology*. 7(4), 547, 1986.

43. Renwick, A.G., Data-derived safety (uncertainty) factors for the evaluation of food additives and environmental contaminants, *Food Additives and Contaminants*, 10(3), 275, 1993.

44. WHO (World Health Organization), Assessing Human Health Risks of Chemicals: Derivation of Guidance Values for Health-Based Exposure Limits, Environmental Health Criteria No. 170, International Programme on Chemical Safety, Geneva, 1994.

45. Skowronski, G.A. and Abdel-Rahman, M.S., Interspecies comparison of kinetic data of chlorinated chemicals of potential relevance to risk assessment, *Human and Ecological Risk Assessment,* 3, 635, 1997.
46. Burke, T.A., The Red Book and the practice of environmental public health: Promise, pitfalls, and progress. *Human and Ecological Risk Assessment*, 9, 1203, 2003.
47. Goldman, L.R., The red book: A reassessment of risk assessment, *Human and Ecological Risk Assessment,* 9, 1273, 2003.
48. RCEP (Royal Commission on Environmental Pollution), Chemicals in Products. Safe-Guarding the Environment and Human Health, Twenty-Fourth Report, London, 2003.
49. Calow, P. and Forbes, V., The UK Royal Commission on Environmental Pollution gives risk assessment a vote of no confidence, *SETAC Globe,* (Nov.–Dec), 30, 2003.

Abbreviations

ATSDR	Agency for Toxic Substances and Disease Registry, DHHS
BAT	Best Available Technology
BPT	Best Practicable Control Technology
CAA	Clean Air Act
CCEHRP	Committee to Coordinate Environmental Health and Related Programs, DHHS
CCR	Consumer Confidence Report
CDC	Centers for Disease Control and Prevention, DHHS
CEQ	Council on Environmental Quality
CERCLA	Comprehensive Environmental Response, Compensation, and Liability Act (also called the Superfund Act)
CPSC	Consumer Product Safety Commission
CWA	Clean Water Act
DHEW	U.S. Department of Health, Education, and Welfare (predecessor of DHHS)
DHHS	U.S. Department of Health and Human Services
DHLS	Department of Homeland Security
DoC	U.S. Department of Commerce
EEA	European Environmental Agency
EIS	Environmental Impact Statement (required by NEPA)
EPA	United States Environmental Protection Agency
FAO	Food and Agriculture Organization, United Nations
FCCC	Framework Convention on Climate Change (United Nations)
FDA	Food and Drug Administration, DHHS
FEMA	Federal Emergency Management Agency, DHLS
FIFRA	Federal Insecticide, Fungicide, and Rodenticide Act
FQPA	Food Quality Protection Act
GAO	Government Accountability Office, nee General Accounting Office (name changed July 7, 2004)
Hazmat	Hazardous materials
IARC	International Agency for Research on Cancer, WHO
ILO	International Labour Organization, United Nations
IPCS	International Programme on Chemical Safety, WHO
MACT	Maximum Available Control Technology
MCL	Maximum Contaminant Level
MCLG	Maximum Contaminant Level Goal
NAAQS	National Ambient Air Quality Standard
NAS	National Academy of Sciences
NASA	National Aeronautics and Space Agency

NCI	National Cancer Institute, NIH
NCP	National Contingency Plan
NCTR	National Center for Toxicological Research, FDA
NEPA	National Environmental Policy Act
NIEHS	National Institute of Environmental Health Sciences, NIH, DHHS
NIH	National Institutes of Health, DHHS
NIOSH	National Institute for Occupational Safety and Health, CDC, DHHS
NOAA	National Oceanographic and Atmospheric Administration, DoC
NRC	National Research Council of the National Academies
OPA	Oil Pollution Act
OSHA	Occupational Safety and Health Administration, U.S. Department of Labor
PELs	Permissible exposure limits
PHS	U.S. Public Health Service, DHHS
PPA	Pollution Prevention Act
PRP	Potentially Responsible Party under CERCLA provisions
QSAR	Quantitative Structural Activity Relationship
RCRA	Resource Conservation and Recovery Act
RELS	Recommended exposure limits
ROD	Record of Decision
SDWA	Safe Drinking Water Act
SIP	State Implementation Plan
TRI	Toxics Release Inventory (required by Title III of CERCLA, as amended)
TSCA	Toxic Substances Control Act
TSDF	Treatment, Storage, and Disposal facility
UN	United Nations
UNCED	United Nations Conference on the Environment and Development
UNEP	United Nations Environment Program
WHO	World Health Organization, UN
WTO	World Trade Organization

Glossary of Key Terms

Absorbed dose   The amount of a substance that penetrates across the exchange boundaries of an organism through either physical or biologic processes after contact (exposure) [1].

Absorption   The process of taking in, incorporation, or reception of gases, liquids, light, or heat [2].

Acute   Occurring over a short time, usually a few minutes or hours. An *acute* exposure can result in short-term or long-term health effects. An *acute* effect happens a short time (up to one year) after exposure.

Administered dose   The amount of a substance given to a human or test animal. Administered dose is a measure of exposure because absorption is not considered [1].

Agency   A government office or department that provides a specific service.

Agent   An entity (chemical, radiologic, mineralogic, or biologic) that may cause effects in an organism exposed to it.

Ambient   Surrounding; pertaining to the air, noise, temperature, etc. in which an organism or apparatus functions [2].

Analytic epidemiologic study   Investigations that evaluate the causal nature of associations between exposure to hazardous substances and disease outcome by testing scientific hypotheses [3].

Anemia   A decreased ability of the blood to transport oxygen; low numbers of red blood cells or hemoglobin.

Anthropogenic   of, relating to, or resulting from the influence of human beings on nature.

Applied dose   The amount of a substance given to a human or test animal, especially through dermal contact. Applied dose becomes a measure of exposure because absorption is not considered [1].

Assessment   The process of determining the nature and extent of hazards and health problems within a jurisdiction.

Background level   A typical or average level of a chemical in the environment. *Background* often refers to naturally occurring or uncontaminated levels.

Biologic indicator   A chemical, its metabolite, or another marker of exposure that can be detected or measured by biomedical testing of human body fluids or tissues to validate human exposure to a hazardous substance.

Biologic monitoring   Measuring chemicals in biologic materials (e.g., blood, urine, breath, hair) to determine whether chemical exposure has occurred in living organisms.

Biologic uptake   The transfer of substances from the environment to living organisms.

Blood lead level   The concentration of lead in a sample of blood.

Body burden   The total amount of a chemical in the body. Some chemicals accumulate in the body because they are stored in fat or bone or other tissues.

Bully Pulpit   a prominent public position (as a political office) that provides an opportunity for expounding one's views [4].

Carcinogen   A substance that can cause cancer.

Carcinogenicity   Capacity to cause cancer.

Census block   Small geographic areas enclosed by visible features such as streets, roads, streams, and railroad tracks, or by invisible borders such as city, town, township, and county limits; property lines; or short, imaginary extensions of streets and roads [5].

Census block group   A geographic block group or tabulation block group. The former is a cluster of blocks having the same first digit of their three-digit identifying numbers within a census tract or block numbering area. A tabulation block group is a geographic block group that may be split to present data for every unique combination of county subdivision, place, American Indian and Alaska Native area, urbanized area, voting district, urban/rural and congressional district shown in the data product [5].

Chromosome   The structure (normally forty-six in humans) in the cell nucleus that is the bearer of genes.

Chronic   Occurring over a long period of time (e.g., more than one year).

Climate change   A condition that can be caused by an increase in the atmospheric concentration of greenhouse gases, which inhibits the transmission of some of the sun's energy from the earth's surface to outer space.

Command-and-control regulation   A regulation that requires polluters to meet specific emission-reduction targets.

Community   A group or social class having common characteristics.

Concentration   The amount of one substance dissolved or contained in a given amount of another.

Confidence interval   An interval of values that has a specified probability of containing a given parameter or characteristic [1].

Contaminant   Any substance or material that enters a system (e.g., the environment, human body, food) where it is not normally found.

Cost-benefit analysis   An economic technique applied to public decision-making that attempts to quantify in dollar terms the advantages (benefits) and disadvantages (cost) associated with a particular policy.

Cost-effectiveness analysis   An analysis that measures the net cost of providing a service as well as the outcomes obtained.

Demographics   The statistical study of human populations.

Dermal   Referring to the skin. *Dermal* absorption means absorption through the skin.

Developing countries   Those countries that are in the process of becoming industrialized but have constrained resources.

Diagnostic test   A laboratory test used to determine whether a person has a particular health problem.

Disease   Illness; sickness; an interruption, cessation, or disorder of body functions, systems, or organs [2].

Disease incidence   The rate of new occurrences of a disease.

Disease surveillance   A data collecting system that monitors the occurrence of specific diseases (e.g., cancer).

Dose   The total amount of radiation or toxicant, drug, or other chemical administered or taken by the organism (adapted from [6])

Dose-response study   A toxicological study of the quantitative relationship between the amount of a toxicant administered or taken and the incidence or extent of the adverse effect [6].

Ecosystem   A biological community of interacting organisms and their physical environment [7].

Effluent   Waste material discharged into the environment.

Emissions   Pollutants released into the air or waterways from industrial processes, households, or transportation vehicles.

Environment   The circumstances, objects, and conditions by which one is surrounded [8].

Environmental contamination   The presence of hazardous substances in the environment.

Environmental equity   The proportionate and equitable distribution of environmental benefits and risks among diverse economic and cultural communities [9].

Environmental health   Comprises of those aspects of human health, including quality of life, that are determined by physical, chemical, biological, social and psychosocial factors in the environment. It also refers to the theory and practice of assessing, correcting, controlling, and preventing those factors in the environment that can.

Environmental justice   Concern about the disproportionate occurrence of pollution and potential pollution-related health effects affecting low-income, cultural, and ethnic populations and lesser cleanup efforts in their communities [10].

Environmental medium   Material in the outdoor natural physical environment that surrounds or contacts organisms (e.g., surface water, groundwater, soil, air) and through which substances can move and reach organisms (adapted from [1]).

Epidemiologic surveillance   The ongoing, systematic collection, analysis, and interpretation of health data essential to the planning, implementation, and evaluation of public health practice, closely integrated with the timely dissemination of these data to persons who need to know.

Epidemiology   The study of the occurrence of disease in human populations.

Ergonomics   An applied science concerned with the characteristics of people that need to be considered in designing and arranging things that they use in order that people and things will interact most effectively and safely [8].

Exposure   The amount of a stressor (e.g., a hazardous substance) that living organisms contact over a defined period of time.

Exposure assessment   Determination of the sources, environmental transport and modification, and fate of pollutants and contaminants, including the conditions under which people or other target species could be exposed, and the doses that could result in adverse effects [10].

Exposure investigation   The collection and analysis of site-specific information to determine whether human populations have been exposed to hazardous substances. The site-specific information may include environmental sampling, exposure-dose reconstruction, biologic or biomedical testing, and evaluation of medical information.

Exposure pathway   The path by which pollutants travel from sources via air, soil, water, or food to reach living organisms (adapted from [10]).

Exposure-response relationship   The relationship between exposure level and the incidence of adverse effects.

Exposure route   The way a substance enters an organism after contact (e.g., inhalation, ingestion, dermal absorption).

Federalism   A kind of government in which power is divided between a central government and independent regional (e.g., states) governments.

Fibrosis   Formation of fibrous tissue as a reparative or reactive process [2].

Fossil fuel   A fuel that is formed in the earth from animal or plant remains.

Fungicide   A substance that kills molds.

Gene   The functional unit of heredity that occupies a specific place or locus on a chromosome [2].

Genotoxicity   An effect on the genetic material (DNA) of living cells that, upon replication of the cells, is expressed as a mutagenic or a carcinogenic event [6].

Geographic information system (GIS)   A computer hardware and software system designed to collect, manipulate, analyze, and display spatially referenced data for solving complex resource, environmental, and social problems.

Global warming   The progressive gradual rise of the earth's surface temperature, thought to be caused by the greenhouse effect and responsible for changes in global climate patterns.

Governance   Administration, establishment, brass, organization   the persons (or committees or departments etc.) who make up a governing body and who administer something.

Government   The act or process of governing; *specif*: authorative direction or control [8].

Greenhouse gases   Gases that can absorb heat in the atmosphere.

Hazard   A factor or exposure that may adversely affect health [11].

Hazard surveillance   A data collecting system that monitors the distribution of specific hazards (e.g., carcinogens).

Health   Health is a state of complete physical, mental, and social well-being and not merely the absence of disease or infirmity [12].

Health education   A program of activities to promote health and provide information and training about reducing exposure, illness, or disease that result from hazardous substances in the environment.

Health investigation   An investigation of a defined population, using epidemiologic methods, that would help determine exposures or possible public health impact by defining health problems which require further investigation through epidemiologic studies, environmental monitoring or sampling, and surveillance.

Health surveillance   The periodic medical screening of a defined population for a specific disease or for biologic markers of disease for which the population is, or is thought to be, at significantly increased risk.

Herbicide   A chemical that kills weeds and other plants.

Hypersensitivity   A greater than normal bodily response to a foreign agent.

Incidence   The rate of development of disease in a population that can be expressed as either incidence density or cumulative incidence. Prevalence refers to existing cases of a health condition in a population, and incidence refers to new cases [13].

Ingestion   Taking food or drink into the body. Chemicals can get in or on food, drink, utensils, cigarettes, or hands from which they can be ingested.

Inhalation   Breathing. Exposure can occur from inhaling contaminants because they can be deposited in the lungs, taken into the blood, or both.

Insecticide   An agent that kills insects.

Interaction   An outcome that occurs when exposure to two or more chemicals results in a qualitatively or quantitatively altered biologic response than that predicted from the actions of the components administered separately.

In utero   Within the womb; not yet born.

In vitro   In an artificial environment, as in a test tube or culture medium.

In vivo   In the living body.

Kyoto Protocol   An international agreement struck by 159 nations attending the Third Conference of Parties to the United Nations Framework Convention on Climate Change, held in December 1997 in Kyoto, Japan, to reduce worldwide emissions of greenhouse gases.

Leukemia   Cancer of the blood-forming tissues.

Media   Soil, water, air, plants, animals, or any other parts of the environment that can contain contaminants.

Metabolism   The sum of chemical changes occurring is tissue. For example, food is metabolized (chemically changed) to supply the body with energy. Chemicals can be metabolized and made either more or less harmful by the body.

Metabolite   Any product of metabolism.

Microgram (μg)   One one millionth of a gram.

Milligram (mg)   One one thousandth of a gram.

Mixture   Any set of two or more chemical substances, regardless of their sources, that may jointly contribute to toxicity in the target population.

Morbidity   Illness or disease.

Morbidity rate   The number of illnesses or cases of disease in a population.

Mortality   The condition of being mortal; death.

National Priorities List (NPL)   EPA's listing of Superfund sites that have undergone preliminary assessment and site inspection to determine which locations pose immediate threat to persons living or working near the release.

Particulate matter   A kind of air pollution that includes soot, dust, dirt and aerosols.

Peer review   Evaluation of the accuracy or validity of technical data, observations, and interpretation by qualified experts in an organized group process [10].

Percentile   Any of the values in a series dividing the distribution of the individuals in the series into one hundred groups of equal frequency.

Picogram (pg)   One one trillionth of a gram.

Plume   An area of chemicals in a particular medium, such as air or groundwater, that moves away from its source in a long band or column. For example, a plume can be a column of smoke from a chimney or chemicals moving with groundwater.

Policy   A definite course or method of action selected from among alternatives and in light of given conditions to guide and determine present and future directions [8].

Politics   The total complex of relations between people living in society [8].

Potentially responsible parties   Persons or organizations liable under CERCLA for the costs of remediating NPL sites.

Precautionary principle   Decisions about the best ways to manage or reduce risks that reflect a preference for avoiding unnecessary health risks instead of unnecessary economic expenditures when information about potential risks is incomplete [10].

Prevalence   The proportion of ill persons in a population at a point in time, expressed as a simple percentage. Prevalence refers to existing cases of a health condition in a population, and incidence refers to new cases [13]

Primary prevention   The prevention of an adverse health effect in an individual or population through marked reduction or elimination of the hazards known to cause the health effects.

Public comment   Invited comment from the general public on agency findings or proposed activities.

Public health   Public health is the process of mobilizing local, state, national, and international resources to solve the major health problems affecting communities [14].

Public health assessment   An evaluation by ATSDR of data and information on the release of hazardous substances into the environment to assess any current or future impact on public health, develop health advisories or other recommendations, and identify studies or actions needed to evaluate and mitigate or prevent human health effects; also, the document resulting from that evaluation.

Pulmonary   Pertaining to the lungs.

Quantitative structure activity relationships (QSAR)   The relationship between the properties (physical and chemical) of substances and their ability to cause particular effects, enter into particular reactions, etc. The goal of QSAR studies in toxicology is to develop procedures whereby the toxicity of a compound can be predicated from its chemical structure by analogy with the known toxic properties of other toxicants of similar structure (adapted from [6]).

Random samples   Samples selected from a statistical population so that each sample has an equal probability of being selected [1].

Range   The arithmetic difference between the largest and smallest values in a data set.

Record of decision An EPA document that discusses the various cleanup techniques that were considered for a site and an explanation of why a particular course of action was selected [15].

Registry A system for collecting and maintaining, in a structured record, information on specific persons from a defined population.

Residual risk The health risk remaining after risk-reduction actions are implemented, such as risks associated with sources of air pollution that remain after maximum achievable control technology has been applied [10].

Risk The probability that an event will occur [11].

Risk assessment The characterization of the potential adverse health effects of human exposures to environmental hazards [16].

Risk communication An interactive process of exchange of information and opinion among individuals, groups, and institutions [17].

Route of exposure The means by which a person may contact a chemical substance. For example, drinking (ingestion) and bathing (skin contact) are two different routes of exposure to contaminants that may be found in water.

Rule The whole or a part of an agency statement of general or particular applicability and future effect designed to implement, interpret, or prescribe law or policy [18].

Rulemaking The agency process for formulating, amending, or repealing a rule [18]).

Screening A method for identifying asymptomatic individuals as likely, or unlikely, to have a particular health problem.

Screening program A program of screening for a health problem, diagnostic evaluation of persons who have positive screening-test results, and treatment for persons in whom the health problem is diagnosed.

Secondary prevention The prevention or slowing of the progression of a health problem attributable to specific hazards through use of education, protective equipment, relocation away from the hazards or other means to avoid contact with the hazard.

Soluble Dissolves well in liquid.

Solvent A substance that dissolves another substance.

Stakeholder An individual or group that has an interest in or will be affected by an action.

Statistical significance A calculated value that infers the probability whether an observed difference in quantities being measured could be due to variability in the data rather than an actual difference in the quantities themselves.

Stressor A chemical, material, organism, radiation, noise, temperature change or activity that stresses an organism's health or well-being.

Superfund Another name for the Comprehensive Environmental Response, Compensation, and Liability Act of 1980 (CERCLA). The term is also used to refer to the Hazard Substance Superfund, the trust fund established by CERCLA.

Surveillance A data-collection system that monitors the occurrence of disease (disease surveillance) or the distribution of hazard (hazard surveillance).

Sustainable development   Development which meets the needs of the present without compromising the ability of future generations to meet their own needs [19].

Synergism   A response to a mixture of toxic chemicals that is greater than that suggested by the component toxicities.

Toxicant   A substance not produced by a living organism that causes a harmful effect when administered to a living organism. See toxin.

Toxicity   The property of chemicals that causes adverse effects on living organisms.

Toxicokinetics   Toxicodynamics; the study of the quantitative relationship between absorption, distribution, and excretion of toxicants and their metabolites [6]

Toxicology   The science that deals with poisons (toxicants) and their effects [6].

Toxics Release Inventory   A publicly available EPA database that contains information on toxic chemical releases and other waste management activities reported annually by certain covered industry groups as well as federal facilities. This inventory was established under the Emergency Planning and Community Right-to-Know Act of 1986 (EPCRA) and expanded by the Pollution Prevention Act of 1990.

Toxin   A toxicant produced by a living organism [6].

Tumor   An abnormal mass of tissue.

Volatile organic compounds (VOCs)   Substances containing carbon and different proportions of other elements such as hydrogen, oxygen, fluorine, chlorine, bromine, sulfur, or nitrogen. These substances easily become vapors or gases. Many VOCs are commonly used as solvents (paint thinners, lacquer thinner, degreasers, and dry cleaning fluids).

Weight-of-evidence   A systemic method for applying biomedical judgment to empirical observations and mechanistic considerations to qualitatively assess the potential toxicity of a substance, singly or in a chemical mixture, for a given target organ or system.

REFERENCES

1. RiskFocus®, Analysis of the Impact of Exposure Assumptions on Risk Assessment of Chemicals in the Environment: Part I, VERSAR, Inc., Springfield, IL, 1990, 123.

2. Hensyl, W.R., ed., *Stedman's Pocket Medical Dictionary*, Williams & Wilkins, Baltimore, MD, 1987.

3. NRC (National Research Council), *Environmental Epidemiology*, vol 2, National Academy Press, Washington, D.C., 1997.

4. Merriam-Webster, *Merriam-Webster Online*. Available at http://www.m-w.com/cgi-bin/mwwod.pl,, 2001.

5. BOC (Bureau of Census), Census of Population and Housing, 1990: Summary Tape File 3 on CD-ROM Technical Documentation, Department of Commerce, Washington, D.C., 1992.

6. Hodgson, E. and Levi, P.E., *A Textbook of Modern Toxicology*, Elsevier, New York, 1987, 274, 361.

7. Oxford Dictionary, Oxford Dictionary On-Line. Available at http://www.askoxford.com/?view=uk, 2005.

8. *Webster's Ninth New Collegiate Dictionary*, Merriam-Webster Publishers, Springfield, MA, 1985.

9. WDOE (Washington Department of Ecology), A Study on Environmental Equity in Washington State, Report No 95-413, Olympia, 1995.

10. CRARM (Commission on Risk Assessment and Risk Management), Framework for Environmental Health Risk Management, U.S. Environmental Protection Agency, Washington, D.C., 1997.

11. Yassi, A. et al., *Basic Environmental Health*, Oxford University Press, Oxford, 2001.

12. WHO (World Health Organization). Available at http//:www.who.int, 2000.

13. Dicker, R.C., Analyzing and interpreting data, in *Field Epidemiology*, Gregg, M.B., ed., Oxford University Press, New York, 1996.

14. Detels, R. and Breslow, L., Current scope and concerns in public health, in *Oxford Textbook of Public Health*, 2nd ed, vol 1, Holland, W.W., Detels, R., and Knox, G., eds., Oxford Medical Publications, Oxford University Press, Oxford, 1991.

15. EPA (U.S. Environmental Protection Agency), Superfund Progress Aficionado's Version, Report PB92-963267, Office of Solid Waste and Emergency Response, Washington, D.C., 1992b, 11.

16. NRC (National Research Council), *Risk Assessment in the Federal Government: Managing the Process*, National Academy Press, Washington, D.C., 1983.

17. NRC (National Research Council), *Improving Risk Communication*. National Academy Press, Washington, D.C., 1989, 20, 21.

18. LII (Legal Information Institute), Subchapter II – Administrative Procedure. Available at http://www.law.cornell.edu/uscode/htm/uscode05/usc_sup_01_5_10_1_30_5_40_11.html, 2005.

19. Brundtland, G.R., *Our Common Future: The World Commission on Environment and Development*, Oxford University Press, Oxford, 1987.

Index

Page numbers in italic refer to figures and tables.